Zoo Animal Welfare

Zoo Animal Welfare unpacks the different concepts of welfare in the zoo and aquarium, considering how enclosure design, nutrition, conservation activities, visitor engagement and daily husbandry all need to have a welfare focus to be relevant to the species being housed. The book provides the practical application of evidence-based welfare theory, centred on the concept of "welfare-focused husbandry", which enables zoo animals to experience positive welfare.

Recognising that welfare is a complex and challenging area of science, the first section of the book provides core introductory material defining welfare in the zoo and looking at how the wild is managed ethically. It then delves into current issues with contributed chapters exploring topics such as environmental enrichment, nutrition, the impacts of human presence in the zoo, conservation welfare, use of social media and population management.

Packed with case studies and exploration of emerging issues in the field (e.g. novel technology for welfare assessment, welfare considerations of overlooked taxa), this book is essential reading for higher-level undergraduate and postgraduate students and academics who wish to study the welfare of captive wild animals, as well as for all those working in animal care and zoo operations. *Zoo Animal Welfare* will empower readers to go "beyond the norm" when considering what zoo animals really need, applying their new knowledge for the benefit of both animals and zoos.

Zoo Animal Welfare

Edited by
Paul Rose, Zoë Newnham and Ian Hickey

CRC Press
Taylor & Francis Group
Boca Raton London New York

CRC Press is an imprint of the
Taylor & Francis Group, an **informa** business

Designed Cover Image: Shutterstock: 2353747319

First edition published 2026
by CRC Press
2385 NW Executive Center Drive, Suite 320, Boca Raton FL 33431

and by CRC Press
4 Park Square, Milton Park, Abingdon, Oxon, OX14 4RN

CRC Press is an imprint of Taylor & Francis Group, LLC

ISBN: 9781032551760 (hbk)
ISBN: 9781032547367 (pbk)
ISBN: 9781003429425 (ebk)

DOI: 10.1201/9781003429425

Typeset in Minion
by Deanta Global Publishing Services, Chennai, India

Contents

Foreword

This is a very welcome book at a time of great change in thinking about what a zoo is and what it should be. The public who come to zoos, and the public who do not, are driving change in zoos, in a similar way to the changes being forced in all areas where humans have impacts on the lives of non-human species; for example, in animal production, plant production, transport, sport and scientific research. There is a rapid increase in public concern about climate change, biodiversity loss and the welfare of individual animals. Zoos, like many other areas of human activity, have been too slow to change and too slow to take opportunities that can increase their importance in the world. People entering a zoo, or seeing a video of animals in a zoo, are rapidly confronted with individual animals whose conditions of living and behaviour raise questions about their welfare. As a consequence, welfare has become the highest priority issue for zoos, with other issues such as concerns about life in the wild and the future of all life in the world not far behind.

The book commences with a thoughtful discussion about what various people mean when they talk about welfare and, in relation to the zoo community, how this has limited their actions. These ideas are developed further in a discussion of how there might be action to facilitate zoo animals moving from having to struggle with difficulties in having their needs met to thriving. The change in thinking about what a zoo is for and the development of knowledge about welfare in a range of species is described in several of the chapters in the book. How the zoo animals perceive environmental quality, nutrition, health and the optimal amount and nature of interaction with humans is all discussed in relation to studies of the welfare of those animals.

While the welfare of animals in a zoo can be scientifically evaluated, once this has been done, this evidence can be utilised when deciding which animals should be kept in a zoo and how animals should be kept. Deciding this is an ethical matter, so a chapter on ethics is an important part of this book. An ethicist may reasonably advise on minimal levels of poor welfare that should be tolerated. Ethical decisions can be very difficult; for example, what degree of poor welfare might be tolerated in order that animals originating in zoos can contribute to conservation programmes?

Practical advice is of key importance when trying to improve the welfare of zoo animals. Hence, information on what zoo keepers can and should do is an important chapter in this book. A further issue considered in the book is for which animals the term welfare applies. It is my view that we should consider the welfare of all vertebrate and invertebrate animals, but that, while we can evaluate the welfare of humans and all other animals, the term does not apply to plants, bacteria, or inanimate objects. Sentient animals are those that have the capacity to have feelings, and while those that can feel pain, etc., should be treated accordingly, the concept of welfare applies to both sentient and non-sentient animals. Some mechanisms that a sentient individual uses when coping with its environment involve emotions and feelings, but others, such as immunological and cellular defence reactions, help to cope and improve welfare but need not involve any feelings or perception. Welfare can be poor because the individual is sick, whether or not it perceives and feels the sickness. The chapters in this book include details of the welfare of invertebrate animals, both those like decapod Crustacea that many scientists now consider to be sentient and those other invertebrates like gastropod molluscs, which are not at present thought to be sentient.

Several chapters of this book encourage consideration of the motivation of animals in zoos. The welfare of all individuals is improved if they have some opportunity to choose what to do and where to be, and if they have control over aspects of their environment. Environmental enrichment is facilitated by choice and is more effective if highly motivated behaviours are promoted. Encouraging this way of thinking is an example of one current area of opportunity for zoos. Zoos can take a lead in society by informing and encouraging the public who visit, physically or online, about how to become less human-centred and to consider the valuable diversity of life. If zoos are seen to focus on the welfare of individuals and hence to respect non-human life, they can promote change in thinking towards a more balanced world where the lives of all animals and all other living species are valued. The idea that most of life on earth is just a resource for people has already caused enormous damage to the world. Zoos can encourage people to consider all aspects of the sustainability of every human activity, including reversing biodiversity decline, utilising resources well, reducing greenhouse gas production, and considering the welfare of every individual.

Donald M. Broom
18th January 2025

Preface: Zoo animal welfare: A journey of faring well into welfare-focused husbandry

PAUL ROSE, ZOË NEWNHAM AND IAN HICKEY

Welfare in the English language is an old word, dating to the Middle English period of 1150 to 1300 and referring to a wish or will ("wel") and going on a journey ("faran"). Welfare literally means being on the right path to feeling well. The Oxford English Dictionary's first written record of its usage is in the works of Benedictine monks from 1347. Shakespeare's works feature the word several times, using it in a context that the modern reader would be familiar with – to feel well or to be well. Therefore, welfare is the compound of an adjective ("well") and a noun ("fare"), and this etymological background is important for how we use the word today when pertaining to non-human animals.

From a scientific perspective, it is essential that we understand how to provide welfare inputs and assess welfare outputs. The benchmark for doing this correctly in zoos and aquariums ("zoos") can be found in the natural history of each species being kept. Welfare assessment must be repeatable, rapid but thorough and (above all) valid and meaningful to the focal species. Generic, descriptive approaches that cannot accurately illustrate species-specific indicators (such as behavioural changes to body language, behavioural expression, resource usage, social interactions or environmental engagement) can lead to poor-quality information being used as evidence for what welfare "looks like" for different species of zoo animals. Indeed, poor-quality evidence is not evidence at all. As welfare scientists and practitioners, we must critique and appraise the literature which, sadly, is not all of good quality and thus not always helpful to in-zoo developments.

As such, *Zoo Animal Welfare* aims to provide the practical application of welfare theory and welfare concepts and is centred on the concept of "welfare-focused husbandry," which requires valid evidence for husbandry practices to enable zoo animals to experience positive welfare outcomes. Academic ideas that may look good on paper do not always translate into useful, relevant, or logistically possible aspects of in-zoo husbandry or animal management regimes. There are many aspects of species' behaviours and behavioural expressions (i.e. body language) that we have yet to fully decipher. The primary sensory modality of many zoo-housed species will differ from how we, as visual primates, take in and process information. What we see will provide a different impact on our welfare compared to what an animal smells, hears, tastes, or touches.

In many cases, evidence on wild animal ecology (including sensory ecology) is limited, further restricting our benchmark for "what are appropriate positive outputs" in the zoo. It is therefore essential that we, as mammals, do not inflict a mammal-centric interpretation on what species responses may mean when we make observations across taxonomic groups in the zoo. Time and energy spent trying to illuminate emotional quotients of species such as the giant African land snail (*Achatina fulica*), the poison dart frog (Dendrobatidae) or the neon tetra (*Paracheirodon innesi*) may not be the best use of time when there are multiple practical challenges to solve relating to captive management for populations such as these, including correct care, health provision, formulation of social groups, climatic variables, enrichment protocols and nutrition. Welfare-focused

Table 1 Examples of measurable indicators that can be used to infer welfare states (not an exhaustive list)

Physical indicators	Behavioural indicators	Psychological inferences
Body condition.	Time-activity budgets that include performance of key welfare indicators (e.g. maximal rumination, appropriate time spent asleep, indicators of diverse foraging activity).	Perceived control over their environment via expression of choices (interaction with environmental variables/ conspecifics) and display of interest in their surroundings.
Food intake.	Ecologically relevant behavioural diversity.	Appropriate coping mechanisms and resilience.
Mass.	Appropriate (for the species) social associations and interactions.	Engaging in activities for benefit beyond a physical reward (e.g. contrafreeloading).
Pelage/plumage/integument colouration and condition.	Temporal, seasonal, and physiological-driven changes in behavioural rhythms.	Positive indicators of behavioural expression (e.g. outputs from qualitative behavioural assessment that reflect ecologically relevant emotions and moods).
Faecal consistency.	Change in "normal" or "expected" behaviour for that individual animal.	Confident and exploratory. Comfortable in engaging with their wider environments.
Indicators of baseline homeostasis such as heart rate, core body temperature, and respiration.	Change in how an individual interacts with people in the zoo (e.g. keepers, visitors).	Positive outcomes from (repeated, replicable) cognitive bias testing.
Clinical signs of disease (e.g. lameness, loss of appetite).	Signs of fear, distress or frustration, or maladaptive indicators such as abnormal repetitive behaviours.	Personality traits (behavioural syndrome) being consistent for that individual.
Senescence and associated health changes.	Behavioural development (learning new behaviours) across different life stages.	Use of environmental enrichment that produces positive behavioural expression.
Growth and developmental changes associated with life stage.	Use of environmental enrichment that promotes adaptive traits or ecologically relevant time budgets.	Apathy, disinterest or other such signs of low mood or lack of care over self or immediate situation.
(Validated) endocrine responses associated with stress (e.g. glucocorticoid concentrations).	Prolonged inactivity, lethargy, or restlessness, or self-removal from normal engagement with conspecifics and environment.	Negatively-valenced behavioural expression (e.g. boredom) noted multiple times when positive outputs would be expected.
Reproductive indicators and outputs (e.g. normal oestrous cycle, viable young produced).	Parenting skills and ability to rear young to maturity or dispersal age.	Interest and engagement in positively-valenced activities, such as courtship, breeding and parenting.
Integrity of, and use of, sensory modalities and the outputs from these (covers all indicators).		
Quality of life assessment protocols (covers all indicators).		

husbandry is a process – once we have husbandry and care as the best standard possible, then we can devote our time and efforts to understanding the more esoteric aspects of how different species present welfare outputs.

Alongside embedding biological evidence into welfare inputs and outputs, this book is grounded in the clear and reasonable definition of welfare from Broom (1986) - "the state of the individual as it attempts to cope with its environment". According to our understanding of welfare as comprising physical, behavioural, and psychological components, this definition enables us to develop ways of assessing, measuring, and inferring what welfare looks like in different species, based on key aspects of the scientific method (Table 1).

We can only identify, and subsequently address, welfare shortfalls for zoo animals with a correct assessment of welfare states. Table 1 aims to provide some context (i.e. how could welfare be inferred?) to the subject or topic-specific chapters that follow. It is essential that welfare measures are validated for each species to ensure that we are using appropriate indicators that are grounded in natural history and behavioural biology. By considering as many potential indicators as we can (such as some of the examples shown in Table 1) and aligning these with our understanding of a species' biology, we can start to home in on what an animal may be trying to tell us. We must avoid the trap of assuming what a species or individual needs based on historical and anecdotal evidence passed down within zoological institutions; so, rather than relying on *"It's always been done this* way", we instead we ask ourselves, *"What would good welfare really look like for this animal?"*

Welfare science evolves rapidly and continually, and how to provide for good welfare can change according to newly available information, improvements in our understanding and associated amendments to policy and regulation. We, the editors of this book, are aware of changes in region-specific zoo legislation that occurred during the writing and preparation of this text – notably the amendments to the UK's Zoo Licencing Act, which sees new requirements coming in to force in 2027. Whilst we were unable to consider and discuss these new Standards for Modern Zoo Practice in our book due to the timeframe of publication of these Standards, we welcome such changes to policy that will ultimately improve how zoos and aquariums enable and promote good welfare for all of their inhabitants.

Along with its sister text, *The Behavioural Biology of Zoo Animals* (Rose, 2022), this book seeks to promote the aims and work of the modern, scientific zoo; those zoological facilities that are generally part of professional accreditation or membership schemes, and that wish to achieve educational, public engagement, research, and conservation impacts from the animal collections they maintain. Extending these well-known "four aims" of the modern zoo to include well-being (Rose & Riley, 2022) should be given clear credence moving forward. We know that the work of zoos can promote what good animal welfare looks like (in the eyes of the visiting public) whilst installing connectivity to nature and the positive mood-boosting elements of interaction with wildlife (Rose & Riley, 2023).

In the chapters ahead, you will read ideas and viewpoints from subject matter experts encompassing a range of welfare-focused topics, all of which aim to expand our understanding of zoo animal welfare (using taxa-specific examples, as well as considering their relevance to zoo operations more holistically) and help us go "beyond the norm" when considering what zoo animals actually need. We have gone beyond academic discussion of what welfare comprises, to providing application of such knowledge so that animals (and zoo operations) can ultimately benefit. From this book, we hope you take away new insights into what positive welfare should look like in the zoo and, ultimately, continue this welfare-focused husbandry journey to creating an even brighter future for all zoo- and aquarium-housed species.

REFERENCES

Broom, D. M. (1986). Indicators of poor welfare. *British Veterinary Journal, 142*(6), 524–526.

Rose, P. E. (Ed.). (2022). *The behavioural biology of zoo animals*. Boca Raton, USA: CRC Press.

Rose, P. E., & Riley, L. M. (2022). Expanding the role of the future zoo: Well-being should become the fifth aim for modern zoos. *Frontiers in Psychology, 13*, 1018722.

Rose, P. E., & Riley, L. M. (2023). Five ways to wellbeing at the zoo: Improving human health and connection to nature. *Frontiers in Psychology, 14*, 1258667.

Editor biographies

Paul Rose gained his PhD in 2018, investigating the behaviour and welfare of captive flamingos using a variety of methodological approaches including social networks' analysis. He is a Senior Lecturer in the Psychology Department at the University of Exeter, UK, where he specialises in animal behaviour. Paul also works for the Wildfowl & Wetlands Trust (WWT) where he manages the Animal Welfare & Ethics Committee, the Academic Lead for the Wetland Learning Hub and is a research associate for WWT's captive animal research programme. Paul is also a lecturer on the MSc Zoo Biology course for University Centre Sparsholt, UK. Paul also delivers an MSc module in biodiversity conservation for Birkbeck College, University of London, UK. As a member of Defra's Zoos Expert Committee, Paul writes and reviews policy pertaining to the running and licencing of zoos and aquariums. Paul is a member of the IUCN SSC Flamingo Specialist Group and the IUCN SSC Giraffe & Okapi Specialist Group. He is a member of the steering committee for the BIAZA Bird Working Group, and a member of the BIAZA Animal Welfare Working Group. Paul is involved with three European (EAZA) Taxon Advisory Groups (Gruiformes, Charadriiformes, Anseriformes) and with the North American (AZA) Waterfowl TAG in an educational advisor capacity. Paul completed his first piece of in-zoo research in 2002 and has been professionally involved in zoo animal behaviour and welfare in an academic and practitioner capacity since 2006. His research predominantly focusses on behaviour and welfare, and evidence for zoo husbandry. He is a Senior Fellow of the Higher Education Academy and a Fellow of the Zoological Society of London. Paul's first book, "The Behavioural Biology of Zoo Animals" was published in 2022.

Zoë Newnham is a senior animal keeper in the Hoofstock Department at Marwell Zoo and a sessional lecturer at University Centre Sparsholt, UK, where she teaches on higher education programmes pertaining to animal management and zoo biology. She is also a partner associate lecturer for the University of Portsmouth, UK. Zoë started her zookeeping career in 2008 at the New Forest Wildlife Park, working with many British and European species and in wildlife rehabilitation for native species. Zoë has a BSc (hons) in Animal Management (2011) and an MSc in Applied Zoo Biology (2023). Zoë's current zookeeping career focuses on ungulates and she has researched the nutrition of captive browsing species for her postgraduate research. She has published work on training of okapi to assist with health and veterinary-related procedures and has contributed to the textbook "The Behavioural Biology of Zoo Animals" writing about how zookeepers use behavioural evidence in their roles. Zoë is a trustee of the zoo-focussed charity "Browse Poster", a professional member of the Association of British & Irish Wild Animal Keepers (ABWAK), a member of the Southampton Natural History Society (SNHS), a member of the Association for the Study of Animal Behaviour (ASAB) and a member of the Universities Federation for Animal Welfare (UFAW).

Ian Hickey is Enclosure Development Manager at Chester Zoo where he looks after the animal enclosure upgrades and enhancements for all species, while representing the Animal & Plant Department throughout the design and construction process for new, large-scale animal exhibits. Ian moved to the UK from Ireland in 2006 to study BSc (Hons) Animal Management at Sparsholt College Hampshire and he subsequently obtained his MSc in Wild Animal Biology from the Royal Veterinary College in 2010. Ian has worked for the British and Irish Association of Zoos and Aquariums (BIAZA) where he delivered their 2010 Elephant Management Audit and spent five years as a zookeeper at Blair Drummond Safari Park and then Chester Zoo (before moving into his current role). His work now involves ensuring that each new development meets the needs of both animals and keeping staff within the spaces being created. Ian is also a consultant for zoo design consultancy Dearadh Zú, having contributed to developments at zoos and safari parks throughout the UK, and is a qualified Accreditation Screener for BIAZA. Ian has contributed the elephant chapter to the textbook "Behavioural Biology of Zoo Animals" and has presented at various zoo research conferences on his undergraduate and postgraduate work.

Acknowledgements

We'd like to thank our families, friends, and all who have been a part of our journey along the way, as well as a giving special thanks to all chapter authors for their invaluable contributions that have made this book what it is.

Although there are shared ideas throughout much of this book, all views are each author's own and may not necessarily reflect that of their affiliated institution(s).

Zoo Animal Welfare glossary

Throughout this book, we discuss animal welfare in zoos, aquariums, safari parks, wildlife parks and other places where ex situ populations of non-domestic species are managed. For brevity, we use the word "zoo" to encompass all these different forms of zoological establishments/institutions.

Animal welfare: An individual animal's state as it attempts to cope with the environment that it is currently experiencing. Welfare comprises of behavioural, physical, and psychological components, and will change as an animal experiences different stimuli. In this book, we consider welfare to encompass Broom's concepts of coping (as an indicator of positive and negative physical, behavioural, and emotional – where measurable – states) alongside Fraser's ideas on biological functioning, affective states, and enabling natural living.

Abnormal repetitive behaviour (ARB): A behaviour performed in an invariant manner, repeated in duration, orientation, and location that may serve no biological function to the individual animal. ARBs are usually not part of a species' wild behavioural repertoire. Stereotypical pacing behaviour, redirected /self-directed goal-orientated behaviours (e.g. feather picking in place of social interaction) and displacement behaviours (e.g. intense vocalisation without context) are all forms of ARB.

Adaptive behaviour: A behaviour whose performance conveys adaptive benefits (e.g. promotes the fitness of the individual) to the animal that is performing it.

Affective state: Transient, subjective experiences that are not hedonically neutral (i.e. can be pleasurable or can be painful). Often used as an overarching term for moods and emotions. An individual's mood is a longer-term experience than an emotion.

Agency: The capacity of an individual animal to engage in voluntary, self-generated, and goal-directed behaviour that they are motivated to perform.

Appetitive behaviour: A behaviour that is highly motivated, whereby the animal has an "appetite" for a particular stimulus (searching and foraging actions, for example). Such a behaviour may continue even if the end goal of the behaviour cannot be achieved (e.g. the animal continues to display foraging patterns even if there is no food present in an enclosure).

Arousal: The activation level of an individual and the intensity of the state they are in.

Back-of-house: The staff or animal area(s) that are part of a zoo or enclosure but that non-zoo personnel (e.g. visitors) do not have access to and usually cannot see.

Behavioural need: The need to perform a specific behaviour irrespective of current environmental conditions (e.g. internal motivation to migrate during a specific season). The behaviour is largely controlled by internal factors that will be present regardless of the individual's current environment or location.

Behaviour training: The active manipulation of an animal's behaviour to achieve a desired outcome. Primarily used in this book when referring to a planned interaction initiated by a caregiver with voluntary cooperation by an animal.

Best Practice Guidelines: The currently accepted "gold standard" of husbandry, housing and management used for a specific species within ex situ populations. Best Practice Guidelines can be regionally specific (e.g. EAZA or AZA husbandry manuals).

Biologically (ecologically) relevant behaviour: A behaviour that is considered normal in the behavioural repertoire of that species. One that enrichment (for example) would seek to promote and would likely be displayed by a free-living animal in the wild.

Caregiver: A person providing care and husbandry services to an animal in captivity. Used interchangeably in this book with "(zoo)keeper", "animal care staff" and "caretaker".

Conspecific: An individual animal of the same species. Primarily used in this book when discussing animals that share the same enclosure space. Different species (allospecific) would occur together in a mixed-species enclosure.

Consummatory behaviour: A behaviour that results in completion of an end goal or reward that satiates the animal. Consummatory behaviour signals the end of an appetitive behaviour.

Contrafreeloading: The animal will exert energy on behaviour to work for a reward even if the reward is readily available within the animal's environment (e.g. the animal will engage in complex foraging actions to gain food from an enrichment device when food is also easily available from a bowl or similar source).

Ethogram: A list of species-specific behaviours, and the descriptions of the mechanical actions of the behaviour, exhibited by an animal. Used for behavioural research methods when recording behaviour in an objective and repeatable manner.

Environmental enrichment: Applications of animal husbandry practice and/or enclosure design features that promote positive welfare by encouraging a wider range of (ecologically relevant) behavioural performances. Enrichment can comprise of nutritional (e.g., scatter feeding), occupational (e.g., puzzle feeders or other problem-solving opportunities that encourage an animal to spend time with the enrichment), sensory (e.g., the use of scents, textures, or tastes to engage an animal's different sensory modalities that in turn promotes behavioural diversity), social (e.g., maintenance of an ecologically relevant social group according to individual animal's life stage) and physical (e.g. structures, furnishings, and devices that the animal engages with to perform different behaviours) ideas. Animal training regimes (e.g. positive reinforcement training) can also be considered a form of enrichment. Different categories of enrichment are continuous and not mutually exclusive.

Ex situ conservation: Conservation of a species within managed populations in conditions away from their native range states and wild places (e.g., in zoos, aquariums, safari parks and similar zoological facilities).

Fission-fusion (society): A fluid social structure whereby a group may split into smaller sub-groups (fission) or merge into larger groups (fusion) as it moves through its environment.

Hedonism: Welfare states are greatest when an animal has many positive stimulations that result in pleasant or enjoyable experiences and few negative stimulations that cause unpleasant or painful experiences.

In situ conservation: Conservation of a species within populations occurring in native range states and wild places where the species would naturally occur (e.g. in national parks, wildlife reserves and refuges, and in unmanaged wild spaces).

Motivational state: Physiological (internal) states within an animal that play an important role in initiating behaviour, selecting actions to perform, and orienting such actions to achieve the animal's desired goals.

One Plan Approach: Integrated conservation management of species between wild and captive populations, which involves the concept of a metapopulation approach (all individuals of a species wherever they occur) to conservation action.

Prosocial behaviour: An action performed by an animal to achieve a benefit to a conspecific or an individual of another species.

Quality of life: Cumulative experience of an animal's psychological and physical well-being. Quality of life (QoL) is complex and hard to assess objectively. Inferences of QoL rely on multiple measures of validated welfare indicators that are scored by trained assessors across specific periods of time.

Qualitative Behavioural Assessment (QBA): A method of animal welfare assessment that describes and records the emotionally expressive qualities observed in an animal's behaviour, posture, orientation, and body language. QBA focusses on behavioural expression as an outward indication of an animal's moods, emotions, and feelings and not on the mechanics of the behaviour being performed.

Valence: The degree to which an affective state is perceived as positive (e.g. rewarding) or negative (e.g. aversive).

Welfare assessment tool: A standardised method of data collection for inferring an individual animal's welfare state, based on assessment of the quality and relevance of inputs (e.g. resources, physical cues, social environment) and the type of outputs (behaviour, emotions, feelings) shown by the animal. These tools are used routinely in a zoo to undertake a welfare audit of individual animals and groups of animals in the zoo's collection. A variety of welfare assessment tools are available and can be bespoken to a specific organisation or zoo regions (e.g. Animal Welfare Assessment Grid, BIAZA Animal Welfare Toolkit).

Wellbeing: Although often used interchangeably with welfare, this book considers wellbeing within a human's perspective of their physical and mental health.

Contributors

Unnar Aevarsson
Zoological Society of London, Regent's Park, London, UK

John E. Andrews
Association of Zoos and Aquariums Population Management Center at Lincoln Park Zoo, Chicago, Illinois, USA

Kyle Banton-Jones
Uxbridge, Ontario, Canada

Jonathan Beilby
The North of England Zoological Society, UK

Nicholas Bishop
Adelaide Zoo, Frome Road, Adelaide, South Australia

Jack Boultwood
Formally at Wild Planet Trust, Paignton, Devon, UK

Lorna Cameron
Equine Research Centre, Hartpury University, Hartpury, Gloucestershire, UK

Lisa M. Clifforde
Zoological Society of London, Regent's Park, London, UK

William Costa
WWT, Slimbridge, Gloucesteshire, UK

Matthew Fiddes
CJ Hall Veterinary Surgeons, East Sheen, London, UK

Ricardo Lemos de Figueiredo
Bristol Zoological Society, Hollywood Mansion House, Bristol, UK

Marianne Freeman
University Centre Sparsholt, Sparsholt, Winchester, UK

Linda Greening
Equine Research Centre, Hartpury University, Hartpury, Gloucestershire, UK

Philip Greenwell
Salce, St. Georges, France

Jessica J. Harley
Knowsley Safari, Prescot, UK
University Centre Reasheath, Rease Heath, Nantwich, UK

Ian Hickey
Chester Zoo, Caughall Road, Upton-by-Chester, Cheshire, UK

Kerry Hunt
SEALIFE Protect, Poole, UK

Robert Kelly
Centre for Research in Animal Behaviour, Psychology, University of Exeter, Exeter, Devon, UK

Jack Lewton
School of Animal and Environmental Sciences, Nottingham Trent University, Nottingham, UK

Andrew Mooney
Dublin Zoo, Phoenix Park Dublin, Ireland

Camille Munday
Zoological Society of London, Regent's Park, London, UK

Campbell Murn
Hawk Conservancy Trust, Andover, Hampshire, UK
School of Biological Sciences, University of Reading, Whiteknights, Reading, UK

Steve Nash
Formally of Wild Planet Trust, Devon, UK

Zoë Newnham
Marwell Wildlife, Winchester, Hampshire, UK

Michelle O'Brien
WWT, Slimbridge, Gloucestershire, UK

Justine K. Partoon
Zoos South Australia, Adelaide, South Australia

Sarah Pellett
Byre Veterinary Surgery, Peterborough, UK

Lisa M. Riley
University of Winchester, Sparkford Road, Winchester, Hampshire, UK

Paul Rose
Centre for Research in Animal Behaviour, Psychology, University of Exeter, Exeter, Devon, UK
WWT, Slimbridge, Gloucestershire, UK

Christina R. Stanley
Animal Behaviour & Welfare Research Group, University of Chester, Chester, UK

Chloe H. Stevens
Animals in Science Department, RSPCA, Parkside, Horsham, West Sussex, UK

Steven A. Trim
Ventera Bio, CoLab, Sandwich, Kent, UK

Jake S. Veasey
Care for the Rare Ltd, C/O Open University, Animal Computer Interaction Laboratory, Kents Hill, Milton Keynes, UK

Mark Whiteside
School of Biological and Marine Sciences, University of Plymouth, Drake Circus, Plymouth, UK

Nathalie Wissink-Argilaga
Chipping Norton Veterinary Hospital, Banbury Road, Chipping Norton, Oxfordshire, UK

Lisa K. Yon
School of Veterinary Medicine & Science, University of Nottingham, Sutton Bonington, Leicestershire, UK

1

Animal welfare in the zoo: The what, why and how

PAUL ROSE, ZOË NEWNHAM AND IAN HICKEY

1.1 INTRODUCTION

Animal welfare is a core component of the aims and operations of modern zoos and aquariums (hereafter zoos). Welfare concepts and definitions abound in the scientific literature, but perhaps more important than continual academic debate on all potential nuances of welfare is to consider what is meaningful to the animal and how good welfare outputs can be enabled by meaningful care. In this book, we extend the original definition of animal welfare from Broom (1986) that considers welfare to be an individual's state with regard to its attempts to cope with its environment (both internal and external), to include the behavioural, physical and psychological influences over wellbeing (and how these will impact an individual's health, fitness, and abilities to survive and thrive). With this extended focus, we can provide living conditions that promote thriving in the zoo (Figure 1.1). Health is also a part of welfare and the link between the two means that, when one is discussing health status, welfare is being considered by default. If an animal is coping with a pathology, this impacts its overall health and, as such, the welfare state which it can experience (Broom, 2006).

Species will struggle to survive in, and cope with, a captive environment if it does not meet their intrinsic biological and behavioural needs. For example, previous husbandry protocols for captive giraffe (*Giraffa camelopardalis*) have been linked to a shortened lifespan in the zoo when animals are fed inappropriate forage, not kept warm, and not provided with a correct concentrate pellet ration to boost available digestible energy. Changes to giraffe husbandry, to ensure animals are fed *ad lib* legume forage supplemented with a formulated browser pellet and are housed inside during low external temperatures, have improved animal health and therefore captive lifespans. These husbandry changes, based on empirical review and quantification of good practice, alongside information on animal biology, ecology and behaviour have provided meaningful markers for welfare assessment of this species. Therefore, whilst welfare is not a "thing" that an animal has, but something that is experienced, it is intrinsic to that animal's state at a specific time (Broom & Johnson, 1993). Improving care so that it has added species'

DOI: 10.1201/9781003429425-1

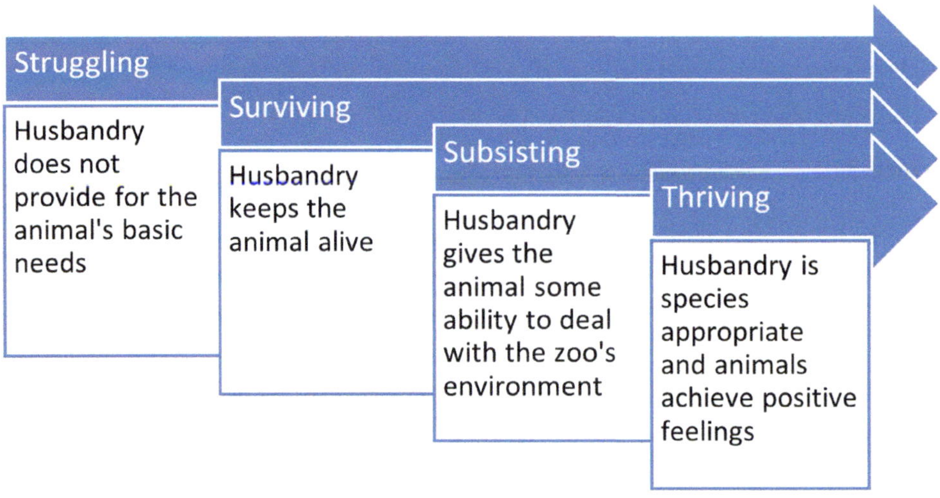

Figure 1.1 The evolution of welfare aims in the zoo. Species will struggle when basic needs are not met, from systems that simply keep an animal alive to modern-day, evidence-based practices that enable species to thrive and fulfil their biological needs. An evidence-based approach of husbandry evaluation is needed to move all species from left to right, as there are still many examples of taxa that are struggling and not yet surviving under human care.

relevance will push the balance of an animal's positive experience to maximum against negative ones. Figure 1.2 illustrates how species-specific knowledge is used to shift husbandry actions and (perhaps just as importantly) includes thoughts and ideas about what species need in the right direction so that they become a solid foundation for positive welfare.

Identification of husbandry issues and associated welfare changes, e.g. lack of forage availability and limited browsing opportunities leading to poor body condition and performance of stereotypic behaviours, can be achieved based on comparison of resource use in the wild state and measurement of behavioural and physical condition of giraffe in other zoos that do provide ad lib forage and browse. This assessment of good practice and of behavioural ecology allows for objective identification of why the species is failing to cope and can provide the baseline to assess improvements to animal behaviour and health when husbandry is altered. Review and assessment at regular intervals of how the giraffes continue to respond to their husbandry is required to ensure that management practices remain supportive of good welfare outputs (Figure 1.2).

Welfare is on a continuum that changes across the course of an animal's day and, in the zoo, this can be caused by social groupings, visitor presence, husbandry routines, veterinary interventions, feeding time, use of enrichment, physiological state, growth and development, past experiences, and available space (Figure 1.3). The more pressures or negative challenges an animal faces, the less they can cope. This is because whatever mechanisms they employ to cope do not reduce the presence of the pressure or challenge (they do not quell the motivation creating a behavioural need) and, consequently, the animal experiences a state of distress or apathy or learned helplessness (depending on individual circumstances). We should be mindful of this continuum when we undertake welfare assessment and remember to monitor welfare at different time periods, different days, and consistently throughout the course of the animal's life in the zoo.

Our ideas of what constitutes animal welfare have changed as we attempt to move individuals from a life worth living to a good life (Mellor, 2016; Yeates, 2011). Environments that provide choice and control and aspects of positivity can help to buffer against the negative stressors that all zoo animals will experience due to the operational nature of a zoological facility. Figure 1.3 illustrates how external factors, many of which can be beyond an animal's control, will impact welfare states at that given time. The presence of animal care staff in the morning is likely to be associated

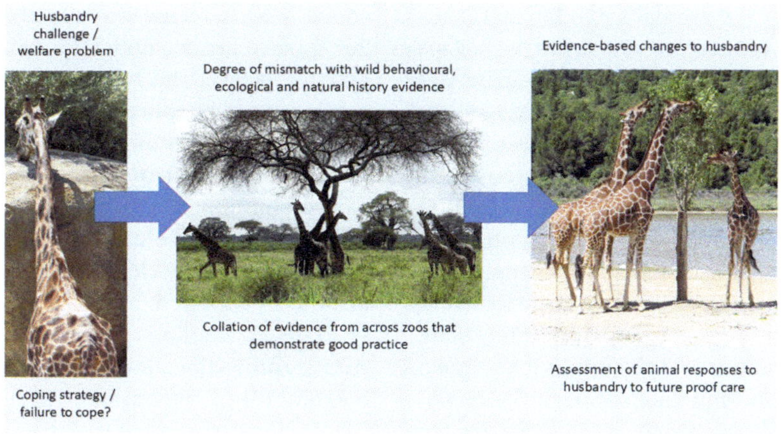

Figure 1.2 Identification of signs that an animal is not coping with current "care". In this case, oral stereotypic behaviour in a zoo-housed giraffe occurs for a significant proportion of a daily time budget and poor body condition indicative of inappropriate nutrition. Critique of why poor welfare indicators are apparent by matching current "care" to that from other facilities and review of wild data to understand ecological and behavioural needs. Measurable changes in performance (i.e., condition, growth, behavioural patterns) correlate with improvements or evolution of management practices. In this instance, provision of browse increases dry matter roughage intake for zoo giraffes to provide opportunities for rumination, social feeding, and increased time spent foraging. Degree of mismatch relates to the extent of any difference between what the animal would do naturally and what is permissible in captivity. Wild giraffe photo credit: G. Kautz.

with positive inputs; inputs being the resources (e.g. access to food, conspecifics or more space) that influence attainment of positive welfare states as the day progresses. The presence of the first visitors to the zoo may go unnoticed by the animals, as there is a latency of positive mood from earlier experiences. The arrival of a veterinary surgeon may cause some alarm to the individuals, but indicators of discomfort (e.g. heightened vigilance or attempts to flee) will reduce if nothing drastic occurs. Visitor presence may drop over lunchtime and the arrival of animal care stuff with enrichment (for example) in the early afternoon may further promote positive states. An unscheduled or

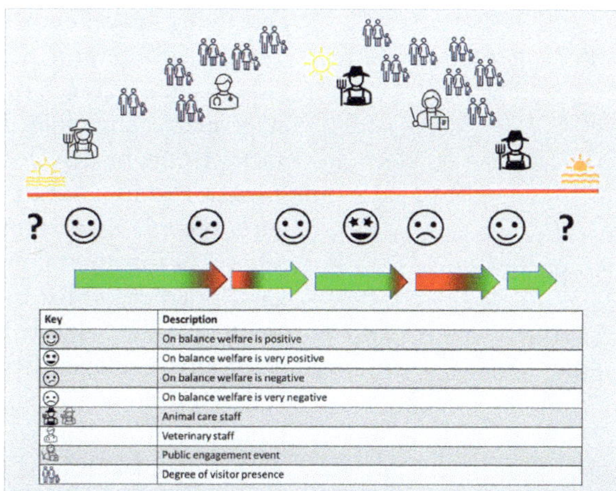

Figure 1.3 The cycle of welfare experience as a zoo animal is exposed to different positive, neutral, and negative stimuli across the course of its day. Green to red arrows represent a change from positive to negative welfare; red to green represents a change from negative to positive welfare. Question marks refer to times of the day when animal welfare may not be measured (e.g. overnight).

seasonal event at the animal's exhibit that causes large crowds to congregate can result in negative welfare states, and these may not dissipate until the last visitors leave the zoo at closing time and keepers fulfil their final rounds of feeding and daily care before the animals are left until the following day. In this hypothetical situation, the animal's routine is important; the familiarity of animal care and the use of husbandry methods to buffer against stressors can help reduce the length of time animals experience negative welfare states. Changes to behaviour and the amount of time an individual spends "not looking right" can be a helpful guide to how animals are feeling. Further, physiological measures (e.g. samples of faecal glucocorticoids) can be taken if an animal's behaviour does not return to what would be expected of that individual within a given timeframe to further support inferences of poor welfare state. A physiological change may be apparent even if the animal's outward activity does not appear much altered. For example, an overt "fight or flight" response may not have occurred when animal care staff enter an exhibit, but reduced time resting or foraging and increased time on vigilance or alert behaviour are subtle cues that a sympathetic adrenal medullary response has occurred. Homeostasis has changed from a "rest and digest" baseline and, if the causal stimulus were to be frequently repeated, and the animal had no way of coping with it, poor welfare could occur.

1.2 WHAT TO REMEMBER ABOUT ZOO ANIMAL WELFARE

Historically, zoos in the Post-War era have centred their key aims and objectives around conservation, education, research and entertainment (Kleiman, 1985), providing extra value to the existence of their living collections by attempting to engage, inspire, protect and teach. It is important to remember that physically and psychologically healthy animals greatly enable zoos to achieve these aims. Into the 21st century, the evolution of the entertainment aim further towards engagement (Rose & Riley, 2022) and the consideration of the wellbeing of the living collection and human components of the zoo as well (Greenwell et al., 2023; Rose & Riley, 2022) place greater obligation on zoos to promote positive welfare and reduce instances of suffering, dependence on coping and lack of opportunities to thrive. This obligation can be realised when zoos readily share what practices work and are transparent and honest about those that do not, whilst seeking potential solutions.

Whilst it is now clear that animal welfare states need to be assessed and evaluated, and measured via a mixed methods approach, what should constitute zoo welfare for the future is that which is meaningful to the animal. As covered by Veasey (2017), measurement of variables or predictors that are not relevant to the individual or its prevailing environment will result in incorrect and inappropriate conclusions as to how the animal is doing in the zoo. Using protocols that enable our improved understanding of the animal's psychological priorities, together with measurement of the overall impact of housing, husbandry, and management decisions, we can promote the relevant physical, behavioural, and psychological aspects of zoo animal care. A key principle of such meaningful assessment is to remember that, whilst the wild is important, it is no "Garden of Eden." Welfare in the wild is not always positive and zoo animals can, on balance and at the end of their lives, have experienced a more positive quality of life, thanks to the dedicated care of zoo staff.

Free-living animals will suffer from many stresses, including competition for food, resources and living space, choice of mate, acceptance into a social group, and possible predation. As individuals age and reach the end of their reproductive stage of life, they may struggle to keep up with a social group, be displaced from a profitable area of habitat, and lose out during agonistic interactions with others. Zoos should promote the positive aspects of zoo animal management and share such good practice, whilst being open and transparent about persistent challenges that cause some species to not thrive. Flexibility in collection planning and consideration of all reasons why a species is being housed is also required. Meaningful reflection on what a species' role is within the zoo, alongside meaningful measurement of how all species are doing, makes the zoo both more welfare-friendly and more ethical.

Therefore, an overall aim of this book is to prompt the reader (and practitioner) to check, recheck and check again that what they are providing to their animals (the inputs that will influence welfare states) is species-appropriate and aligned to the needs of the individual animals themselves. "Ask the animals – they will tell you what they need" must be a grounding principle for modern welfare-focussed husbandry.

1.3 WHY IS ZOO ANIMAL WELFARE AN IMPORTANT "NEW" AIM OF THE ZOO?

The idea of managing captive populations of wild species is contentious. Added value is given to a zoo's animal collection when welfare is good, and species can feel some degree of autonomy over their current state and environment. This is because many examples of stereotypic or unnatural behaviour are obvious to a zoo's visitors and therefore distracting from educational and scientific messages about the animals. Likewise, animals that can display choice and control over how they organise their lives will promote the aims of the zoo. Animals in this situation will also enable the zoo to realise a fifth "welfare aim" because the professional lives of the animal care and operational staff will be made easier by knowing that their animals are not stereotyping to deal with the challenges of captivity. If we review autonomy as a key end point of positive welfare, using Broom's original welfare definition as a starting point, we can identify where to intervene to enable our animals to be more autonomous. From Figure 1.4 (Rose & Lewton, 2025), we can see how the animal's state, its attempts at coping and the environmental factors that determine coping (internal and external) all have a role to play in how much self-control an animal has over life in the zoo.

If we consider that an animal's behaviour is a reflection of how they perceive the world (behaviour is defined as the observable response that an animal gives to external and internal stimuli) then psychological and physical states will be influenced by behaviour, and vice versa. An animal will attempt to cope better in the zoo (and therefore thrive rather than just survive) if they have control over the factors suggested in Figure 1.4. Likewise, when modern zoo enclosures give options over where to be, what to use, who to socialise with, what resources are available and where to retreat and seek shelter, animals can choose to control what they do. Finally, positive challenges in the form of temporal and social complexity (e.g. unpredictable, yet welcome, events across a 24-h period or a diverse and varied social group) can promote beneficial outcomes (feelings and emotional states) and behavioural development that further allows animals to be resilient. The degree to which positive welfare can be achieved via control, choice and challenge is down to the animal's environment. And the zoo environment is a hybrid of animal-centred features and those needed by various human elements of the zoo. Therefore, when planning out husbandry and housing, all stakeholders (including the key element of zookeepers who represent the captive species themselves) should be involved in developing and enabling the evolution of species care.

1.3.1 "CARE" OR CARE?

Care is an important word to spend some time discussing. We often talk about how the care provided can impact an animal's abilities to experience positive welfare states, and especially how its emotional state can remain healthy. But how do we know we are moving away from "care" to actually caring about what we provide? Species will struggle with captive "care" when this is inappropriate for their needs, and one should question if such incorrect care can even be considered care at all. If husbandry and housing is substandard and not fit for purpose, are we actually caring for our zoo's animals? Real care is achieved by integrating a compassionate approach to animals and to animal care staff, by looking back (but not staring) at where we have come from and by horizon scanning (but not trying to predict) the future.

As outlined in the British & Irish Association of Zoos and Aquariums (BIAZA) Animal Welfare Toolkit, Figure 1.5 illustrates that our early caring about animals had a key aim of stopping cruelty. Animals were or had to have already suffered before intervention commonly took place. Now, the aim of caring is to prevent suffering before it occurs and therefore the outcome of caring is to ensure animals can thrive and not just be kept alive. Originally, caring simply meant reducing or removing negative factors whereas now, caring is

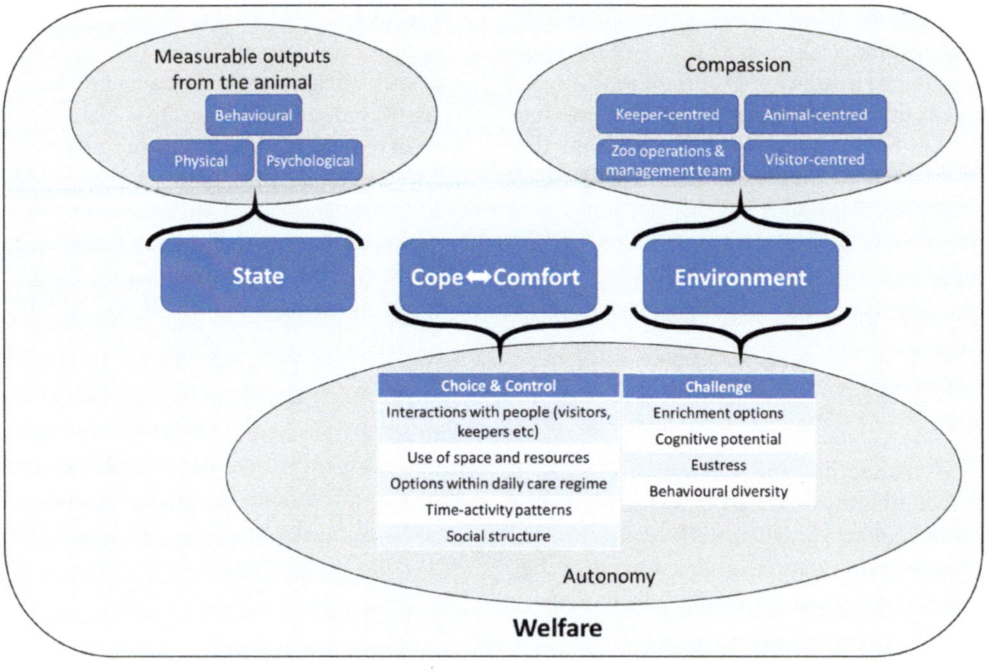

Figure 1.4 Taken from Rose and Lewton (2025), using the Broom welfare definition as a basis, and then considering how control, choice, and challenge can be added to ensure that coping strategies are positive and animals can thrive in a managed environment, and can experience comfort. Assessment of physical, behavioural, and psychological states can be used to underpin the positive impacts of enabling control and choice and providing positive challenge. Examples in the Control, Choice, and Challenge boxes are not exhaustive. "Environment" considers the internal (e.g. animal-centred elements) and external (e.g. impacts of people or stimuli potentially beyond the animals' control) conditions an animal will be experiencing that will have an impact on the attainment of positive welfare. Compassion is also important as correct care of the animals improves the working life of zoo animal management staff.

focus,sed on promoting positive experiences. And by positive experiences, we also include positive challenges. Animals are never expected not to be stressed, but the stress they experience must have a beneficial output for them to develop from and become more behaviourally flexible. This change in our caring focus to give attention to an animal's mental states has set new goals for the modern zoo, both in terms of how animals are presented and how the animals' responses to captivity are measured, interpreted and explained. For example, if a zoo cares about its animals and has invested in a new enclosure, why does the visitor still observe abnormal repetitive behaviours? Of course, there are a myriad of reasons as to why captive wild animals will perform abnormal behaviours, but it is an obvious signal to the visitor that something is not right for that animal. And the visitor could think, rightly or wrongly, that the care provided is not good. These are continual challenges for the zoo, and whilst solutions to ease obvious welfare issues have been implemented, there is still a need to do more. This conundrum, to promote an ideal message to the visitor yet still dealing with negative indicators of welfare, is beautifully summarised in Gray (2017).

> At the zoo you find much changed. The old cages are gone, replaced with new habitats. Fewer animals occupy larger spaces. Conservation messages have replaced zoological trivia...But many things remain the same: the school groups still pour through the gates, parents share stories and ice-cream with their children, and tigers pace.

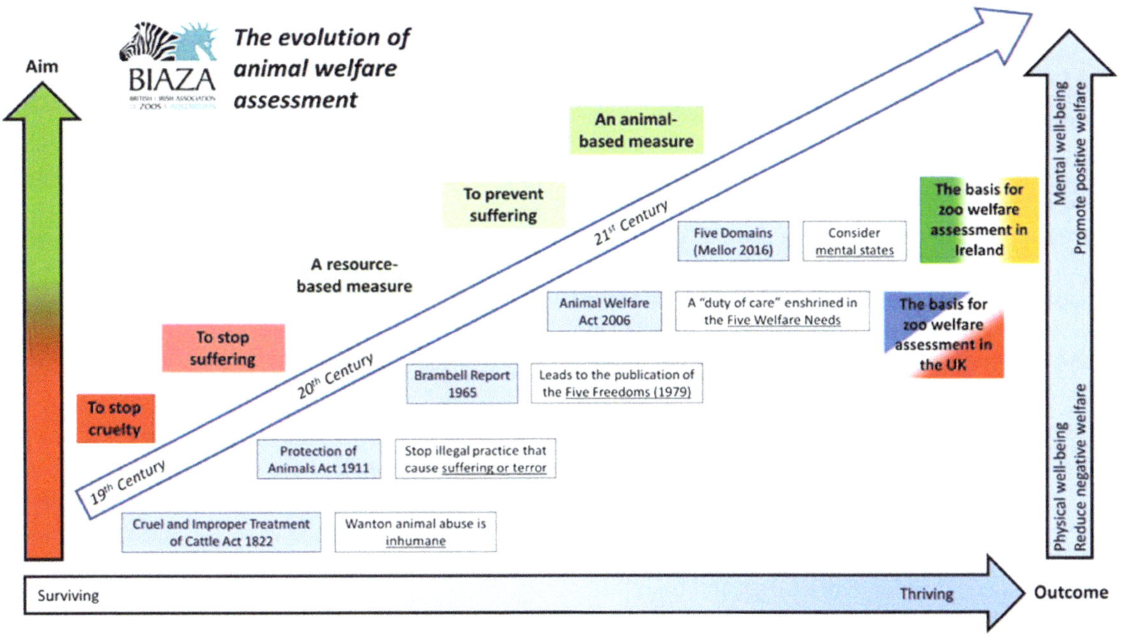

Figure 1.5 Taken from Harley and Clark (2019) the evolution of the thoughts and ideas behind animal welfare and how it can be assessed and protected in Great Britain and Ireland from the 19th Century to the present day. The first piece of animal welfare legislation, passed in 1822 solely to stop cruel practices, and this formed the basis of later legislation, theory and /or policy that attempts to prevent suffering, promote correct housing and husbandry, and to provide for an animal's emotional wellbeing too.

Scientific research can be used to predict what species are like to struggle in captivity and stand a higher probability of exhibiting abnormal repetitive behaviours (Clubb & Mason, 2003; Mason, 2010; Mason & Clubb, 2004). These research outputs can enable targeted interventions (e.g. with specific changes to housing and husbandry) as the link between "being captive" and exhibiting poor welfare has been established and verified – for example, Mellor et al. (2021). Perhaps an ideal is for an evolution of collection plans which are feasible and realistic. Zoos could develop these plans to manage species that are more suitable to ex situ care and provide support, expertise and financing of conservation work in situ for species that have been shown to cope less well in ex situ facilities. For example, support of mountain gorilla (*Gorilla beringei beringei*) in situ conservation efforts that are promoted by ex situ populations of lowland gorillas (*G. gorilla gorilla*).

Anticipatory behaviours are those that show an animal's sensitivity to a reward and may help to indicate the animal's perception of its own wellbeing (Watters, 2014). Study of such actions helps zoo staff understand what is perceived as valuable to an animal. However, caution is needed when determining anticipation as an overall positive output; if the animal spends much of its time in an expectant state and its time budget is comprised of anticipatory actions (e.g. actions that occupy the animal until the reward is provided) then an unnatural or inappropriate behavioural repertoire is being presented to zoo visitors and animal satiation may be unfulfilled. Distraction behaviours, that allow an individual to cope with the delay before a reward, are common across species (Brucks et al., 2022). Therefore, if the individual is attempting to distract itself from a situation that would otherwise be experienced negatively, and it can only cope with this by performing an unusual behaviour, then anticipatory actions (and any links to underlying positive welfare) need careful consideration. For example, when the anticipatory behaviour takes up a large part of the individual's time activity budget and continues even after the anticipated event has concluded.

1.4 HOW DO WE PROVIDE FOR MEANINGFUL POSITIVE WELFARE STATES?

Using data on species' responses to captivity can allow evidence-based husbandry refinements that, over time, improve welfare experiences. And moving care of some species from ex situ to in situ frees up space and resources for other taxa that may have a better role to play in the zoo. However, just as the wild is not a positive welfare paradise, so the wild may be very much reduced or non-existent for some of these more challenging-to-care-for species. For example, if climatic change continues at the pace predicted, then quality foraging and breeding habitats for polar bears (*Ursus maritimus*) will be severely restricted or eliminated completely. The immense size of a polar bear's range and their daily travelling distances have previously been identified as factors that predispose captive individuals to abnormal weaving, bobbing and pacing actions (Clubb & Mason, 2003). Yet, alterations to enclosure design and layout, targeted and ecologically relevant enrichment and technological advances in how key elements of "the wild" are replicated have seen advancements in polar bear quality of life in the zoo. An idealised view of "the wild" as the best place for an animal to be may not always hold true when scanning the metaphorical horizon. Perhaps more important for the polar bear in the zoo is novelty and environmental engagement, rather than an empty quantity of space. Measures of personality and behavioural expression, for example via use of a Qualitative Behavioural Assessment (QBA) methodology (Wemelsfelder, 2007) used to describe body language in a repeatable and objective manner, would allow for individual welfare scores of emotional health of captive polar bears under different housing styles.

QBA is helpful as zoos attempt to infer what animals may be thinking and feeling at a given time. Many zoos and other captive animal organisations reach for the Five Domains (Mellor et al., 2020) as a model for welfare assessment. Consideration is given to the four physical domains (the resources provided), and these are assessed to understand their impact on the outputs expressed by the animal (the fifth or mental domain). Valid description and definition of the mental domain is important. It is easy to fall into a trap of "dressing up" behaviour as behavioural expression. The mechanics of an action, the animal's state (long duration) and event (short duration) behaviours are different to the body language and expression of personality, emotion, mood and feeling that are expressed during a mechanical behaviour. Measurement of the fifth domain is challenging and there is, at the time of writing, limited information available on how to do this reliably, repeatably and without any subjectivity. Wider application of QBA across different species seems to be one way forward, and has been shown to be a useful approach for different taxonomic groups (Rose & O'Brien, 2020). Such methods for welfare assessment must be validated to ensure they are appropriate for the species being focussed upon, and they must generate reliable data to ensure valid husbandry advancements can be made. Use of automated tools or apps designed to collect behavioural data may be useful, if applied correctly, but scrutiny is needed to see whether such generic forms of data collection represent a validated means of determining welfare states. As questions arise over the utility of popular welfare frameworks (Hampton et al., 2023), a species-specific, more focussed approach to care is likely key to moving zoo animal husbandry and management, and the chances of attaining positive welfare states, forward (Rose, 2023).

Providing opportunities for positive challenge in the form of eustress, or beneficial stress (Selye, 1975), that gives extra behavioural outputs through the form of enhanced cognitive functions or behavioural diversity (Meehan & Mench, 2007; Villalba & Manteca, 2019) can allow zoo-housed individuals some options to express positive psychological and emotional states. Understanding of the value of a "reward" (e.g. access to enrichment, to conspecifics, or to different areas of an enclosure) would further help advance what is meaningful to measure when defining physical, behavioural, and psychological indicators of welfare. To enable this, an overall focus on positive measures, and not just negative behaviours, is required. This allows further understanding of the signs suggestive of appropriate, relevant housing and husbandry rather than signpost poor welfare that requires an immediate action or change to animal management. This focus on positive indicators is further explored in Chapter 6 (Environmental enrichment and promoting motivated behaviours).

"Naturalness" and the importance of the performance of species-appropriate behaviour is something that still needs to be at the forefront of welfare assessment and interpretation. Whilst direct comparison of wild traits with those of zoo animals is not always appropriate for a complete and valid welfare assessment (Veasey et al., 1996), our understanding of welfare challenges in the zoo is better informed when we consider evolutionary aspects of anatomy, physiology and behaviour. For example, Veasey et al. (1996) suggest that increased time spent grazing by captive giraffes may be of little concern; however, we now know that grass consumption by giraffes wears down teeth, causes rumen blockage and exposes animals to dangerous parasitic infections (Arnold et al., 2017). Therefore, even seemingly benign changes in the time spent on behaviour in captivity need evaluation against wild evidence, the available literature and records that document any potential consequences of this action. Well-managed mixed species' enclosures, where grazing species are housed alongside browsing species, can help reduce exposure of browsing species to parasites (Garretson et al., 2009). Building ecological knowledge into feeding areas and practical food delivery allows species to consume food in a natural manner, further separating foraging individuals from potential parasites. Therefore, meaningful measurement of welfare is linked to meaningful change to housing and husbandry; a scientific review coupled with a practical activity that provides animals with rewards, opportunities for satiation, as well as positive challenges that further enhance their resilience in a human-created environment.

1.5 CONCLUSIONS

Broom's (1986) definition is still a suitable foundation for inferring, interpreting and measuring welfare inputs and welfare outputs in the zoo, especially when we consider the physical, behavioural and psychological components of welfare and how these are influenced by the unique environment of the zoo. The need to provide for a species' behavioural needs is clear, but this must be done in a balanced and objective way using both wild information and evidence from ex situ facilities (Box 1.1). Therefore, to futureproof the "what, why and how" of zoo animal welfare, we need to:

- Further collaborate on, consider, and construct ways of meaningfully assessing all of the five domains of welfare across different zoo-housed species.
- Continue/start to monitor behaviour for all species that we house in captivity, even for those where behaviour may appear tricky to define and interpret. It is essential that we gather information to describe ethograms for all captive species and share these widely with all the interested parties. Without useful and accurate descriptions of behaviour and time-activity budgets, an understanding of what different welfare states may present as will be unavailable.
- Validate physiological measures of welfare outputs (e.g. metabolite concentrations of circulated hormones that are released in response to stress, excitement or comfort) at the species-specific level that can be reliably correlated with behaviour patterns.
- Measure, apply, evaluate, and reassess husbandry practices at the species-specific level to ensure their suitability and future-proofing.
- Share knowledge and information across species where general ecology may be similar, where something well known in one species might help to interpret "how is my animal doing?" for another species. Such sharing of information can provide the basis for more detailed, species-specific observation or study to refine husbandry and care.
- Consider how we subdivide stress. The physiological response to a stressor is the same regardless of whether that stressor is positive (e.g. excitement/ "eustress") or distressing. To simplify that a stressor is only a stimulus that elicits a change to homeostasis and, if prolonged or not biologically relevant, that it can be distressing, may allow further discussion and interpretation on the types of stressors that animals experience in the zoo.

The application of this background, "what, why and how" welfare information, is further expanded upon in the "applications" section of this book. With this, we can utilise our knowledge of

BOX 1.1: Increasing capacity and embedding welfare-friendly practices into day-to-day care. Some useful sources that i) explore cross-taxa welfare challenges and potential solutions, or ii) explain and outline initiatives in capacity building on species' husbandry needs and welfare requirements, or iii) evaluate the usefulness of evidence on ecology and behaviour to enable us to interpret, and therefore manage, welfare states and quality of life in the zoo.

- Blackett, T. A., McKenna, C., Kavanagh, L. & Morgan, D. R. (2017). The welfare of wild animals in zoological institutions: are we meeting our duty of care? *International Zoo Yearbook*, *51*(1), 187–202.
- Kagan, R., Carter, S. & Allard, S. (2015). A universal animal welfare framework for zoos. *Journal of Applied Animal Welfare Science*, *18*(sup1), S1–S10.
- Melfi, V. & Hosey, G. (2011). Capacity building for better animal welfare. *International Zoo Yearbook*, *45*(1), 274–281.
- Mellor, E. L., McDonald Kinkaid, H. K., Mendl, M. T., Cuthill, I. C., van Zeeland, Y. R. A. & Mason, G. J. (2021). Nature calls: Intelligence and natural foraging style predict poor welfare in captive parrots. *Proceedings of the Royal Society B: Biological Sciences*, *288*(1960), 20211952.
- Riley, L. M. (2018). Conserving behaviour with cognitive enrichment: A new frontier for zoo conservation biology. In M. Berger & S. Corbett (Eds.), *Zoo animals: Husbandry, welfare and public interactions* (pp. 199–264). New York, NY: Nova Science Publishers.
- Rose, P. E. (2018). Ensuring a good quality of life in the zoo. Underpinning welfare-positive animal management with ecological evidence. In M. Berger & S. Corbett (Eds.), *Zoo animals: Behavior, welfare and public interactions* (pp. 141–198). New York, NY: Nova Science Publishers Inc.
- Rose, P. E., Brereton, J. E. & Gardner, L. (2016). Developing flamingo husbandry practices through workshop communication. *Journal of Zoo and Aquarium Research*, *4*(2), 115–121.
- Yon, L., Williams, E., Harvey, N. D. & Asher, L. (2019). Development of a behavioural welfare assessment tool for routine use with captive elephants. *PLoS One*, *14*(2), e0210783.

behavioural needs, meaningful welfare measures, species differences in enrichment design, enclosure construction and layout, animal training and human interactions, and nutrition. Hopefully, this will, in time, mean that, when the crowds of visitors pour through the zoo's gates and engage with conservation and educational messaging, the tigers will no longer be pacing.

REFERENCES

Arnold, N., Cain, A., Rowland, R., Steele, S., Martel, C., Burton, M. & Bissell, H. (2017). *The influence of giraffe behavior on parasite load: Impact of husbandry modifications at Busch Gardens Tampa Bay.* Paper presented at the Twelfth Conference on Zoo and Wildlife Nutrition, Frisco, USA.

Broom, D. M. (1986). Indicators of poor welfare. *British Veterinary Journal*, *142*(6), 524–526.

Broom, D. M. (2006). Behaviour and welfare in relation to pathology. *Applied Animal Behaviour Science*, *97*(1), 73–83.

Broom, D. M. & Johnson, K. G. (1993). *Stress and animal welfare* (Vol. 993). Cham, Switzerland: Springer Nature.

Brucks, D., Petelle, M., Baldoni, C., Krasheninnikova, A., Rovegno, E. & von Bayern, A. M. P. (2022). Intra-and interspecific variation in self-control capacities of parrots in a delay of gratification task. *Animal Cognition*, *25*(2), 473–491.

Clubb, R. & Mason, G. J. (2003). Captivity effects on wide-ranging carnivores. *Nature*, *425*(6957), 473–474.

Garretson, P. D., Hammond, E. E., Craig, T. M. & Holman, P. J. (2009). Anthelmintic resistant Haemonchus contortus in a giraffe (Giraffa camelopardalis) in Florida. *Journal of Zoo and Wildlife Medicine*, *40*(1), 131–139.

Gray, J. (2017). *Zoo ethics: The challenges of compassionate conservation.* Ithaca, NY: CSIRO Publishing.

Greenwell, P. J., Riley, L. M., Lemos de Figueiredo, R., Brereton, J. E., Mooney, A. & Rose, P. E. (2023). The societal value of the modern zoo: A commentary on how zoos can positively impact on human populations locally and globally. *Journal of Zoological and Botanical Gardens*, *4*(1), 53–69.

Hampton, J. O., Hemsworth, L. M., Hemsworth, P. H., Hyndman, T. H. & Sandøe, P. (2023). Rethinking the utility of the Five Domains model. *Animal Welfare*, *32*, e62.

Harley, J. J. & Clark, F. E. (2019). *Animal welfare toolkit*. London, UK: BIAZA.

Kleiman, D. G. (1985). Criteria for the evaluation of zoo research projects. *Zoo Biology*, *4*(2), 93–98.

Mason, G. J. (2010). Species differences in responses to captivity: Stress, welfare and the comparative method. *Trends in Ecology & Evolution*, *25*(12), 713–721.

Mason, G. J. & Clubb, R. (2004). Pacing polar bears and stoical sheep: Testing ecological and evolutionary hypotheses about animal welfare. *Animal Welfare*, *13*, 33–40.

Meehan, C. L. & Mench, J. A. (2007). The challenge of challenge: Can problem solving opportunities enhance animal welfare? *Applied Animal Behaviour Science*, *102*(3–4), 246–261.

Mellor, D. J. (2016). Updating animal welfare thinking: Moving beyond the "five freedoms" towards "a life worth living". *Animals*, *6*(3), 21.

Mellor, D. J., Beausoleil, N. J., Littlewood, K. E., McLean, A. N., McGreevy, P. D., Jones, B. & Wilkins, C. (2020). The 2020 five domains model: Including human–animal interactions in assessments of animal welfare. *Animals*, *10*(10), 1870.

Mellor, E. L., McDonald Kinkaid, H. K., Mendl, M. T., Cuthill, I. C., van Zeeland, Y. R. A. & Mason, G. J. (2021). Nature calls: Intelligence and natural foraging style predict poor welfare in captive parrots. *Proceedings of the Royal Society B: Biological Sciences*, *288*(1960), 20211952.

Rose, P. E. (2023). Identifying essential elements of good giraffe welfare - can we use knowledge of a species' fundamental needs to develop welfare-focussed husbandry? *Journal of Zoological and Botanical Gardens*, *4*(3), 549–566.

Rose, P. E. & Lewton, J. (2025). Key concepts for enhancing zoo animal welfare: Coping, comfort, choice, control, challenge and compassion *Journal of Applied Animal Welfare Science*, in press.

Rose, P. E. & O'Brien, M. (2020). Welfare assessment for captive Anseriformes: A guide for practitioners and animal keepers. *Animals*, *10*(7), 1132.

Rose, P. E. & Riley, L. M. (2022). Expanding the role of the future zoo: Wellbeing should become the fifth aim for modern zoos. *Frontiers in Psychology*, *13*, 6649.

Selye, H. (1975). Stress and distress. *Comprehensive Therapy*, *1*(8), 9–13.

Veasey, J. S. (2017). In pursuit of peak animal welfare; The need to prioritize the meaningful over the measurable. *Zoo Biology*, *36*(6), 413–425.

Veasey, J. S., Waran, N. K. & Young, R. J. (1996). On comparing the behaviour of zoo housed animals with wild conspecifics as a welfare indicator. *Animal Welfare*, *5*, 13–24.

Villalba, J. J. & Manteca, X. (2019). A case for eustress in grazing animals. *Frontiers in Veterinary Science*, *6*, 303.

Watters, J. V. (2014). Searching for behavioral indicators of welfare in zoos: Uncovering anticipatory behavior. *Zoo Biology*, *33*(4), 251–256.

Wemelsfelder, F. (2007). How animals communicate quality of life: The qualitative assessment of behaviour. *Animal Welfare*, *16*(2), 25–31.

Yeates, J. W. (2011). Is 'a life worth living' a concept worth having? *Animal Welfare*, *20*(3), 397–406.

2

The complexity of conceptualising and defining animal welfare

LISA M. RILEY

2.1 ZOO ANIMAL WELFARE IN CONTEXT

The welfare of zoo-housed animals is a hot topic. As societies globally grapple with the justification of keeping wild animals in captivity (Gray, 2017), animal welfare (hereafter 'welfare') is often the deciding factor in a complex ethical argument. While zoo animals are (technically) protected from predation, starvation, dehydration, and disease, welfare is more than the maintenance of life and legitimately involves more than the scientific measurement of welfare state, requiring that we consider societal views and the representation or place of animals in society. Zoos exemplify this as the public access zoos daily and form opinions about the animals and enclosures seen. The characteristics, indicators, and measures that the public holds important should also be considered important by the zoo industry, even if just to communicate welfare messages to visitors. There is perhaps no better example of this multi-stakeholder approach to welfare than the emergence of animal welfare science as an academic discipline, a field which informs our very concept of welfare across all animal contexts, including zoos.

Animal welfare science emerged from the Brambell Report in 1965 – full name, "Report of the Technical Committee to Enquire into the Welfare of Animals Kept Under Intensive Livestock Husbandry Systems" – which followed a government inquiry into the way farm animals are kept in intensive production systems. It was Ruth Harrison's book, *Animal Machines* (Harrison, 1964) that documented the experiences of animals in intensive farming and generated the public concern which led to the inquiry. After the Second World War, intensive farming practices in the UK and globally gained popularity as efficient conversion of food to carcass was required to satisfy hungry post-war nations. This took animals out of the fields into the realm of the invisible, as animals in intensive systems are kept at high stocking density indoors to facilitate growth and reproductive output. The general population questioned intensification little and mostly adopted an out-of-sight-out-of-mind philosophy. Harrison's book provided emotive and enlightening descriptions of intensive farming, and the subsequent social awareness

DOI: 10.1201/9781003429425-2

stimulated the UK government to hold the inquiry that was led by Professor F. W. Rogers Brambell.

The Brambell Report made multiple recommendations regarding what should be considered as 'welfare', including health, behaviour, stress, and emotional state, and concluded all animals in intensive systems should have 'sufficient freedom of movement to be able without difficulty, to turn around, groom itself, get up, lie down and stretch its limbs' (p.13). 'Freedom' was later explored by the Farm Animal Welfare Council (FAWC), and the Five Freedoms welfare assessment framework was established in 1979 and later published in refined form (FAWC, 1993). The Five Freedoms were subsequently transposed into the Five Welfare Needs in the Animal Welfare Act 2006 (England) as a duty of responsibility to ensure the welfare of a protected animal (which includes zoo vertebrates). The Five Welfare Needs are:

- The need for a suitable environment
- The need for a suitable diet
- The need to be able to exhibit normal behaviour patterns
- The need to be housed with, or apart from, other animals
- The need to be protected from pain, suffering, injury and disease

These needs are used extensively in the zoo industry and beyond; they form a major component of accreditation guidance and zoo licensing and underpin the mission statements and modern roles of the zoo – Well-being, Research, Engagement, Conservation, Education (Rose and Riley, 2022). Yet the Five Welfare Needs do not tell us what welfare is; rather, they tell us a possible way in which good welfare could be achieved (by meeting every need). To determine the former, we need to consider the welfare concept.

2.2 WELFARE CONCEPT

If reading this book, you are interested in zoo animal welfare, and you likely have your own definition of welfare. But if you asked your colleagues, fellow students, or consulted the literature, you would probably find multiple definitions of welfare that differ from yours. While stating your definition is essential, seeking a universal definition of welfare is perhaps a misdemeanour, distracting from the application of evidence-based husbandry (Melfi, 2009) to create opportunity for good welfare. However, a helpful and necessary step, often overlooked, is the consideration, as in the Brambell Report, of what welfare involves. This book therefore needs an overarching concept of welfare (Figure 2.1) into which definitions and indicators of welfare (specific, indirect measures of welfare, e.g. rumination time, cortisol/glucocorticoid metabolite concentration, growth rate, expression of joyfulness) fit.

2.2.1 THE COMPONENTS OF ANIMAL WELFARE

Welfare, as noted in Chapter 1, is a continuum; good welfare and poor welfare are not separate concepts, rather the components of welfare vary in degree or type along a continuous line between (very) good and (very) poor welfare. With dozens of potential indicators, a model which categorises indicators into welfare components is needed and this will guide both welfare definition and assessment. Fraser (2003, 2008) proposed a welfare concept that remains progressive and inclusive. While exploring a multidisciplinary approach to welfare and examining why opposing views can be reached regarding the same animal's welfare, Fraser proposed the differing welfare perspectives that people can combine to make a more complete welfare concept, that essentially comprises three components: 1) biological functioning; 2) affective state; 3) naturalness (Figure 2.1b). This concept is particularly relevant to zoo animals who need excellent health to remain reproductively viable and fulfil the zoo's conservation aims, need to experience a range of affective states to regulate their physiological, psychological and social wellness, and, being wild (non-domesticated) animals, need a life aligned to their evolution and behavioural biology (Rose, 2022) which allows for the conservation of both genetics and behaviour; conserving behavioural skills and flexibility for survival in the natural ecological niche (Riley, 2018). For example, a zoo-housed primate requires autonomy over their feeding, locomotion and (to a degree) their social interaction (the 'autonomy approach', Riley, 2022). This aligns with their behavioural biology

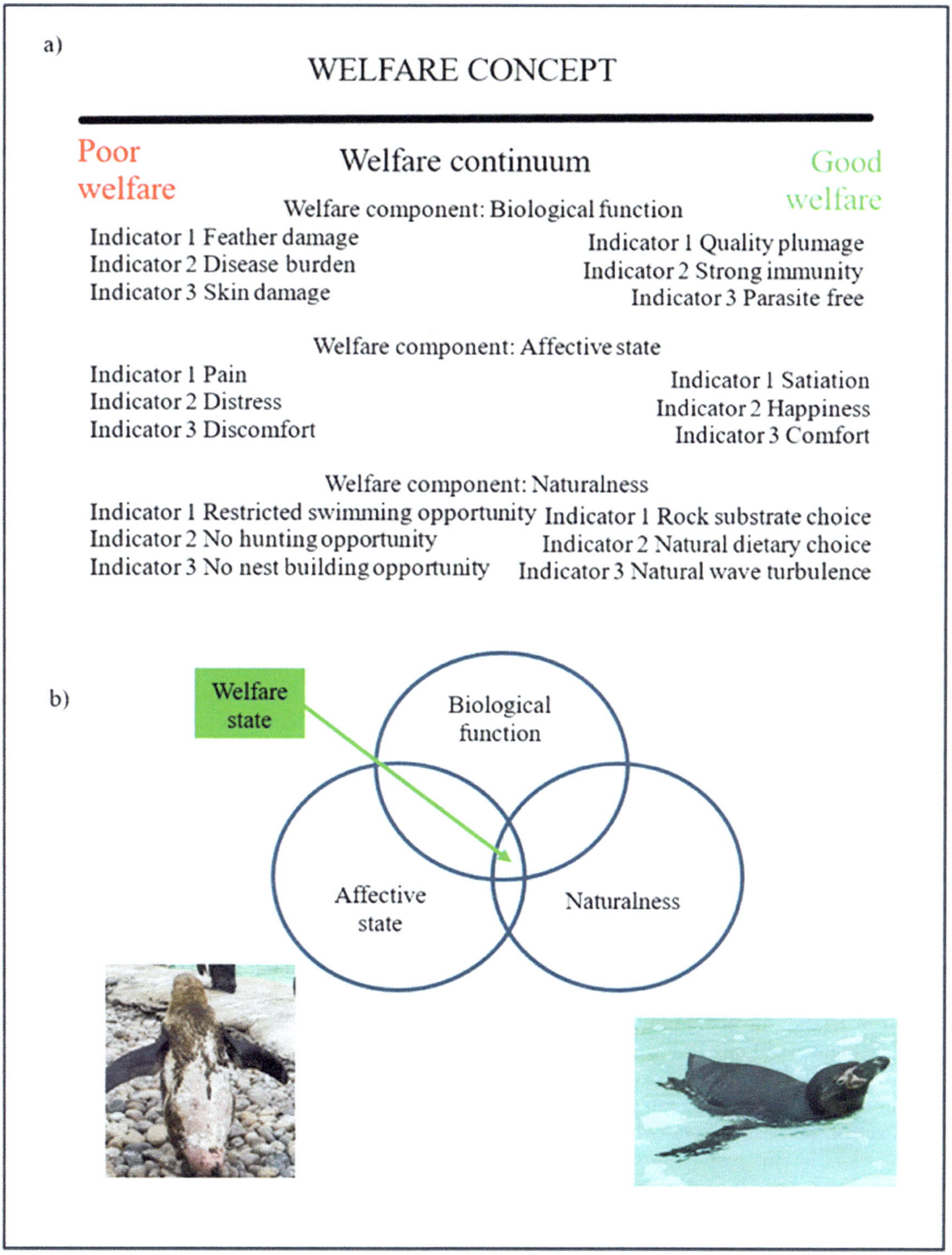

Figure 2.1 Welfare theory and definition. a) the separate components of welfare with example indicators of both poor and good welfare; b) how welfare components merge to evidence the welfare concept. Animal example using Humboldt penguins (*Spheniscus humboldti*).

regarding foraging, food preparation and social interaction, the major activities of primates in the wild and in which they invest significant cognition, time and resource. In the zoo environment, provision of learning opportunity, a diverse and varying physical and nutritional environment and biologically relevant social organisation are essential if primates are to thrive, having what Green and Mellor (2011) would describe as 'a good life' where 'the balance of salient positive and negative experiences is strongly positive' (p. 267).

2.2.2 BIOLOGICAL FUNCTION

Biological function refers to the health, growth, reproductive output and productivity of animals; in other words, how well their physical biology is working. Arguably, productivity (carcass quality, milk yield, egg yield, for example) is less relevant to zoo animals but, in totality, biological functioning, inclusive of physical health, is inherently part of zoo animal welfare. For some, welfare *is* health. Human health is defined in the constitution of the World Health Organisation (WHO) as 'a state of complete physical, mental and social well-being and not merely the absence of disease or infirmity'. Early in the development of animal welfare science, and pre the 1990s, when discussion of animal emotions was rare, health was widely the focus of welfare discussions and research. Hughes' (1976) definition of animal welfare aligns with that of human health, that welfare is '...a state of complete mental and physical health, where the animal is in harmony with its environment' (which technically defines *good* welfare, rather than welfare). While this definition importantly identifies the role of the current environment in which the animal is housed in relation to that animal's welfare state, complete mental and physical health does not, by default, lead to good welfare, as even in the absence of clinical pathology, animals can experience negative affective states, such as boredom and frustration. Dawkins has for decades proposed that health should be the major concern when discussing welfare, stating 'the essence of good welfare is therefore being currently healthy and also having good prospects for future health' (Dawkins, 2021, p. 7).

While it is undeniable that an animal who has sustained an injury or who is in a disease state will experience some biological compromise and, therefore, may experience negative welfare states, it is oversimplistic to consider that either health encompasses all of welfare or that complete mental and physical health can be achieved or maintained indefinitely. An animal with good health may experience negative welfare states, and conversely, an animal with illness can still experience positive affective states, fulfil behavioural motivations and thus, despite being ill, can still have *a good life*. For example, a zoo ungulate with hoof issues, such as an okapi (*Okapi johnstoni*) with excessive sole pad wear (Dadone and Foxworth, 2023) may still have good welfare, present with normal gait, natural behaviour and no noticeable clinical signs of ill-health; however, in the future, as the sensitivity of the foot tissues increases, the animal will likely start to experience pain and this could result in an overall negative welfare state. The issue here should be the potential of the situation to cause the animal to experience poor welfare in the future, and that warrants intervention to mitigate potential risk to welfare. This is not to say that health is not an important component of welfare; it is, firstly due to the far-reaching potential effects of illness on the physiology and psychology of the animal; secondly, for the threat to life from ill-health or injury; and thirdly, because the often obvious, directly observable symptoms of ill-health may be the only noticeable indicators of poor welfare. Biological functioning indicators are particularly important when individuals are not identifiable from the social group (in a shoal of fish, for example) and when the behaviour of the species and/or their affective states are not well understood, making valid conclusions about welfare states based on behaviour and/or affective states difficult to achieve. Thus, the biological functioning of an animal should always be considered if assessing welfare, but not in isolation

2.2.3 AFFECTIVE STATE

Affective state refers to the emotional and mood-based experiences of an animal that are transient, subjective and characterised by hedonic value – the extent to which the affective state is pleasurable, or not (Mendl et al., 2010; Mendl and Paul, 2020). While emotions are typically considered short-term experiences elicited by a specific internal or external stimulus, moods are long-term and generalised as they do not result from specific stimuli (Paul et al., 2005; Mendl et al., 2010).

Such emotions and moods affect decision-making, behaviour and physiology and have adaptive value and thus an effect on fitness and survival.

The importance of affective states to welfare cannot be understated and was first acknowledged in a scientific context in the Brambell Report (1965, p. 9), which stated 'Welfare ... embraces both the physical and mental well-being of the animal. Any attempt to evaluate welfare, therefore, must take into account the scientific evidence available concerning the feelings of animals that can be derived from their structure and functions and also from their behaviour.' When the Brambell Report was published, scientific evidence of non-human affective states was in its infancy and observable, objective evidence scarce, therefore. While affective states are often studied in laboratory or domestic animals, zoo animals should neither be considered exceptional nor excluded in their ability to experience pain, joy, fear, contentment, and a range of other affective states. Zoos hold a diverse range of species and while brain structure and function (and therefore the capacity to experience affective states) of mammalian and avian zoo animals may be comparatively well understood. Fish, reptiles, amphibians, and invertebrates are typically overlooked when considering affective states.

While we may not know the exact range or extent of affective states in a given species, with the introduction of the Animal Welfare (Sentience) Act 2022 in England and established recognition of animal sentience in EU law for example, we must acknowledge the possibility (some might say likelihood) of all animals being, to greater or lesser extent, sentient and thus able to experience affective states.

In relation to welfare, it may not be essential to determine specifically what affective states are or are not being experienced by the animal, as the *core affect* of the emotion or mood should be prioritised. Mendl et al. (2010) conceptualised the *dimensional approach* of the study of affective states. This approach assesses two dimensions of affective state: arousal and valence. Arousal simply refers to the intensity of the affective state which varies on a continuum of low-to-high arousal, and valence refers to the animal's perception of the affective state as positive (pleasurable, reward-based) or negative (unpleasant, punishing). By considering the animal's demeanour (the holistic expression of the animal as it performs behaviour, Wemelsfelder, 2007) affective state's core affect can be determined and thus welfare state interpreted. If the animal's demeanour indicates a positively valenced affective state, welfare can be interpreted as potentially good; if the animal's demeanour indicates a negatively valenced affective state, welfare can be interpreted as potentially poor, without needing to specifically label what affective state is being expressed.

As animal welfare science progresses, affective state is increasingly incorporated into welfare concept. For some, welfare *is* (and has always been) an affective state. Duncan (1993, 2004) considers welfare as a subjective experience and therefore considers welfare to be about how an animal feels, thus Duncan would consider good welfare is '...the absence of negative subjective emotional states [suffering]....and probably with the presence of positive subjective emotional states that are usually called pleasure' (p. 88). Yet due to their core affect, affective states are both influenced by and influence the wider biology of the animal. If we consider the positively valenced 'joy', this experience is rewarding and therefore encourages behaviour change as the animal will be motivated to repeat the behaviour associated with the feeling of joy. This can positively regulate serotonin (neurotransmitter) production and uptake and deactivates the Hypothalamic-Pituitary-Adrenal (HPA) stress response, thus leading to physiological changes also. It would seem oversimplistic to therefore consider affective state in isolation when discussing the welfare concept, but similarly if we are investigating welfare, it would be absurd to omit affective state. Affective state is vital for, but not sufficient, if conceptualising welfare.

2.2.4 NATURALNESS

Fraser (2003, 2008) considers naturalness, or 'natural living' as an equally important component of welfare concept. In Harrison's *Animal Machines*, the author describes the caging of intensively reared poultry as so removed from the animal's natural way of living that it causes the animals to be unhealthy and miserable and is thus detrimental to their welfare. In a lifetime of works Bernard Rollin has consistently referred to the nature of animals, which he labels 'telos'. The notion of telos was discussed by Aristotle who described telos in his Four Causes. Aristotle used telos in relation to plant and animal biology as well as human ethics. Aristotle describes telos as the end goal or purpose of human behaviour, for example, to eat is to satiate hunger, to research animal husbandry is to satisfy a thirst for knowledge. Rollin applied this to animals describing the natural basis of animal behaviour and affective state, postulating 'it is plausible to suggest that happiness resides in the satisfaction of the unique set of needs and interests, physical and psychological, which make up what I have called the telos, or nature of the animal in question. Each animal has a nature which is genetically and environmentally constrained, from which flow certain interests and needs, whose fulfilment or lack of it matter to the animal' (Rollin, 1989, p. 203).

Wild animals in captivity retain their natural motivations to hunt, forage, maintain social relationships, establish territory and avoid predation, and every animal is the result of speciation – the genetic isolation of a species from a common ancestor as a result of adaptation to environmental change. Therein, the telos of an animal aligns with the animal's evolution and environmental niche – the anatomy, physiology, behaviour, affective states and interests of the animal represent adaptations of that animal for particular purpose(s), which Rollin (2020) simply describes as the "pigness" of a pig, "cowness" of a cow (p1695), and being able to achieve said purpose is relevant to welfare. Thus, to achieve good welfare a zoo animal must be able to actualize its biological and psychological nature; it must be able to evidence its *speciesness*.

Rose (2022) explores the importance of biologically relevant zoo animal husbandry that aligns with the behavioural biology of zoo animals, noting 'the challenge for the zoo is to provide an environment that enables a wild animal to experience environmental fluctuations and challenge in a way that promotes the performance of advantageous behaviours that deliver fitness benefits' (p. 8). It is therefore essential that the welfare concept includes naturalness, even if ultimately considered in terms only of natural behaviour. Thus, both the behavioural actions of an animal (the mechanistic movement of body/body parts, e.g. state behaviour which is relatively long lived such as grooming, and event behaviour which is instantaneous such as a threat face display, or yawn) and the behavioural expression (demeanour, mood and emotion) as well as the physiological basis of both behavioural actions and behavioural expression are important considerations when assessing naturalness, the welfare of zoo animals.

In this book, we therefore consider that welfare comprises biological functioning, affective state and naturalness and this guides how we discuss, define, assess and prioritise welfare throughout the book.

2.3 DEFINING AND FRAMING WELFARE IN THIS BOOK

Now that welfare concept is understood and we accept that welfare is comprised of biological functioning, affective state and naturalness, we seek a definition that encompasses these components. As mentioned in Chapter 1, this book embraces a long-standing definition used extensively in the animal management industry: "The welfare of an individual is its state as regards its attempts to cope with its environment" (Broom, 1986). For some this may be an outdated choice, while others will find the familiarity of the definition easy to process. For those who are sceptical, we seek to reassure you that our take on Broom's definition throughout this book is progressive and inclusive and considers welfare from the perspective of Ohl and Putman (2018, p. 45), 'welfare describes an internal state of an individual, as experienced by that individual'. We consider in this book therefore that welfare is the perception of an individual animal (consciously or subconsciously) regarding their current state of being (Figure 2.2). Our approach recognises, as Broom (1996) emphasised, humans do not give an animal good welfare, we merely provide an animal with the opportunity to perceive good welfare given the environment provided to the animal (physical, nutritional, sensory, social, cognitive) and wider husbandry under which that animal is maintained. Thus, for the purpose of this book, the 'state' referred to in Broom's definition is the animal's own perception of how they are doing.

The endorsement of a coping-based concept of welfare will potentially cause controversy. Coping suggests that the animal in question has to adapt to a substandard environment, and as long as adaptation occurs (no matter the cost), the animal

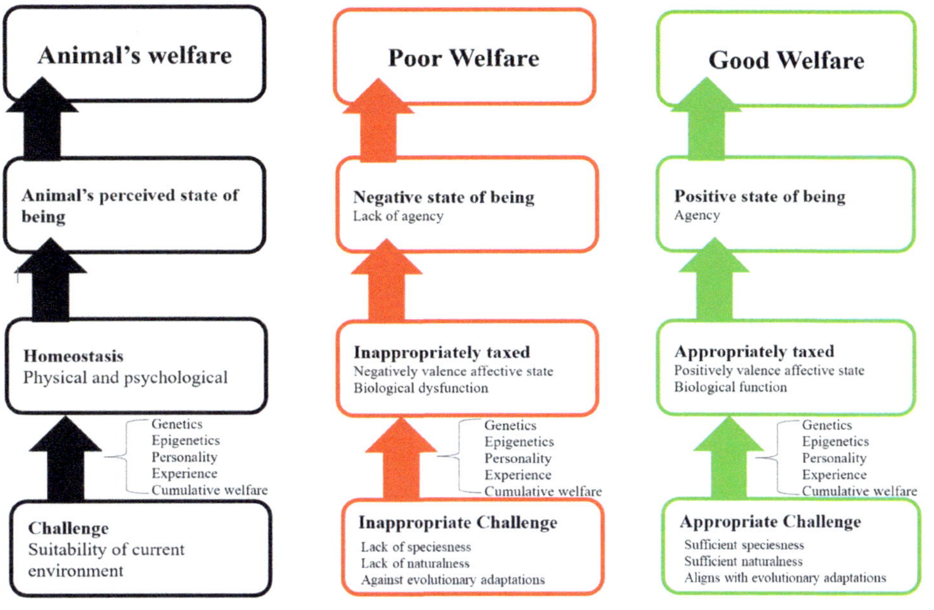

Figure 2.2 The welfare concept is evidenced, and the pathways to both poor and good welfare are shown. The effects of challenge are mediated by the individual characteristics of the animal (genetics, epigenetics, personality, experience, and cumulative welfare) that determine the individual's welfare state by taxing the animal's homeostasis, leading to either positive outputs or negative valence and interruption of biological function.

will have good welfare (Broom, 1991). Because of the way 'coping' is used in everyday language – *the strategies enacted to deal with a difficult* situation – welfare may incorrectly be considered a state only achieved in adversity, which pushes the thresholds of acceptable welfare to a low level and focuses attention on suffering and negative welfare states. For 30 years post-publication of the Brambell Report, animal welfare scientists focused on negative welfare states. There is a legal, ethical, and moral obligation to intervene if an animal is experiencing negative/poor/compromised welfare, with the aim of providing an opportunity for the animal's welfare to improve. No such obligation exists if the animal is experiencing good welfare, and good welfare is not merely the absence of negative experiences, such as distress, suffering, frustration, and pain. This way of thinking builds upon the WHO definition of human health stated previously – that health is not merely the absence of disease or infirmity. Appleby and Sandøe (2002) suggest that to experience good welfare, an animal will need to experience pleasure in addition to an absence of negative affective states (fear, pain, etc.). For good welfare to be achieved, an animal must experience wellness, satiation, a range of other positive affective states, and good health.

In the literature there is increasing recognition of the need to assess welfare using both positive and negative indicators, with the need for 'explicit' inclusion of positive welfare outcomes in any assessment framework (Yeates and Main, 2008). Broom (1999) considered measurement of welfare should jointly involve indicators of poor welfare (e.g. disease, reduced longevity, inhibited physiological processes, behavioural pathology) and indicators of good welfare (e.g. extent to which strongly preferred behaviours can be expressed, physiological and behavioural indicators of pleasure). To incorporate this into Broom's original welfare definition, we can reframe coping as homeostasis (Figure 2.2). Homeostasis refers to the physical and psychological self-regulatory processes that maintain the functionality of a biological system. The system as a collective whole has a narrow range of optimal processing and a wider range of tolerance limits below or above the range of optimal processing. We will consider challenge to be any condition (external or internal) that affects homeostasis, which equally includes positive and negative challenges. A negative challenge moves homeostasis away from optimal, may lack biological relevance, and may chronically push homeostasis to near the tolerance levels or exceed those tolerance levels. Examples would include exposure to extreme weather conditions, sustained threat of a predator, direct and sustained competition from a conspecific, depression, or other negatively valenced affective states. Positive challenge moves homeostasis away from optimal, typically has biological relevance, and may result in acute departure from optimum that is recoverable. Examples would include surmountable obstacles to food acquisition, sexual competition that leads to copulation, working to achieve physical comfort, or other positively valenced affective states. The challenge should be biologically relevant to incorporate naturalness, and ultimately the cycle of biologically relevant challenge and homeostatic regulation can be considered biological functioning. The affective states of the animal will be impacted by and will impact the way the challenge is perceived and the exact homeostatic processes that are up- or down-regulated. By reframing coping slightly, welfare is considered holistically as a continuum of positive to negative states, and this encourages a more inclusive assessment of welfare to focus equally on indicators of positive and negative welfare states.

The definition of welfare provided here is to guide discussions throughout the book. It is not a universal definition (nor does it aim to be), but it is operationalisable and, when reframed slightly from Broom's original definition, it can be considered progressive and holistic for defining welfare as a perceived state of being, focusing welfare on an individual rather than a group of animals, considering all three components of the welfare concept (biological functioning, affective state and naturalness) and by considering equally both positive and negative welfare states. While chapter authors and readers may prefer other definitions, this should not hinder our exploration or assessment of zoo animal welfare.

2.4 INTEGRATING RESOURCE-BASED INPUTS AND ANIMAL-CENTRIC OUTPUTS FOR ASSESSING WELFARE

Broom's (1986) welfare definition considers environmental challenge. Broom clearly meant this to include internal and external environments – the animal's bodily (physical and psychological) environment and the space/time environment in which the animal is housed (Broom, 1991, 1999). However, a misconception that an animal's environment determines welfare state quickly became ingrained in research and practice. For example, throughout the zoo industry, there is a misconception that provision of 'environmental enrichment' (EE) will create a positive welfare opportunity for the animals who interact with that EE. In fact, the items we provide to zoo animals are *potentially* enriching – they provide challenge to the animal who can work in a goal-oriented way to overcome that challenge and achieve the goal. The zookeeper provides a biologically relevant opportunity for choice and control (Meehan and Mench, 2007); the animal can choose a course of action when they are motivated to do so and gain some control over their space/time environmental provisions. The animal can perceive these opportunities independently as boring, frustrating, irrelevant, engaging, exciting, or joyful, for example. Exactly what they perceive is determined by the individual and is something beyond human control or direct influence.

The individual's perceived state will be determined by a combination of factors including:

- Genetics. Some species are more likely to perform specific stereotypies. Mason et al. (2007) found that carnivores are more likely to develop stereotypic pacing. This makes sense given carnivores' behavioural biology; carnivores would journey long distances to establish and maintain territory and hunt prey. The frequency of pacing in zoo-housed carnivores is significantly predicted by home range size and daily journey length (Clubb and Mason, 2007).

- Epigenetics. Animals not only inherit genes, they also inherit epigenetic biomarkers that can up- or down-regulate gene expression (and protein production therefore), relevant to welfare. While epigenetics in zoo animals is an emerging area of research, increasing literature from laboratory animal science suggests that rodents inherit biomarkers of stress – markers that alter myelination of genes coding for glucocorticoid receptor for example (Zhang et al., 2013).
- Personality. Personality traits are stable over time and are heritable. With the development of personality scales in more species, it may be possible to use personality profiling in the zoo to predict welfare state in response to particular environmental challenges (Tetley and O'Hara, 2012).
- Prior experience. Rearing history and thus early life experience can condition a zoo animal to have poor or good welfare. Premature weaning, isolation rearing, nursery rearing, hand rearing, and impoverished early life environments predict the development of abnormal repetitive behaviours (ARBs) across multiple taxa including primates, carnivores, elephants, birds, and, if inactivity is considered an ARB, reptiles (Rose et al., 2017).

These factors, in addition to the current environment in which the animal is kept, determine how an animal perceives their state of being. Given every individual is unique (they have a unique combination of genetics, epigenetics, personality, and rearing history) the way in which an animal interacts with their current environment is also unique. Any given animal in a group will have a different experience of their current environment, therefore. By defining welfare as a state belonging to an individual, we acknowledge this and advocate for the consideration of the individual to be central to welfare assessment, meaning the assessment of welfare must, when individuals are identifiable, be made at the individual level rather than the group level. In this way we also encourage consideration of agency. Agency is the tendency of an animal to gather knowledge, develop skill and retain behavioural flexibility by engaging actively with their environment (Špinka and Wemelsfelder, 2011). Impoverished environments will limit agency and cause an animal to experience negative welfare. Consideration of environmental opportunities for the agency will be beneficial when assessing welfare and the inclusion of such opportunities into zoo husbandry plans is encouraged. It is morally and ethically correct to ensure that what is provided for an animal in the zoo meets their biological needs and ensures that they can cope, develop, learn and adapt to their environment (and any challenges) accordingly. As such, Box 2.1 considers the role of an ethical approach to animal welfare for all individuals housed in zoos and aquariums. As outlined in Box 2.1, integration of key concepts from animal-based ethics into the measurement and assessment of animal welfare inputs and outputs can help us better understand the cumulative experiences of an animal over time and perhaps help practitioners, or animal care staff, or veterinary staff determine quality of life. Cumulative (affective) experience over time is increasingly recognised as important to an animal's current welfare state (Reimert et al., 2023). The emotions and moods experienced by an animal have a cumulative effect on their welfare state. By focusing on individual welfare, there is increased opportunity to consider cumulative experience and this book advocates for the inclusion of cumulative experience into welfare assessment. This may be achieved by Quality of Life assessment, that objectively considers the balance of positive to negative experiences of an individual over a specified time period.

This book's definition of animal welfare further requires a shift in our approach to welfare assessment as it advocates for better integration of species-appropriate resource-based input measures alongside enhanced consideration of relevant and valid animal-centric output measures. Resource-based measures focus on the external environmental provisions given to animals, for example the number of feeds per day, the type of dietary presentation and the provision of a varied, nutritious diet. Animal-centric output measures focus on the animal, for example the time spent feeding, diversity of feeding behaviours shown and consideration of how satiated the animal is by their diet. Animal-centric measures are increasingly used in industry and WAZA (World Association of Zoos and Aquariums) now requires member organisations

BOX 2.1: The interface between animal welfare and animal ethics

Animal welfare and animal ethics are different subjects born from science and philosophy, respectively. For some zoo-focused people, welfare will be at the forefront of their minds and daily practice, while ethics may rarely be considered. Indeed, ethics may be seen as the enemy, an argument for why zoos should not exist or why they should be heavily criticised. Similarly, for an animal ethicist, a book on zoo animal welfare may simply mask the fundamental questions like *should zoos really exist?* Welfare may be seen as the scapegoat for caging wild animals; if welfare is good enough, consideration of the rights of animals to live free lives may be ignored or deemed inconsequential. For some, the existence of zoos will always be ethically wrong, whereas for others, the existence of zoos will be essential and warranted if species are to be conserved and the public encouraged to care for wild animals and their natural habitats. With such stark and conflicting views, it may seem viable to keep welfare and ethics compartmentalised; however, when strategizing our welfare definition, it seemed remiss to omit ethics and prohibitory to the progression of welfare application at the zoo. In this chapter's discussions on welfare concept, definitions and assessment, we have considered agency, the autonomy approach, recognised animals as individuals and considered the telos of an animal (their personality, emotions and perception of being). We are therefore discussing not just welfare, but the characteristics used to argue for the moral consideration of animals and their protection. The world we build for an animal in captivity should not simply allow the animal to survive, but to live a good life. This application of *Quality of Life* represents a potential way to unite both welfare-based and ethics-based approaches to animal care and affords the opportunity to reexamine and reinterpret some of the established and opposing welfare-based and ethics-based views surrounding the keeping of wild animals in captivity. By considering not the divisions but the commonality between the disciplines of ethics and welfare – that of respect for animals – we can progress the theoretical and applied fields of welfare science further than before. By seeking to respect the individuality of every animal, their intrinsic worth would be realised, and their welfare improved. By simply asking *what is best for this individual animal,* we start to examine both their welfare and their needs concurrently, and evidence an approach to animal management built on respect. Respecting an animal as the individual being they are, focuses our attention firmly on the state of being of that animal and this must surely be the future of zoo animal welfare.

to assess welfare using output-based measures as part of its Animal Welfare Goal (WAZA, 2023).

The Five Freedoms (FAWC, 1993) focuses on negative welfare states and tends to encourage environmental and husbandry assessment rather than welfare assessment. While familiar to most in the zoo industry, the Five Freedoms encourages group assessment rather than individual assessment where environmental provisions are considered equal for all group members. The Five Domains welfare assessment framework (Mellor and Reid, 1994; Mellor and Beausoleil, 2015; Mellor, 2017) is directly applied to every individual and requires four physical domains (behaviour, environment, nutrition, health) to be related to a fifth mental domain. In theory, the Five Domains framework better encompasses all three welfare components under our concept of welfare. However, it must be recognised that the practical application of the Five Domains framework can be challenging due to the subjective scoring of the four physical domains and subjective interpretation of the mental domain.

While Mellor (2017) considers assessment to require assemblage of an expert panel to quantify welfare by scoring each domain, very little formal guidance on inferring mental state has been published, and this raises questions of validity and reliability that is attracting criticism (Hampton et al., 2023). More research is required regarding the application of the Five Domains, for example to determine when welfare may be compromised. Let us consider a familiar example – hunger. It is normal for animals to feel hunger, as this emotion regulates (in part) feeding behaviour. Yet, how hungry does the animal need to be before a negative mental state is inferred and welfare considered compromised/poor?

Both the Five Domains and the Five Freedoms have practical limitations and whilst it may be unachievable to reach consensus over which framework is more appropriate, the inclusion of input- and output-based measures within welfare assessment is essential. Output measures allow inferences of welfare state, input measures allow potential environmental/ nutritional/ social/ behavioural deficits to be identified. Appropriate husbandry is the essential foundation to good welfare but not necessarily sufficient to ensure welfare is always positive.

2.5 CONCLUSIONS

The multidimensional concept of animal welfare proposed by Fraser informs us that welfare involves biological functioning, naturalness and affective state. In this book, we use this concept as an overarching guide to discussing and assessing welfare. The definition of welfare we propose, though controversial to some, is an interpretation and slight reframing of Broom's coping definition which firmly places emphasis on the individual and their perception of how they are doing currently. Overall, this book will demonstrate an inclusive (every individual matters) and progressive (consideration of affective state equally with biological functioning, consideration of both positive and negative welfare state) approach to welfare. While chapter authors, and readers, may differ in their chosen definitions of welfare in comparison to Broom's reframed definition, this should not hinder discussions on how zoos can provide husbandry which allows an animal to perceive good welfare, nor should it discourage welfare assessment, which should include input- and output-based measures. While there will likely always be some disagreement between zoo practitioners, students, academics and members of the public regarding what welfare is or how to assess it, all can universally agree that zoos should prioritise welfare across all operations and strive to provide for the needs of every animal in their care, so that all can experience good welfare.

REFERENCES

Appleby, M. C., & Sandøe, P. (2002). Philosophical debate on the nature of well-being: Implications for animal welfare. *Animal Welfare*, 11(3), 283–294.

Broom, D. M. (1986). Indicators of poor welfare. *British Veterinary Journal*, 142(6), 524–526.

Broom, D. M. (1991). Animal welfare: Concepts and measurement. *Journal of Animal Science*, 69(10), 4167–4175.

Broom, D. M. (1996). Animal welfare defined in terms of attempts to cope with the environment. *Acta Agriculturae Scandinavica, Section A – Animal Science*, 27, 22–28.

Broom, D. M. (1999). The welfare of dairy cattle. In Aagaard, K. (Ed.). *Proceedings of the 25th International Dairy Congress, Aarhus 1998. III Future Milk Farming* (pp. 32–39). Aarhus: Danish National Committee of I.D.F, Denmark.

Clubb, R., & Mason, G. J. (2007). Natural behavioural biology as a risk factor in carnivore welfare: How analysing species differences could help zoos improve enclosures. *Applied Animal Behaviour Science*, 102(3–4), 303–328.

Dadone, L., & Foxworth, S. (2023). Footcare for zoo ungulates (zoo hoofstock trim program). In Miller, R. E., Lamberski, N., & Calle, P. P. (Eds.). *Fowler's Zoo and Wild Animal Medicine Current Therapy, Volume 10* (pp. 303–312). St Louis, MO: WB Saunders.

Dawkins, M. S. (2021). *The Science of Animal Welfare: Understanding What Animals Want*. Oxford, UK: Oxford University Press.

Duncan, I. J. (1993). Welfare is to do with what animals feel. *Journal of Agricultural and Environmental Ethics*, 6(2), 8–14.

Duncan, I. J. (2004). A concept of welfare based on feelings. In Benson, G. J. & Rollin, B. E. (Eds.). *The Well-Being of Farm Animals: Challenges and Solutions* (pp. 85–101). Oxford, UK: John Wiley & Sons.

Farm Animal Welfare Council (FAWC). (1993). *Second Report on Priorities for Research and Development in Farm Animal Welfare*. London, UK: DEFRA.

Fraser, D. (2003). Assessing animal welfare at the farm and group level: the interplay of science and values. *Animal Welfare*, 12(4), 433–443.

Fraser, D. (2008). Understanding animal welfare. *Acta Veterinaria Scandinavica*, 50(1), 1–7.

Gray, J. (2017). *Zoo Ethics: The Challenges of Compassionate Conservation*. Clayton, Australia: Csiro Publishing.

Green, T. C., & Mellor, D. J. (2011). Extending ideas about animal welfare assessment to include 'quality of life' and related concepts. *New Zealand Veterinary Journal*, 59(6), 263–271.

Hampton, J. O., Hemsworth, L. M., Hemsworth, P. H., Hyndman, T. H., & Sandøe, P. (2023). Rethinking the utility of the Five Domains model. *Animal Welfare*, *32*, e62, 1–10.

Harrison, R. (1964). *Animal Machines: The New Factory Farming Industry*. London, UK: Vincent Stuart.

Hughes, B. O. (1976). Behavior as an index of welfare. In *Proceedings of the fifth European Poultry Conference, Malta*, pp. 1005–1018.

Mason, G., Clubb, R., Latham, N., & Vickery, S. (2007). Why and how should we use environmental enrichment to tackle stereotypic behaviour? *Applied Animal Behaviour Science*, 102(3–4), 163–188.

Meehan, C. L., & Mench, J. A. (2007). The challenge of challenge: can problem solving opportunities enhance animal welfare? *Applied Animal Behaviour Science*, 102(3–4), 246–261.

Melfi, V. A. (2009). There are big gaps in our knowledge, and thus approach, to zoo animal welfare: A case for evidence-based zoo animal management. *Zoo Biology*, 28(6), 574–588.

Mellor, D. J. (2017). Operational details of the five domains model and its key applications to the assessment and management of animal welfare. *Animals*, 7(8), 60–80.

Mellor, D. J., & Beausoleil, N. J. (2015). Extending the 'five domains' model for animal welfare assessment to incorporate positive welfare states. *Animal Welfare*, 24(3), 241–253.

Mellor, D. J., & Reid, C. (1994). Concepts of animal well-being and predicting the impact of procedures on experimental animals. In Baker, R., Jenkin, G., & Mellor, D. (Eds.). *Improving the Well-Being of Animals in the Research Environment* (pp. 3–18). Sydney, Australia: Australian and New Zealand Council for the Care of Animals in Research and Teaching.

Mendl, M., Burman, O. H., & Paul, E. S. (2010). An integrative and functional framework for the study of animal emotion and mood. *Proceedings of the Royal Society B: Biological Sciences*, 277(1696), 2895–2904.

Mendl, M., & Paul, E. S. (2020). Animal affect and decision-making. *Neuroscience & Biobehavioral Reviews*, *112*, 144–163.

Ohl, F., & Putman, R. (2018). *The Biology and Management of Animal Welfare*. Dunbeath, UK: Whittles Publishing.

Paul, E. S., Harding, E. J., & Mendl, M. (2005). Measuring emotional processes in animals: The utility of a cognitive approach. *Neuroscience & Biobehavioral Reviews*, 29(3), 469–491.

Reimert, I., Webb, L. E., van Marwijk, M. A., & Bolhuis, J. E. (2023). Towards an integrated concept of animal welfare. *Animal*, 17(S4), 100838.

Rogers Brambell, F. W. (1965). *Report of the Technical Committee to Enquire into the Welfare of Animals Kept Under Intensive Livestock Husbandry Systems*. London, UK: Her Majesty's Stationery Office.

Riley, L. M. (2018). Conserving behaviour with cognitive enrichment: A new frontier for zoo conservation biology. In Williams, E., Bremner-Harrison, S., & Ward, S. J. (Eds.). *Zoo Animals: Husbandry, Welfare and Public Interactions* (pp. 199–264). New York, NY: Nova Science Publishers, Inc.

Riley, L. M. (2022). The behavioural biology of primates. In Rose, P. E. (Ed.). *The Behavioural Biology of Zoo Animals* (pp. 49–65). Boca Raton, FL: CRC Press.

Rollin, B. E. (1989). *The unheeded cry: Animal consciousness, animal pain and science*. Oxford, UK: Oxford University Press.

Rollin, B. E. (2020). Ethical and societal issues occasioned by xenotransplantation. *Animals*, 10(9), 1695–1705.

Rose, P. E. (Ed.). (2022). *The behavioural biology of zoo animals*. Boca Raton, FL: CRC Press.

Rose, P. E., & Riley, L. M. (2022). Expanding the role of the future zoo: Wellbeing should become the fifth aim for modern zoos. *Frontiers in Psychology*, 13, 1018722–1018733.

Rose, P. E., Nash, S. M., & Riley, L. M. (2017). To pace or not to pace? A review of what abnormal repetitive behavior tells us about zoo animal management. *Journal of Veterinary Behavior*, 20, 11–21.

Špinka, M., & Wemelsfelder, F. (2011). Environmental challenge and animal agency. In Appleby, M. C., Olsson, A. S., & Galindo, F. (Eds.). *Animal Welfare* (2nd ed., pp 27–43). Wallingford, UK: CABI.

Tetley, C. L., & O'Hara, S. J. (2012). Ratings of animal personality as a tool for improving the breeding, management and welfare of zoo mammals. *Animal Welfare*, 21(4), 463–476.

WAZA. (2023). *Animal Welfare Goal*. https://www.waza.org/priorities/animal-welfare/2023-animal-welfare-goal/

Wemelsfelder, F. (2007). How animals communicate quality of life: The qualitative assessment of behaviour. *Animal Welfare*, 16(S1), 25–31.

Yeates, J. W., & Main, D. C. (2008). Assessment of positive welfare: A review. *The Veterinary Journal*, 175(3), 293–300.

Zhang, T. Y., Labonté, B., Wen, X. L., Turecki, G., & Meaney, M. J. (2013). Epigenetic mechanisms for the early environmental regulation of hippocampal glucocorticoid receptor gene expression in rodents and humans. *Neuropsychopharmacology*, 38(1), 111–123.

3

Zoo ethics vs. welfare – complementary or conflicting?

LINDA GREENING AND LORNA CAMERON

3.1 INTRODUCTION

At the time of writing, there were 131 members of the British and Irish Association of Zoos and Aquariums (BIAZA), over 400 member institutions of the European Association of Zoos and Aquariums, based in 47 countries throughout Europe, Western Asia and beyond, and 235 zoos and aquariums accredited to the Association of Zoos and Aquariums (AZA) across 13 countries. It is estimated that there are over 10,000 zoos worldwide (Abbott & Tan-Kantor, 2022). In its 75th year, the International Union for Conservation of Nature (IUCN) celebrated the major contribution of zoos to conservation efforts. Specifically, the organisation released a position statement that "zoos are essential", recognising how they set the standards for keeping animals under human care. However, the IUCN also recognised that some institutions had yet to reach the level of welfare that leaders in the field were promoting (Rodriguez & Brooker, 2023). The lack of standardised approaches to optimising welfare globally is only part of the juxtaposition of zoos. Ethically, the post-humanist argument, that animals have fundamental rights and should not be confined to zoos where they are deprived of their liberty (Milan, 2023), is pitted against the laudable role that zoos play in education and conservation.

Unequivocally, zoos are places frequented by many people (Nekolný & Fialová, 2018). Although the public has an opportunity to learn about species they would not normally see, this is usually within an informal learning environment (Godinez & Fernandez, 2019) and the impact of zoos on visitor knowledge about animals and conservation can be variable (McNally et al., 2024). And, due to the variable standards of welfare, the question becomes "what is the public learning?" relative to the practice of keeping animals in captivity. Zoos are as much subject to social licence to operate as any industry, such that there is a growing need to understand the opinions and perspectives of visitors, rather than visitors being passive receivers of what they see. Transparency becomes increasingly important to build public confidence and trust, as much as knowledge. Good examples of this include zoos that provide information signs around animals/enclosures explaining how behaviour may

DOI: 10.1201/9781003429425-3

have been affected by medical procedures. Zoos and accrediting associations need to increase communication and marketing around accreditation schemes (Warsaw & Sayers, 2020). More recently, there have been calls for zoos to encompass another dimension to their aims, one that conserves/promotes animal welfare and engages with human wellbeing (Rose & Riley, 2022).

AZA states that the trademark of modern zoos and aquariums is their commitment to the scientific advancement of contemporary animal management, husbandry, veterinary care and conservation through research (Cannataro et al., 2024). In 2023, AZA reported an expenditure of $356.7 million on field conservation across 125 countries, and a further $39.7 million spent on research across 56 countries. Nearly half of the research focused on animal care, health and welfare (48%), with 70% of these projects studying mammalian species. Meanwhile, the Universities Federation for Animal Welfare (UFAW) actively supports research to advance zoo animal welfare. Zoo animal research has clearly improved our knowledge and understanding of captive animal populations, with increasing numbers of publications and outputs year on year (Rose et al., 2019). However, there are gaps in our knowledge because research focuses on anthropocentrically selected captive environmental variables (e.g. digital technology in smart environments, Webber et al., 2022) and a continuing trend where knowledge about animal welfare is generated from very few taxa.

So how do we know what is right and what is wrong when it comes to animal welfare? Ultimately, welfare comprises three central components for every individual: biological functioning, affective state, and naturalness (Fraser, 2003, 2008). As a characteristic of the animal, welfare is not something that is given to it – for example, by a human caregiver (Broom, 1991). And yet, zoos house collections of animals for which human caregivers are still trying to work out how to optimise animal welfare. And even though zoos are keeping animals in captivity, they are morally justifiable because of the virtuous things they do with their collection of animal assets. This chapter aims to explore some key ethical concepts and explore whether, in the context of zoos, concepts of morality and welfare conflict or complement one another.

3.2 CONTEXTUALISING ETHICAL VIEWPOINTS

We agree that animal welfare is fundamentally the experience of each individual animal relative to the environment and the challenges presented. When animals are in captivity or are managed by humans, societal attitudes and ethical viewpoints around animals become a key determinant of animal welfare (Haddy et al., 2023). Broadly, ethics is a branch of philosophy concerned with the study of moral principles, considering what is right and what is wrong and how people should behave. There are three key ethical philosophies, and several theories, under which the reader may implicitly or subconsciously operate. This chapter will explore some of these relative to what is known about zoos and animal welfare. In all the examples below, when we apply ethical viewpoints to zoos, we refer to zoos as a collection of people involved in the moral decisions and actions of an organisation. It is fair to expect that zoos are organisations that employ individuals who share the organisations concept of corporate social responsibility.

3.2.1 INTRODUCING ETHICAL VIEWPOINTS

Virtue ethics emphasizes the value of moral character, requiring rationality and agency (Hursthouse & Pettigrove, 2023). A virtue is an excellent trait of character; the individual is seen as a morally good or admirable person who does the right thing. Previously we described zoos as being virtuous, because of the charitable and benevolent things these organisations do for the good of animals and people. However, to possess perfect virtue is rare, and the virtuous can still fall foul of blind spots. Virtuous individuals (or indeed organisations) are also less likely to act in a way that is beneficial, that is to know when to be honest, compassionate or courageous, without moral or practical wisdom. In this sense, sayings such as "To know what you do not know is best" (Lao Tzu) and "When you know better, do better" (Maya Angelou) are important. Knowledge around animal welfare has exponentially increased, evidenced through the growing number of publications in the field, in the last fifty years. It is unfair to say that zoos were managing animal collections badly prior to this because there was little motivation to call out practices that were bad. It is fair to say however that since the times when zoos were menageries, they have to some extent developed practical wisdom by increasingly recognising what is virtuous and acting accordingly. There are several forms of virtue ethics that combine virtue and practical wisdom. For example, eudaemonist virtue ethicists promote traits that contribute towards happiness or well-being, agent-based virtue ethicists look for exemplary motivations, and target-centred virtue ethicists focus on the right action within context. Zoos need to reconsider their virtuous actions in terms of the Social License to Operate (SLO) debate and move beyond existing exemplary motivations around conservation and education to take the right action. At this time, adhering to target-centred virtue ethics might encourage zoos to consider conserving/ promoting animal welfare and engaging with human well-being as the right action to take.

Deontology states that the morality of an action is determined by whether it is right or wrong (Alexander & Moore, 2024). Judgement takes place against existing moral rules which are absolute and do not change regardless of the situation. Deontology is sometimes referred to as rule-based ethics, and the word is derived from the Greek 'deon' (duty) and 'logos' (science). The moral obligation behind the action may arise from religious law, or personal, or cultural values. Kantianism is a duty and deontologically based philosophy, applicable to all human beings (Johnson & Cureton, 2024). It states that actions are morally right if guided by a universal moral law and a sense of duty. It also advocates for human rights, emphasizing that humans are worthy of respect and should not be treated as objects. In terms of animal rights, the deontological viewpoint that no animal should be killed regardless of the benefit to humans was initially proposed by Tom Regan (1975). Compared to virtue ethics, which seems highly subjective and context-driven, deontology offers a very black and white approach to what is right and what is wrong that appeals to many. However, this viewpoint fails to consider that the right action in accordance with a predefined rule may not result in the most advantageous or beneficial outcome. And who sets these rules? Both these factors can affect the widespread agreement about moral viewpoints and judgements required for acceptance of deontological philosophy. Even where the right action is to follow relevant legislation, those laws may only advocate for minimal welfare standards. A consequentialist approach may afford individuals an opportunity to think outside the box and innovate around practices that promote optimal welfare, for example through positive experience and agency (Rose & Lewton, 2025).

Consequentialism is a moral philosophy that judges the rightness or wrongness of an action by the goodness of its outcomes (Sinnott-Armstrong, 2023). It encourages individuals to be thinking about acting according to whatever makes the world better in the future. The rightness or goodness of an act depends on the outcomes as opposed to the act itself or anything that happens before it. Classic utilitarianism is a form of consequentialism which claims that an act is morally right if that act maximises the good within its outcome, with welfarism as one of its defining elements (MacAskill et al., 2023). In this sense, the educational and recreational status of zoos maximises the good for humankind. The moral viewpoints hold that there are guiding principles in what is right and what is wrong but that these are not rules to be followed if their outcomes are deleterious rather than advantageous. The arguments against consequentialism and utilitarianism include that, so long as the outcome is beneficial to well-being without discrimination or disaggregation, any kind of act can be justified as being good. The more common concept of act-consequentialism helps to negate this argument, in that the outcome of an act by an individual (or organisation) is ranked according to its overall goodness, its goodness for the individual (or organisation), and its goodness for others besides the individual.

It is worth noting that ethical egotism, an act-consequentialist theory, ranks an act on how good the outcome is for the individual (or organisation), even if the outcome makes things worse for everyone else involved. In a discussion around ethical behaviour, it would be important to highlight the detriment of the selfish act as part of the ranking process. In his essay "Narveson on egoism and the rights of animals" (Regan, 1977), the author highlights how egoism, particularly rational egoism, excludes beings that are incapable of entering into rational agreements including animals, without consideration for how they may suffer as a result. Regan is a deontologist who, as part of the animal rights movement, is committed to the total abolition of the use of animals in science, commercial animal agriculture, and commercial/sport hunting and trapping. He posits that the system allowing us to view animals as resources is fundamentally wrong because this leads us into ignoring their needs. Many see the animal rights movement as an extreme or fundamentalist viewpoint because of its abolitionist nature (Francione & Garner, 2010). For example, in a world where it was accepted that animals had rights, animal ownership and zoos would cease to exist. But of course, it is impossible to say whether animals would fare better in current society if they were not kept in zoos. General disbelief in animal rights occurs for a variety of reasons, for example some think that animals act purely on instinct whilst humans engage in rational thought, and others believe that animals do not behave morally, nor do they belong to a moral community. In this sense, the use of animals comes from the central tenet of consequentialism; whatever we do with animals maximises the good somehow e.g. feeding the population through animals in agriculture or providing pleasure for humans in zoos (Learmonth, 2020). In between these two extreme viewpoints are animal welfarists, who believe it is morally acceptable for humans to use animals. Whilst welfarists would prefer to eliminate unjustifiable pain inflicted on animals, they support measures that incrementally reduce suffering amongst animals (Chiesa, 2016). From a consequentialist perspective, this approach also maximises the good but highlights how it is possible to hold different ethical viewpoints on different things. A Hybrid View is any moral theory that endorses treatment according to different principles to different classes of individuals (Fischer et al., 2023). In 1974, Robert Nozick proposed utilitarianism for animals but rejected it for humans, favouring Kantianism instead (Killoren & Streiffer, 2020). This Hybrid View was used to explain the positions about the use of animals if they were replaced by humans, considering for example the concept of human zoos or the ownership of human pets.

Alongside these ethical principles sit a variety of cognitive biases, which explain why we sometimes selectively accept information whilst overlooking evidence that contradicts them. As seen in Table 3.1, several cognitive biases exist that are useful to those dealing with zoo animals because it is often difficult to recognise one's own biases; therefore, it is difficult to reflect on how these can impact our decisions, and the decisions of others, based on judgements. Please note that the list presented in Table 3.1 is not exhaustive.

One final point is that ethical actions are different to the law which functions as a system of rules overseen by political authority to maintain order within society. Illegal activity is often easier to call out than unethical activity. An illegal act is something which is against the law, whilst unethical behaviour is something that is considered morally wrong. Illegal activity usually results in legal consequences, whilst the deciding agent on unethical behaviour largely comes down to the individuals' conscious. Legislation therefore can help pragmatically guide society towards the right behaviour. Globally, animal welfare legislation differs markedly highlighting how different levels of value are placed upon the consequence of animals as conscious sentient beings (UFAW, 2024). Zoos adhere to practices as required by legislation, however this deontological approach to animal welfare may involve adhering to practices that advocate for only minimal animal welfare standards. Moral relativism tells us that judgements are only true or false relative to some standpoint, but that no standpoint can be proven objectively superior to another (Westacott, 2021). As such, we should refrain from passing moral judgements on practices of characteristics of cultures other than those we identify with. For example, it may be that zoos keep animals relative to the growth of the field of animal welfare science in the country they are situated in. Of course, this viewpoint is in direct contrast to the idea of moral absolutism which posits that actions are inherently right or wrong according to universal moral laws. This is certainly the case for legislation but again, which such variation amongst countries, the need for moral relativism appears obvious in the consideration of zoo animal welfare.

3.3 APPLYING ETHICAL PHILOSOPHIES IN THE CONTEXT OF ZOOS

The following examples aim to enable contextualisation of the theory that has been defined within this chapter. With a disclaimer, these are presented with as much balance as possible but are not meant

Table 3.1 Examples of identified types of biases

Type of bias	Definition	Example
Confirmation bias	Seeking and selecting information that confirms beliefs, whilst ignoring (consciously or not) contradictory evidence.	Keeping an animal according to the way you were taught but not seeking out new information or ideas on keeping animals.
Availability heuristic	Basing decisions or judgements on the information that is most readily available from memory (even if it is incorrect), which can be influenced by currency and emotionality.	Anthropocentric thinking which involves the tendency to use human analogies as a basis for reasoning. Anthropomorphism involves characterising animals/objects as possessing human traits/emotions/intentions.
Clustering illusion	Overestimating the importance of clusters in large samples of random data (phantom patterns).	Failing to use statistical tests to determine significant probabilities.
Bias blind sport	The tendency to believe oneself to be less biased than those around them, or to be able to identify biases in others.	False confidence in decisions or judgements about other people.
Third-person effect	People believe that others are more likely to be persuaded by a message, thereby leading to inappropriate action.	Holding a belief that others are more impressionable than oneself means the individual has not leant into the sources that are influencing the decisions they are making.
Choice-supportive bias	When people choose something that they feel positive about, even if it is flawed.	A type of forage is provided by a human caregiver because they think it is best, even though it is not the preferred forage type/ some herbivores don't eat it.
Zero-risk bias	In general, humans thrive on certainty and will eliminate risk to reduce the chance of harm.	Keeping animals separate to avoid risk of injury.

to be definitive answers to the problems that zoos face.

At the time of writing, London Zoo appeared in the news due to the birth of six species of endangered small mammal (see https://www.bbc.co.uk/news/articles/c4gmlvmg79yo, accessed 02/03/2025). This type of activity can be seen as ethical and positive for zoos. It is virtuous because zoos have highlighted a need to preserve endangered animal species. However, this type of captive breeding programme likely does not address the reasons behind the need for it, potentially highlighting several biases behind the actions of zoo organisations. For example, confirmation bias would involve only looking at positive news stories about the births. From an animal rights perspective, if we were to apply the clustering illusion, we could go so far as to ask why breed into zoos when there are so many deaths? This last point has taken data on the number of animal deaths in zoos to mean captivity is the cause, but it is known that many species kept in captivity have longer average life spans than free-living counterparts.

Enriching environments that provide greater space and opportunities for exploration are optimal, next to not having animals in captivity at all. But a construction site is no place to keep animals, such that the animal's routine and location are disrupted during any building. In this situation, the greater good for the animals is certainly removing them from the building site but this is also to reduce the risk to the animal which would be at cost to the zoo.

The act of treating animals well, such as allowing nocturnal creatures to sleep during the day without manipulating dark/light cycles, is part of our duty of care. However, displays where animals are asleep might result in poorer experiences for humans, which may not be seen as virtuous within an anthropocentric society (Haddy et al., 2023). In this sense, an ethical approach of consequentialism, where the outcome is best for humans, would certainly trump the rights of an animal in a human-centric society. Increasing human activity is the reason for decreasing biodiversity, and modifications amongst/ impoverishment of surviving wild animal populations (Paquet & Darimont, 2010) such that it can become increasingly difficult to hold a line of virtue around animal welfare when it is disregarded by many. Public pressure is also felt increasingly through the SLO where public consensus, based around perceptions of an entity, holds a "moral authority" over their operations (Breakey et al., 2025). Four types of SLOs are proposed: 1) social legitimacy linked to social acceptance for consequential stakeholders, those who belong to most affected group(s); 2) local-values legitimacy according to whether operations align with local values; 3) overall moral legitimacy, determining whether the operation is morally entitled to be performed; 4) moral authority, where social legitimacy is a necessary condition of the operation's overall moral legitimacy (Breakey et al., 2025). These four types of SLOs draw on all the ethical principles described already highlighting the complexity of dealing with public perception.

3.4 THE CHALLENGE OF PUBLIC PERCEPTION

Zoological collections rely on the support of the public through attendance and viewing of the animals to fund their existence and as such invite public scrutiny on a daily basis, unlike the majority of animal industries. This unrivalled scrutiny leads to judgement and in a continuously evolving societal outlook, these judgements will change and adapt over time. This has led to dilemmas surrounding "compassionate conservation" (Clay & Visseren-Hamakers, 2022), recognising the individual animal's interests, whilst still meeting conservation goals and being "seen" to be doing the best for animals within a collection. Opinions on the worth of zoos and the conditions of the animals are often made by members of the public with limited general animal management knowledge, let alone the species-specific needs of the animals on their current views. This often leads to erroneous, or at the very least, ill-informed welfare assessments of the individual animals. The inaccuracy of these assessments is, however, irrelevant as public opinion does not have to be correct to be powerful. As society changes and develops, the moral compass moves, shifts and evolves into new, widely held beliefs impacting any organisation's SLO. In its most simple terms, SLO is the public perception of an industry, organisation or activity and many previously acceptable uses of animals for human gain have become subject to changing public perceptions, such as equestrian sports (Heleski, 2023), wildlife use (Hampton & Teh-White, 2019) and agriculture (Hampton et al., 2020). The need to maintain, retrieve or create SLO is imperative to any organisation and the zoo community has embraced this challenge. At about the same time as "menagerie" collections' SLO were being called into question in some countries, so was the relevance and ethics of the utilisation of animal performers in circus environments (Kapera & Kapera, 2022), although there remain geographical and cultural differences in public perception. Zoological collections grasped this challenge and "shifted" their aims, roles and moral compass to envelope conservation, species recovery and education, being rewarded with societal approval in this changing world. To maintain their role in society and the acceptance of their existence, to promote conservation and education, zoos must be seen to be trustworthy and credible (Spiriti et al., 2022) and positive public perception is imperative (Carr & Cohen, 2011).

3.4.1 BALANCING ETHICAL CONSIDERATIONS, WELFARE ISSUES AND PUBLIC PERCEPTION

This need to maintain positive public perception can often be at odds with animal welfare decisions, such as captive breeding requirements for conservation (Norman and Brando, 2024), species-specific enclosure design which meets both the animals' needs and the viewing requirements of the public (Shaw et al., 2022) and transport of animals between collections. There exists a trichotomy in any management decision for zoo animals which is complex and challenging to balance: the needs of the individual animal, the conservation mission of the collection and the perception of the public as to the outcome of these management choices (Figure 3.1).

This decision-making process will also be further influenced by the ethical standpoint of the organisation and the individuals involved. Often these diverse factors can lead to compromise but should be informed by the guiding principle that it is not enough to simply avoid poor welfare for the animals; positive experiences should be promoted (Melfi, 2009). This is often further complicated in under-researched species or in species little understood by the zoo visitor. In the natural environment, animals will encounter challenging experiences such as predation or the search for nourishment and, as such, experience arousal or anxiety, often lacking within their captive environments. The recognition of this lack of arousal has led to advancements in enclosure design to include environmental enrichments that can promote, at least, a semblance of naturalistic behaviours. This has led to an increased public expectation of enriched, naturalistic enclosures; however, these enrichments, although challenging, are still controlled and chosen by those responsible for the animal's environment and management. This lack of control or choice with an individual animal's environment may lead to further welfare challenges as the concept of agency is missing. Individual animals have agency when they are engaged in self-generated, voluntary, goal-directed behaviour that they are motivated to perform (Littlewood et al., 2023) leading to positive affective engagement. Some argue that without agency, or even freedom (Browning and Veit, 2021) animal welfare cannot be assured; however, we could argue that it depends on "freedom from what?". Surely, freedom from hunger and thirst is positive; however, freedom from choice may well be construed as negative. The impact of enclosure design on this aspect of agency is an example of the challenge of balance; a naturalistic, enriched enclosure must balance challenge, agency, choice, hyper and hypo stimulation (Norman & Brando, 2024).

Additionally, zoo animal management will also include procedures that may be necessary for the physical health of the animal but may not be pleasurable for the individual, posing an ethical dilemma. Recent developments in the use of preconditioning training (Clay & Visseren-Hamakers, 2022) to reduce the novelty of predicted management tasks in captive animals have been promoted.

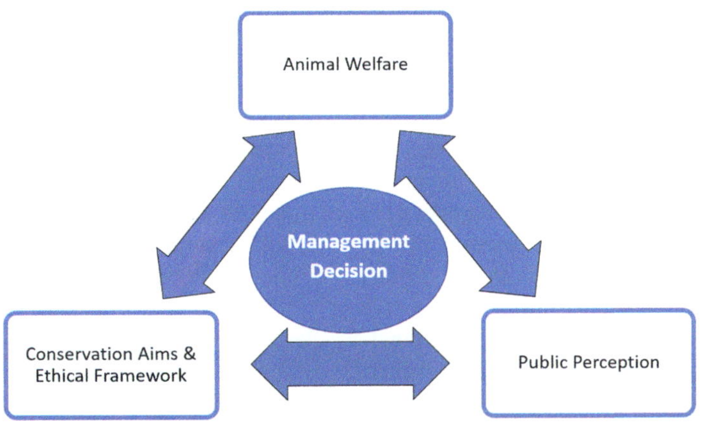

Figure 3.1 The trichotomy of management decision-making processes

This pre-conditioning, however, does not allow for choice in the animal's interactions with humans but may lessen the negative impacts of these necessary encounters. Acceptance or lack of resistance is not an indication that the animal would choose to submit themselves to an activity. In required procedures, however, acceptance or lack of resistance may be the only option available when ethical considerations come into conflict with the complex welfare assessment of the individual animal in question. Ethics and welfare may be either conflicting or complementary, dependent on the situation, but may ultimately always lead to compromise. If this compromise is made with the best interests of the individual animal in mind and is made by knowledgeable, considerate, and empowered staff, these decisions should maintain the welfare of each individual throughout their life span.

3.5 CONCLUSIONS

Ethics welfare and reputation appear to contradict more than complement. For example, although there is increasing research in the field of animal welfare in zoos, this is limited and typically species-centric. Greater consideration is required for the full range of species within zoological collections, and better transparency around the lived experience of individual animal needs is required to enable to public to embrace why they may not be on display or active for their entertainment at that time. The future of zoos is important because of the work they do to address and prevent extinction. Whilst keeping animals in captivity does not address the root problem of an extinction threat (e.g. deforestation, habitat loss), it is largely because this is beyond the capability of zoos to do so. Education of the public is critical to redirect the SLO lens away from the issue of animals in captivity and to explore the reasons behind the needs for conservation strategies in zoos. Increasing awareness of mental states with the Five Domains Model, in terms of choice and agency, highlights how we could further adopt (credible and validated) knowledge gained from relevant theoretical and empirical study of affective states. Increasingly, zoos may wish to shift their focus on animal welfare away from avoiding unpleasant experiences and towards an inclusion of pleasant experiences (Reimert et al., 2023). Zoological collections face the challenge of public scrutiny more often than many other animal industries and should make sure they are clear about their decision-making processes, giving full consideration of their ethical standpoint and the welfare of each individual animal within their care.

REFERENCES

Abbott, M. & Tan-Kantor, A. (2022). Accounting for zoo animals: It is a jungle out there. *Australian Accounting Review, 32*(1), 91–105.

Alexander, L. & Moore, M. (2024). Deontological Ethics. In: E. N. Zalta & U. Nodelman (Eds.) *The Stanford Encyclopedia of Philosophy* (Winter 2024 Edition). https://plato.stanford.edu/archives/win2024/entries/ethics-deontological/. (Access Date 02/03/2025).

Breakey, H., Wood, G. & Sampford, C. (2025). Understanding and defining the social license to operate: Social acceptance, local values, overall moral legitimacy, and 'moral authority'. *Resources Policy, 102*, 105488.

Broom, D. (1991). Animal welfare: Concepts and measurement. *Journal of Animal Science, 69*(10), 4167–4175.

Browning, H. & Veit, W. (2021). Freedom and animal welfare. *Animals, 11*(4), 1148.

Cannataro, G., Grow, S., George, L. & Hart. K. (2024). 2023 Annual report on conservation and science highlights. Association of Zoos & Aquariums (AZA). https://assets.speakcdn.com/assets/2332/aza_arcshighlights_2023_final_web.pdf. (Access Date 02/03/2025).

Carr, N. & Cohen, S. (2011). The public face of Zoos: Images of entertainment, education and conservation. *Anthrozoös, 24*(2), 175–189.

Chiesa, L. E. (2016). Animal rights unraveled: Why abolitionism collapses into welfarism and what it means for animal ethics. *The Georgetown Environmental Law Review, 28*, 557.

Clay, A. S. & Visseren-Hamakers, I. J. (2022). Individuals matter: Dilemmas and solutions in conservation and animal welfare practices in zoos. *Animals, 12*(3), 398.

Fischer, B., Palmer, C. & Kasperbauer, T. J. (2023). Hybrid theories, psychological plausibility, and the human/animal divide. *Philosophical Studies*, *180*, 1105–1123.

Francione, G. L. & Garner, R. (2010). *The Animal Rights Debate: Abolition or regulation?* Columbia University Press, New York, NY.

Fraser, D. (2003). Assessing animal welfare at the farm and group level: The interplay of science and values. *Animal Welfare*, *12*(4), 433–443.

Fraser, D. (2008). Understanding animal welfare. *Acta Veterinaria Scandinavica*, *50*(1), 1–7.

Godinez, A. M. & Fernandez, E. J. (2019) What is the zoo experience? How zoos impact a visitor's behaviors, perceptions, and conservation efforts. *Frontiers in Psychology*, *10*, 1746.

Haddy, E., Burden, F., Raw, Z., Rodrigues, J, B., Zappi Bello, J. H., Brown, J., Kaminski, J. & Proops, L. (2023). Belief in animal sentience and affective owner attitudes are linked to positive working equid welfare across six countries. *Journal of Applied Animal Welfare Science*, *28*, 1–19.

Hampton, J. O. & Teh-White, K. (2019). Animal welfare, social license, and wildlife use industries. *The Journal of Wildlife Management*, *83*(1), 12–21.

Hampton, J. O., Jones, B. & McGreevy, P. D. (2020). Social license and animal welfare: Developments from the past decade in Australia. *Animals*, *10*(12), 2237.

Heleski, C. R. (2023). Social license to operate–why public perception matters for horse sport–some personal reflections. *Journal of Equine Veterinary Science*, *124*, 104266.

Hursthouse, R. & Pettigrove, G. (2023) Virtue Ethics. In: E. N. Zalta & U. Nodelman (Eds.) *The Stanford Encyclopedia of Philosophy* (Fall 2023 Edition). https://plato.stanford.edu/entries/ethics-virtue/. (Access Date 02/03/2025).

Johnson, R. & Cureton, A. (2024). Kant's Moral Philosophy. In: E. N. Zalta & U. Nodelman (Eds.) *The Stanford Encyclopedia of Philosophy* (Fall 2024 Edition). https://plato.stanford.edu/entries/kant-moral/. (Access Date 02/03/2025).

Kapera, I. & Kapera, A. (2022). Use of animals in entertainment-an outline of the problem. *Geography and Tourism*, *1*(10), 17–24.

Killoren, D. & Streiffer, R. (2020). Utilitarianism about animals and the moral significance of use. *Philosophical Studies*, *177*, 1043–1063.

Learmonth, M, J. (2020). Human–animal interactions in zoos: What can compassionate conservation, conservation welfare and duty of care tell us about the ethics of interacting, and avoiding unintended consequences? *Animals*, *10*(11), 2037.

Littlewood, K. E., Heslop, M. V. & Cobb, M. L. (2023). The agency domain and behavioral interactions: Assessing positive animal welfare using the Five Domains Model. *Frontiers in Veterinary Science*, *10*, 1284869.

MacAskill, W., Meissner, D. & Chappell, R. Y. (2023). Elements and Types of Utilitarianism. In: R. Y. Chappell, D. Meissner, and W. MacAskill (Eds.) *An Introduction to Utilitarianism*. https://www.utilitarianism.net/types-of-utilitarianism. (Access Date 02/03/2025).

McNally, X., Webb, T. L., Smith, C., Moss, A. & Gibson-Miller, J. (2024). A meta-analysis of the effect of visiting zoos and aquariums on visitors' conservation knowledge, beliefs, and behavior. *Conservation Biology*, *39*(1), e14237.

Melfi, V. A. (2009). There are big gaps in our knowledge, and thus approach, to zoo animal welfare: A case for evidence-based zoo animal management. *Zoo Biology*, *28*(6), 574–588.

Milan, G, O. (2023). The legacy of zoos? Moral reasons against considering Zoos as heritage. In: Á. López López, G. J. Quintero Venegas, and C. Kline (Eds.) *Tourism, Heritage and Commodification of Non-human Animals*, CAB International, Egham, UK, pp. 34–46. https://doi.org/10.1079/9781800623309.0003.

Nekolný, L. & Fialová, D. (2018). Zoo tourism: What actually is a zoo? *Czech Journal of Tourism*, *7*(2), 153–166.

Norman, M. & Brando, S. (2024). The concept of agency, animal wellbeing, and the practical realities of ex situ breeding programs in zoos and aquariums. *Journal of Zoological and Botanical Gardens*, *5*(4), 563–578.

Paquet, P. C. & Darimont, C. T. (2010). Wildlife conservation and animal welfare: Two sides of the same coin?. *Animal Welfare*, *19*(2), 177–190.

Reimert, I., Webb, L. E., Van Marwijk, M. A. & Bolhuis, J. E. (2023). Towards an integrated concept of animal welfare. *Animal*, *17*, 100838.

Regan, T. (1975). The moral basis of vegetarianism. *Canadian Journal of Philosophy*, *5*(2), 181–214.

Regan, T. (1977). Narveson on egoism and the rights of animals. *Canadian Journal of Philosophy*, *7*(1), 179–186.

Rodriguez, A. & Brooker, H. (2023). IUCN species survival commission acknowledges vital contributions of botanic gardens, aquariums and zoos, to wildlife conservation. https://iucn.org/news/202310/iucn-species-survival-commission-acknowledges-vital-contributions-botanic-gardens. (Access Date 02/03/2025).

Rose, P. E., Brereton, J. E., Rowden, L. J., Figueiredo R. L. & Riley, L. M. (2019). What's new from the zoo? An analysis of ten years of zoo-themed research output. *Humanities & Social Science Communications*, *5*(128), https://doi.org/10.1057/s41599-019-0345-3.

Rose, P. E. & Lewton, J. (2025). Key concepts for enhancing zoo animal welfare: Coping, comfort, choice, control, challenge, and compassion. *Journal of Applied Animal Welfare Science, 28*(3), 497–514. https://doi.org/10.1080/10888705.2024.2440891

Rose, P. E. & Riley, L. (2022) Expanding the role of the future of the zoo: Wellbeing should become the fifth aim for modern zoos. *Frontiers in Psychology*, *13*, 1018722.

Shaw, M. N., McLeod, E. M., Borrie, W. T. & Miller, K. K. (2022). Human positioning in close-encounter photographs and the effect on public perceptions of zoo animals. *Animals*, *12*(1), 11.

Sinnott-Armstrong, W. (2023). Consequentialism. In: E. N. Zalta & U. Nodelman (Eds.) *The Stanford Encyclopedia of Philosophy* (Winter 2023 Edition). https://plato.stanford.edu/archives/win2023/entries/consequentialism/. (Access Date 02/03/2025).

Spiriti, M. M., Melchiori, F. M., Dierkes, P. W., Ferrante, L., Bandoli, F., Biasetti, P. & de Mori, B. (2022). Development of a tool for assessing the reputation of zoos: The Zoo Ethical Reputation Survey (ZERS). *Animals*, *12*(20), 2802.

Universities Federation for Animal Welfare (UFAW). (2024). The importance of science to animal welfare. https://www.ufaw.org.uk/about-ufaw/ufaw-and-animal-welfare. (Access Date 02/03/2025).

Warsaw, D. & Sayers, J. (2020). The influence of animal welfare accreditation programmes on zoo visitor perceptions of the welfare of zoo animals. *Journal of Zoo and Aquarium Research*, *8*(3), 188–193.

Webber, S., Cobb, M. L. & Coe, J. (2022). Welfare through competence: A framework for animal-centric technology design. *Frontiers in Veterinary Science*, *9*, 885973.

Westacott, E. (2021). Moral relativism. *The Internet Encyclopedia of Philosophy*. https://iep.utm.edu/moral-re/. (Access Date 02/03/2025).

4

Confirming the credibility of animal care: Evidence-based husbandry and husbandry-based evidence

PAUL ROSE

4.1 INTRODUCTION

A fundamental way of providing an animal with the best possible chance of experiencing positive welfare (physically, behaviourally, and psychologically) is to ensure that the husbandry, housing, and management are species-appropriate and individual animal relevant. The provision of this daily care is the most essential element of a zoo's daily operations, and any captive species (an individual or population) will not achieve positive welfare experiences when the key, basic needs of this species are not being met. Zoo animals will have the opportunity to express and experience positive welfare outputs because their daily care is good. Experiencing best practice husbandry indicates the zoo is already providing for the animal's physical and behavioural needs; therefore, the zoo can now work towards delivering positive psychological welfare experiences too. Consideration and application of evidence-based approaches to zoo husbandry were apparent in the 1950s, with the pioneering work of Hediger (1950); he collected information from across zoos and on wild animals too, to understand their needs in captivity and to also show that wild animals face constraints, ecologically and environmentally, by pressures from within their ecosystems. These first steps of the evidence-based approach gained further traction in the 1970s with the progression of "behavioural engineering" (what we now call environmental enrichment), led by Portland (now Oregon) Zoo (Markowitz et al., 1978). Ensuring that, whilst a wild system cannot ever be fully replicated, animals have every opportunity to behave in a manner as if they were in the wild based on species-specific enrichment embedded into husbandry or from an enriching enclosure. The relevance of environmental enrichment to zoo animal welfare and the evidence behind enrichment is covered in detail in Chapter 5, but it is worth mentioning here as a milestone on the road to wider acceptance that zoo husbandry must be supported by evidence and not anecdote. The dramatic increase in interest in evidence for husbandry, and what we still need to know about zoo animal care, occurred in the late 2000s (Melfi, 2009). As the husbandry protocols,

DOI: 10.1201/9781003429425-4

housing specifications and management plans of more species become evidence-based, such biologically relevant husbandry (see Chapter 1 for further definition and interpretation of this term) itself becomes the evidence for the specifics of care that must be provided. Zoos and their animal care staff are then promoting opportunities for a good life within, and good welfare experiences at, the zoo for all animals in the collection.

> This work is the first attempt by a trained zoologist to collect information on the available knowledge required for, and the responsibilities involved in the maintenance of a zoological garden... Detailed evidence is given in support of the theories developed; evidence based on the author's own studies of animals captive in zoological gardens and wild in many parts of the world.
>
> *Dr Heini Hediger, Wildtiere in Gefangenschaft (Wild Animals in Captivity), January 1950.*

Seminal pieces of research have identified the knowledge gap that has existed (Melfi, 2009), and continues to exist (Binding et al., 2020; Rose et al., 2019) in our expertise in species-specific care. Collaboration and coordinated research action, and the application of research outputs that are evidence-based and demonstrate best practice, help to improve welfare and the sustainability of zoo populations for particular taxonomic groups (Roller et al., 2021; Scherer et al., 2022). We can continue to embed fundamental evidence from the wild into our animal management plans to align what species have evolved to do with what they are able to do in the zoo. For example, when we consider enclosure sizes and quality space (Clubb & Mason, 2003), the way in which food is provided to enhance positive behavioural diversity (Mellor et al., 2021) or the size of social groups to promote breeding (Mooney et al., 2023).

Evidence-based husbandry can be considered as a method that takes quality, validated and current scientific information and applies it to the practicalities of zoo animal care (Figure 4.1). This evidence comes from global husbandry surveys to identify what specific aspects of care work best for a species or taxonomic group (Brereton & Rose, 2022; Harley et al., 2021; Hummel et al., 2006; Lewis et al., 2010; Marshall et al., 2016; Rose & Roffe, 2013; Rose & Rowden, 2020; Rowden & Rose, 2016), as well as from data collection on key aspects of natural history, evolutionary ecology, life history and behaviour that need to be integrated into captive management (Ferrie et al., 2014; Moise et al., 2023; Troxell-Smith & Miller, 2016). "The wild" is often considered an important first port of call for the gathering of evidence on what a species is and what it does that can be translated into the zoo to improve the relevance of animal care; but it is important to remember that not everything an animal does in the wild is needed in the zoo, and not all aspects of the wild are welfare-positive. It may be challenging to fully explain and interpret the motivations behind all behaviours performed by free-living species (Howell & Cheyne, 2019) and therefore we should consider the outputs of such behaviours (positive behavioural expression, mood and emotion) to determine what behaviours must be enabled in the zoo.

Zoo science outputs are increasing across regions and topics (Binding et al., 2020; Escribano et al., 2021; Hosey et al., 2019; Loh et al., 2018; Rose et al., 2019) and this directed research output can be useful to add clarity to species-specific requirements, e.g. answering questions to provide clarity on what species need (Rose et al., 2014), as well as to advance management across a wide range of organisms that may be kept in a similar manner (Da Silva et al., 2019). Further research is needed to assess the suitability of husbandry across more taxonomic groups, however. Figure 4.2 identifies only small increases in the number of husbandry-focused papers (with husbandry as either an original aim or a specific outcome of the research) across different years of a decade-long period. In some instances, any increase in number is only a small proportion of the total number of papers across a wide range of topics investigated. Advances in how zoo scientists and researchers can disseminate empirical evidence on husbandry more readily are demonstrated by the European Association of Zoos and Aquaria (EAZA)'s *Journal of Zoo & Aquarium Research* specific section for "Husbandry Reports" (Journal of Zoo & Aquarium Research, 2023) and in *Zoo Biology's* "Research Review with Case Studies", "Husbandry Report", and "Technical Report" paper submission categories (Zoo Biology, 2023).

Figure 4.1 Steps involved in generating evidence-based husbandry include the review of quality (i.e. peer-reviewed, objective) source material; the collection of empirical evidence (i.e. from direct observation, experimental study, or survey), the cycle of assessment and evaluation of husbandry and housing provision; and the implementation of best practice care based on this bank of evidence. The cycle would continue to ensure that evidence is still relevant, based on further review of quality sources.

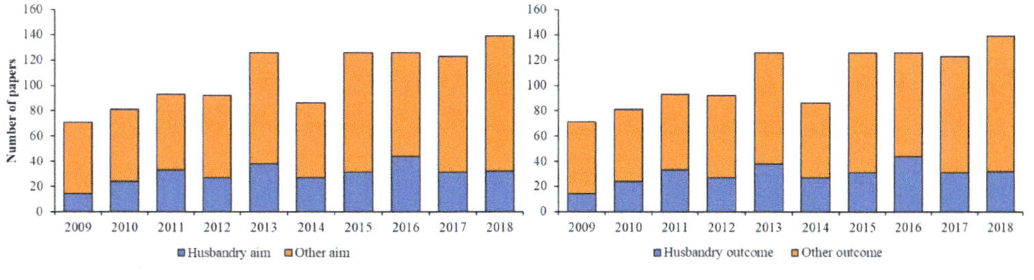

Figure 4.2 Data taken from Rose et al. (2019) showing that for a sample of 1,063 papers available on the Web of Science© database, published from 2009 to 2018, overall zoo research outputs increased during this time period. It is difficult to identify a trend in the number of papers with a specific husbandry aim (i.e. the point of the research was to investigate husbandry) or husbandry outcome (i.e. the relevance of the findings was to inform husbandry and care), but husbandry as a topic is investigated more frequently at the end of this decade compared to the start.

Animal welfare is evidently going to be impacted by the husbandry experienced by the individual animal at a given time. Therefore, the more information that can be gathered on husbandry practices for that species (to enable objective review and evaluation), the more likely baseline conditions for good welfare will be provided. The increasing number of best practice guidelines that are being written by zoo organisations, for example, those available on the EAZA website, show the application of empirical study to zoo husbandry developments. Increasing the interest of students, academics, research scientists, and zoo personnel in answering husbandry-based questions, using the scientific method, will only expand the number of species that will benefit from the application of these best practice approaches. The scientific method centres on making observations and then creating testable predictions, collecting data to test predictions, drawing conclusions and evaluating them, and then revisiting the original observation to see if further study is required. It is imperative for those promoting zoo welfare that interested and qualified parties can be employed in the collection, analysis, and evaluation of husbandry-focused data for as many species as possible. Without such data collection, housing and husbandry protocols, and species' population management practices will not advance.

4.2 CREATING HUSBANDRY-BASED EVIDENCE

As more facilities use an evidence-based approach to animal care, and welfare states improve (including emotional or psychological quotients of welfare), the good practice employed by zoos becomes the evidence needed to show how zoo animals must be kept and maintained. Evolving evidence-based husbandry into husbandry-based evidence is the pinnacle of gold standard care in the zoo; a way of understanding why a husbandry process works well because of the outputs it provides (i.e. excellent animal health, behaviour, and welfare). Longitudinal monitoring of behaviour and comparing results to a published benchmark to quantify change in animal welfare states help to provide evidence of how species respond favourably to daily husbandry and care. For example, a study of a herd of Asian elephants (*Elephas maximus*) across 2018 and 2019 in a large UK zoo showed marked improvements in behavioural indicators of welfare (Finch et al., 2020) when compared to benchmark values from a 2008 research project conducted across all UK elephants by Harris et al. (2008). Changes to elephant husbandry and how the social group was managed to provide increased opportunity for interactions and social stability manifested in reduced incidences of stereotypic behaviour, improved social cohesion, and increased time foraging (Finch et al., 2020). This combination of social stability and husbandry change has been beneficial to both elephants and their keepers. Here, within this example, the implementation of the advanced elephant husbandry model and knowing how it is reviewed and assessed regularly becomes the evidence that all zoos need to follow to enhance positive welfare outputs in their elephants. The original benchmark for elephant welfare improvement came from evidence within the literature; now the fact that husbandry "works" for these animals is the evidence that supports how to achieve positive welfare states.

Figure 4.3 identifies an example of how evidence-based husbandry (the information needed on the species, its environment, and natural history itself) can become husbandry-based evidence when the outputs from good husbandry (how the animal's behaviour, its physical condition, and use of space) become the evidence needed to show why *X* husbandry approach is required by all zoos for that species.

This approach can be undertaken across all species housed in captivity, even those where, traditionally, welfare assessment was considered more challenging. By measuring how individuals display specific traits or characteristics that can only occur when the resources and the environment as a whole are suitable, a proxy for baseline good welfare is achieved. For example,

- Attainment of appropriate carotenoid-based colouration in species that use such pigments for signalling, e.g. red-eyed tree frog (*Agalychnis callidryas*). Enhanced social behaviour, breeding outputs, and viability compared to animals fed on carotenoid-limited diets.
- Opportunities to background match for cryptic species that increase the use of overall available space compared to when such species are housed with limited opportunities to background match.
- Aposematic species, e.g. *Dendrobates* sp. frogs, are on show and more visible due to ecologically relevant environmental replication within their vivaria. Increased time spent visible and confidence in being "on display" compared to individuals housed in environments that have reduced ecological relevance.
- Enhanced shoaling, movement, and exploration of aquaria in school-dwelling fish, e.g. cardinal tetras, *Paracheirodon axelrodi*, as the group size and volume of an aquarium are large enough to promote important social activities.
- A stable social group with reduced aggression, good body condition, and enhanced longevity has been achieved for grazing tropical marine fish such as surgeonfish (Acanthuridae). Multiple foraging (grazing) sites are available across the fish's aquarium (Figure 4.4).

Figures 4.3. and 4.4 both identify how information for good husbandry (the evidence-based approach)

Figure 4.3 Taking information from ecology, natural history, and behaviour for appropriate husbandry (evidence-based husbandry) and indicators that husbandry can evidence optimal care (husbandry-based evidence) using a poison dart frog as an example. Key pieces of evidence on the left of the diagram make husbandry plans suitable. Animal responses to husbandry (on the right of the diagram) become the husbandry evidence that shows all holders of a species how to better manage these animals.

evolves into husbandry-based evidence for zoos to follow to promote good welfare states in specific species being maintained ex situ. The responses of such species to their care, i.e. the information from husbandry, are what must be disseminated across all facilities and holders of these populations so that standardised approaches to promoting good welfare and best practice care can be achieved. Increasing research outputs and building multi-zoo studies from individual institution case studies expand the reach of husbandry-based evidence to improve welfare more broadly.

4.3 ENSURING THE VALIDITY OF EVIDENCE

An essential foundation to the evidence-based approach, which will ultimately ensure any husbandry provided is as relevant and applicable as possible, is rigorous following of the scientific method. Validity and integrity of data collection approaches and being open, honest, and unbiased in how methods are described, results are analysed, and findings are critiqued and appraised. The wider replication crisis noted in publication outputs of several scientific disciplines erodes trust and confidence in the collection and application of evidence (Boulesteix et al., 2020; Maxwell et al., 2015). Zoo-based research is already under scrutiny due to the logistical challenges of implementing robust experimental designs (Kuhar, 2006); e.g. due to small sample sizes and potential pseudoreplication, perceived lack of generalisability across individuals and situations, perceived limited useful comparison of zoo-housed animals to wild animals, and a need to perform repeated measures on the same animals resulting in a potential lack of independence. Therefore, adhering to the scientific method and being true, accurate, and objective in the explanation and evaluation of research designs and outputs is even more crucial. Figure 4.5 illustrates key points for those collecting evidence for, and evidence from, husbandry to follow to ensure the relevance and honesty of their research output.

Examples of how zoo-based husbandry research can move towards being more objective, unbiased, and open, as well as avoid subjectivity, ambiguity, and lack of replication include:

Figure 4.4 Good husbandry is reflected in the behaviour and physical attributes of the species being maintained. In this example, a yellow tang (*Zebrasoma flavescens) is* in excellent body condition (left), has vivid colouration, and displays healthy eyes and fins. These features are indicative of species-appropriate water chemistry, filtration, heating, lighting, water flow, enrichment, enclosure furnishings, social group, and daily care (including diet). Middle: husbandry-based evidence for good yellow tang welfare maps key biological needs (e.g. multiple opportunities for grazing on algae) onto fish responses to care. Aquaria that provide a healthy growth of algae, with numerous opportunities for grazing, and a husbandry regime that provides a suitable marine herbivore diet promote good health for this species of fish. Right: a fish in poor, thin condition, with pale colouration, ragged fins, and erosion around the head and eye. Such symptoms are evidence of incorrect husbandry and therefore fish care needs to be re-evaluated with the utmost urgency and changes implemented.

Repeatability

- All methods are described fully to allow for accurate replication for future data collection.
- The title and aims are clear, not misleading, or hard to comprehend.
- Sample sizes, location of data collection and all interventions are fully described.
- Results can be replicated if study is repeated in its entirety.

Unbiased review of sources

- Credible use of citations from a wide range of published literature.
- Review of papers that are independent of the authors of the study they are used in.
- Use of empirical study to support claims, rather than review articles.
- Clear and unambiguous supporting citations for all claims and arguments.

Biological significance

- Consideration of the biological relevance of any findings, as well as statistical significance and effect size.
- Even small changes in numerical data can be statistically significant, but is there any real change to the animal's behaviour, welfare etc?
- Truthfulness of the measure (e.g., does behaviour change truly equate to better welfare?)

Valid analyses

- Appropriate use of descriptive and inferential statistics suitable for any data collected and that can reliably test predictions.
- Clear explanation of all statistical testing for each dependent variable under scrutiny.
- Relevant choice of any inferential testing, avoidance of over-analysis and "significance fishing".

Figure 4.5 Removing any repeatability or replication issues in animal welfare research by following steps that enable all those who read published work to fully replicate all aspects of an intervention, case study or research project. Key points on how to be repeatable, how to use sources to support aims and arguments correctly, how to ensure that outputs have biological significance, and why it is essential to fully consider the choice and application of descriptive and inferential testing.

- Avoiding publication in predatory journals (Grudniewicz et al., 2019) that can lack academic rigour in the review and criticism of methods and approaches, but that provide an apparently honest platform for content to be disseminated.
- Avoiding over-use of self-citations and a lack of thorough review of the wider literature. Limited to no global reach in the literature reviewed that does not consider all authors in all regions.

- A reliance on review articles to support overarching or generalised statements when empirical papers are also available, which provide data-based support for arguments and claims.
- An over-analysis of data in an attempt to find significance. Considering weak or hard to define general trends as potentially significant outputs.
- Use of inappropriate inferential analyses that do not account for small sample sizes (Plowman, 2008) or repeated measures (Rose & Riley, 2021).
- Limited pilot studies or proof of concept approaches to check that data collection methods are applicable or relevant to data collection for a specific question or species.
- Lack of description of full result outputs, including no reporting of degrees of freedom, model fit, or test statistic values to evidence the relevance of inferential analyses to the collected data set.
- Over-stretching the findings of case studies, and lack of testing across different institutions to pool data and determine wider generalisability.
- Limited biological relevance of results. Findings may be of statistical significance, but is any perceived change or difference between conditions actually biologically relevant to the animal(s)?
- Lack of fluent explanation of data analysis choices. Statistical testing is not fully described so that the reader can understand why a specific test was chosen, what output variable and predictors were included, and what aim the testing hoped to achieve.
- Limited explanation and description of aims, objectives and descriptions so the reader is unsure about the background to the experimental design and why this investigation into zoo husbandry and welfare is needed.
- Writing an unclear title that lacks clarity in explaining what the paper is all about at the expense of a quirky or catchy "journalistic" title.

Creating our evidence bank for husbandry, which ultimately allows for the positive responses to such husbandry (from the animals) to become evidence in their own right, cannot happen without the use of the scientific method. The need to ask a question of or make a husbandry/welfare observation; to perform some research into this area of investigation; to draw up some predictions that can be tested; to collect data and to analyse these data to test the predictors; and to then report any conclusions (significant or otherwise) on animal husbandry; and to consider further questions and examination of husbandry and welfare based on these outputs is fundamental.

4.4 HEDIGER INTO THE 2050s AND BEYOND: ENSURING HUSBANDRY REMAINS RELEVANT

Using information from wild animals, their habitats and how they interact with them, is not a new way of evolving zoo animal management for the better. As Hediger has shown us at the start of this chapter, the more we understand about the natural history and evolutionary ecology of the species we manage, the better informed we are on their needs and requirements. Zoos into the 21st Century and beyond will be asked to care for more species of conservation concern – species where the wild is no longer a reality. It is essential that we learn as much as we can about species in the zoo that we are familiar with, so that these can become a proxy for good care and management for "new to the zoo" species and those in a desperate state in the wild (Kerr, 2021; Rose, 2021). For these familiar species to be relevant and valid proxies, the nature of their care (how appropriate it still is) needs to be regularly considered. Whilst zoos strive to save species of conservation concern, those species that are still common (but that have a place in the zoo) are also important. Review of the management of common species will ultimately keep them abundant and sustainable in the zoo too and provide useful information for those species that are less common.

Previous research has highlighted bias in the study conducted on zoo animal husbandry, with a preference towards the study of charismatic mammalian species, such as primates (Melfi, 2009; Rose et al., 2019). Whilst some research into such charismatic mammals is still relevant and needed, broadening the scope of research to start understanding how other species respond to care would help advance welfare across all zoo taxa. Birds, herptiles, fish and invertebrates are still complex and highly functioning organisms, clearly aware of their surroundings. Even though they are more removed from mammalian taxa that we have more of an affinity with, these non-mammal species need to become more of a focus for those researching husbandry in the zoo. Examples of husbandry evidence that could be prioritised in our research questions include:

- Species whose behaviour may be hard to define or categorise due to their activity or ecology.
- Wholly nocturnal or crepuscular species.
- Social or group-dwelling species where individual behaviours are highly influenced by their social environment
- Species perceived as "boring" or inactive in their daily activity patterns
- Over-looked mammalian species that may be considered less complex (e.g. small carnivores or rodents).
- Domestic species housed in the zoo, e.g. in Children's' farms or petting zoos
- Housing of zoo animals in quarantine or veterinary facilities.

Dissemination of case studies that have focussed on such species or situations, and wide availability to their findings in an open access manner is required to help answer similar challenges that may be occurring in many different institutions around the globe. If the keepers at one zoo are struggling with a specific aspect of tortoise husbandry, the chances are the same questions are being asked by keepers of the same species in other zoos in other locations. Perhaps a Conservation Evidence approach to zoo publication and distribution of information is therefore required?

Conservation Evidence is an online, open-access publication that documents research into species and habitats; linking together the animal and the environment with the outcome of supporting, as best possible, management decision that relate to these species and habitats (https://www.conservationevidence.com/). The Conservation Evidence website (as of May 2023) clearly states that "We summarise the documented evidence for the effectiveness of conservation actions." And this is exactly the tool that zoos need to succeed in providing across-the-board relevant husbandry and the potential for positive welfare. To encourage further publication of key studies, that are of high quality and applicability, in a manner that is accessible to all, collaboration and a conjoined effort by all zoo-focussed researchers could build a similar tool that would provide this evidence for ex situ husbandry and be a gatekeeper of the quality of such evidence. Zoo animal welfare can move forwards in huge steps when progressive and involved stakeholders get together to consider what changes to animal care are needed and why. By pooling resources and asking directed questions for specific species or taxonomic groups, we would be able to collect evidence more rapidly and use collective expertise to review the relevance of such evidence more efficiently. Box 4.1 provides a framework for transition from evidence-based husbandry to husbandry-based evidence and how this process can encourage buy-in and engagement from all parties interested in a species' presence in the zoo.

It is up to the future Hedigers of the zoo world to take up such a call-to-arms, to drive forward the evidence-based approach and to further distribute the positive responses to such evidence, from the animals that experience it, so that husbandry-based evidence is seen as the gold standard for positive animal welfare for 21st Century zoos.

BOX 4.1: Creating a research programme to generate husbandry-based evidence

1. Identify the current need for evidence for a specific species.
2. Identify the challenges associated with positive animal welfare states due to current housing, husbandry, and management.
3. Identify examples of good practice in zoos and ex situ facilities that could be measured or observed to understand why they are appropriate.
4. Identify and keep species' needs from wild, natural history information.
5. Gather interested and engaged stakeholders in that species' persistence and presence in the zoo.
6. Collate thoughts and ideas on the most practical and robust way to measure husbandry and species' responses to it.
7. Collect data across facilities. Review and consider findings.
8. Implement changes and repeat observation and evaluation of species' responses.
9. Document aspects of husbandry that evidence positive welfare.
10. Disseminate good practice around all holders and encourage the use of such husbandry-based evidence.
11. Reflect, review, re-assess evidence on a regular basis to reduce "keeper drift" from best practice guidance and to ensure that practice keeps pace with scientific discovery and outputs.

4.5 CONCLUSIONS

Husbandry for all species housed in the zoo and aquarium must be based on evidence and not anecdote. This is clearly illustrated in the large number of quality scientific papers that have made observations of practice, implemented changes, and reviewed the animals' responses. Animal welfare in the zoo is intrinsically linked to the quality of their care. This is the key foundation to the attainment of positive welfare states. Animals will not cope with poor husbandry, or that which is incorrect for their needs; therefore, without evidence for housing and husbandry, in-zoo welfare will not advance or improve for any species in ex situ populations. This approach to husbandry and welfare which relies on the collection and use of evidence of relevant throughout an animal's entire life, and concerns past experiences as well as present engagement and interaction with care and management. Collaboration and cooperation on research projects, a sound understanding of the scientific method, and an objective approach that implements a robust experimental design will mean that more and more zoo-housed species can benefit from husbandry evidence, and then their responses overall become the key information for how to manage a particular taxon in an ex situ facility.

REFERENCES

Binding, S., Farmer, H., Krusin, L. & Cronin, K. (2020). Status of animal welfare research in zoos and aquariums: Where are we, where to next? *Journal of Zoo and Aquarium Research*, *8*(3), 166–174.

Boulesteix, A.-L., Hoffmann, S., Charlton, A. & Seibold, H. (2020). A replication crisis in methodological research? *Significance*, *17*(5), 18–21.

Brereton, J. & Rose, P. E. (2022). An evaluation of the role of 'biological evidence' in zoo and aquarium enrichment practices. *Animal Welfare*, *31*(1), 13–26.

Clubb, R. & Mason, G. J. (2003). Captivity effects on wide-ranging carnivores. *Nature*, *425*(6957), 473–474.

Da Silva, R., Pearce-Kelly, P., Zimmerman, B., Knott, M., Foden, W. & Conde, D. A. (2019). Assessing the conservation potential of fish and corals in aquariums globally. *Journal for Nature Conservation*, *48*, 1–11.

Escribano, N., Ariño, A. H., Pino-del-Carpio, A., Galicia, D. & Miranda, R. (2021). Global trends in research output by zoos and aquariums. *Conservation Biology*, *35*, 1894–1902. https://doi.org/10.1111/cobi.13757.

Ferrie, G. M., Alford, V. C., Atkinson, J., Baitchman, E., Barber, D., Blaner, W. S., Crawshaw, G., Daneault, A., Dierenfeld, E. S. & Finke, M. (2014). Nutrition and health in amphibian husbandry. *Zoo Biology*, *33*(6), 485–501.

Finch, K., Sach, F., Fitzpatrick, M., Masters, N. & Rowden, L. J. (2020). Longitudinal improvements in zoo-housed elephant welfare: A case study at ZSL Whipsnade Zoo. *Animals*, *10*(11), 2029.

Grudniewicz, A., Moher, D., Cobey, K. D., Bryson, G. L., Cukier, S., Allen, K., Ardern, C., Balcom, L., Barros, T. & Berger, M. (2019). Predatory journals: No definition, no defence. *Nature*, *576*(7786), 210–212.

Harley, J. J., O'Hara, L. & Rose, P. E. (2021). A global survey of current zoo housing and husbandry practices for fossa: A preliminary review. *Journal of Zoological and Botanical Gardens*, *2*(3), 388–405.

Harris, M., Sherwin, C. & Harris, S. (2008). *The welfare, housing and husbandry of elephants in UK zoos* (DEFRA Science and Research Project WC05007). London, UK: Department of Food, The Environment and Rural Affairs.

Hediger, H. (1950). *Wild animals in captivity*. London, UK: Butterworth-Heinemann.

Hosey, G., Harley, J. & Ward, S. (2019). Research and research training in BIAZA zoos and aquariums. *Journal of Zoo and Aquarium Research*, *7*(4), 210–217.

Howell, C. P. & Cheyne, S. M. (2019). Complexities of using wild versus captive activity budget comparisons for assessing captive primate welfare. *Journal of Applied Animal Welfare Science*, *22*(1), 78–96.

Hummel, J., Zimmermann, W., Langenhorst, T., Schleussner, G., Damen, M. & Clauss, M. (2006). *Giraffe husbandry and feeding practices in Europe. Results of an EEP survey.* Paper presented at the 6th Congress of the European Association of Zoo and Wildlife Veterinarians, Budapest, Hungary.

Journal of Zoo & Aquarium Research. (2023). Submissions. EAZA. https://jzar.org/jzar/about/submissions. (Access Date 20/04/2023).

Kerr, K. C. R. (2021). Zoo animals as "proxy species" for threatened sister taxa: Defining a novel form of species surrogacy. *Zoo Biology*, *40*(1), 65–75.

Kuhar, C. W. (2006). In the deep end: pooling data and other statistical challenges of zoo and aquarium research. *Zoo Biology*, *25*(4), 339–352.

Lewis, K. D., Shepherdson, D. J., Owens, T. M. & Keele, M. (2010). A survey of elephant husbandry and foot health in North American zoos. *Zoo Biology*, *29*(2), 221–236.

Loh, T.-L., Larson, E. R., David, S. R., de Souza, L. S., Gericke, R., Gryzbek, M., Kough, A. S., Willink, P. W. & Knapp, C. R. (2018). Quantifying the contribution of zoos and aquariums to peer-reviewed scientific research. *Facets*, *3*(1), 287–299.

Markowitz, H., Schmidt, M. J. & Moody, A. (1978). Behavioural engineering and animal health in the zoo. *International Zoo Yearbook*, *18*(1), 190–194.

Marshall, A. R., Deere, N. J., Little, H. A., Snipp, R., Goulder, J. & Mayer-Clarke, S. (2016). Husbandry and enclosure influences on penguin behavior and conservation breeding. *Zoo Biology*, *35*(5), 385–397.

Maxwell, S. E., Lau, M. Y. & Howard, G. S. (2015). Is psychology suffering from a replication crisis? What does "failure to replicate" really mean? *American Psychologist*, *70*(6), 487.

Melfi, V. A. (2009). There are big gaps in our knowledge, and thus approach, to zoo animal welfare: A case for evidence-based zoo animal management. *Zoo Biology*, *28*(6), 574–588.

Mellor, E. L., McDonald Kinkaid, H. K., Mendl, M. T., Cuthill, I. C., van Zeeland, Y. R. A. & Mason, G. J. (2021). Nature calls: intelligence and natural foraging style predict poor welfare in captive parrots. *Proceedings of the Royal Society B: Biological Sciences, 288*(1960), 20211952.

Moise, R. I., Eccles, G. R. & Mettke-Hofmann, C. (2023). Enclosure background preferences differ between sexes and color morphs in the Gouldian finch. *Animals, 13*(8), 1353.

Mooney, A., Teare, J. A., Staerk, J., Smeele, S. Q., Rose, P. E., Edell, R. H., King, C. E., Conrad, L. & Buckley, Y. M. (2023). Flock size and structure influence reproductive success in four species of flamingo in 540 captive populations worldwide. *Zoo Biology*. https://doi.org/10.1002/zoo.21753.

Plowman, A. B. (2008). BIAZA statistics guidelines: Toward a common application of statistical tests for zoo research. *Zoo Biology, 27*(3), 226–233.

Roller, M., Müller, D. W. H., Bertelsen, M. F., Bingaman Lackey, L., Hatt, J. M. & Clauss, M. (2021). *The historical development of juvenile mortality and adult longevity in zoo-kept carnivores* (0733-3188). Wiley Online Library.

Rose, P. E. (2021). Evidence for aviculture: Identifying research needs to advance the role of ex situ bird populations in conservation initiatives and collection planning. *Birds, 2*(1), 77–95.

Rose, P. E., Brereton, J. E., Rowden, L. J., Lemos de Figueiredo, R. & Riley, L. M. (2019). What's new from the zoo? An analysis of ten years of zoo-themed research output. *Palgrave Communications, 5*(1), 1–10.

Rose, P. E., Croft, D. P. & Lee, R. (2014). A review of captive flamingo (Phoenicopteridae) welfare: A synthesis of current knowledge and future directions. *International Zoo Yearbook, 48*(1), 139–155.

Rose, P. E. & Riley, L. M. (2021). Conducting behavioural research in the zoo: A guide to ten important methods, concepts and theories. *Journal of Zoological and Botanical Gardens, 2*(3), 421–444.

Rose, P. E. & Roffe, S. M. (2013). A case study of Malayan tapir (Tapirus indicus) husbandry practice across 10 zoological collections. *Zoo Biology, 32*(3), 347–356.

Rose, P. E. & Rowden, L. J. (2020). Specialised for the swamp, catered for in captivity? A cross-institutional evaluation of captive husbandry for two species of lechwe. *Animals, 10*(10), 1874.

Rowden, L. J. & Rose, P. E. (2016). A global survey of banteng (Bos javanicus) housing and husbandry. *Zoo Biology, 35*(6), 546–555.

Scherer, L., Bingaman Lackey, L., Clauss, M., Gries, K., Hagan, D., Lawrenz, A., Müller, D. W. H., Roller, M., Schiffmann, C. & Oerke, A. K. (2022). The historical development of zoo elephant survivorship. *Zoo Biology, 42*(2), 328–338.

Troxell-Smith, S. & Miller, L. J. (2016). Using natural history information for zoo animal management: A case study with okapi (Okapia johnstoni). *Journal of Zoo and Aquarium Research, 4*(1), 38–41.

Zoo Biology. (2023). *Author guidelines*. Hoboken, NJ: John Wiley & Sons. https://onlinelibrary.wiley.com/page/journal/10982361/homepage/forauthors.html. (Access Date 20/04/2023).

5

Control and choice, and positive challenge…a route to ideal zoo welfare?

PAUL ROSE, PHILIP GREENWELL AND JACK LEWTON

5.1 INTRODUCTION

Good welfare for animals requires attention to both their physical health and their psychological needs – moving away from straightforward "Five Freedoms" welfare approaches to those that assess inputs and outputs, giving a more complete understanding of an individual's welfare. In the 21st century, a growing focus on using the affective states of animals (i.e. their feelings and emotions) as welfare indicators means that good welfare can be linked to evidence of positive mental states (e.g. play, exploration) validated with measures of behavioural diversity or time-activity and physiological measures (e.g. glucocorticoid metabolite concentrations that can link to the stress response). Consequently, the field of animal welfare science is increasingly emphasising the significance of evaluating welfare outcomes to provide insights into the day-to-day experiences of animals. To promote such an assessment of welfare, evidence-based, species-specific inputs provide the surest foundation for zoo-housed animals to be enabled and empowered to experience positive welfare outputs (Rose, 2023), including positive affective states that come from pleasurable moods and emotions. Examples of inputs include following of best (better) practice care guidelines, providing a dynamic enclosure with environmental complexity, using positive reinforcement training to enable animals to be involved in their daily care, and providing opportunities for choice to be on- or off-show (to zoo visitors) when desired.

For many species, where assessment of the mental domain of the Five Domains Approach may be hard or impossible, understanding how animals respond to the options provided to them in their environment gives us a secure footing when inferring that "welfare may be good." These responses can come from opportunities to experience choice and control and overcome challenges within the environment while helping to shape welfare

DOI: 10.1201/9781003429425-5

outputs for individual species based on how the animals demonstrate their preferences and meet the demands of the challenge provided.

When species-specific inputs are effectively utilised, animals can experience control and choice over their environment and benefit from feelings of autonomy and agency (Figure 5.1). In the context of zoo management, promoting control and choice can significantly enhance the cumulative quality of life experiences for the animals. Additionally, animals that experience positive challenges (i.e. ones that they can solve and gain benefits from) receive mental and physical stimulation and can explore new behaviours as they respond to the outside world in a novel way. Enclosure modification, environmental enrichment, or removing banal elements of routine in feeding schedules are examples of positive challenges. This creates interest (in the animal's immediate situation) and builds resilience from the experiences of dealing with and rectifying short-term stressors. This chapter explores how these key concepts of choice, control, and (positive) challenge are applied in zoos, the benefits they offer, and the trials involved in implementing them effectively.

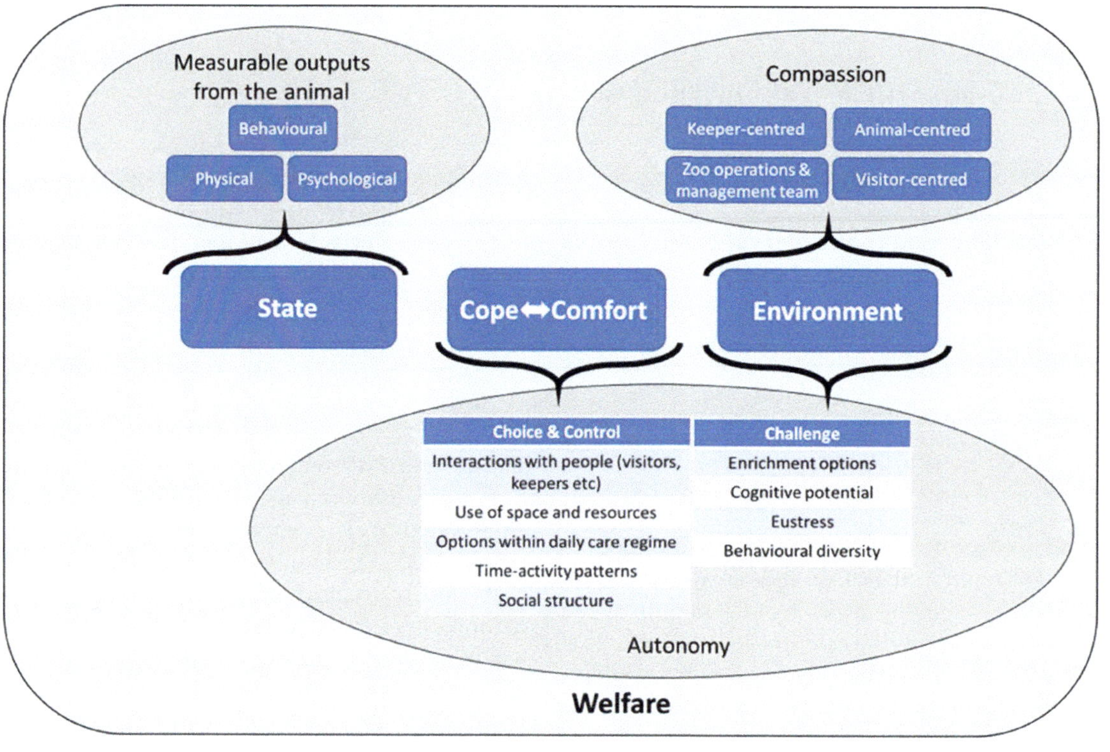

Figure 5.1 Taken from Rose and Lewton (2025). Using the key elements from Broom (1986) definition of welfare (state, cope, and environment) and expanding what they are to consider further influences over welfare experience in the zoo. States are measurable entities where data can be quantitative or qualitative, providing inferences on the animal's fitness and feelings. Coping (negative connotations) can be balanced against comfort (positive connotations), and these welfare outputs will be influenced by the degree of choice and control that the animal has over its immediate situation. The environment that the animal lives within is under the influence of the zoo and is a balance of needs – the animal's, the zoo's visitors, the keeping staff, and the will of the zoo's management. A compassionate management approach enhances human well-being and animal welfare. Challenge, which builds in behavioural resilience, can be incorporated into the animal's environment (e.g. via the provision of ecologically relevant enrichment) and, together with choice and control, leads to feelings of autonomy. The extension of state, coping, and environment by including comfort, choice, control, challenge, and compassion should provide a more complete picture of the prevailing welfare state.

5.2 CONCEPTS OF CHOICE, CONTROL AND CHALLENGE

Control and choice are distinct but related concepts that can be utilised to significantly contribute to animal welfare. Control refers to an animal's ability to influence its environment through its actions, whereas choice involves providing animals with options to express their preferences (for example, over what they do and where they do this). Both concepts are rooted in the idea that animals are capable of experiencing a range of emotions and motivations. Therefore, notions of sentience are also important here – to experience and be aware of feelings and sensations that can be pleasurable or aversive, to learn from such experiences, and to have abilities in evaluating actions and consequences (of self and others) (Broom, 2019). By expressing a choice or preference, an animal is clearly aware of something that it would like to do or engage with. Providing control and choice can reduce stress (Ross, 2006), thus preventing boredom and promoting natural behaviours that can be crucial for improved animal welfare (White et al., 2003). In the wild, animals constantly make choices about where to go, what to eat, and how to interact with others. Replicating these opportunities in zoos helps to meet a species' psychological and physical needs more effectively, as opportunities for more positive interactions with the environment or with caregivers, are possible. Even for often under-studied species, such as arachnids, providing wider habitat complexity that increases opportunities for novel interaction with the animal's environment may promote better welfare (Stalter et al., 2024).

Whilst comfort is important, animals can benefit from appropriate levels of challenge (Meehan & Mench, 2007). We consider positive challenges to be biologically relevant problems or tasks to solve that engage an animal's cognitive, behavioural, and physical abilities. Physical tasks can be used for body weight management, to promote muscle development, and improve motor functions, whilst mental tasks can practise memory, planning, decision-making, and problem-solving. This positive form of stress, termed "eustress", which arises from solving a challenge, contributes to enhancing physical and psychological health. The positive outcomes of controlling a situation and choosing to engage with eustress, resulting from positive challenge, are illustrated by Figure 5.2.

Eustress, in the context of animal welfare, refers to a positive or beneficial form of stress (Villalba & Manteca, 2019). A stressor that motivates an animal to "do something" and this "doing" ultimately builds resilience, enables coping, and solves a problem that leads to beneficial gain (learning, development) from the experience. Examples could include finding food via the use of a puzzle feeder and thus also experiencing cognitive exercise, manipulating enrichment objects to test hand-eye coordination, or engaging in social interactions at a specific time of the year for a specific purpose or intent. Ultimately, eustress is caused by challenges that are manageable, stimulating and ecologically relevant to that species; this promotes alertness, focus, and motivates the performance of natural behaviours that have a fitness or adaptive benefit to the individual or population. The overall positive outcome should be an improvement in the animal's welfare.

Eustress differs from distress, as something that is distressing is caused by overwhelming or chronic stressors which lead to feelings of anxiety, fear, or frustration (Moberg, 1999). When animals have the ability to make choices and influence their environment, they are better equipped to handle challenges, reducing the likelihood of frustration. Encouraging environmental interactions that develop behavioural responses and problem-solving abilities is especially relevant to conservation outcomes (Griffin et al., 2000; Shepherdson, 1994), giving better assurances that animals are equipped for any potential life in the wild that is independent of human care (Reading et al., 2013). Aligning key aspects of enclosure design, habitat recreation and provision of suitable enrichment not only improves behavioural diversity and welfare outlook but also adds more value to how that population supports the zoo's key aims.

In the zoo, distress can be caused by cramped or barren enclosures, a lack of enrichment, social isolation for species that require social stimuli, exposure to loud noises and no opportunity for an animal to remove themselves from being "on view". Chronic and unchanging abnormal repetitive behaviours are often a common symptom of distress, and a lack of choice and control, with such behaviours being widely studied in zoo-housed animals. Zoos need to consider how to work

Figure 5.2 Definitions of choice, control, challenge and the interplay between them that can lead to good welfare. (A) A giraffe (*Giraffa camelopardalis*) chooses to engage in an optional public feeding session, demonstrating **choice**. (B) An Asian elephant (*Elephas maximus*) participates in an optional health check with a keeper, showcasing **control** over the interaction. (C) A red panda (*Ailurus fulgens*) enjoys a reward after choosing to participate in a training task, highlighting both **choice** and **control**. (D) A meerkat (*Suricata suricatta*) engages with a puzzle feeder, demonstrating **challenge** as it solves the puzzle to access food. **(E)** Ring-tailed lemurs (*Lemur catta*) interact and socially forage on a puzzle feeder, illustrating both **challenge** and **choice** in their problem-solving and social behaviours.

eustress into daily husbandry and management plans so that animals have both comfort and positive challenge during their lives in the zoo (Figure 5.3).

5.3 IMPLEMENTING CHOICE AND CONTROL IN ZOOS

Published scientific research shows the importance of understanding what is meant by control and choice when it comes to promoting agency and ensuring that animals are autonomous in their lives and abilities in decision making (Englund & Cronin, 2023). When we develop our collection and application of valid evidence to inform husbandry, such evidence-based husbandry then becomes husbandry-based evidence (Rose & Lewton, 2025), in terms of the positive outlooks that we see in the zoo. Here, in this part of the chapter, we provide some practical ideas for including choice, control and challenge in zoo husbandry and management focussing on:

- Enclosure and exhibit design
- Environmental enrichment
- Social grouping
- Husbandry training

As is covered in other chapters in this book, the design of animal enclosures plays a pivotal role in providing control and choice. Enclosures that mimic natural habitats and offer diverse features allow animals to engage in a variety of behaviours, and designing complex enclosures that mimic natural environments can provide ongoing challenges for zoo animals (Rose et al., 2024). This includes creating multi-dimensional spaces with various substrates, vegetation, and microhabitats. If such habitats change over time, with new features introduced regularly, the enclosure remains vibrant, dynamic, and stimulating for the inhabitants. For example, rotational exhibits that circulate different species through them can provide benefits for the zoo and animals (White et al., 2003). For the zoo, this approach offers advantages such as space efficiency and enhanced visitor experiences, whilst for the animals, it offers a dynamic environment

Manageable and relevant challenge

Provide challenges that are appropriate for the species and individual animal. If too difficult or overwhelming, distress and fear may result.

Positive result

The behaviour that is promoted is fulfilling and ultimately satiates the animal. The animal learns something that can be applied in the future.

Balancing routine and variety

Zoos need to ensure a balance between predictability (routine so animals are comfortable with daily care) and environmental variation that develops new skills, opportunities for experiencing novelty, and chances of building resilience.

Figure 5.3 Managing challenges in zoos to return positive results (i.e. welfare enhancement and enhanced overall quality of life).

with the choice of interactions. Rotational exhibits can include multiple species if all their ecological needs are met, and they are behaviourally compatible. For example, at Longleat Safari Park in the UK, a large drive-through enclosure known as the "Big Game Park" can be shared by southern white rhinoceros (*Ceratotherium simum simum*), Bactrian camel (*Camelus bactrianus*), scimitar-horned oryx (*Oryx dammah*), ostrich (*Struthio camelus*), and Ankole cattle (*Bos taurus*), with species being rotated as per the zoo's and the animal's needs (Figure 5.4). In such a mixed species display, however, even in a substantive area, animals still need to be given the freedom to remove themselves from visitors or from conspecifics as and when they choose. Simple changes to daily husbandry, such as access to 'on-show' or 'off-show' locations, provide a chance for animals to decide if they wish to be on view to the zoo's visitors, and this aspect of enclosure choice has been an emerging priority in zoos. Evidence shows that, across taxa, zoo animals exhibit less stress when given the choice over where to spend their time (Anderson et al., 2002; Owen et al., 2005; Ross, 2006). Providing the animals with access to their full living space (when appropriate) is a great example of exercising choice.

Alongside of enclosure design, environmental enrichment is also an important aspect of choice, control, and challenge (see Chapter 6 for more). Regularly alternating different types of enrichment prevents habituation and keeps the environment engaging (Kuczaj et al., 2002). Enrichment can also be tailored to an individual animal's preferences and needs, and this is especially important when considering how enrichment is used to reduce the performance of stereotypies, for example. The dynamic environment that enrichment creates in an enclosure is a vital tool for zoos, as this empowers animals to make choices, control the pace of their day, and ultimately contributes to improving physical and psychological welfare. Correctly applied, biologically relevant enrichment is a way that animals can gain new information which can prepare them for the future (Veissier et al., 2024). As such, enrichment is an essential framework for

Figure 5.4 The "Big Game Park" at Longleat Safari Park, UK - a multi-species rotational exhibit providing animals with a choice of interspecies social interactions. The image shows southern white rhinoceroses with Ankole cattle and scimitar-horned oryx in the background within the same enclosure.

building resilience and giving more autonomy to the daily lives of zoo animals.

As an example, using nutritional enrichment techniques that give animals control over their feeding experiences can promote appetitive foraging activities that ultimately result in the performance of consummatory behaviours, and thus lead to satiation. However, consistent feeding schedules can also be important, for example when management routines are supported by the reliable presence of food. Some animals may become stressed by any absence of food, and not knowing when food will next become available may also result in abnormal changes to behaviour. It is therefore important to assess the implementation of any flexible feeding strategies to ensure that they are promoting desired outcomes and that they are not, ultimately, reducing an animal's welfare experience. A balance is required, however, between offering reliable resource access and building in variation to encourage behavioural diversity and flexibility in a species-specific way.

Social animals benefit greatly from the ability to choose their social partners and interactions (Rose & Croft, 2015). In zoos, this can be facilitated by housing animals in compatible social groups and providing spaces where they can be together, directly or indirectly, or even retreat from others if they choose (Figure 5.5). This choice reduces the chance of aggressive encounters through forced interactions and maximises positive social interactions. Positive interactions may be important for an animal's development as they provide opportunities to play and learn from each other. Encouraging social play can stimulate the physical and cognitive development of individuals and strengthen social bonds between individuals that engage in play.

Positive reinforcement training is a powerful method for giving animals control over their environment and interactions with caregivers (Laule & Whittaker, 2007). By rewarding desired behaviours, animals learn that their actions can influence outcomes, which can reduce stress and improve cooperation during husbandry procedures. Training also empowers zoo animals to participate in their own care. For instance, when animals are trained to present body parts for routine health checks, to enter transport crates voluntarily, or to participate in medical procedures such as injections, they are more in control of the situation because they have a choice to take part. Such training supports a compassionate approach to animal welfare (Rose & Lewton, 2025) as zookeepers also benefit from engaging in a less stressful, more equitable relationship with the animals they are responsible for.

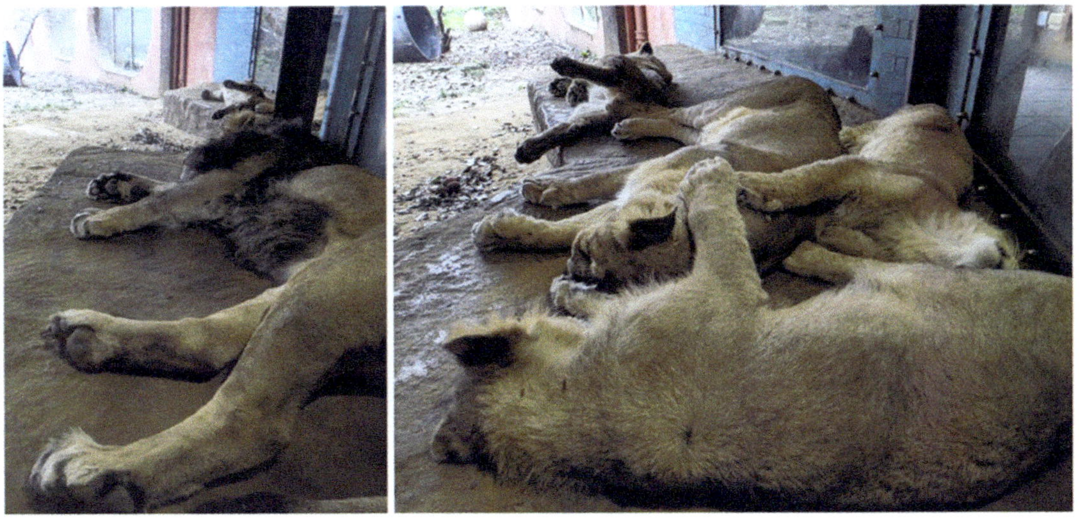

Figure 5.5 Enough room to pile on yet still be apart. A valued resting area for lions (*Panthera leo*) allows individuals to socialise in their own way without quarrelling over who goes where.

5.4 PROMOTING AUTONOMY THROUGH CHOICE AND CONTROL

The examples noted above provide animals with a sense of autonomy. Autonomy refers to the feeling of independence and self-governance; therefore, allowing animals to make choices in their daily activities promotes some degree of independence. Some authors argue that the fact that animals are captive removes their dignity because they are not free to make choices (Browning & Veit, 2021). With evidence-based care that is provided in an environment that recreates important habitat features and is dynamic and engaging, we suggest that animals can experience independence and control over their feelings, emotions, motivational states, and needs. Enabling animals the decision over when to rest, when to explore, or when to engage with enrichment items further leads to satiation. Knowledge of the features of an enclosure, as well as where favoured resources are located, can buffer zoo animals against unpredictable stressors (e.g. the behaviour of different visitors when at the enclosure). The animals can control their interaction with different visitors by knowing that they have safe, comfortable retreat areas to seek refuge in. Using measures such as "giving-up densities" (Troxell-Smith et al., 2017) that can assess preference and avoidance of resources in particular enclosure locations provides data on how animals choose to use their space in relation to other contexts (e.g. weather, visitor presence, time of day). Environmental complexity and heterogeneity can be created within the three-dimensional space of the enclosure (Figure 5.6), enabling different resources to be accessed in different ways.

Opportunities for choice and control often lead to increased physical activity and a more diverse suite of natural behaviours, which are crucial for maintaining good health. For example, as demonstrated in Troxell-Smith et al. (2017), animals that can choose to feed at different foraging patches exhibit more natural movement patterns and behaviours. Thus, control ("I wish to feed now") and choice ("I'd like to feed here and not there") contribute significantly to an animal's psychological welfare. Engaging in preferred activities and having the autonomy to make decisions can lead to positive emotional states, reducing the risk of boredom, lethargy or apathy. Different species have unique needs and preferences; what works for one species might not be suitable for another. Similarly, individual animals will also have different personalities that impact how they perceive different features of their environment as positive and negative (even the things that we, as humans, want to be perceived as positive, such as enrichment). Understanding the natural history and behaviour of each species is critical for designing effective interventions that provide choice and control, but also when considering how disruption or variability can be presented to provide challenging stimuli

Figure 5.6 A large cat enclosure with considerable environmental heterogeneity and variation in how the inhabitants can use the space provided. However, stereotypic behaviours can still be seen, even in complex recreations of habitat such as this (see https://youtu.be/KE3KD7jZErQ for this example). Understanding stimuli that are bigger than the enclosure (climate, time of day, an animal's motivational state or anticipatory urge) is required to objectively judge the responses of all individuals to the space they are provided with.

5.5 ASSESSING THE IMPACT OF CHOICE, CONTROL AND CHALLENGE

Implementing control and choice requires ongoing monitoring and evaluation to ensure that interventions are effective and beneficial. Regular welfare assessments, behavioural observations, and feedback from zookeepers are essential for refining practices and addressing any issues that arise. Each species and individual has unique preferences that may change over time. And so, continuous monitoring ensures animals are experiencing positive affective states and are provided with opportunities without causing stress or frustration long after implementation. This in itself may become a challenge to monitor, as some animals may have lagged responses or exhibit stress during out-of-zoo hours. Some zoos monitor their animals 24 hours a day with surveillance systems, and recent advances in Artificial Intelligence are being explored to monitor animal behaviour in captive settings (Rohan et al., 2024). This may prove a useful tool for continuous monitoring of zoo animals and assessment of their reactions to any choices or challenges they are provided with. Using behaviour-specific methods, such as applying social network analysis to conservation-focused questions or practices (Lewton & Rose, 2021), can provide further information on individual animal responses and how these are shaped by the opportunities they have to experience choice and challenge, or to feel in control of their current situation.

Research has shown that species do prefer to have a choice, but individual differences in the context of their choices are important (Perdue et al., 2014). Therefore, assessment of the challenges we introduce is crucial to ensure that animals are being stimulated and engaged without experiencing undue frustration or stress. To the simplest level, this can be thought of as "how are they compared to their baseline?". Baseline behavioural activity can be calculated by monitoring the frequency and durations of behaviours under normal conditions. Some behaviours may be of interest; for example, a challenge may be introduced with the goal of reducing abnormal repetitive behaviours. Additionally, physiological parameters can be measured under baseline conditions and subsequent changes monitored. Once we have an idea of what is "normal" for that individual we can start to compare aspects of behaviour and physiological indicators. For example, a decrease in pacing (repetitive behaviour associated with stress) and an increase in exploration time may be seen as positive behavioural indicators of welfare for most animals. Whereas an increase in cortisol concentration and elevated heart rate variability are often physiological indicators of stress or excitement and thus can have very different welfare inferences. Assessing these indicators at different intervals after introducing a challenge can be used to measure their short-term and long-term effects. By employing these methods, zookeepers can ensure that animals are receiving positive challenges. Like in all aspects of animal welfare assessment and measurement, being valid in how measures are taken is essential. Box 5.1 provides some further discussion of what it means to employ a "validated measure of animal welfare" and why using validated welfare assessment tools is essential to making evidence-based changes to husbandry and management that we have confidence in.

BOX 5.1: Considerations for what it means to use a validated animal welfare measure

We asked all the authors of this book to provide their ideas on what the statement "validated measure of animal welfare" means to them. We provide a summary of their responses, as well as a word cloud (Figure 5.7) to illustrate common descriptors, below.

A validated measure is one that evaluates a relevant and important aspect of an animal's well-being, and that it (however it is being assessed) accurately captures (reflects) that animal's actual status for that aspect of well-being. The most important consideration for validating a measure is to have a relevant and accurate benchmark against which to evaluate or assess the results for that measure. It's also important to consider the period of time that the measure is intended to reflect (i.e. is it just for that instant in time, or is it intended to reflect a longer time period), and this should depend on the question being asked (i.e. fit the appropriate measure to the appropriate question).

A process or procedure that has been shown to describe, reliably and objectively, an animal welfare metric and how this metric changes over time or in response to husbandry methods. For example, foot scores for ungulates. Reliability and repeatability (especially over long timeframes) are also imperative. The validation process of an animal welfare measure must be easily understood by a variety of people across multiple settings.

A validated welfare measure accurately reflects the specific aspect of welfare it is intended to measure, and it is reliable and repeatable. Ensuring biological relevance is key. Animal welfare is about an animal's experience and its ability to cope with its environment. A biologically relevant measure directly connects to the animal's physiological or behavioural responses, which in turn reflect its internal state. With that said, I also feel that the process should not be focussed on what is easy to quantify, but rather on ensuring that the measure reflects what matters to the animal. This means that validation can be extremely challenging because welfare includes the subjective experience. I think measures that try to capture this are important even when full validation is not possible. For example, we know from the literature that 'choice and control' is important to animals, and providing choice has been linked to a reduction in stereotypic behaviour. Asking questions such as, "What opportunities exist for the animal to have choice and control within its environment?" are important – even if each of those potential choices has not been validated to a reduction of abnormal behaviour or an increase in positive behaviours, or a lack of choice and outcome is not assessed in the species of interest.

So, some measures may not be fully validated but should not necessarily be omitted from the process.

A validated animal welfare measure is one that has been rigorously tested for validity of the measurement through both, content precision (how well the measure represents welfare) and criterion validity (does it correlate with a valid measure/or how accurate is it to the gold standard value). Overall, this should lead to construct validity – are we measuring what we think we are measuring i.e. if we interpret a behaviour, are we correct in the interpretation (like the example of pacing (abnormal (negative) or anticipatory (positive)). So, we need several measures to make that judgement. A combination of convergent measures is probably the most important consideration to ensure a valid measure.

Biological validation is also essential to ensure that meaningful results are generated that are applicable to the identification and measurement of welfare states in that particular species. Beaulieu (2024) provides a useful review of these concepts.

Comparison against any already validated parameter, when possible; however for many of our species, these simply do not exist. I find this is easier for the health domain for instance, but much harder for behaviour. I am comfortable with using a model species (e.g. domesticated species) and then using with a degree of caution in the application. Science tells us to then measure the indicator against a controlled variable, but again, this is often not possible for our species due to constraints with animals and environmental management, or ethical boundaries. I think that using individual measures and comparing/contrasting individuals can help us here. I believe that the most practical thing that we can do is to use multiple indicators to assess the same aspect of welfare, and then look for relationships and trends.

Essential to any definition is the need for objective measurement alongside clear criteria. And a main consideration is to produce consistent and reproducible results against the specific construct being measured, with species, context and duration being key issues.

A set of criteria that can be objectively measured with reproducible results that are linked to the health or well-being of the species being studied.

A validated measure must be proven to be robust and reproducible, and is linked to a clearly defined health or well-being status such as the grimace scales in rodents and lagomorphs (NC3Rs, 2022).

A replicable and reliable means of measuring (species-specific) welfare that is repeatable across time and by different people, independent of observer skill and bias. For example, I think I would naturally gravitate to measures that focus on testable areas, such as hormone levels, faecal analysis for parasites, feather structure and dermatitis, prevalence of feather picking and environmental humidity etc., more than behavioural observations that are more subjective and dependent on external variables such as weather, food, people, conspecifics etc. Ideally, there would be much prior evidence that the approach has been undertaken before, that the method was ethical and did not compromise welfare/behaviour. For example, if the measure was used as a means to improve welfare or change a behaviour, what were the outcomes?

Figure 5.7 Word cloud of key themes that this book's authors feel to be more important when considering what "valid animal welfare measure" means.

5.6 A CONTROL, CHOICE AND CHALLENGE CASE STUDY: PROVIDING FOR PARROTS

Colourful, raucous and familiar parrots continue to be a staple of zoo exhibits. However, unless displayed well in appropriate enclosures that encourage positive activity and diverse behavioural performance, these animals can be sedentary, easily bypassed by visitors, or engage in abnormal repetitive behaviours such as stereotypical feather plucking (Mellor et al., 2021). Zoos vary in their approaches to housing parrots; these approaches can be based on the overall theme of the collection, available space, inherited infrastructure, level of popularity or endangerment, and general interest in the taxa from management and curatorial staff. Within the order of Psittaciformes, a great deal of behavioural, social and ecological variation exists, and treating the group as homogeneous is patently unacceptable. To treat polyamorous species such as eclectus parrots (*Eclectus* sp.) or vasa parrots (*Coracopsis* sp.) in a similar manner to monogamous species such as African grey parrots (*Psittacus erithacus*) or the larger macaws (*Ara* sp.) would result in inappropriate husbandry and thus diminished welfare. The importance of mate choice for pair compatibility is significant, as the killing of the female by the male bird can be a real risk in incompatible pairs. Provision of a suitable number of mates either for polyamory, colony nesting, or mate-choice selection can be complicated for zoos to manage, as multiple birds are needed (and this being an expense to available space, time, and diversity of other species housed in the zoo's animal collection).

Similar considerations need to be taken for colonial and cooperative breeding species: introvert versus extrovert and neophobic versus neophilic species, for example. African grey parrots, well known for shyness, may benefit greatly from having access to out-of-view areas more so than extrovert species. Having an area to retire to may instil greater confidence in shyer species, as they know they can remove themselves from view or conflict with other aviary occupants. The ability to choose can be of intrinsic importance to many species, as demonstrated by empirical work on primates (Perdue et al., 2014), and therefore the opportunity for choice may be more valuable than the options that are actually presented.

When considering choice, control, challenge, and agency for captive parrots, it merits looking at wild ethograms for a particular species, or for closely related species, to determine fundamental behavioural activities and how time is budgeted across these. For example, up to 75% of waking hours can be spent on foraging activities in wild parrots (Mellor et al., 2021), suggesting that the search for, manipulation, and consumption of food matter is of high value for these animals. Species that eat food items that require time spent on

Figure 5.8 Wild scarlet macaws (*Ara macao*) foraging in the rainforest of Costa Rica illustrate the complexity and dynamism of this environment and why these parrots spend such a large portion of their day foraging and travelling to foraging locations. The social aspect of flocking (information sharing and opportunities for affiliation and social exchange) is also apparent in these photos. Photo credit: R. Lemos de Figueiredo.

manipulation in the wild are suggested to be more prone to feather-damaging behaviours in captivity (Mellor et al., 2021). Within this framework of "foraging activity", there will also fall flight and other locomotion, vigilance, and socialisation behaviours, presenting a rich suite of behaviours overall (Figure 5.8). Captivity denies or severely limits these activities, with food often presented in one single location, at set times, in an easily accessible and consumable manner, and with little annual variation in food type. This style of husbandry is based on old avicultural practices, derived from farming applications that preferred speed and ease of keeping animals. Research is showing that parrots are incredibly complex animals, and not only show significant variation between species but also within regarding personality traits and behaviours. Brain size (encephalisation), in relation to body mass, also influences welfare in parrots. Mellor et al. (2021) suggest that more intelligent parrot species require increased cognitive stimulation. Therefore, emerging technologies can help support these behavioural needs, such as by the use of touch screen devices, complex puzzles, and challenging training activities (Kleinberger et al., 2024).

As naturalistic enclosures become the norm in zoos, husbandry regimes must adapt as well. Many parrot species may become aggressive during the breeding season, and this can restrict the use of walk-through enclosures or mixed aviaries during these times for these species. One of the reasons that 'traditional' aviaries have external access to feed and water bowls is to reduce the risk of aggressive encounters with breeding birds. It can therefore be complicated to improve choice and control in feeding in some housing setups, for example, and parrot keepers may also be pressed for time to ensure that higher welfare options are constantly presented. One manner of navigating breeding aggression may be to hold bachelor groups of birds; thereby, a single-sex colony may then be kept in a free-flight, multi-species aviary in relative harmony. However, the trade-off is that species then do not have the ability to engage in natural reproductive activities that are in themselves highly enriching. The reason why the parrot is being kept by the zoo should therefore be considered. Breeding stock required for conservation or population sustainability initiatives may be best housed in off-show aviaries that provide an environment suitable for mate choice, nesting and hatching of young, whereas birds not needed for breeding can be displayed to visitors in a more dynamic and engaging fashion.

Encouraging agency in captivity can present as follows: options for evading other aviary members, feeding from multiple feed sites, accessing food in a manner that presents challenge, retiring from view when desired, and performing natural mate selection and reproductive behaviours (Figure 5.9). Supporting parrots to have agency helps retain or develop skills or behaviours that may be important for future use, certainly if birds are to be used to supplement depleted wild populations and other conservation programmes. For example, captive yellow-shouldered Amazon parrot (*Amazona barbadensis*) chicks kept in pre-release flights prior

Ecological relevance of enclosures Contents/enrichment permits diversity of activities and behaviours. Engagement in new and emerging technologies to improve welfare.	**Species specific husbandry** e.g. vasa/eclectus vs. greys, Amazons, macaws. Colonies kept where feasible and safe, or birds are housed within sight/sound of conspecifics.	**Reproduction** Free-mate choice, bachelor groups, breeding vs non-breeding aviaries. Young spends appropriate time with parents to learn key skills. Nest boxes can be modified by pair.
Education Signposting and application of high zoo bird welfare in relation to pet birds via signage. Consider important information from wild populations.	**Considerations for a Gold Standard in parrot welfare**	**Enclosure design** Height, width, floors, shade, water, perches, weather etc. built around welfare needs. Enclosures designed to maximise hygiene and safety, with secluded areas away from visitors.
Recognition of emotional and intellectual "uniqueness" Pair bonds, PTSD, longevity, grief, senescence etc. Treat like primates due to intelligence and sociality and complex ecology/habitat.	**Distribution of food resources** Presentation of food in a complex and engaging manner to replicate feed/foraging behaviours. Seasonal variation of food items. Flowers, foliage, bark and other non-traditional feedstuffs available.	**Physical and psychological health checks** These are managed appropriately. Training and enrichment provided to enhance welfare, particularly for free-flight, display birds.

Figure 5.9 Key considerations to improve welfare in captive parrots to a gold standard across zoological facilities.

to release were fed local food that required high degrees of manipulation, such as cactus fruits left with spines on (Sanz & Grajal, 1998). Enabling free flight and the diversity of movements, learned experiences, and motor control that go with navigating between, amongst, over and under obstacles when in the air manifests as individual parrots suitable for wild release (Brightsmith et al., 2024). Both examples show the importance of naturalism in housing and husbandry – to create opportunities for wider environmental engagement that provide experiences of challenges that need to be overcome.

5.6.1 PERFECTION FOR PARROTS…?

Bioparc Zoo de Doué-la-Fontaine in the Pays de la Loire region in France is renowned for its unique setting. Situated on the site of a former quarry, the Bioparc's exhibits are linked by old underground service tunnels through the rock, while the enclosures are built around, over, or through the remains of the quarry. Of particular note, given its scale and diversity of species, is the "La Grande volière sud-américaine" (South American aviary). This aviary is constructed within the quarry, with 'islands' of land left to provide both roof support and nesting sites, and the walls comprising raw stone. Into this exhibit are mixed waterfowl and wading birds, Chilean flamingos (*Phoenicopterus chilensis*), raptors, Humboldt penguins (*Spheniscus humboldti*), and multiple parrot species. To witness large flocks flying through these canyons is a remarkable experience.

The exposed stone walls of the aviary are a focus for species that inhabit or use cliff-faces in the wild (Figure 5.10), and for the resident flock of Patagonian conures (*Cyanoliseus patagonus*) this is possibly the most naturalistic habitat that could be replicated in a zoo. This species is also known as the burrowing parrot, as they excavate nesting chambers in cliff faces in their natural range – a behaviour generally difficult to replicate in zoos. Some zoos have circumnavigated this by building fake cliffs and inserting pipes to replicate burrows

Figure 5.10 Two aspects of the South American Aviary at Bioparc Zoo de Doue-la-Fontaine are noteworthy. Nesting cavities can be seen excavated in the walls. Bricks have been used to create nesting chambers in larger holes. Perching sites are multiple and vary in dimensions. Mixed species allow for intra- and interspecific interactions, and multiple members of the same species permit free-mate choice.

with nesting chambers at the rear for nest inspection, an admirable effort to replicate habitat and educate people on the natural history of the species. In Doué-la-Fontaine, however, birds have the choice of nest site selection, their proximity to neighbours, and to engage in nest excavation, construction, and defence against usurpers. These are all activities that may reinforce pair bonds, keeping body and mind active, as well as providing fascinating viewing for the public. As a terrestrial feeder, the conures feed both from the feeders provided but also actively search for fallen grains on the floor, thereby avoiding other parrots on the feeders and replicating natural foraging behaviour. Substrate varies throughout the aviary, with lawn, gravel, turned earth, running water, and wild vegetation in different areas, providing choice and challenge in feeding behaviours. The scale of the aviary also permits large numbers of individuals of a species to be held, which supports free mate choice and selection. Individuals can associate with preferred partners and affiliates. Flight is unrestricted; indeed, movement between areas is only possible by flying from nesting sites to perches to feeding areas and so on. Thus, these parrots experience choice, control, challenge, and a high degree of autonomy (that one hopes keeps them comfortable in their environment) on a daily basis.

5.7 CONCLUSIONS

In this chapter, we have discussed how promoting control and choice is a vital aspect of enhancing animal welfare in zoos. As our understanding of animal welfare continues to evolve, incorporating control and choice will grow in prominence and emphasis. By providing an environment that allows animals to express natural behaviours and make choices about the flow of their daily lives, zoos can significantly improve welfare at an individual, species, and population level. Validated measures of welfare that consider how animals respond when they are provided with choice, control, and (positive) challenges are key. Without understanding how animals react to opportunities for choice and control, we will not understand their relevance to promoting good welfare outcomes.

Figure 5.11 The relationship between control and choice, and challenge, along with coping, comfort, and compassion, and how these influence welfare outcomes for zoo animals.

Providing positive challenges and experiencing stressors that can be rectified is essential for promoting resilience and ensuring an animal's physical and mental welfare is robust. By encouraging natural behaviours, enhancing cognitive function, and supporting physical health, positive challenges enrich the lives of zoo animals and help them thrive when managed under human care. Implementing these practices requires thoughtful enclosure design, ongoing welfare assessment of the individual and the animal group, and a deep understanding of each species' needs, ecological background, and natural history. As zoos continue to evolve and prioritise animal welfare, the integration of positive challenges and stressors will remain a key component of compassionate and effective animal care. We feel that Figure 5.11 best sums up our approach to these important "Cs" of good zoo animal welfare, and we can further understand how to deliver choice, control, and challenge by answering the questions in Box 5.2

BOX 5.2: Further research questions for understanding, providing and ensuring the relevance of choice, control and challenge for zoo and aquarium animals

- For individuals of a particular species, what choice is a good choice? What evidence do we have that the choice provided is meaningful?
- How do we know an animal has control? What are the signs and signals presented that indicate some degree of autonomy?
- How can we effectively measure eustress in zoo and aquarium animals, distinguishing it from distress and validating positive outcomes from it?
- What does a resilient individual of a particular species present as? What behavioural, psychological, and physical clues can be valid identifiers of resilience?
- What types of environmental, social, or cognitive stimuli reliably elicit eustress in different species, and how do we keep these experiences novel?
- How does promoting eustress (from positive challenge) influence the long-term welfare and behaviour of zoo animals with regard to building resilience and positive behavioural diversity?
- What are the physiological and behavioural indicators presented by a species that show a biological challenge is beneficial rather than harmful?
- How can biological challenges be adapted to individual animal needs and species-specific behaviours that have wider importance (e.g. in conservation programmes)?
- What are the unintended consequences of introducing novel biological challenges in controlled environments?
- How can cross-institutional collaborations increase opportunities for animal choice and control in captivity by developing new ways of providing these options or validated methods to measure outputs?
- What are the barriers to implementing shared enrichment programmes across zoos, and how can those barriers be overcome for specific species?
- How do zoo animal "giving up densities" vary with the type of resource (e.g. food, enrichment devices) that they are engaging with?
- What external or extraneous factors influence the decision of an animal to abandon the use of a resource within a specific location of their enclosure?
- How do "giving up densities" inform us about perceived predation risk, competition, or resource quality in captivity?
- Why do some animals continue to exhibit abnormal repetitive behaviours despite being housed in enriched, complex environments?
- Are certain types of enrichment less effective at reducing abnormal behaviours in specific species?
- Can we develop an effective assessment of enrichment programmes to always ensure that enrichment is biologically relevant (Brereton & Rose, 2022) and thus actually enriching for that species?

REFERENCES

Anderson, U. S., Benne, M., Bloomsmith, M. A. & Maple, T. L. (2002). Retreat space and human visitor density moderate undesirable behavior in petting zoo animals. *Journal of Applied Animal Welfare Science, 5*(2), 125–137.

Beaulieu, M. (2024). Capturing wild animal welfare: A physiological perspective. *Biological Reviews, 99*(1), 1–22.

Brereton, J. & Rose, P. E. (2022). An evaluation of the role of 'biological evidence' in zoo and aquarium enrichment practices. *Animal Welfare, 31*(1), 13–26.

Brightsmith, D. J., Biro, C., Mendes, H. F. & Woodman, C. (2024). Free flight training as a tool for psittacine reintroductions. *Birds, 5*(3), 522–542.

Broom, D. M. (1986). Indicators of poor welfare. *British Veterinary Journal, 142*(6), 524–526.

Broom, D. M. (2019). Sentience. In J. Chun Choe (Ed.), *Encyclopedia of animal behavior* (2nd ed., Vol. 1, pp. 131–133). Cambridge, MA: Academic Press.

Browning, H. & Veit, W. (2021). Freedom and animal welfare. *Animals, 11*(4), 1148.

Englund, M. D. & Cronin, K. A. (2023). Choice, control, and animal welfare: Definitions and essential inquiries to advance animal welfare science. *Frontiers in Veterinary Science, 10,* 1250251.

Griffin, A. S., Blumstein, D. T. & Evans, C. S. (2000). Training captive-bred or translocated animals to avoid predators. *Conservation Biology, 14*(5), 1317–1326.

Kleinberger, R., Cunha, J., McMahon, M. & Hirskyj-Douglas, I. (2024). *No more angry birds: Investigating touchscreen ergonomics to improve tablet-based enrichment for parrots.* Paper presented at the Proceedings of the CHI Conference on Human Factors in Computing Systems, New York, NY.

Kuczaj, S., Lacinak, T., Fad, O., Trone, M., Solangi, M. & Ramos, J. (2002). Keeping environmental enrichment enriching. *International Journal of Comparative Psychology, 15*(2), 127–137.

Laule, G. & Whittaker, M. (2007). Enhancing nonhuman primate care and welfare through the use of positive reinforcement training. *Journal of Applied Animal Welfare Science, 10*(1), 31–38.

Lewton, J. & Rose, P. E. (2021). Social networks research in ex situ populations: Patterns, trends, and future directions for conservation-focused behavioral research. *Zoo Biology, 40,* 493–502.

Meehan, C. L. & Mench, J. A. (2007). The challenge of challenge: Can problem solving opportunities enhance animal welfare? *Applied Animal Behaviour Science, 102*(3–4), 246–261.

Mellor, E. L., McDonald Kinkaid, H. K., Mendl, M. T., Cuthill, I. C., van Zeeland, Y. R. A. & Mason, G. J. (2021). Nature calls: Intelligence and natural foraging style predict poor welfare in captive parrots. *Proceedings of the Royal Society B: Biological Sciences, 288*(1960), 20211952.

Moberg, G. P. (1999). When does stress become distress? *Lab Animal, 28*, 422–426.

NC3Rs. (2022). Grimace scales. The National Centre for the Replacement, Refinement and Reduction of Animals in Research. https://nc3rs.org.uk/3rs-resources/grimace-scales. (Access Date 24/12/2024).

Owen, M. A., Swaisgood, R. R., Czekala, N. M. & Lindburg, D. G. (2005). Enclosure choice and well-being in giant pandas: Is it all about control? *Zoo Biology, 24*(5), 475–481.

Perdue, B. M., Evans, T. A., Washburn, D. A., Rumbaugh, D. M. & Beran, M. J. (2014). Do monkeys choose to choose? *Learning & Behavior, 42*(2), 164–175.

Reading, R. P., Miller, B. & Shepherdson, D. (2013). The value of enrichment to reintroduction success. *Zoo Biology, 32*(3), 332–341.

Rohan, A., Rafaq, M. S., Hasan, M. J., Asghar, F., Bashir, A. K. & Dottorini, T. (2024). Application of deep learning for livestock behaviour recognition: A systematic literature review. *Computers and Electronics in Agriculture, 224*, 109115.

Rose, P. E. (2023). Identifying essential elements of good giraffe welfare - can we use knowledge of a species' fundamental needs to develop welfare-focussed husbandry? *Journal of Zoological and Botanical Gardens, 4*(3), 549–566.

Rose, P. E. & Croft, D. P. (2015). The potential of social network analysis as a tool for the management of zoo animals. *Animal Welfare, 24*(2), 123–138.

Rose, P. E., Freeman, M., Hickey, I., Kelly, R. & Greenwell, P. J. (2024). Considering what animals "need to do" in enclosure design: Questions on bird flight and aviaries. *Birds*, *5*(3), 586–603.

Rose, P. E. & Lewton, J. (2025). Key concepts for enhancing zoo animal welfare: Coping, comfort, choice, control, challenge and compassion *Journal of Applied Animal Welfare Science*, in press.

Ross, S. R. (2006). Issues of choice and control in the behaviour of a pair of captive polar bears (Ursus maritimus). *Behavioural Processes*, *73*(1), 117–120.

Sanz, V. & Grajal, A. (1998). Successful reintroduction of captive-raised yellow-shouldered amazon parrots on Margarita Island, Venezuela. *Conservation Biology*, *12*(2), 430–441.

Shepherdson, D. (1994). The role of environmental enrichment in the captive breeding and reintroduction of endangered species. In P. J. S. Olney, G. M. Mace, & A. T. C. Feistner (Eds.), *Creative conservation: Interactive management of wild and captive animals* (pp. 167–177). Dordrecht: Springer Netherlands.

Stalter, L., Dorleus, T., Milone, N., Sincage, J., Skurski, M. & Leeds, A. (2024). Tarantula welfare may be improved with greater environmental complexity: A preliminary behavioral study with Brazilian black tarantulas (Grammastola pulchra). *PLoS One*, *19*(12), e0314501.

Troxell-Smith, S. M., Whelani, C. J., Magle, S. B. & Brown, S. (2017). Zoo foraging ecology: Development and assessment of a welfare tool for captive animals. *Animal Welfare*, *26*(3), 265–275.

Veissier, I., Lesimple, C., Brunet, V., Aubé, L. & Botreau, R. (2024). Review: Rethinking environmental enrichment as providing opportunities to acquire information. *Animal*, *18*(9), 101251.

Villalba, J. J. & Manteca, X. (2019). A case for eustress in grazing animals. *Frontiers in Veterinary Science*, *6*, 303.

White, B. C., Houser, L. A., Fuller, J. A., Taylor, S. & Elliott, J. L. (2003). Activity-based exhibition of five mammalian species: Evaluation of behavioral changes. *Zoo Biology: Published in Affiliation with the American Zoo and Aquarium Association*, *22*(3), 269–285.

6

Environmental enrichment and promoting motivated behaviours

PAUL ROSE AND NICHOLAS BISHOP

6.1 INTRODUCTION

It has long been proposed that when captive animals are provided with resources that they are most motivated to interact or engage with, welfare can be improved (Dawkins, 1983). Environmental enrichment (EE) is one way of improving animal welfare by providing such important resources. EE is now commonplace in many zoos and aquariums (Beer et al., 2023) and is featured routinely in the best practice husbandry guidelines that are being written for specific species of captive wild animals. Originally termed "behavioural engineering", EE arose in the 1970s as a way of addressing challenges associated with the performance of inappropriate or unwanted behaviours of zoo-housed species (Markowitz et al., 1978). Considered something that was an optional extra to provide on top of basic care, over time, EE is becoming an essential part of daily husbandry (Shepherdson et al., 1999) although there is still wide disparity in its usage across institutions globally. "To be provided routinely but to not become routine" is a key mantra of best practice approaches to enrichment. For when EE becomes routine (the same programme of activities, the same objects provided, the same additions to the enclosure given), it becomes predictable and loses its important role in stimulating wider, positive, behavioural diversity. Although an animal does not always have to engage with a type of EE for it to have a welfare benefit (Decker et al., 2023) (i.e. the presence of the form of enrichment can itself be comforting or positive in some way for the animal), it is important that zoo animal care staff observe how EE is used or utilised to keep its relevance for the species it is provided to.

EE supports many of the aims of the modern zoo, not just those related to animal welfare. The use of EE increases the conservation relevance of animals in zoos by improving their fitness or behavioural flexibility, or it can enhance the educational benefits of zoo animals by promoting natural behaviours that engage visitors with biodiversity (Shepherdson et al., 1999). Understanding intrinsic motivation for behaviour and what the adaptive function of the behaviour is enables us to provide enrichment that can be a causal factor for a specific, adaptive action. Such joined-up thinking is especially important for animals in conservation programmes whereby EE would be used to promote responses to challenges, hone the ways in which animals respond to aversive stimuli, and encourage the development of behaviours that

DOI: 10.1201/9781003429425-6

make resource access, predator detection, or the use of cues from a social group (as a few examples) more efficient. How do we ensure that EE includes everyday responses to stressors, delivered sensitively and evaluated accurately, that keeps animal behavioural repertoires appropriately functional and "fit for purpose"? A question for budding zoo behavioural scientists to answer!

As we understand more about the evolutionary history of the species that we keep, their anatomical and physiological adaptations, we can better provide EE or adapt what was EE into daily care. In most cases, a key aim of EE is the promotion and performance of wild-type activity patterns and naturalistic time budgets. Attitudes towards what constitutes EE are changing as zoo operations and animal care protocols become more evidence-based. For example, the use of cut tree branches ("browse") for ungulate species that are obligate browsers in the wild (e.g. the giraffe, *Giraffa camelopardalis*) is now known to be an essential element of daily care (Figure 6.1), and not something that is given as a "treat" on irregular occasions (Rose, 2023). The body condition, behaviour patterns, health status, and lifespan of captive giraffe, living under current conditions, are all markedly improved now that browse is given as an essential and not as a treat.

Similarly, social enrichment originates from a correctly maintained population structure, and as zoos have gained more knowledge on how many animals to keep and in what numbers, the animals within these social groups gain more behavioural and welfare benefits from a correctly managed population size. For example, guidelines for the minimum flock size of captive flamingos (Phoenicopteriformes) used to recommend 20 birds "for good welfare" and 40 birds to encourage breeding (Pickering et al., 1992), but more recent findings show that flock sizes of over of a 100 or more individuals should be maintained to encourage full performance of courtship and breeding behaviours (Mooney et al., 2023). We use past evidence to support current practice, and as we review and evaluate such practice, we find that new evidence needs to be collected as our knowledge of what is correct for a species evolves. Numerous species-specific best practice guidelines have an enrichment section within them, but as more anecdotes turn into empirical evidence, there is a strong case for renaming such resources as "better practice" guidelines to reflect their nature as living documents.

Figure 6.2 shows that behavioural challenges should be identified from a welfare assessment and that enrichment can be a tool to give the animal

Figure 6.1 Attitudes towards what enrichment is and what its value is to the animals it is provided for should change over time. For example, browse (cut tree branches) provided to captive giraffes was formerly considered something that would be nice for them to have on occasion; then browse was classified as a form of enrichment, and now we know that browse is essential for daily husbandry and care.

more control over their situation to alleviate the problem or issue. Perhaps in this case pacing or stereotypic swaying in a polar bear (*Ursus maritimus*) occurs during the early afternoon, and after reviewing biological and ecological information on this species, it is identified that the enclosure does not provide sufficient novelty during times of heightened arousal and activity. An enrichment device is created, tested, and provided to the animal. The animal's response is measured both in terms of immediate behavioural change and reaction to the enrichment, and the latency of an enriching effect (how many times does the animal re-engage with the enrichment or how much more exploration occurs even when the enrichment is not used?). Further welfare assessment confirms the positive benefits of the enrichment on the polar bears' behavioural and psychological welfare states.

6.2 THINGS THAT ARE ENRICHING

Enrichment should come from key aspects of husbandry and housing, and not only by other resources provided on top of daily care – it starts with a design that responds to what the species evolved for. Whilst some authors suggest that enrichment is too simple to be truly effective (Vicino et al., 2022) and that using enrichment to advance husbandry is indicative of a failure to advance zoo animal care, we need to be mindful that, for many species, husbandry is still being trialled and evidence for best practice care is limited to non-existent. Therefore, enrichment is vital to transitioning to best practice care and the promotion of positive welfare impacts. Even apparently simple forms of enrichment can provide meaningful behavioural and physical outcomes for individual animals (Figure 6.3).

Hay (forage) is an essential element of zoo elephant husbandry. The collection, processing and consumption of hay is in itself enriching because it supports the overall health of the animal (from dietary, metabolic, gut microbiome and dental perspectives). Creative means of providing such forage then add more enriching elements to daily husbandry. Current guidelines for elephants recommend "topfeeder" (i.e. elevated in nets and on poles) provision of forage rations for zoo-housed animals across all adult and sub-adult life stages and sexes to strengthen muscles and encourage physical development (Bolechova et al., 2020). Providing forage in this manner simulates wild feeding strategies and time spent gathering forage is extended and can become a social activity. Therefore, multiple benefits are accrued by all

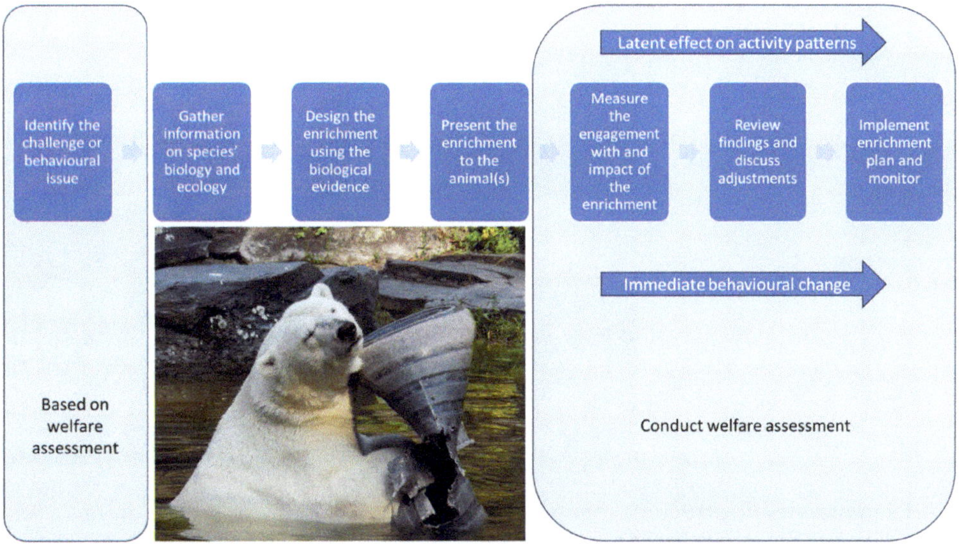

Figure 6.2 Guiding principles for the design, implementation, and review of species-specific EE in modern-day zoos.

Figure 6.3 Three different zoos, two species of elephant (*Elephas maximus, Loxodonta africana*), different individual life stages and sexes. All animals are benefitting from an enriched method of feeding a daily staple (their forage ration). Feeding elephants from height encourages physical development (enhanced musculature and strength), is more challenging for the individual and therefore diversifies foraging behaviours and encourages problem solving.

individuals in an elephant herd from the change from ground feeding of forage to the use of elevated hay nets or feeding poles. And whilst this may appear simplistic, designing and implementing changes to husbandry are complex at many operational levels (Figure 6.4). Manufacturing and implementation of feeding poles, engineering and installation of winches and automated technology to raise and lower hay nets, placement of poles around an enclosure and within indoor housing, changes to substrate and topography to encourage physical activity all require background research, expenditure of time from personnel, and financial outlay for construction and maintenance means that potentially "simple" enrichment can be anything but.

When people think of EE, images of boomer balls for wrestling, sand pits for digging, and termite mounds for fishing may spring to mind. Yet EE is more than just specially crafted devices or

Figure 6.4 Examples of infrastructure change and use of engineering and technology to create enriched feeding opportunities for zoo-housed elephants include the use of tree trunks within indoor housing, automatic winches to raise and lower hay nets when needed, and building outdoor structures in multiple feeding locations to encourage animals to move around their space. This is a logistical and financial challenge.

toys provided in a limited number at certain times. Enrichment comes from the wider environment around the animal. The idea that "It's not just the getting, it's the getting there…" is important to remember when providing relevant EE. Animals are learners, evolved to operate within their environments for desired outcomes. When this environment is biologically relevant and suitable for the species, opportunities to work for worthwhile consequences become more apparent. This concept, of an individual animal being prepared to expend energy to access a resource when the same resource can be freely available elsewhere, is now termed contrafreeloading (de Jonge et al., 2008). The principles of contrafreeloading relate to the theories of behavioural economics (Mason et al., 1998), which is in essence "what will animals work for?" How valuable is a resource to an individual and therefore how much time and energy are they prepared to expend on obtaining it? Enrichment that has a relevant impact on behaviour and welfare aims to promote the idea of contrafreeloading, as this provides the animal with an outlet for the display of a wider range of natural behaviour patterns.

A sound knowledge of appetitive (highly motivated, searching activities) and consummatory behaviours (that cause satiation) – for full definitions please see the Glossary – can be crucial to the development and design of effective enrichment strategies that promote desired behavioural outcomes, as this knowledge will inform us what animals are prepared to work for and why. Evidence from wild data supports the design and implementation of EE regimes that encourage contrafreeloading by distributing or presenting resources in a manner that individuals could experience in the wild. Evaluating the furnishings, substrates, and habitat features recreated within a species' enclosure adds further enrichment opportunities, including opportunities for contrafreeloading. For example, well-managed and planted waterbodies in the aviaries of wading birds encourage blooms of aquatic invertebrates that can be hunted by the occupants of the aviary. Mixed species housing also provides an array of interaction opportunities between and across species that diversify the performance of behaviours within the same enclosure. Further thought and consideration of how the enclosure (and husbandry) is, in itself, a form of EE should be a future direction for zoo research and developments in zoo operations.

Figure 6.5 illustrates different animal behaviours that show how the wider environment of the zoo is enriching and why giving zoo animals extra choice and control, and a degree of autonomy over what they do, promotes positive behaviours that further enrich their daily lives. Not all the concepts illustrated in this figure might be considered enrichment per se, but giving thought to each when enclosures are planned and designed, and when social groups are constructed and populations are managed, ultimately ensures a more enriching and interesting life for the inhabitants. The following (not exhaustive) list of enclosure features or elements from daily care could be further investigated to understand animal responses to them and what any overall enriching influence is:

- Shade and movement of the sun across an enclosure during the day;
- People (caregivers and other staff) and associated indicators of daily husbandry events;
- Training;
- Conspecifics (associations and interactions);
- Time away from other animals;
- Opportunities to be inside;
- Opportunities to be visible outside;
- Hiding places and use of refuges.

Regardless of the intervention, or the form of EE provided, what is essential is valid and replicable measurement and evaluation of the animal's responses to it. An objective and rational interpretation of behavioural data means that the correct and most plausible explanation for any effect of EE on behaviour will be reached. For example, not all increases in activity are always beneficial, and not all indirect or passive uses of EE (for example, an individual animal might rest or sleep next to an enrichment item rather than actively use it in locomotory behaviours) mean the EE intervention has failed. To illustrate why sound interpretation is key, highly social pinyon jays (*Gymnorhinus cyanocephalus*) display elevated rates of stereotypic behaviour when housed alone, and abnormal behaviour markedly reduces in performance when birds are pair housed (Wolff & Stevens, 2024). This example shows the importance of quantifying and describing what is meant by changes in activity. In this case, jays reduced activity and spent more

Figure 6.5 Examples of how enrichment is provided holistically from the opportunities of social engagement, resource allocation, space and cover, and climatic variability within an enclosure. From the top. Social state: A shoal of silver monos (*Monodactylus argenteus*) can swim as one large, synchronised unit, whereas a lion (*Panthera leo*) can rest separately from his harem of lionesses. Reproductive state: Nest building and associated courtship displays in a pair of Abdim's storks (*Ciconia abdimii*) show that their zoo's housing and husbandry promote the performance of reproductive activity, and conversely, at the end of the breeding season, a pair of mallards (*Anas platyrhynchos*) has the space to moult out of summer plumage into new autumn courtship colouration. Resource use: Animals can access multiple valued resources, e.g. basking spots, but can be alone if territorial (e.g. a male pool frog, *Pelophylax lessonae*, in breeding condition) or in a group (e.g. yellow-bellied sliders, *Trachemys scripta scripta* all basking together to feel more secure). No social groupings are forced by resource availability being limited. Space and cover: Animals can seek refuge when needed, like this emperor goose (*Anser canagicus*) utilising a meadow of long grass, or they have the open space for unimpeded movement, such as around a coral reef for an aquarium-housed clown triggerfish (*Balistoides conspicillum*). Sun or shade: Enclosures provide access to direct sunlight and shaded spots at different times of the day, and inhabitants can orient themselves accordingly depending on their motivation to (e.g.) sunbathe, as in this hooded vulture (*Necrosyrtes monachus*), or slumber in a cool corner, such as these warthogs (*Phacochoerus africanus*).

time resting and preening when social enrichment was provided. Therefore, EE has still promoted better welfare, even though birds are less mobile, as biologically relevant behaviours are now being displayed in place of the former reliance on abnormal or displacement activity that suggested a lack of coping with the former prior.

6.3 PROMOTING WELFARE WITH ENRICHMENT

For zoo visitors to see the benefits of EE, it is important that it is explained to them at the animal's enclosure, especially if the enrichment appears unnatural (Packer et al., 2018). Concepts of "functional substitution" are based on using EE to recreate specific habitat features that a species has evolved to directly interact with (Robinson, 1998). The whole habitat does not need to be completely rebuilt in the zoo; the focus is on providing the ecologically relevant aspects of the habitat that the species being kept needs to utilise to survive and thrive. Therefore, understanding the benefits of EE, i.e. by choosing an enrichment programme that is evidenced as being relevant for that species (Brereton & Rose, 2022), can support development of enclosures and animal habitats to embed enrichment into the animal's wider physical environment.

From this chapter so far, it is clear that enrichment directly impacts behavioural states. Individual animals will respond differently to enrichment, therefore demonstrating different behavioural responses that can be used to measure impacts of enrichment. As welfare comprises physical, psychological, and behavioural components, so enrichment can be used as both an influence over and a measure of such factors. Thus, measuring engagement with EE will help zoos fulfil animal welfare assessment criteria (i.e. that stipulated in zoo accreditation or membership policies), and can be particularly relevant to determining the impact of inputs (the things given to the animals in the zoo, such as EE) on emotional or psychological outputs (feelings, moods, body language and expression).

There should be a latent effect to EE that lasts after the animal has used up or directly interacted with the EE device, procedure, or intervention (Figure 6.6). This means that the change to the animal's behaviour (e.g. increased foraging or exploration of the environment) continues after the EE has "vanished" or the animal returns to the site or source of the EE and gains further stimulation later in the day. Of course, this approach will not work for all forms of EE, and passive engagement with EE is still important (Decker et al., 2023) but latency should be considered (and ideally measured) for EE interventions that aim to promote wider behavioural diversity or activity.

Positive (and biologically relevant) behaviour is key. EE should not just promote activity for the sake of it. The key to making EE beneficial to the animal is to understand what the outcome of the enrichment is – what behaviour will be changed/promoted/diversified from the presence of the enrichment item or the enriched environment? If the enrichment creates an immediate change in behaviour that is positive, can this be sustained over a timeframe that is relevant to the individual/species at a time of day when whatever activity is being promoted would naturally occur? Depending on the desired behavioural outcome, the competencies of the individual, and the role of EE in daily husbandry and management, EE needs to present some form of challenge. For individuals in populations destined for in situ reintroduction, more challenge is required (and perhaps the rewards given have different values) to prepare animals for a life in the wild. Otherwise, in this instance, EE has limited value in replicating the wild and the problems that animals will encounter.

As mentioned earlier, EE aims to promote the performance of wild-type time-activity budgets. For individuals that display large proportions of their time engaged in abnormal repetitive behaviours, EE can be a means of giving them more control over their environment to help diversify the activities displayed. Similar tools and interventions can be used across species (Figure 6.7) to enable similar behavioural outcomes. It is essential to remember to never prevent an abnormal repetitive behaviour from occurring but to address the internal motivation for its performance and to understand the background history of the animal that influences why such behaviour occurs. Providing an animal with outlets to develop important naturalistic behaviours that will ultimately fill the place of the abnormal behaviour in the animal's daily activity pattern highlights the role of EE in managing stereotypies and promoting good welfare (Figure 6.8).

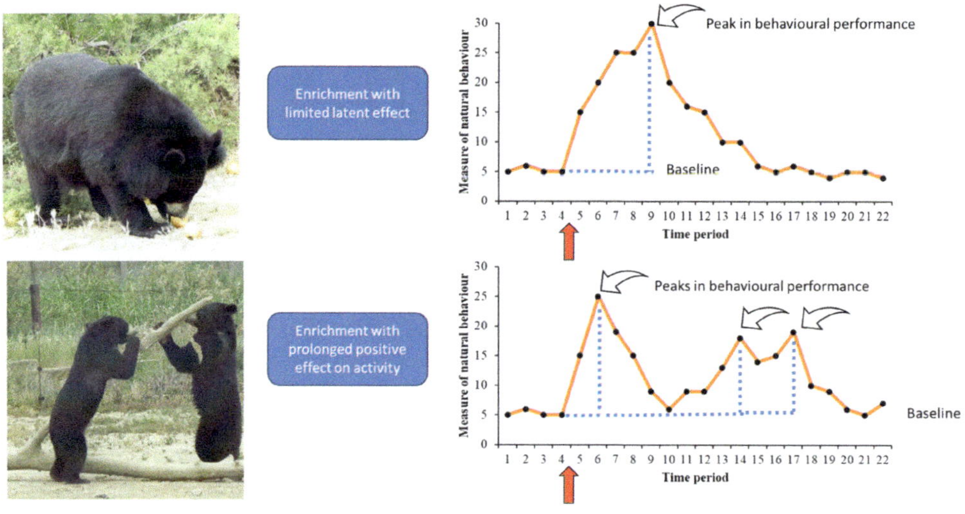

Figure 6.6 Evaluating the longer-term impacts of EE to assess any latentt effect well beyond the presence of the original stimuli in the animal's environment. For example, a limited latent effect seen when an animal is given a scatter feed of treats in one uniform area of their enclosure (top) that rapidly increases activity and exploration but only for a short period of time. Compared to an object in the enclosure that is mobile yet requires effort to move, which can be engaged with by multiple individuals on different spatial levels (bottom), interest is therefore interest is maintained over a longer period of time. Evaluating the efficacy of EE relies on prior data regarding an animal's baseline activity or performance of a behaviour that we need to change or alter, so that peaks (i.e. positive changes to this behaviour) can be objectively assessed.

6.4 WHAT DOES BIOLOGICALLY RELEVANT ENRICHMENT LOOK LIKE IN PRACTICE?

It is essential to operationalise theory into practice, so that animals truly benefit from EE aligned to evolutionary biology. Examples of enrichment across taxa that provide multiple behavioural, welfare and health benefits can be found in Table 6.1.

Figure 6.7 The same tool used to promote beneficial activity and biologically relevant behaviours in one species can be adapted for another with minimal effort. For example, a feeding device that a warthog pushes along to dispense pelleted food can make a forage feeder that encourages a giraffe to spend longer gathering lucerne.

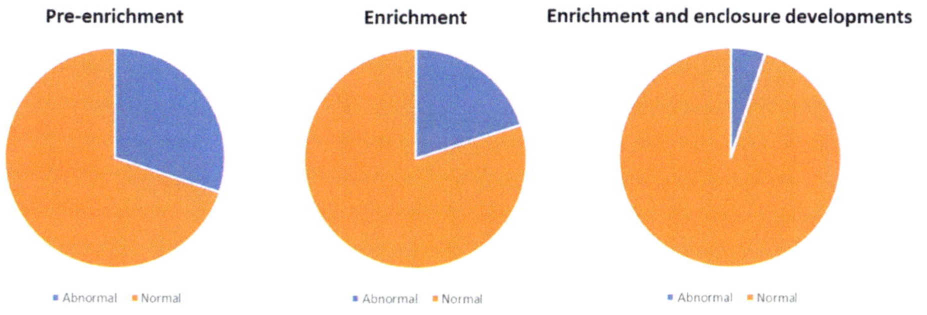

Figure 6.8 The main principles of behaviour management include enrichment as part of a wider range of interventions to reduce the performance of abnormal behaviour in an individual animal's time-activity budget. Including enrichment alongside measures such as enclosure modification, changes to husbandry routines and feeding schedules, and the use of training, all encourage an animal to increase the performance of positive, biologically relevant behaviour patterns that ultimately outcompete any need for stereotypy or similar abnormal activities.

Table 6.1 Examples of enrichment across taxa that provide multiple behavioural, welfare and health benefits, with video links to each.

Enrichment example	Description	Video link
Exercise benefits of elevated feeding in Asian elephants.	Hanging hay nets provide opportunities for challenge, behavioural dexterity, and a wider range of physical activity. Providing elevated hay increases overall foraging opportunities and thus improves time-activity budgets.	https://youtu.be/b-trCw-XITk
Multi-faceted enrichment outputs (nutritional, sensory, physical, occupational) from scatter feeding a spectacled bear (*Tremarctos ornatus*) in a log pile.	A bear moves around branches and logs whilst searching for food. It is important to remember that enrichment categories are not mutually exclusive, and in this example, a nutritional enrichment (scatter feeding) of a bear in a log pile is also enriching across multiple other dimensions. The value of the original form of enrichment (the scatter feed) in this scenario is likely to remain long after all the food items have been consumed.	https://youtu.be/Aw9FbOArJWo
"Bubbler" to move water around a pool and maintain water quality also provides filter feeding opportunities for Caribbean flamingos (*Phoenicopterus ruber*).	Flamingos filter feed around an underwater "bubbler" in their pool. Species that would naturally spend a large period of their time searching for, and collecting, food in the wild can be prone to disturbed behaviour patterns in captivity. Flamingos, for example, can be more inactive compared to wild counterparts (Brereton & Rose, 2019), so the use of "bubblers" to aerate and circulate water keeps their waterbodies healthy whilst enabling more opportunities for natural filter feeding behaviour.	https://youtu.be/aScBGonUmtU

(*Continued*)

Table 6.1 (Continued)

Enrichment example	Description	Video link
Simulation of a hunt for a pack of African hunting dogs (*Lycaon pictus*).	A pack of African hunting dogs chases a fake prey item across their enclosure. Carnivorous mammals can suffer from poor welfare in zoos due to their need to locate, talk and chase down prey (Kroshko et al., 2016). Creative means of simulating prey detection, stalking and chasing enables outlets for such important behavioural needs, whilst being excellent opportunities for exercise and social cohesion.	https://youtu.be/N36RVxz4Aqo
Sand is a dynamic substrate, when managed so that it does not compact, and remain ideal for promoting foraging and exploratory behaviours.	A domestic hen (*Gallus gallus domesticus*) scratches through a sandy substrate to locate food items. A fundamental feature of the enclosure, when managed correctly (raked and turned regularly to stop grains from compacting), becomes a source of enrichment. Such a situation improves behavioural outputs and associated welfare states within minimal financial and time costs to zoo staff.	https://youtu.be/mQFDpleasD8
When a social group provides enriching benefits for all individuals.	Schooling behaviour in a large group of freshwater aquarium fish. Individuals that gather in groups for safety and security, and display collective movement, must have these elements of their social system recreated in captive conditions. Fish that live in large groups will use more space, be more active, and display a wider range of behaviours when the social environment provided is relevant to them. Therefore, in itself, the shoal is a key form of enrichment for them.	https://youtu.be/jqlY_kEm0SE

Extra information on the key principles of environmental enrichment for zoo and aquarium animals can be found in:

Carlstead, K. & Shepherdson, D. (2000). Alleviating stress in zoo animals with environmental enrichment. In G. P. Moberg & J. A. Mench (Eds.), The biology of animal stress: Basic principles and implications for animal welfare (p. 337). Wallingford, UK: CABI.

de Azevedo, C. S., Cipreste, C. F. & Young, R. J. (2007). Environmental enrichment: A GAP analysis. Applied Animal Behaviour Science, 102(3–4), 329–343.

Kleiman, D. G., Thompson, K. V. & Baer, C. K. (2010). Wild mammals in captivity: Principles and techniques for zoo management. Chicago, IL: University of Chicago Press.

Rose, P. E. (2022). The behavioural biology of zoo animals. Boca Raton, FL: CRC Press.

Shyne, A. (2006). Meta-analytic review of the effects of enrichment on stereotypic behaviour in zoo mammals. Zoo Biology, 25(4), 317–337.

Although providing EE is now commonplace, zoos could do more to enthuse their visitors about how the positive messages from seeing EE in practice could be used in their daily lives. This case study (Box 6.1) by Nicholas Bishop from Zoos South Australia provides some food for thought in this area of public engagement.

BOX 6.1: "Do Try this at Home" - taking enrichment home from the zoo

Words have power. Just a handful can change the course of history. In 1943, the cowboy Curly McClean proclaimed the bright golden haze on the meadow, signalling the start of Rogers and Hammerstein's musical Oklahoma! These words are significant because they heralded the concept of the integrated musical: stories about real people with relatable challenges, and not just a float for a Broadway kick line. These words are also compelling because they express wonder at the beginning of a new day, where all the elements come together to engage with nature - both an invocation and an invitation.

As a musical theatre tragic and a passionate nature lover, they resonate greatly with me. Additionally, as Animal Behaviour and Creative Programs Manager at Zoos South Australia, they motivate an enduring curiosity about human-nature connectivity and how we can improve this link for the behavioural benefit of all. Words can get us to take notice, and I'm especially interested in the ones that help us take action; the pairing of both declaration with application is key to influencing our visitors towards positive behaviour change.

At Zoos South Australia, the Nature Theatre endeavour is sharpening a focus on how storied demonstrations of reinforcement-based enrichment may improve the lives of animals beyond our zoo borders. To grasp this mission, it's worth considering the power of the tales we tell; stories have an intrinsic relevance to human progress and our zoo stories are vibrantly close to their original context and purpose, as their deep-time potency and power have always been about how to understand the world around us. Crucially, learning is a change in behaviour due to experience (Chance, 2003) and as all behaviour necessarily relies on an environmental setting, we see our evolutionary progress as one inextricably linked with nature.

Storytelling and screenplay coach Lisa Cron is fascinated by the neuroscience of story and embeds its power deep in the practical realities of the human survival model:

> "We turn to story to navigate reality. Story, as it turns out, was crucial to our evolution – more so than opposable thumbs. Opposable thumbs let us hang on; story told us what to hang on to (Cron, 2012)." Story is therefore more than entertainment: it's a biological boon to the road-mapping of individual and collective progress.

At Zoos South Australia, we've been collaborating on how to tap into this ancient urge. Generally, our question is "How can we use our social licence to improve animal welfare in the community?" This motivates the demonstration of methods within programmed engagement that educate as they entertain, leading to the specific and core concept, both theme and invitation: "Do try this at home!" In this context, we introduce the concept of our species as "Humanimals" – when we see ourselves as a part of nature, rather than apart from nature. All are combined and showcased in our "Wildshow" production, delivered in the Variety Children's Zoo domain in a purpose-built amphitheatre. We collaborate with animal colleagues that we refer to as 'co-learners' across a variety of presentations. Using training methods that focus on positive reinforcement, the team aims to share behavioural strategies that open up pathways for learning within an options-rich environment.

Our guiding markers are intrinsic to the Applied Behaviour Analysis query: that behaviour has a function and relies on thoughtful antecedent arrangements that make the right behaviour easier and consequences that make it more worthwhile. In this context, enrichment is understood and promoted as providing a diversity of opportunities that replicates the randomised richness that in-situ contexts provide. Provision of enrichment devices, demonstration of how they're made and delivery within the show context are all part of the engagement strategy. Further, emphasis on cognition, sensory richness and animal agency is layered into our narrative.

Wildshow programme development features a variety of models, and all can be presented as individual sharing events. Training positively demonstrates positive reinforcement-based techniques with our goats, guinea pigs, rabbits, quokkas, and possums that can be applied to foster cooperative care rather than coercive compliance with animal co-learners at home. We establish basic foundations for learning by demonstrating recall, target, and station techniques – useful for replication on the home front with dogs, cats, budgies, ferrets, and goldfish. Enrichmentoring Wildshow Workshops use everyday articles to engage senses, expand cognition and promote problem solving. For example, "snuffle mats" are a favourite demonstration to encourage more complex foraging behaviour for guinea pigs. Design a Domain develops internal and external spaces that enhance welfare, we encourage observation of "Choice Voice" – how animals show their preferences through body language – and this helps us to think about how we provide relevant spaces for animal needs. In the Wildshow context, creating a microenvironment encourages participation from young humanimals as they assist the presenter to create the ultimate "Rabbitat".

Crucially, there is also a bonus benefit to all this: the creation and sharing of enrichment strategies are in themselves enriching for humanimal carers. As a corollary, the "Do try this at home!" programme additionally enriches our lives as animal care professionals, with each concept giving rise to further creativity to improve our welfare outreach. Ultimately, these five words express the same invitation as that first lyrical line of Oklahoma!: involve yourself more closely in the marvellous gifts of nature surrounding you. On from this, we remember that science brings us fresh wonders every day and also equips Humanimals to support the agency of other species that share our lives.

Here's to open pathways, interspecific enrichment and ongoing welfare – for all of us.

6.6 CONCLUSIONS

EE is a fundamental and essential part of the daily zoo animal husbandry and management. EE needs to be routine in its provision but not routine in what is provided. Animal care staff can be routine in their EE interventions over time by planning enrichment in cycles and delivering EE after a certain time period (relevant to the species) when it would appear as novel and interesting. To continue to be useful to the animals in the zoo, caregivers, zoo operations staff, and other stakeholders need to document what the aim of the EE is (the required behavioural outcome) and develop the EE intervention accordingly (Figure 6.9). Environmental enrichment may be one of the most studied of behavioural topics in zoological organisations. However, the evidence behind what EE works best and why can still be lacking. Identifying the need for EE to be implemented and then observing and collecting data to see how it is received by the animals supports the evidence-based approach and can be integrated into publications, such as best practice guidelines. Further work needs to focus on how animals engage with enrichment but not in an overtly active manner, for example, when the presence of enrichment is positive even if the animal does not "do" anything directly with it. Zoo animal welfare scientists need to grapple with the thorny question of what animals want and need, and how does planned, appropriate, scheduled enrichment provide for these wants and needs. Providing a range of EE and using concepts such as preference tests or cognitive bias tests may help stakeholders develop EE that is both species-specific and relevant to upholding positive welfare for specific individual animals in the zoo. Alongside of this, more work is needed on the value of stressors to the lives of zoo animals, and how we think about the use of such stressors as part of an appropriately enriched existence. We would be doing our animals a disservice if we only ever delivered positive experiences. Enhancing functional behaviour capacity includes learning around aversive stimuli, experiencing negative states, and being supported to decipher how to successively emerge from dealing with (and meeting the challenge of) a stressor with an expanded behavioural repertoire.

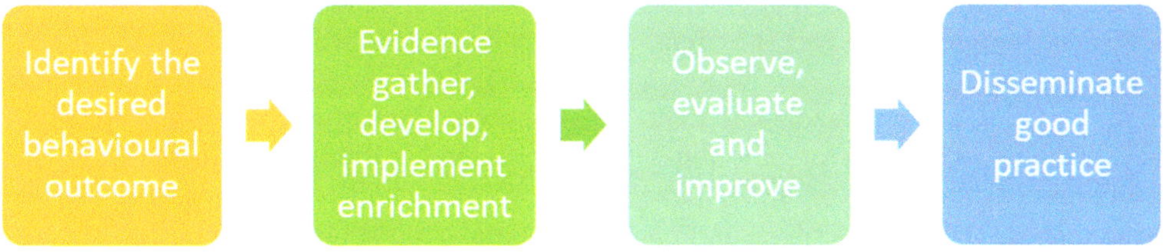

Figure 6.9 Ultimately, achieving measurable welfare outputs from EE requires an understanding of what the goal or function of the enrichment is, and then developing the enrichment device or programme with this behavioural outcome in mind. Once EE has been provided, its usage and impact need to be reviewed and measured before information on its use and relevance is imparted to others within the zoo's animal care teams (and more widely to other interested parties).

For many species housed under human care, enrichment is used to increase and enhance the physical, sensory, and social environment for individuals and animal groups. Although enrichment is commonly provided, more work is needed to evaluate its impact and to understand any short- and longer-term efficacy and benefits to behaviour and welfare. Key to the development, implementation and outputs of a successful enrichment programme is for all those involved to know what the goal of the enrichment is, and to plan how the enrichment will work for the animal based on knowledge of species biology and ecology. Extending the value of enrichment beyond the zoo's boundaries is possible, as demonstrated in the case study in this chapter, and such initiatives should become more commonplace in zoo education and outreach activities. As all species have evolved to perform behaviour in response to environmental conditions, to provide a benefit to them at a specific time, so enrichment should consider the behavioural needs and motivational states that are intrinsically linked to when, how, and where behaviour will be displayed. Providing enrichment that is useful and tailored to the individual, the species and the population is a continual challenge in the zoo, but the more evidence that is collected on its application, coupled with our increasing knowledge of species' biology, will ultimately mean that EE can increase in its animal welfare and husbandry relevance.

We finish this chapter with a human perspective – that use of EE is also enriching to animal care staff. The value of zoo staff being involved in evidence collection for EE is an important point to emphasise as this builds capacity and enhances their understanding of the species in their care. A sense of being worthwhile in enhancing animal lives and contributing to improvement is crucial to the motivational states that sustain commitment to their role and to the development and application of EE for the species they are responsible for.

REFERENCES

Beer, H. N., Shrader, T. C., Schmidt, T. B. & Yates, D. T. (2023). The evolution of zoos as conservation institutions: A summary of the transition from menageries to zoological gardens and parallel improvement of mammalian welfare management. *Journal of Zoological and Botanical Gardens*, 4(4), 648–664.

Bolechova, P., Clauss, M., de Man, D., Galeffi, C., Hofman, S., Kappelhof, J., Kfir, G., Kjellson, B., Kölpin, T., Lawrenz, A., Lüders, I., McKenzie, A., Mul, C., Oerke, A.-K., Pluhackova, J., Sach, F., Schaftenaar, W., Schiffmann, C., Schmidt, H., Sos, E. & Versteege, L. (2020). *EAZA Best practice guidelines for elephants* (2nd ed.). Amsterdam, The Netherlands: EAZA.

Brereton, J. & Rose, P. E. (2022). An evaluation of the role of 'biological evidence' in zoo and aquarium enrichment practices. *Animal Welfare*, 31(1), 13–26.

Brereton, J. E. & Rose, P. E. (2019). Comparing the behaviour of wild and captive flamingos: An evaluation of published data on time-activity budgets. *Flamingo*, 2, 34–49.

Chance, P. (2003). *Learning and behavior.* Melbourne, Australia: Thomson/Wadsworth.

Cron, L. (2012). *Wired for story: The writer's guide to using brain science to hook readers from the very first sentence*. New York, NY: Ten Speed Press.

Dawkins, M. S. (1983). Battery hens name their price: Consumer demand theory and the measurement of ethological 'needs'. *Animal Behaviour*, 31(4), 1195–1205.

de Jonge, F. H., Tilly, S.-L., Baars, A. M. & Spruijt, B. M. (2008). On the rewarding nature of appetitive feeding behaviour in pigs (Sus scrofa): Do domesticated pigs contrafree-load? *Applied Animal Behaviour Science*, 114(3), 359–372.

Decker, S., Lavery, J. M. & Mason, G. J. (2023). Don't use it? Don't lose it! Why active use is not required for stimuli, resources or "enrichments" to have welfare value. *Zoo Biology*, 42(4), 467–475.

Kroshko, J., Clubb, R., Harper, L., Mellor, E., Moehrenschlager, A. & Mason, G. J. (2016). Stereotypic route tracing in captive Carnivora is predicted by species-typical home range sizes and hunting styles. *Animal Behaviour*, 117, 197–209.

Markowitz, H., Schmidt, M. J. & Moody, A. (1978). Behavioural engineering and animal health in the zoo. *International Zoo Yearbook*, 18(1), 190–194.

Mason, G. J., McFarland, D. & Garner, J. (1998). A demanding task: Assessing the needs of captive animals. *Animal Behaviour*, 55, 1071–1075.

Mooney, A., Teare, J. A., Staerk, J., Smeele, S. Q., Rose, P. E., Edell, R. H., King, C. E., Conrad, L. & Buckley, Y. M. (2023). Flock size and structure influence reproductive success in four species of flamingo in 540 captive populations worldwide. *Zoo Biology*, 42(3), 343–356.

Packer, J., Ballantyne, R. & Luebke, J. F. (2018). Exploring the factors that influence zoo visitors' perceptions of the well-being of gorillas: Implications for zoo exhibit interpretation. *Visitor Studies*, 21(1), 57–78.

Pickering, S. P. C., Creighton, E. & Stevens-Wood, B. (1992). Flock size and breeding success in flamingos. *Zoo Biology*, 11(4), 229–234.

Robinson, M. H. (1998). Enriching the lives of zoo animals, and their welfare: Where research can be fundamental. *Animal Welfare*, 7(2), 151–175.

Rose, P. E. (2023). Identifying essential elements of good giraffe welfare - can we use knowledge of a species' fundamental needs to develop welfare-focussed husbandry? *Journal of Zoological and Botanical Gardens*, 4(3), 549–566.

Shepherdson, D. J., Mellen, J. D. & Hutchins, M. (1999). *Second nature: Environmental enrichment for captive animals*. Washington, DC: Smithsonian Institution Press.

Vicino, G. A., Sheftel, J. J. & Radosevich, L. M. (2022). Enrichment Is simple, that's the problem: Using outcome-based husbandry to shift from enrichment to experience. *Animals*, 12(10), 1293.

Wolff, L. M. & Stevens, J. R. (2024). Less activity means improved welfare? How pair housing influences pinyon jay (Gymnorhinus cyanocephalus) behavior. bioRxiv, 2024.2002.2007.579343.

7

Enclosure design, space use and animal welfare

PAUL ROSE AND IAN HICKEY

7.1 INTRODUCTION

A ubiquitous feature of all zoological establishments, regardless of their type or focus, is that animals live within human-created environments that are of a specific size and shape. Such managed environments will ultimately restrict where animals can go and what they can do, irrespective of how much space is provided. While spatial restriction may not be a problem for all species, the quality of the space provided and the variety of resources within will ultimately dictate animal welfare states. For example, wide-ranging carnivores experience novelty within their home range on a daily basis. Because of the scale of home range size compared to how much they travel, these animals are more likely to suffer from poorer welfare in a restricted zoo enclosure (Bandeli et al., 2023). Conversely, species with a smaller and more defined territory that is fully travelled, explored, and inspected daily show fewer indicators of poorer welfare when housed in zoo enclosures. Therefore, it is imperative that zoo enclosure design is grounded in the fundamental needs of the species being housed, and that due care and consideration are given to how animals will use and access space, both on-exhibit and in "back of house" areas, which also need to be designed in a way sympathetic to species ecology. These considerations extend beyond an animal keeper's working day; housing design and enclosure features need to provide for the crepuscular, cathemeral, and nocturnal activities of species, as these behaviour patterns can be just as diverse (and some more so) than diurnal ones (Duggan et al., 2016; Rose et al., 2018; Wilson et al., 2006). Ultimately, whenever an enclosure is planned, designed, and constructed, animal welfare must be a priority (Richardson, 1999).

Since the creation and opening of the modern zoo in the early 19th century, zoos have been aware of the impact of housing on species' ability to survive, breed, and thrive in the zoo (Rees, 2011). Rigid bars and cages for containment have been replaced by immersive designs that blur boundaries between a viewing audience and the animals. Bringing the animal's environment into areas dedicated for visitors may enhance the experience of observing the animals and the associated conservation messaging related to the exhibit (Moss

DOI: 10.1201/9781003429425-7

& Pavitt, 2019). Immersing zoo guests and visitors in the animal's world allows them appreciate the wider environment that a species has evolved to inhabit. It is understandable, therefore, that there is a growing trend within zoos and aquariums to refer to enclosures as "habitats", but we must exercise caution when using such terms, particularly if we are serious about educating visitors on the natural world and the complexities of ecological interactions.

A habitat is a natural system that is unpredictable and full of a complex web of interactions between species. In contrast, zoo enclosures are static, under human control, and may house only one species. Any change to the habitat areas created in an enclosure needs to be instigated by humans and is not under the control of biotic or abiotic factors. Therefore, if zoos wish to be honest and open with their visitors, the use of the word habitat should be carefully considered regarding how an enclosure or exhibit is formally described, named, and branded. Enclosures can replicate habitat areas and resources used by a species in the wild, but functional replication of key areas of a habitat that elicit natural behaviour performance (Robinson, 1998) and enhance welfare is not the same as calling an entire enclosure a habitat.

7.2 CREATING A USEABLE SPACE FOR ZOO SPECIES

For those who first started to evaluate and review how zoo animals respond to the space of an enclosure, its size, and its dimensions, it became apparent that the quality of the space provided can be more important than the quantity of space a species has access to (Hediger, 1950). Quality space refers to enclosure features, furnishings, and substrates, as well as the area provided for the animal, that are useful, relevant and engaging to the inhabitants (Richardson, 1999). Animals housed in enclosures that provide them with quality space are more likely to display naturalistic time activity budgets as they will be engaging with resources that fulfil their biological needs (Fàbregas et al., 2012). An open expanse of empty space provides an illusion of freedom, but a species may prefer to engage with specific features, structures and substrates within an enclosure that reduces the overall amount of space within but provides complexity and environmental variation essential to their biological functioning.

While zoo enclosures may need to house a mixture of different species, designing, building, and maintaining enclosures that promote good welfare for specific single-species housing can be complex enough; individuals within these populations will grow, develop, and change over the course of their lifespans. These spaces should be designed to cater for an individual's needs throughout all their life stages from adolescence to adulthood, and consider the potential for additional physical and psychological needs of health-compromised individuals or geriatric individuals (Figure 7.1).

Such needs are even greater when considering socially orientated species that will share the same space with other individuals, as each will have their own specific needs. For example, big cats (Felidae) such as lions (*Panthera leo*) are drawn to elevated vantage points within an enclosure, and such structures are often made with just one or two levels, thus minimising the concurrent use by all individuals in the pride, often requiring animals to jump to access the levels. Not only do we need to consider the needs of the social group and allow enough elevated spaces to cater for more individuals (a concept further explored in section 7.5), but we should also consider the potential for reduced mobility in older individuals. For example, within section 7.5, we show an example of how such enclosure structures can be created to provide equitable resource access for all group members that ultimately facilitates social order, promotes wild-type behaviour, and maximises positive welfare outputs. Lions also operate under a fission-fusion social structure, with prides in the wild often splitting, resulting in subgroups being formed (Mbizah et al., 2019). Housing designs should consider such complex social structures too, allowing for the potential needs of individuals to be split into separate groups during various life stages. Additional housing considerations are also needed for social species where males live largely solitary lives in the wild. For example, Asian elephant (*Elephas maximus*) bulls should have access to separate housing away from the main herd (Hartley et al., 2019). Practical challenges then arise when designing bull housing that is far enough from the herd but is still close enough to allow for connected pathways

Figure 7.1 An example of enclosure changes to promote good welfare in captive flamingos (Phoenicopteriformes). Reconstruction of original pool edges (left) to create a shallow, "beach style" incline (right) facilitates easy access between water and land for older birds in the flock. Such types of life-stage-appropriate enclosure change are particularly important for species such as these, that can live for potentially eight decades under human care.

that can be used by the bull to merge with the herd for breeding. When considering breeding, we must also consider the birthing process for all species we are designing for, and how we might incorporate those needs within designs. As elephants give birth to live young standing up, modern enclosures should have a sand substrate which concurrently meets a variety of other biological needs. This particularly helps during the birthing process by providing a softer substrate for the new calves when they fall to the floor. Other mammals, however, such as big cats and bears (Ursidae), will naturally give birth in secluded dens in the wild, away from conspecifics. For those species, we need to design such spaces into potentially quieter parts of an enclosure away from the noise and view of visitors, with the capability of being separated from conspecifics if needed. By considering species biology and incorporating such features within designs from the outset, we can minimise the need for any enclosure alterations that may impact on gestation.

While single-species housing can present many design difficulties to overcome, mixed species exhibits (MSE) often prove to be even more challenging to design. As zoos create more complex exhibits that house complex and dynamic groups of different taxa and attempt to replicate the types of species co-habitation that may exist in wild habitats, it is essential that welfare is considered for all species and individuals housed. An enclosure may be designed to provide the appropriate resources and environmental conditions for all species, and the practical challenges of catering for their different needs may be overcome. There can, however, still be external variables that impact on co-habitation, and designers need to consider these variables. The presence of visitors, for example, may affect the behaviour of certain species or individuals, so viewing spaces should be designed to ensure minimal impact and allow for all species to access their required resources while feeling secure, a concept further discussed in Chapter 9.

To ensure that welfare is not compromised by any element of such multi-species mixing, or that there are no compromises in the enclosure design when housing a variety of species together, regular observation of the inhabitants of an MSE is required (Figure 7.2). Methods that assess niche overlap and how mixed species communities aggregate around resources within an enclosure may be useful for constructing harmonious groups of different species (Brereton & Rose, 2023). Similarly, measuring how individuals use resources within an enclosure should be undertaken to check the evenness of enclosure usage – especially for enclosures where resources ("zones") are not all of a uniform size (Plowman, 2003) or when a species' enclosure is altered, changed, or redeveloped (Ross et al., 2009). The size and shape of resources or zones within an enclosure, plus the number of resources or zones designated, will impact on any assessment of species' enclosure usage and zone occupancy (McConnell et al., 2022). Consideration of social traits (i.e. a need to be within a social group will drive occupancy of a specific zone) also needs to be considered as, for some species, being with conspecifics within a specific biologically relevant zone may be important to the performance

Figure 7.2 A complex multi-species aviary (left) and a habitat-specific aquarium for one species of fish (right) both come with their animal husbandry and welfare opportunities and challenges. Monitoring of individual animal behaviours, space usage, and resource access is required to ensure that such enclosures remain relevant and usable to the inhabitants.

of highly motivated behaviours (Rose et al., 2018a). Combining assessment of behaviour (e.g. calculation of time-activity budgets or time spent on behaviours indicative of positive welfare) alongside the time spent within particular areas of an enclosure provides valuable information on how species are comfortable and remain contented with the environment created for them. Therefore, it is important to consider the right method of data collection or the most appropriate index for calculating enclosure zone usage to get a valid and credible estimate of how suitable the enclosure is for individual, species, or population needs.

7.2.1 WHAT ABOUT THE WILD?

The environment recreated within zoo housing (habitat features that animals will engage with) must have its basis in the wild ecology of the species. The concept of "functional replication" has already been mentioned, and the features/substrates/furnishings that we create and provide within an exhibit can only be functional if they are based on information from species' ecology and knowledge of their wild habitat. Collecting such information is more straightforward for single-species enclosures that can be designed, themed, and fitted out with the specifics of that taxon in mind. But what about for more complex, multi-species enclosures? Can we look to the wild for evidence of how mixed-species aggregations will work? Potentially, to some degree; although Des Pallieres and Rose (2023) suggest that this is complex and challenging, and estimating why species are together at specific times can be hard to judge. Certainly, the wild is helpful for deciphering what type and dimension of space to provide (e.g. vertical, horizontal) as data on where individuals are most commonly found within their habitat can be used to recreate such elements of the habitat's structural and spatial variety. Gibbons (Hylobatidae) show a preference for life at higher elevations compared to lower levels (Anderson, 2014) and, when provided with the option to be higher up, natural behaviour patterns become more diverse and outdoor space is used more frequently. This study notes the importance of what is termed "a species' natural desires" to the design and build of welfare-positive zoo enclosures.

For those studying species' wild ecology, niche selection and occupancy can be used to inform habitat management and conservation plans to ensure that the habitat conserved is relevant for that species to persist and thrive within the ecosystem. Such information can be translated into the design and build of an enclosure – providing highly valued resources in areas that animals can access in the same way as free-living individuals would. Wild animals inhabit two basic areas of their habitats: a fundamental niche and realised niche (Soberón & Arroyo-Peña, 2017). A species' fundamental niche refers to areas that have the right balance of resources suitable for a species to exist without the presence of limiting factors, whereas a realised niche is a location in which you actually find the organism living, breeding, and surviving, based on an ability to cope with limiting factors (e.g. predation, disease, environmental fluctuation). Can this be used to determine quality space in the zoo and what species need to engage with? Could a realised niche be considered the areas a species finds most suitable for spending time in, and the fundamental niche be all areas of the enclosure, some of which are only inhabitable when specific stimuli or pressures have eased or been removed for the day (e.g. visitor presence)? Figure 7.3 illustrates this concept of niche ecology in the zoo in practice.

Ultimately, the overarching objective of any zoo enclosure is to convey the message that animals are not separate from their environment (Fekete & Szilágyi, 2021) and therefore naturalistic zoo design that incorporates ecological features, complexity, and variety not only provides important outlets for psychological well-being but also leaves a lasting, positive impression on zoo visitors too. Still, more traditional, functional, and species-specific housing designed to maximise welfare can still maintain visitor interest and extend dwell times at viewable enclosures (Smart et al., 2021) so designers need not be hampered by the constraints of providing a fully naturalistic environment should budget or development opportunities be limited.

7.3 QUALITY SPACE IS IMPORTANT

As already mentioned, animals in zoos live in environments that are controlled by humans. Therefore, their social and physical environments can differ from what would be experienced under natural conditions. Such differences may be caused by logistical challenges and feasibility, or they may be caused by management needs. If the enclosure is a compromise, because the fundamental needs of a species are hard or impossible to meet, and if key aspects of its habitat cannot be replicated due to their scale, complexity, or uniqueness, are our conservation and management efforts better engaged elsewhere? Zoos need to reflect and ask themselves whether, if the housing provided to a species can only ever be a compromise, it is it morally and ethically right to try and maintain that species in the zoo in the first place? Are there better and more appropriate ways of positively impacting on that species' existence and persistence without trying to house it in a substandard ex situ environment? For many species, however, we can still provide the right environment that meets most, if not all, of a species' biological needs while still creating an engaging exhibit for visitors (Figure 7.4). If we have a greater knowledge of a species' behavioural needs and the flexibility of its behaviour patterns

Figure 7.3 A large, complex enclosure for western lowland gorillas (*Gorilla gorilla gorilla*), photo left, provides a range of replicated habitat features that the gorillas can use at different times of the day. Limited public views mean most of the enclosure is without external pressures and therefore the whole exhibit is a realised niche. Within walk-through exhibits, photo right, specific areas of the enclosure are designed for visitor passage and therefore at certain times of the day will be less desirable spaces for the animals and, as in the case of this pheasant pigeon (*Otidiphaps nobilis*), are spaces to transition through to useful, realised niches within the enclosure. Such resources can be considered fundamental niches – areas that the enclosure inhabitants could use once external pressures have gone (i.e. visitors have gone home for the day).

– understanding behaviours must be provided for to prevent poor welfare (Kelly et al., 2025) – we can design and build enclosures with greater certainty of their biological relevance for the inhabitants.

While it is possible to provide the right environment that meets most, sometimes all, of a species' biological needs (Figure 7.4), recreating aspects of a habitat and providing resources in a naturalistic manner is easier for some species, and in some systems, than it is in other species and systems. For example, the correct life support systems, lighting, and water flow required for maintaining a marine coral reef tank all provide a final product that is so close to nature that the animals would not know they were contained in a glass vessel. To quote Rees (2011), "A management goal of zoos should be to provide an environment in which the animal, if it could know, would not know that it was in captivity." One would imagine that we stand a better chance of achieving such a goal for a common clownfish (*Amphiprion ocellaris*) than we do for a killer whale (*Orcinus orca*) (Figure 7.5). Use of the term "habitat" is likely more truthful for the aquarium than for the killer whales.

7.3.1 ALIGNING WELFARE MEASURES WITH ENCLOSURE USAGE

When considering a species' requirements in enclosure design, we must ask the question, "What does the animal need to do?". Although it sounds obvious, this simple question, if investigated and answered truthfully, will direct designers towards the most suitable environments where all space is usable and fulfilling. Instead of assuming what useful space means to a species based on either historical information or anecdote, considering what is actually meaningful to that species can direct us towards a more appropriate enclosure design (Rose et al., 2024). Owls (Strigiformes) for example, are species commonly kept in relatively small aviaries when compared to the space given for other similarly sized birds, and often in exhibits that neither allow sufficient space for flight nor provide much natural light. Are such designs considered best practice or simply "acceptable practice" because we do not know enough about how these species live? How do owls in the wild utilise patches of sunlight during the daytime? Do they require habitat variety within an enclosure the same as other birds of prey? We must question the validity of information used as the basis for current housing to check

Figure 7.4 An example of a naturalistic, ecologically relevant enclosure for Sunda gharials (*Tomistoma schlegelii*) in a large European zoo's immersive tropical rainforest house. The enclosure enables control and choice due to the complexity of the environment that has been recreated. The animals can access different resources according to their needs, wants, and physiological states.

Figure 7.5 Managing the wild to its fullest extent is possible for species where replication of all important biotic and abiotic factors can be performed in a zoo's exhibit (in this example, directly recreating a marine coral reef with sea anemones for a shoal of common clownfish). However, the same may not be true for a group of captive killer whales living in a uniform pool of defined depth and boundaries that do not provide the same degree of social and environmental complexity as that experienced by free-living counterparts.

whether it is fit for purpose. In this avian example, we investigate whether flight is truly a meaningful behaviour for these animals, so that instead of designing aviaries because a bird species might not "need" to fly, we first attempt to answer the question, do they "want" to fly?

Therefore, the way that animals use their space should be considered part of the assessment process of welfare scoring, as it will allow us to determine the usefulness of such spaces. Zookeepers and those in charge of daily care can build up experience of how each individual animal they are responsible for uses their space and engages with resources. Any change to where an individual is found, or how they move around the enclosure, should warrant further investigation. For example, if an animal is spending more time than usual in an area of an enclosure that is further from, or out of, visitor view, we need to question the motivation. Is it simply that they are choosing to be further from visitors and, if so, why? Could it be that there are more available resources in that space that the animal is perceiving to be more desirable than usual? Has keeper activity changed or has husbandry routine been recently altered? Might the animal therefore be anticipating feeding (or similar) while achieving a better back-of-house view from that space within the enclosure? Such questions allow for more focused and evidence-based welfare assessment that aligns with critique and appraisal of space use and enclosure suitability

7.3.2 APPLYING TECHNOLOGY TO WELFARE-POSITIVE ENCLOSURE DESIGN

As zoos move forwards in how research is conducted, the use of artificial intelligence (AI) and bespoke technology developed in collaboration with computer scientists can help gather data from the animal's perspective to improve the design and advancement of zoo enclosures and enrichment devices. Indeed, it is likely that any future version of a book pertaining to zoo animal behaviour and welfare will feature the use of novel technology (and how it can assist zoos with their operations) even more heavily than this. It is perhaps telling that the research body for the Association of Zoos & Aquariums (AZA) in North America is called the "Research and Technology Committee" (AZA, 2024), emphasising the role of tech in data collection and application at the zoo. For example, expanding on how tech can support implementation and integration of zoo animal welfare assessments, with post-occupancy evaluation of enclosures and the appraisal of how effective environmental enrichment is. Open-access AI could also help with science communication and how the zoo tells the stories of its work to its visitors, explaining the complexities of enclosure design and maintenance to a wider audience in an accessible and creative manner. It is essential, however, that all outputs from AI are fact-checked and critiqued by an actual person (i.e. an expert in the field) before wider dissemination.

Working towards an animal's entire enclosure being enriching, so that any added enrichment opportunities are not the main way that biologically relevant behaviours are promoted, should be a key goal of all stakeholders (those involved in animal care, operation, enclosure design, research, and animal health). Furthermore, new technology can also help design and implement more holistic research methods that measure the effect of an enclosure (its design, or alteration) on animal behavioural and physiological responses, as well as the perception of the enclosure by zoo visitors (and how the exhibit impacts them). In this way, future enclosure design research will put the welfare of animals at its heart and provide evidence for how to display species in a manner that can truly be considered engaging and educational. Such an approach aligns with the broader move of re-establishing a zoo's entertainment aim as one engagement (Rose & Riley, 2022) – engagement with the animals themselves and with the natural world more broadly, and with key conservation and sustainability messages (Greenwell et al., 2023) that zoos like to advertise and promote.

7.4 THE CHALLENGE WITH CHOICE – PRACTICAL DESIGN TECHNIQUES TO OPTIMISE WELFARE

It is challenging in itself to design enclosures that provide for all the biological requirements of a species within an engineered environment. Doing so in a way that does not restrict practical management requirements of such spaces, as well as meet the needs of other zoo stakeholders, is an even greater challenge. Choice underpins attainment of good animal welfare states in captivity (Owen et al., 2005; Ritzler et al., 2023; Rose & Lewton, 2025) and the ultimate aim of modern enclosure design should be to give an animal complete choice and control within its environment. But how can we balance giving an animal the freedom to choose where it wants to go and what it wants to do, in an engineered environment that needs to be both actively managed by caregivers and an engaging experience for visitors? For some smaller species, such as reptiles and amphibians, we can create engineered biomes large enough to mimic their natural ecosystems, with the correct landscaping, temperature and humidity that requires minimal hands-on management by caregivers. Housing larger species, particularly those held in zoos with different climates to their natural ranges, in such environments would be impractical. There are, however, several design strategies that can be utilised to overcome these challenges, whilst ensuring that visitor and staff needs are not compromised.

Possibly the greatest challenge of modern zoo enclosure design occurs at the interface of animal welfare and visitor experience. The creation of a space where an animal can freely choose where it wants to spend its time will ultimately conflict with the expectation from visitors to see animals (Kelling et al., 2014). While thoughtful interpretation material can be used to convey the importance of providing an animal with spaces to retreat from view, a visitor-facing zoo needs its animals to be visible as much as is practicable. Some institutions will address this challenge by locking animals outside for the majority of the day in an effort to ensure the animals stay "on show", in those facilities where indoor viewing opportunities are lacking. This directly limits the choice of space usage and will ultimately lead to reduced welfare, as we know that captive animals fare better when given more choice of where to spend their time and have more control over their environment. For example, Smith et al. (2023) found that, when provided with voluntary access to alternative spaces, tigers (*Panthera tigris*) displayed more biologically relevant behaviours and a decrease in stereotypic behaviours. Thus, enclosures should comprise a complex variety of spaces to choose from and, where zoos are geographically located in climates that differ significantly to that of the species' natural distribution, this should include indoor spaces with artificial temperature provision.

To reduce the chance of animals being non-viewable should they choose to go to these indoor spaces, an obvious solution may be to ensure that visitor viewing is carried through into these spaces.

However, these spaces are likely to be smaller than external areas, resulting in reduced distance between visitors and animals and increased potential for stress. While animals will often choose to go into these spaces to escape inclement weather, they may also see indoor areas as spaces to retreat should they want additional seclusion. A simple design technique to maintain these areas as the animals' 'safe' spaces, while still providing visitors with viewing opportunities, is to create windows with smaller apertures. In zoos where immersive theming is used, these windows can be designed to maintain the exhibit's aesthetic, such as gaps between artificial rock or timber-cladded walls (Figure 7.6).

The heating of indoor spaces also comes with practical challenges not related to animal welfare. While enclosure design has historically focused on the three pillars of requirement: 1) Animal needs, 2) Visitor needs, 3) Staff needs (Kelling et al., 2014) – a fourth requirement is becoming increasingly important to the zoo design industry – *sustainability*. This is the need to consider energy usage to manage both monetary and environmental impacts. Providing animals the choice to freely move between indoor and outdoor spaces requires unobstructed access through an open door or slide, but such gaps in a building can make it impractical to heat an indoor space efficiently. As the size of an animal increases, the loss of heat through these apertures can increase significantly. While there are potential technological strategies that could provide automation on doors as animals approach, these devices do not come without their risks to animal safety or operational reliability and can be considerably costly. A simple design technique to overcome this can be the addition of PVC strips or similar materials, which can retain the building's heat while allowing an animal free movement (Figure 7.7). Consideration however should be given to animals that may need a soft introduction to the use of these strips, which can be done by gradually installing each strip one after the other, as the animal becomes habituated. Strips should be of a thickness that is not too heavy for the species to push through, while not being too light that they could be lifted by gusts of wind and cause stress or fright to the resident animals.

Choice can also be impacted by daily operational and husbandry practices. Historically, zoos restricted access to outdoor spaces overnight, particularly for species typically classed as 'dangerous', to ensure additional safety while the facility is unstaffed. A modern zoo enclosure should, be designed to mitigate any such concerns, providing fully suitable containment where animals can be safely housed 24/7. This will allow the choice to go outdoors overnight, which is of particular importance for nocturnal and crepuscular species, but also diurnally orientated species that are known to feed or be active during nighttime hours, such as elephants (Horback et al., 2014) and common hippopotamus (*Hippopotamus amphibius*) (Tennant et al., 2018).

Daily husbandry practices may also result in caregivers segregating animals to specific areas

Figure 7.6 Viewing windows into an African painted dog (*Lycaon pictus*) exhibit seen from both sides, showing that standard window sizes can be themed to create small apertures from visitor spaces, creating more privacy for animals and allowing a more interesting viewing experience at varying eye levels.

Figure 7.7 PVC and rubber strips can be used on doorways for various species of different shapes and sizes. Left: ring-tailed lemur (*Lemur catta*) enclosure. Middle: Sumatran tiger (*Panthera tigris sumatrae*) enclosure. Right: giraffe (*Giraffa camelopardalis*) house.

for tasks such as feeding or behaviour training. Positive reinforcement training has become common practice in zoos for many species and often involves animals being moved into dedicated spaces to work with their caregivers. Behaviour training should always be a choice-based activity, however, dictated by animals as to whether they take part in the process or not. Where possible, training areas should therefore be designed to be incorporated into an animal's daily space, whether this is indoors or outdoors, to allow individuals the full choice of whether or not to engage with their human caregivers. Where hierarchies or dominance behaviours may make this difficult, other staff could encourage an animal's conspecifics away from the training area by feeding or engaging in another activity. Allowing visitors to watch husbandry-focused behaviour training sessions can increase educational opportunities (Whitehouse-Tedd et al., 2020) so incorporating features such as 'training walls' in areas of the enclosure that are viewable by visitors can be beneficial. If this is not practical or desirable, however, designers can consider incorporating CCTV into these spaces and show these sessions on a monitor in visitor areas, where educational talks could still be undertaken.

Many zoos use an "in/out approach" whereby animals are locked inside a building when the outdoor space needs to be serviced or when caregivers need to place food within the enclosure. While food provision is a vital part of animal husbandry, this practice can directly impact an animal's daily experience and remove control over its environment. One solution to this, for species for which keepers cannot safely share the same space, is to use remote feeding systems. For species that forage from the ground, scatter feeding is a common way of providing food in a manner that requires the animal to search and find its food. Rather than entering the animal's space to place food, this can be either manually scattered from outside the enclosure or, more appropriately, to reduce anticipatory behaviours, delivered by automated feeders at randomised time intervals. For predatory species such as the lion and other similar carnivores, meat is often placed in locations where the animal needs to expend energy and use its problem-solving skills. This is often in the form of 'feeding poles', where caregivers attach meat to the top of timber posts, requiring the animal to climb or jump to retrieve it. This method can result in positive health benefits even for tigers (Law & Kitchener, 2020) – a species not known for adopting climbing as a hunting strategy. This same principle can be provided by a remote method using zip lines to deliver a carcass from a safe distance outside of the enclosure to designated points within the animal's space. This method allows animals to be left with free access to all spaces of their enclosure and can simulate prey location, selection, and hunting actions. Consideration, however, should be given to the potential for more dominant individuals in a group to control access to these food delivery locations; thus, parallel feeding methods may also need to be employed should food control by dominant individuals negatively affect the food intake of others.

Such dominance and hierarchical behaviours should also be considered when providing resources or other sources of choice within an enclosure. For example, big cat enclosures often include heated rocks or platforms that give animals a higher vantage point and a warm surface for resting, which can be of particular benefit for enclosures in climates cooler than those of the species' natural ranges. As with food resources, individuals may dominate these features, removing the choice for subordinate individuals to use them. It is important, therefore, to design these features large enough to allow maximum occupancy or provide several of them throughout the enclosure. Even on a budget, such features can be created that allow all individuals within a group to use them simultaneously (Figure 7.8). We do however need to strike a balance between providing individuals enough choice to use resources, but not so much that we completely remove the requirement of hierarchical behaviours that may be necessary for the species' growth and experience. We must also consider the full extent of 'choice' to use such resources. These features are often designed next to, or near, viewing windows to encourage the animals into a space that is easily viewable by visitors. If we only provide these resources near viewing spaces, though, are we really giving the animals a choice? Individuals may be drawn to use these resources but may not necessarily be comfortable being in such close proximity to visitors, so we need to consider providing a number of similar options, in varying areas of an enclosure.

Social structure and dominance also play a role in restricting movement throughout an enclosure, particularly for areas that are separated by doors or slides. Where only one door may be separating spaces, or doors are too close together, an animal may be reluctant or even unable to travel through should a more dominant individual be in the immediate area. Doors should therefore be spaced out as much as possible to allow alternative routes for individuals to pass without stress. Having at least two of these access points between spaces also allows a circular circuit to be maintained, which lowers the risk of an individual being trapped should they ever be chased between spaces by conspecifics. Figure 7.9 illustrates how these circular routes can be employed in a hypothetical enclosure layout for lions, along with some of the other design solutions discussed in this section.

An abundance of factors can restrict the presence of choice and control for captive animals. Zoo managers often face their own challenges when it comes to prioritising, having to consider multiple and often conflicting needs, such as those of the zoo as a business, its visitors, the zoo's staff, and the animals being cared for. However, using species-relevant ecology alongside thoughtful design strategies, the conflict of needs between these different stakeholders can be greatly reduced.

Figure 7.8 Big cats are often drawn to elevated spots within their enclosures, and these features should be created to account for all individuals' needs within those spaces. Left: A tiger surveys its surroundings atop a complex, naturally-constructed rocky outcrop, with a choice of various places to lie or sit. Right: A timber structure for lions which, while not natural-looking in its construction, can allow for multiple individuals within a pride to use the structure at the same time, including individuals with limited mobility such as cubs or geriatric adults (photo: Team Building with BITE).

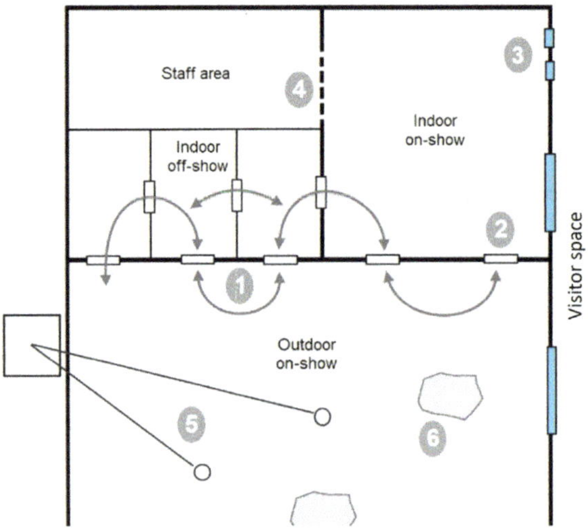

1) **Circulation –** Multiple access and egress points within an enclosure are encouraged, to maximise choice of movement and minimise the risks of dominant animals blocking access points if in close proximity.
2) **PVC strips –** In zoos with colder climates than a species' natural environment, it is essential that we artificially heat as many spaces in an enclosure as possible. Strips made from rubber, PVC or similar material allow slides/doors to be kept open, maximising the choice of space use and minimising heat loss from indoor spaces.
3) **Reduced viewing –** Creating spaces where captive animals do not feel surrounded or overlooked by visitors is essential. Small viewing windows can be employed in indoor spaces, for example, to allow animals a feeling of security while still providing a viewing experience for visitors.
4) **Training wall –** The participation of husbandry-focussed behaviour training should always be through choice. These areas or 'training walls' should therefore be incorporated into the animals' usual daily spaces and could be positioned where visitors can view these husbandry practices if desired.
5) **Remote feeding –** For species such as big cats that are not safe for keepers to share the same space with, food provision often requires animals to be shut within one part of their enclosure, allowing keepers to bring food into another space. Remote feeding mechanisms such as zip lines can be used for such species, to remotely and safely move food such as pieces of meat into the enclosure. Designers should consider too potential hierarchies or dominance behaviours within species such as lions, and provide multiple delivery points to reduce instances of one or more individuals dominating food.
6) **Raised platforms / heated rocks –** While these features can be placed near viewing points to encourage animals closer to visitors, it should not be an 'all-or-nothing' choice for animals, of whether to use such a feature or to move away from visitors. Alternative options should therefore be available in less-overlooked areas of an enclosure.

Figure 7.9 A hypothetical enclosure design for lions, showing several design strategies that can be used to maximise freedom of choice and space use within the full environment.

By using simple and cost-effective practical solutions when designing enclosures, we can alleviate these pressures and ensure that these spaces promote an abundance of opportunities for animals to experience choice and control over their environment. The design ideas discussed above to maximise choice are based around current best practice animal management strategies and husbandry guidelines seen throughout modern zoos, but we should always be striving to go further and consider what animals need beyond the norm in the modern zoo. Chapter 24 discusses this idea further and what the future of enclosure design may resemble.

> Nature, the natural world and behaviours should always be the first approach. A zoo's site conditions should inform design decisions and suitability of species… but not everything can always be achieved due to budget, management, and space. We therefore need to look at species-specific behaviours that would best serve the species both cognitively and physically.
>
> *Jonathan McLoughlin, Chartered landscape architect and director of zoo design consultancy Dearadh Zú*

7.5 CONCLUSIONS

The enclosures and exhibits that house living animals are the fundamental feature of all zoos and aquariums – their main attraction, their key reason for existing, and the most important asset for fulfilling their aims. Consequently, any attempt at inferring welfare states needs to start with an evaluation and assessment of the spaces and environments provided for each and all of the zoo's inhabitants. All species of zoo animals will display behavioural traits that will indicate whether they are comfortable within their enclosure. These behavioural traits may not always be obvious or extravagant (withdrawal, lethargy, and lack of engagement with resources can all indicate inappropriate housing), so animal care staff and designers must be aware of a species' ecology and how it has evolved to live when evaluating housing, enclosure design, resource provision, and amount of space provided. Designing an enclosure that promotes the highest standards of welfare will always present challenges for the needs of other stakeholders within the zoo. For example, how do we give an animal the full choice of what it wants to do and where it wants to go, while ensuring that visitors can see it? Collaboration is key to overcoming these challenges, and the best enclosure

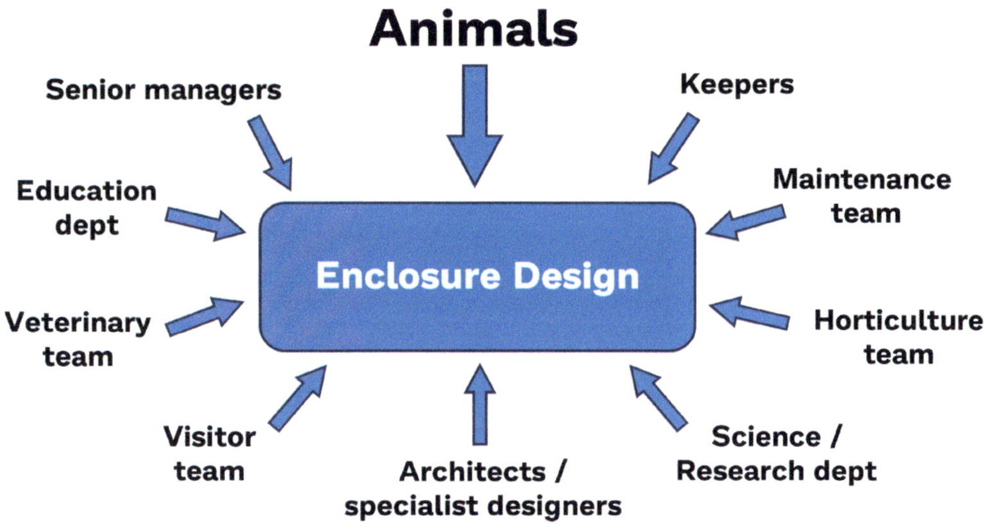

Figure 7.10 An example is presented here of the range of potential stakeholders that need consulting throughout the process of designing an enclosure. Collaboration is key to maximising the benefit of these spaces, especially for their future animal residents. Ultimately, the animal (who did not ask to be in the zoo) must be given the most consideration.

designs are developed through consultation with all relevant end users (Figure 7.10).

Future zoo enclosure design needs to consider how visitors perceive animals, alongside the functional nature of the enclosure to enable animal care staff to provide daily husbandry. Spatial requirements need to be maximised so that meaningful variety is provided within an exhibit, which, in turn, will provide the animals with opportunities to display a wide range of welfare-positive behaviours and thus impact positively on the visitors viewing them (and the messages they leave the zoo with).

There are many complexities and conundrums surrounding the best way of presenting animals in the zoo. Traditional, functional-style housing will hold visitor interest – think the crowds that will congregate around a penguin pool, for example. Yet what is interesting to the visitor? What messages are being presented and how engaging are the animals in this setting? Zoo designers and researchers should grapple with the question – "Is an artificial environment, which does not resemble the animal's natural habitat, fundamentally bad for animal welfare if it provides what animals need?" We need to consider what we are displaying and why we are displaying it. The messages provided by the enclosure (in terms of connection with nature) can be different from promoting positive welfare for the animals within. An artificial environment, full of structural diversity and opportunities to perform wild-type time budgets, is going to be much more fulfilling for the animals than a recreation of their natural habitat that is not accessible, useful, or available to them. Concepts of quality space, rather than artificial replication of habitat features that animals simply cannot use, are still true today as they were back in the 1990s when first mooted (Richardson, 1999). Going back to ecology, all stakeholders involved in zoo enclosure design should ask themselves (before any sketch is drawn, ground is broken, or brick is laid), "what does the animal actually need from, and what do they need in, their environment?"

For those wishing to know more about zoo enclosure design and the practical challenges of managing captive populations in complex spaces, Box 7.1 provides suggestions for further reading.

BOX 7.1: Recommended further reading on the design and construction of zoo enclosures

Salzert, W. (2016). *Making zoos attractive: 12 basic rules of exhibit design*. Munster, Germany: Schüling Verlag.

Thomas, W. D. & Maruska, E. J. (1996). Mixed-species exhibits with mammals. In D. G. Kleiman, M. E. Allen, K. V. Thompson, & S. Lumpkin (Eds.), *Wild mammals in captivity* (pp. 204–211). Chicago, IL: University of Chicago Press.

van Vliet, E. (2015). *Exhibiting zoo animals*. Munster, Germany: Schüling Verlag.

Veasey, J. & Hammer, G. (2010). Managing captive mammals in mixed-species communities. In D. G. Kleiman, K. V. Thompson, & C. Kirk Baer (Eds.), *Wild mammals in captivity: Principles & techniques* (Vol. 2, pp. 151–161). Chicago, IL: Chicago University Press.

REFERENCES

Anderson, M. R. (2014). Reaching new heights: The effect of an environmentally enhanced outdoor enclosure on gibbons in a zoo setting. *Journal of Applied Animal Welfare Science*, *17*(3), 216–227.

AZA. (2024). Research and Technology Committee. Association of Zoos & Aquariums, https://www.aza.org/research_and_technology_committee. (Access Date 24/04/2024).

Bandeli, M., Mellor, E. L., Kroshko, J., Maherali, H. & Mason, G. J. (2023). The welfare problems of wide-ranging Carnivora reflect naturally itinerant lifestyles. *Royal Society Open Science*, *10*(9), 230437.

Brereton, J. E. & Rose, P. E. (2023). The Zone Overlap Index: A new measure of shared resource use in the zoo. *Zoo Biology*, *42*(6), 811–817.

Des Pallieres, C. G. & Rose, P. E. (2023). Two's company, three species is a crowd? A webcam-based study of the behavioural effects of mixed-species groupings in the wild and in the zoo. *PLoS One*, *18*(4), e0284221.

Duggan, G., Burn, C. C. & Clauss, M. (2016). Nocturnal behavior in captive giraffe (Giraffa camelopardalis)—a pilot study. *Zoo Biology*, *35*, 14–18.

Fàbregas, M. C., Guillén-Salazar, F. & Garcés-Narro, C. (2012). Do naturalistic enclosures provide suitable environments for zoo animals? *Zoo Biology*, *31*(3), 362–373.

Fekete, O. B. & Szilágyi, K. M. (2021). Contemporary zoo design and the historical treasure of garden art: Tiergarten Schönbrunn. *Journal of Landscape Architecture*, *16*(3), 82–93.

Greenwell, P. J., Riley, L. M., Lemos de Figueiredo, R., Brereton, J. E., Mooney, A. & Rose, P. E. (2023). The societal value of the modern zoo: A commentary on how zoos can positively impact on human populations locally and globally. *Journal of Zoological and Botanical Gardens*, *4*(1), 53–69.

Hartley, M., Wood, A. & Yon, L. (2019). Facilitating the social behaviour of bull elephants in zoos. *International Zoo Yearbook*, *53*(1), 62–77.

Hediger, H. (1950). *Wild animals in captivity*. London, UK: Butterworths Scientific Publications.

Horback, K. M., Miller, L. J., Andrews, J. R. & Kuczaj, S. A. (2014). Diurnal and nocturnal activity budgets of zoo elephants in an outdoor facility. *Zoo Biology*, *33*(5), 403–410.

Kelling, N. J., Gaalema, D. E. & Kelling, A. S. (2014). A modified operational sequence methodology for zoo exhibit design and renovation: Conceptualizing animals, staff, and visitors as interdependent coworkers. *Zoo Biology, 33*(4), 336–348.

Kelly, R., Freeman, M. & Rose, P. E. (2025). What behavior is important behavior? A systematic review of how wild and zoo-housed animals differ in their time-activity budgets. *Frontiers in Ethology, 4*, 1517294.

Law, G. & Kitchener, A. C. (2020). Twenty years of the tiger feeding pole: Review and recommendations. *International Zoo Yearbook, 54*(1), 174–190.

Mbizah, M. M., Valeix, M., Macdonald, D. W. & Loveridge, A. J. (2019). Applying the resource dispersion hypothesis to a fission–fusion society: A case study of the African lion (Panthera leo). *Ecology and Evolution, 9*(16), 9111–9119.

McConnell, H., Brereton, J., Rice, T. & Rose, P. E. (2022). Do birds of a feather always flock together? Assessing differences in group and individual zoo enclosure usage by comparing commonly available methods. *Journal of Zoological and Botanical Gardens, 3*(1), 71–88.

Moss, A. G. & Pavitt, B. (2019). Assessing the effect of zoo exhibit design on visitor engagement and attitudes towards conservation. *Journal of Zoo and Aquarium Research, 7*(4), 186–194.

Owen, M. A., Swaisgood, R. R., Czekala, N. M. & Lindburg, D. G. (2005). Enclosure choice and well-being in giant pandas: is it all about control? *Zoo Biology, 24*(5), 475–481.

Plowman, A. B. (2003). A note on a modification of the spread of participation index allowing for unequal zones. *Applied Animal Behaviour Science, 83*(4), 331–336.

Rees, P. A. (2011). *An introduction to zoo biology and management.* Oxford, UK: John Wiley & Sons.

Richardson, D. M. (1999). *Staged scene or quality space.* Paper presented at the Conservation centres for the new millennium: Proceedings of the 5th International Symposium on Zoo Design, Paignton Zoo.

Ritzler, C. P., Lukas, K. E., Bernstein-Kurtycz, L. M. & Koester, D. C. (2023). The effects of choice-based design and management on the behavior and space use of zoo-housed Amur tigers (Panthera tigris altaica). *Journal of Applied Animal Welfare Science, 26*(2), 256–269.

Robinson, M. H. (1998). Enriching the lives of zoo animals, and their welfare: Where research can be fundamental. *Animal Welfare, 7*(2), 151–175.

Rose, P. E., Brereton, J. E. & Croft, D. P. (2018a). Measuring welfare in captive flamingos: Activity patterns and exhibit usage in zoo-housed birds. *Applied Animal Behaviour Science, 205*, 115–125.

Rose, P. E., Freeman, M., Hickey, I., Kelly, R. & Greenwell, P. J. (2024). Considering what animals "need to do" in enclosure design: Questions on bird flight and aviaries. *Birds, 5*(3), 586–603.

Rose, P. E. & Lewton, J. (2025). Key concepts for enhancing zoo animal welfare: Coping, comfort, choice, control, challenge and compassion *Journal of Applied Animal Welfare Science*, in press.

Rose, P. E., Lloyd, I., Brereton, J. E. & Croft, D. P. (2018). Patterns of nocturnal activity in captive greater flamingos. *Zoo Biology, 37*(5), 290–299.

Rose, P. E. & Riley, L. M. (2022). Expanding the role of the future zoo: Wellbeing should become the fifth aim for modern zoos. *Frontiers in Psychology, 13*, 6649.

Ross, S. R., Schapiro, S. J., Hau, J. & Lukas, K. E. (2009). Space use as an indicator of enclosure appropriateness: A novel measure of captive animal welfare. *Applied Animal Behaviour Science, 121*(1), 42–50.

Smart, T., Counsell, G. & Quinnell, R. (2021). The impact of immersive exhibit design on visitor behaviour and learning at Chester Zoo, UK. *Journal of Zoo and Aquarium Research, 9*(3), 139–149.

Smith, K. D., Snider, R. J., Dembiec, D. P., Siegford, J. M. & Ali, A. B. (2023). Effects of a modern exhibit design on captive tiger welfare. *Zoo Biology, 42*(3), 371–382.

Soberón, J. & Arroyo-Peña, B. (2017). Are fundamental niches larger than the realized? Testing a 50-year-old prediction by Hutchinson. *PLoS One, 12*(4), e0175138.

Tennant, K. S., Segura, V. D., Morris, M. C., Snyder, K. D., Bocian, D., Maloney, D. & Maple, T. L. (2018). Achieving optimal welfare for the Nile hippopotamus (Hippopotamus amphibius) in North American zoos and aquariums. *Behavioural Processes, 156*, 51–57.

Whitehouse-Tedd, K., Spooner, S. & Whitehouse-Tedd, G. (2020). Making training educational for zoo visitors. In V. A. Melfi, N. R. Dorey, & S. J. Ward (Eds.), *Zoo animal learning and training*. https://doi.org/10.1002/9781118968543.ch10.

Wilson, M. L., Bashaw, M. J., Fountain, K., Kieschnick, S. & Maple, T. L. (2006). Nocturnal behavior in a group of female African elephants. *Zoo Biology: Published in Affiliation with the American Zoo and Aquarium Association, 25*(3), 173–186.

8

Welfare is what you eat: Zoo nutrition and positive animal welfare

PAUL ROSE

8.1 INTRODUCTION

Feeding zoo animals is one of the most essential jobs that caregivers need to provide. Since the turn of the 20th century, zoos have realised the importance of good nutrition for animal health but have often struggled to deliver a quality, appropriate diet to the species in their care. An appropriate diet would be one that meets the nutritional needs of a species at each specific life stage. This appropriate diet supports normal growth and development, as well as meets any physiological challenges (e.g. reproductive events or alterations to homeostatic needs caused by season, such as moulting or hibernation) and enables the performance of important behavioural outputs. Logistical complexities surrounding the sourcing of relevant foodstuffs, a lack of information on wild diets and feeding behaviour, specialised anatomy, physiology, and morphology evolved by a species in a specific niche that cannot be replicated in the zoo, and challenges of appropriate dietary presentation and palatability are all problems that zoos and aquariums have to grapple with when attempting to construct and feed a diet that is nutritionally correct for all species in their living collection.

If welfare is concerned with an individual's perceived state while coping with their immediate environment (Broom, 1986) and with the behavioural, physical, and psychological influencers and mediators of such welfare state, then nutrition clearly has a large part to play in the attainment of positive welfare. Behavioural outputs will be promoted or diminished by how food is presented, and an animal's physical and psychological health will be impacted by the quality of food provided (e.g. freshness, energetic content), the balance of macro- and micro-nutrients within each dietary ingredient, the palatability of food, and the calorific characteristics of the diet too. In some species, the food provided to the animal is also feeding specific aspects of its digestive physiology (e.g. the feeding of browsing ungulates is providing structural fibre and cellulose for ruminal microorganisms to break down and digest), and therefore, individual animals are unlikely to thrive and experience positive welfare if any such symbiotic or commensal relationships are not supported by their zoo diet.

DOI: 10.1201/9781003429425-8

Appropriate diet and the correct intake of nutrients, vitamins, minerals, and energy are integral to animal health and therefore welfare. The correct diet does more than simply keep an animal alive; it enables positive mental states to be experienced, it initiates and satiates behavioural performance, and it influences signalling, social communication, and reproduction. Immune response, growth, development, attainment of sexual maturity, fertility, fecundity, senescence, and overall longevity are all influenced by the quality, quantity, and appropriateness of food and feeding. For example, work on the gut microbiome in species ranging from echidnas (Perry et al., 2022), to elephants (Keady et al., 2021), from marine fish (Egerton et al., 2018) to monkeys (Frankel et al., 2019) has identified the importance of zoo diets to microfloral health and diversity, and associated links to population sustainability, conservation, and good animal welfare.

8.2 THE HISTORICAL BACKGROUND OF ZOO ANIMAL NUTRITION AND WIDER WELFARE CONCERNS

The history of zoo nutrition is a story of transformation and enlightenment, reflecting the evolving understanding of animal welfare and the recognition that correct, species-specific, and life stage-specific nutrition is one of the cornerstones of effective husbandry. Over many decades, zoos have transitioned from simple provision of a basic or "makeshift" diet for survival to embracing a science-driven, evidence-based approach to nutrition and diet formulation that considers species-specific requirements, natural foraging behaviours, anatomy, physiology, and morphology, and overall health requirements (including body condition). This shift in ethos was driven by pioneering zoo and academic personnel who recognised the importance of nutrition as a support for the physical and psychological health of animals in the zoo.

The evolution of zoos has been, in part, driven by changes in societal attitudes towards the keeping of captive wild animals and the reasons why they are being housed. Initially as places of curiosity and entertainment, exhibiting animals as living curiosities without comprehensive consideration or understanding of their needs, biology, or welfare. During the 19th century, as ethical debate intensified, burgeoning legislation started to protect animals from harm, and animal behaviour and evolutionary biology because established as empirical sciences, there was a growing realisation that zoo-housed animals require more than just shelter and basic food to survive and thrive. Changes in the perspective of zoo animal care were driven by concerns over frequent poor animal health, reduced longevity, and evident behavioural distress or disturbance in captive conditions.

The advent of zoo-specific publications, such as the (now defunct) International Zoo Yearbook in 1960, enabled a platform for academic study into zoo nutrition to be distributed widely. Whilst not full articles, this first volume of the International Zoo Yearbook contains several select examples of nutritional developments to improve zoo animal care and welfare. These are an attempted mimicking of natural summer diets in moose (*Alces alces*) using banana fruits and skins at Assiniboine Park Zoo (Voss, 1960), the analysis of common hippopotamus milk (*Hippopotamus amphibious*) at Overton Park (now Memphis) Zoo to help develop successful hand-rearing protocols (Gray, 1960), a review of pelleted diets for captive Galliformes at London Zoo (Yealland, 1960), and horsemeat substitution to improve the nutritional quality of diets for captive carnivores at National Zoo in Washington DC (National Zoological Park, 1960).

Many zoological institutions have also played instrumental roles in improving zoo animal nutrition via investment in analytical facilities to conduct proximate analysis – the basic laboratory procedures for determining the nutritional and energetic contents of feed (McDonald et al., 2022) – and in the training of specialised nutritional staff that have both helped pave the way for wider construction, availability and procurement of species-specific zoo diets. Software used for diet formulation and analysis, such as the widely used "Zootrition" package (Zootrition, 2021) that was developed in the early 2000s by the Wildlife Conservation Society (Crissey, 2001), has been a further valuable tool for the evaluation and assessment of nutritional elements and energetic values in both complete diets and specific ingredients, enabling objective review of what animals are actually eating. The explanation of suitable methods used for intake and digestibility studies has enabled wider calculation of what animals are ingesting, digesting, and assimilating from the food that they are being given. The evolution and application of these tools and comparative methodologies have improved animal welfare by enabling nutritional analysis to be more commonplace and accessible in many more zoos, with outputs from such procedures benefiting large numbers of animals.

Key factors that resulted in greater focus on appropriate zoo nutrition include:

- Scientific advancements across the late 19th and early 20th centuries that ultimately contributed to techniques, methods and approaches that produced data that are more reliable and valid. As human medical science evolved, there was an associated knock-on effect with veterinary medicine. As our understanding of the importance of diet to human health improved, the same knowledge was transferred over to animals. For example, knowledge of "good fats" such as omega-3 fatty acids that protect humans from various diseases has been incorporated into the formulation and processing of animal feeds to provide similar protective benefits.
- Evolution of animal welfare as a movement and concept. A growing shift in societal attitudes towards animal welfare meant that animals were being considered as sentient beings with complex needs (DeMello, 2021), prompting zoos to re-evaluate their approach to animal care and husbandry. Ensuring that zoo animals are provided with a suitable diet has aligned with greater societal scrutiny of zoo animal welfare.
- Recognition of natural behaviour patterns and motivational states. An understanding that animals in captivity needed opportunities to engage in natural behaviours, including those linked to diet (e.g. foraging and feeding) and consequently, the development of diets to encourage these behaviours contributing to better animal health (Clauss & Dierenfeld, 2008; Dierenfeld, 1997).
- Longevity and population sustainability. As the large, mainstream zoos became more focused on conservation post-World War II, the need to protect longevity and reproductive success of species in the living collection became paramount. Zoos evolved into centres for conservation and animal care, and their personnel better understood that healthy animals were more likely to bred and were more likely to produce viable young (Rabb, 2004). Ultimately, recognising that different, for each species, physiological states and different life stages require different nutritional inputs has improved breeding success and the viability of offspring.

Information from domestic animal nutrition has proved somewhat useful in the past for kick-starting the development of zoo diets. However, it is now widely accepted that comparison of the numerous different species in the zoo (and their dietary needs) to that of a handful of domestic livestock and fowl is largely unhelpful. A trial and error approach to zoo nutrition is also no longer appropriate as a means of finely achieving the diet that will promote animal health and welfare (Dierenfeld, 1997). As highlighted by Leus and Macdonald (1997), a wealth of nutritional information on wild boar (*Sus scrofa*) and the domestic pig (*S. s. domesticus*) has been of little usage for captive dietary developments for the myriad of other (wild) suid species that occur in zoological collections and may be of conservation concern. Wild data on ecology and behaviour, on diet choice and selection (such as that presented in Figure 8.1) are required to inform decisions on diet presentation and formulation, and on how to deliver environmental enrichment that is nutritionally based and focussed on the promotion of species-specific feeding and foraging behaviour. Just because all three species illustrated in Figure 8.1 look like pigs, if this argument was used to provide a generic diet and enrichment to promote generic rooting behaviour, none of the species would be particularly well catered for in the zoo.

Overall, the history of zoo nutrition showcases a shift in attitudes on how animals living in the zoo are perceived; from providing basic sustenance to embracing a science-based approach that prioritises species-specific dietary requirements based on key elements of species' biology, ecology, and behaviour. The more we can gain empirical information on diet choices and nutritional needs, the more support will be provided to positive animal welfare from a zoo diet and feeding perspective. This, in turn, upholds population sustainability and conservation outcomes as individual animals

Figure 8.1 Assessment of suid skull and snout anatomy and morphology provides valuable information on foraging ecology that should be translated into how zoo diets are formulated and presented. As reviewed in Leus and Macdonald (1997), from left: Babirusa (*Babyrousa* sp.) are the suid species least likely to root with their snouts as they prefer to consume fruits and leaves and will actively stand on their hind legs to reach growing food. Centre: red river hogs (*Potamochoerus porcus*) use the bridge of their snout, rather than the tip, for rooting, indicating that this species is adapted to foraging in the soil of moist forested environments. Right: the wide, strong rhinarium and tusks of warthogs (*Phacochoerus* sp.) enable digging and rooting in hard, dry earth that allows these species to forage in grassland and other arid environments during challenging times of the year. These examples illustrate the specific relationship between the evolution of morphological traits, habitat structure, and feeding and foraging behaviours.

will be healthier, will experience appropriate longevity, and be more likely to produce viable young that will contribute to the subsistence of future ex situ populations (Figure 8.2).

8.3 BEHAVIOURAL ECOLOGY AND NUTRITION

The consumption of food is an essential requirement for all animal species to ensure continued maintenance of homeostasis and, ultimately, survival. Therefore, species may invest a significant proportion of their daily activity in feeding and foraging behaviours, and their behavioural ecology will have evolved to enable the collection of food in the most optimal manner. Behavioural ecology is defined as the evolutionary basis of animal behaviour driven by environmental pressures (University of Cambridge, 2023) and information pertaining to it is crucial for the effective planning, preparation, and presentation of diets to zoo animals. Understanding the natural behaviours (e.g. time spent foraging), ecological roles (e.g. selectivity of specific dietary items), and feeding habits (e.g. seasonality of food choice) of animals in their native habitats provides valuable insights into their nutritional requirements and dietary preferences, as well as evidence for how food should be presented in the zoo (Figure 8.3).

Figure 8.3 illustrates how the foraging behaviour of a green-winged macaw (*Ara chloropterus*) should be evaluated to ensure that a captive individual can experience positive welfare based on how food is provided in the zoo. Research has demonstrated the link between knowledge of foraging ecology, captive husbandry, and indicators of poor welfare across different species of parrot (Mellor et al., 2018, 2021). A macaw's appetitive behaviours concerning foraging manifest as the identification and selection of food items, and the manipulation of such food items with the beak. Providing outlets for food manipulation (e.g. whole nuts and seeds) that keep birds occupied and that promote dexterity and behavioural diversity will enable captive macaws to experience positive welfare states. All wild animals engage in specific behaviours related to foraging (the searching for and location of food), manipulation and processing of food, and chewing and digestion of selected food items. Many species will spend a large proportion of their time engaged in feeding, foraging, and food handling behaviours. If, because of how diet is provided in the zoo, foraging and feeding times are reduced and the proportion of time normally assigned to feeding and foraging is diminished, animals may fill this time with other, abnormal, or undesirable behaviours. A reduction in the amount of time

Health	Welfare	Longevity	Sustainability
• Appropriate physical condition • Enhanced immune function • Normal gut function	• Output for natural behaviour • Opportunities for enrichment • Buffers against stress	• Matches or exceeds that from wild animals • Development and growth as normal	• Supports conservation aims • Improves fecundity and fertility • Enhances offspring survival

Figure 8.2 Examples of key measurable outputs in the welfare of captive wild animals when nutrition, appropriate to the individual and species, is provided.

available for feeding and foraging does not eliminate or reduce the animal's internal motivation for foraging, and therefore an outlet for this motivation, to reach eventual satiation, is still required. The key principles outlined in Figure 8.3 do not just apply to macaws, but to all species housed in the zoo. Table 8.1 illustrates examples across different taxonomic groups of how restricted opportunities for behavioural performance linked to feeding and foraging can result in a welfare compromise.

Investigation of the digestive physiology of wild specimens can yield important information on what food to provide in the zoo to best support health and longevity. For example, comparative research on the morphology of giraffe stomach chambers by Hofmann and Matern (1988) has been integral to our understanding of why giraffes

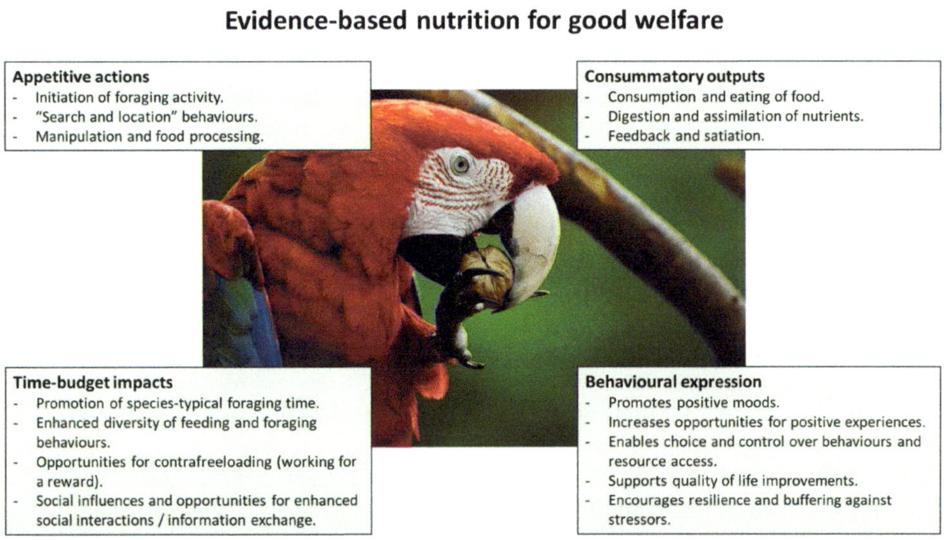

Figure 8.3 Evidence for species-specific nutrition in the zoo comes from our understanding of appetitive (searching or locomotory actions) and consummatory (end goal actions that provide satiation) behaviours, as well as the amount of time a species spends on feeding and foraging (key optimal foraging strategies) and the expression of behavioural actions that are indicative of positive welfare.

are challenging to feed in captivity. This research has shown that the surface area of the rumen and reticulum in wild giraffes is densely covered with long papillae that increase the surface area available for absorption and assimilation of volatile fatty acids (the by-products of microbial fermentation). These authors documented that when the same investigation occurred in captive animals, the papillae were dramatically atrophied or missing entirely, thus restricting the giraffe's abilities to extract energy and nutrients from their diet. Hofmann and Matern (1988) concluded that a lack of leafy forage intake and use of concentrate (pellet) rations for captive giraffes has fundamentally changed the structure of the giraffe's stomach chambers and this has resulted in knock-on effects on the ability to meet energetic demand, homeostatic need, and therefore lifespan. Thanks to comparative wild-captive studies such as this, we know that giraffes must be fed lucerne (legume) hay, have a measured amount of concentrate ration according to body weight and age, and have daily access to cut tree branches to increase structural fibre intake (EAZA, 2006; Hummel et al., 2006; Valdes & Schlegel, 2014).

Free-living animals may consume a wide variety of food items to meet their nutritional needs. Brown bears (*Ursus arctos*), for example, have a widespread distribution and are successful in multiple different habitats because of their extraordinarily diverse diet. Conversely, some species are very selective in how they feed, consuming food items within a very restricted foraging niche, for example, selection of *Eucalyptus sp.* leaves by koalas (*Phascolarctos cinereus*). Knowledge of the degree of specialisation of a species' foraging ecology is essential to supporting long-term good welfare in the zoo. Species that are facultative feeders, such as the aforementioned brown bear, will be less likely to suffer from nutritional deficiencies as they are not tied to the consumption of a specific foodstuff containing a particular nutrient. The dog family (Canidae) is an example of facultative carnivores. Meat is preferred but not essential to sustain life. Whereas obligate carnivores (e.g. members of the cat family (Felidae) and their need for the essential amino acid taurine that cannot be synthesised in the body and is derived from animal protein) will suffer fatal consequences if this is not provided to them in an accessible manner in their diet. Zoo diets can only be nutritionally beneficial (for excellent health) and behaviourally relevant (to promote welfare) if they are grounded in evidence from a species' evolution, anatomy, and physiology. An ability to consume and digest many different food types, for example, might make dietary preparation easier but can lead to obesity in captivity, and therefore body condition should be regularly assessed.

Behavioural ecology information helps zoos select the most appropriate restricted or diverse range of foods that provide the necessary nutrients, vitamins, and minerals that a species requires for optimal health. Providing appropriate diets reduces distress because performance of feeding behaviours can be closely linked to psychological wellbeing as internal motivation to forage is satiated. Replicating natural feeding behaviours is an essential tool for zoo environmental enrichment programmes. An animal's physical heath also benefits by the presence of an appropriate diet. Not only because the diet is suitable for the animal's nutritional and energetic needs, but also because lethargy and boredom, which are potential predictors of obesity, are reduced as animals engage in complex and diverse behavioural repertoires related to foraging and food collection. Therefore, zoo animals are less likely to become overweight and suffer from associated health issues. The more opportunities that zoos can provide for a captive species to actively forage and manipulate food, dependent on dietary niche of course, the more positive welfare benefits will accrue (Figure 8.4).

Dietary palatability is especially important but can be overlooked. A zoo-formulated diet may appear correct on paper, but if this diet is not accepted by the animals it is provided to, and said animals do not eat it, then there are no health and welfare benefits. It is important that the work that goes into nutritional analysis of dietary ingredients, to then formulate a complete diet, is not derailed by a diet being produced in a manner that is unpalatable. Otherwise, all the science and innovation that goes into nutritional analysis will be wasted. As indicated in research on a diverse array of species – from prototype complete diets for aardvarks, *Orycteropus afer* (Davies et al., 2019), to assessing preferences of cut tree branches (and their leaves) for black-and-white colobus, *Colobus guereza kikuyuensis* (Kirschner et al., 1999), and to total diet ingestion by captive giraffes that demonstrated the likelihood of increased dietary consumption with

Table 8.1 Examples of behavioural challenges caused by restricted foraging opportunity in zoo-housed species.

Species	Nutritional challenge/behavioural restriction	Welfare implications
Lowland gorilla (*Gorilla gorilla*)	Limited ingestion of leafy greens and structural fibre results in regurgitation and reingestion behaviours.	Performance of abnormal repetitive behaviour. Damage to teeth, and upper gastrointestinal tract due to regurgitation of stomach acid (Hill, 2018).
Giraffe (*Giraffa camelopardalis*)	Limited browsing opportunities on tree branches and feeding of grass hay in place of legume (e.g. lucerne) limited opportunities for rumination.	Rumen acidosis, oral stereotypic behaviours. Stratification of rumen contents leading to energy deficits and potential reduce longevity (Clauss et al., 2002).
Amazon parrot (*Amazona sp.*)	Restricted opportunities for performance of foraging and food manipulation behaviours increase chances of abnormal repetitive feather picking/plucking.	Increasing environmental complexity, including naturalistic foraging opportunities reduces prevalence and severity of feather plucking (Meehan et al., 2003)
Lesser flamingo (*Phoeniconaias minor*)	Feeding captive flamingos on a specific pellet to maintain their characteristic pink plumage is essential. However, such pellet is a high value food, fed in a limited number of areas in an enclosure. Wild flamingos feed over wide open spaces and maintain individual distances between birds to reduce competition.	Feeding on pellet in bowls or troughs increases aggressive encounters and reduces time spent foraging. Increased aggression between birds may cause undue stress, impact on ingestion of carotenoid pigments and reduce a bird's chances of collecting sufficient food (Rose & Soole, 2020).
Komodo dragon (*Varanus komodoensis*)	Lack of challenge (e.g. stimulation of stalking, hunting and dispatch behaviours) due to unimaginative in-zoo feeding methods and diet presentation. Restricted opportunity for activity within the overall three-dimensional space of the dragon's enclosure.	Reduced muscle tone and increased inactivity (resulting in poorer body condition and physical health) due to reduced behavioural diversity. Nutritional enrichment and creativity in food (prey) presentation benefits health and wellbeing (Waterman et al., 2021).
Perhaps for a majority of amphibian, fish, and invertebrate species in captivity…	Limited understanding of behaviour patterns and motivational needs to forage and collect food.	Reduced choice and control within the environment. Lack of knowledge of the relevance of dietary substitutes to supporting long term health and behaviour. A need for further research application to understand the importance of nutrition to welfare and behaviour patterns.

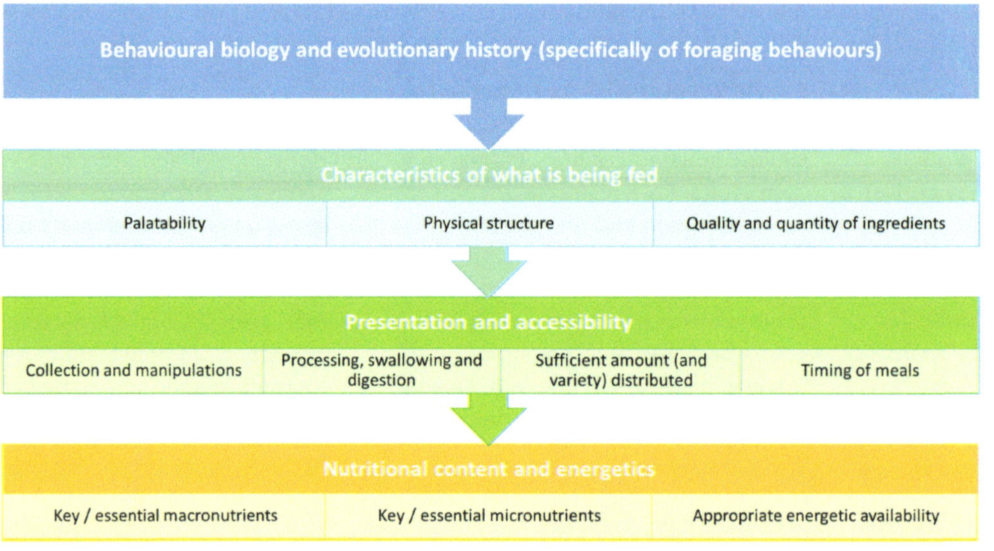

Figure 8.4 A flowchart of key considerations when formulating the most appropriate diet for specific species of non-domestic animals housed in the zoo. Behavioural biology information is fundamental, as the way in which a species has evolved to forage and consume food will form the basis for all aspects of husbandry and care that include nutrition. The presentation of the animals' diet and its accessibility to the individual includes key facets of behavioural biology such as food storage and caching behaviours, stalking, hunting and dispatch of prey, use of tools, and experiences with a specific food item that may dictate how an animal responds to it. The whole process of positive welfare via nutrition is then supported by the actual nutritional content of the animal's diet, which must be correct for each species in the zoo.

improved overall palatability (Clauss et al., 2001) – all evidence the need to evaluate what food looks like, tastes like and feels like when zoo diets are designed. The sensory stimuli provided by a diet are important. More research should be conducted on what feed looks, smells, and tastes like (to each zoo-housed species that it has been designed for) to ensure appropriate updates that will benefit health and welfare in the long term.

Ultimately, use of behavioural ecology knowledge to support zoo diets and by association good welfare can be summarised as:

- Diet Composition. Understanding the natural diet of a species provides insights into the types of foods and nutrients they require. Zoos must use this information to select appropriate food items and formulate balanced diets that closely resemble the animals' wild diets and their nutritional needs across life stages.
- Feeding Mechanisms. Knowledge of species' feeding mechanisms, such as hunting strategies or digestive processes, informs how food should be presented. For example, species that use tools to extract insects from tree bark will benefit from food items hidden within objects in their enclosure to promote (and ultimately conserve) these behavioural actions.
- Feeding Times. Observing when wild animals are most active in foraging will guide the timing of feeding in zoos. Aligning feeding times with natural time-activity budgets supports a species' circadian rhythms and enhances overall welfare by aligning when species wish to be active to when food is available to them.
- Nutritional Requirements. Anatomy and physiology influence what a species will select in its overall diet. Herbivores need diets rich in fibre to support their complex digestive systems, carnivores require diets high in quality protein, and rodents require opportunities for chewing and gnawing to keep their teeth in excellent condition.
- Growth and Development. Understanding growth rates and developmental stages for specific species helps design diets that cater to an individual's changing nutritional needs for all aspects of life.

- Dental Structure. A species' dentition provides insights into their natural diets. Specific tooth shapes, sizes, and growth rates align to the processing, mastication, and digestion of particular types of food.
- Specialised morphological and anatomical adaptations. For example, bill shape and size, hand structure for gripping, or eye position for judging distance during hunting all provide valuable clues on what must be provided in zoo diets as well as how such diets should be presented.

It is essential to realise the niche specificity that has driven the evolution of both anatomy and morphology (to equip an animal to forage and feed successfully) and of the behaviour patterns that enact specific feeding and foraging strategies. Even for apparent "simplistic" species, such as the sea anemone illustrated in Figure 8.5, complex interactions between an individual of the species and their environment are essential for the attainment of nutrition and energy. Sea anemones, in captivity, that are not housed in a replicated marine environment (e.g. with biologically relevant heating, lighting, water chemistry, and water movement) will lose the symbiotic Zooxanthellae within their tentacles and start to bleach white, eventually succumbing to starvation and dying. This is a welfare problem in the same way that the other species in Figure 8.5 will experience if they, too cannot have key aspects of their foraging strategy and specific nutritional needs met in a captive environment. The reindeer without appropriate forage to stimulate ruminal microorganism activity and thus loses body condition; the scarlet ibis without carotenoid pigments to enhance plumage colour therefore turns a pale version of a wild appearance and can suffer from impeded mate choice and captive breeding outputs, and the copperband butterfly that cannot collect food of a specific size in substantial quantities and consequently fares poorly and ultimately experiences a drastically shortened lifespan, all suffer from poor health and reduced welfare and their place in the zoo's collection loses relevance and opens the institution itself up to criticism and accusation of poor animal care.

8.4 NUTRITION AS A PART OF DAILY, CORRECT, ANIMAL HUSBANDRY

It sounds obvious to suggest that the feeding of the zoo's animals are an essential part of day-to-day animal care. Yet, it is important to remember that providing the wrong type of food, or providing food in an incorrect way, or giving an inappropriate diet is not good care and does not support species-specific husbandry and welfare needs (Young, 1997). Wrong types of feed can be dietary ingredients that are readily consumed by the animal yet do little to promote health and behavioural outcomes and can be considered as "empty calories". Feeding bread to ducks, for example. Providing food in an incorrect manner can include providing discrete meal-based timings of food to species that have evolved to "trickle feed" and therefore forage on small amounts of food throughout the day. Many ungulates and coral reef fish forage in this "little but often" style. An inappropriate diet may be one based on outdated or now defunct information, and consequently, best practice nutritional advice is not being adhered to. For example, the feeding of browsing ungulates on grass hay rather than on legume forage. The tapir (*Tapirus* sp.) is fed a large ration of cultivated fruits and vegetables that cause explosive fermentation in the hindgut, the giraffe that exists on a concentrate pellet diet plus grass hay which ultimately blocks its rumen and causes an energy deficit, or the tiger (*Panthera tigris*) that is fed purely on skeletal muscle and experiences a nutritional imbalance due to a lack of vitamins and minerals that would be provided by whole-carcass feeding, are all being fed. But they are not being fed correctly or in a species-appropriate way. All zoos, and all zoo personnel, in fact, every stakeholder in a species' future in ex situ populations, needs to consider nutritional advice and scientific evidence, analyse, and evaluate diets accordingly, and review such diets regularly to prevent "keeper drift" (e.g. not following prescribed best practice over the longer term) to ensure diets remain supportive of good animal health and welfare. For example, keeper drift in measurement and presentation of captive giraffe diets at a large European zoo affected rates of abnormal behaviour performance

Figure 8.5 Examples of how behavioural ecology underpins correct nutrition (and therefore positive welfare and good animal health) in the zoo. Top left; the copperband butterflyfish (*Chelmon rostratus*) has a distinctive elongated snout with a small mouth evolved to extract invertebrates from crevices in the coral reef. Top right; the bright red plumage colour of the scarlet ibis (*Eudocimus ruber*) is produced by carotenoid pigments (notably astaxanthin) metabolised from the crustacean component of their diet. Bottom left; reindeer (*Rangifer tarandus*) have specialised microorganisms in their rumen to digest lichen, which forms a large proportion of their wild diet. Bottom right; sea anemones, such as this bubble-tip anemone (*Entacmaea quadricolor*), host symbiotic dinoflagellates, called Zooxanthellae, in their tentacles that, via photosynthesis, provide the anemone with energy.

as well as dietary intake and amount of leftovers (Cox et al., 2019). Without oversight from curators and animal managers, such dietary drift can have important and negative implications for zoo animal health and well-being.

Zoo animal husbandry and management encompass a complex array of practices aimed at ensuring the health and overall well-being of animals under human care. Among the multifaceted aspects of this endeavour, providing species-specific, nutritionally optimal diets stands out as a cornerstone of responsible and effective zoo management. This chapter has examined why offering appropriate diets tailored to the specific nutritional needs of each species is essential in promoting optimum health, the performance of natural, biologically relevant behaviours, and successful conservation efforts within zoological institutions. Species-specific diets enable zoo animals to express natural behaviours related to foraging, feeding, and food manipulation. Therefore, the provision of correct nutrition is, in itself, enriching. Nutritional enrichment, categorised as a specific form of enrichment by Bloomsmith et al. (1991), can come from the everyday feeding of the staple diet of an animal, if this diet is provided in a way that stimulates important food-seeking and manipulation behaviours. Offering foods that mimic a wild diet encourages animals to engage in activities that challenge their cognitive and physical abilities and thus provide outlets for appetitive actions that lead to consummatory outputs and satiation. This approach prevents boredom and reduces apathy and inactivity by promoting mental stimulation (enhances cognitive abilities) and physical exercise (that maintains muscle tone, supports normal body condition, and protects against obesity).

Different animal species have evolved to depend on specific nutrients derived from their natural diets. Providing an incorrect diet can lead to nutritional deficiencies that compromise immune function, reproductive success, and overall vitality (Hadfield et al., 2006). For example, vitamin deficiencies can result in disorders that range from internal, systemic impacts (e.g. compromised immune responses to poor bone health) to outwards changes in appearance (e.g. pelage, plumage or integument colour) (Donoghue, 2006; Radwinska & Zarczynska, 2014; van Zeeland & Schoemaker, 2014), Figure 8.6. Correct nutrition is a crucial factor in ensuring reproductive success within zoo populations. Female mammals require increased energy and nutrients for both maintenance of their own homeostasis plus the maintenance of a successful pregnancy, lactation, and nurturing of offspring. Female birds, amphibians and reptiles will change their diets leading up to egg laying to ensure that key elements (e.g. calcium) are ingested to support production of viable eggs. If the zoo is unaware of these needs, then inadequate nutrition will lead to reproductive failures, low birth rates, and compromised neonatal health, thereby hindering the genetic diversity and viability of captive populations.

Husbandry that provides species-specific diets contributes to reduced distress and disease susceptibility. When an animal's nutritional needs are met, its immune system is better equipped to combat pathogens and cope with the physiological impacts of stressors. Conversely, incorrect or inappropriate diets can weaken immune responses and increase susceptibility to poor health or infection (Venesky et al., 2012). Any outbreaks of disease within captive population that may have fully or partly resulted in poor health could hamper conservation and negatively impact population sustainability goals for that species. Providing the correct diet in the long term makes economic as well as good welfare sense. Initial outlay on dietary analyses and financial investments into food preparation and diet formulation keeps animal health optimal and reduces fees and costs associated with veterinary treatment of the effects of poor diets.

Ultimately, zoos have an ethical responsibility to provide the best possible care for the animals under their stewardship. Offering species-specific diets aligns with public expectations of responsible animal management. Proper diets contribute to positive public perceptions, fostering support for conservation and education efforts. If well-being is a key goal of the modern zoo, as many believe it should (Rose & Riley, 2022), then investment in the procurement and provision of correct diets becomes more important. Animal care cannot be caring if diets are incorrect, and species are suffering because of ignorance regarding their nutritional needs. Similarly, husbandry cannot advance if zoos are aware of a species' nutritional needs but still attempt to manage a species using a makeshift diet when the actual nutritional needs of said species are impossible to meet. This approach devalues the conservation, education, research, and well-being aims of the zoo and compromises the welfare of individual animals.

The role of correct animal nutrition information in the development and upholding of species-specific husbandry that meets the core aims of the modern zoo can be summarised as follows:

- Physical health and mental well-being. Proper nutrition is essential for maintaining the health and overall positive welfare states experienced by zoo animals. Animals in captivity require a diet that provides all the necessary nutrients, vitamins, and minerals to support their physiological needs. Poor nutrition leads to health issues, including malnutrition, obesity, and compromised immunity, which devalue a species' place in the zoo's living collection.
- Species-specific requirements. Different species have varying nutritional needs based on their evolutionary history, ecological niche, and physiological adaptations. Zoo nutritionists must continue to work to replicate the diet these animals would consume if they were wild as closely as possible. Providing species-appropriate nutrition helps ensure that animals receive the correct balance of nutrients, which is essential for their long-term health and productivity.
- Removing use of "best of a bad option". Some diets can be hard to replicate in zoos, and therefore it has been acceptable to use poor substitutes, e.g. the use of grass hays for browsing ruminants, even though such hay wears down teeth, is unpalatable, and is poorly digested (Clauss & Dierenfeld, 2008). Implementing research findings and continued efforts to collect evidence for husbandry (and the dissemination of such knowledge) will help to remove poor animal welfare practices by explaining to all caregivers why such practices do not support health and welfare (even if they appear cheap and easy to implement).

Figure 8.6 An example of the challenges and potential solutions to correct zoo animal nutrition and promote behaviour and welfare. From left: a male scarlet-chested sunbird (*Chalcomitra senegalensis*) in a captive environment displaying excellent condition and a bright red chest. Middle: another male scarlet-chested sunbird presenting with a yellow chest, an external sign of an incorrect diet and poor health. Right: provisioning captive sunbirds with multiple nectar feeders (in this instance being used by a female bird), plus the use of live insect foods, access to UVA and B light spectrums, supplementary heating, and bathing opportunities, reduces stress, improves body condition, and restores plumage colour. The challenge of adequate nutrition for captive nectivorous birds, with a highly specialised feeding strategy, is further detailed in Peaker (1990).

- Behavioural enrichment. All zoo animals require mental and physical stimulation to prevent boredom, lethargy, and stress in captivity. A part of species-appropriate nutrition involves providing food in ways that encourage natural foraging behaviours, problem-solving, and physical activity. Embedding nutritional enrichment holistically with daily husbandry and feeding promotes positive behavioural diversity.
- Prevention is better than cure. Improper nutrition leads to a range of health problems that are preventable, e.g. dental disease, metabolic disorders, and digestive imbalance. By carefully planning and monitoring diets, zookeepers can help prevent these issues and ensure that the animals under their care do not need to undergo veterinary interventions caused by incorrect nutrition.
- Reproduction and conservation breeding. Successful breeding is crucial for species conservation efforts, whether captive breeding aims for continued sustainability in the zoo or conservation breeding for eventual release into the wild. Proper nutrition plays a pivotal role in ensuring that animals are in optimal reproductive condition. Nutritionists must work with animal care staff to develop specific diets that support reproduction and the health of parents and offspring to ensure conservation goals are viable.
- Longevity. Zoo animals live longer than their wild counterparts, sometimes markedly so (Föllmi et al., 2007). Correct nutrition will significantly influence the longevity of zoo animals and an individual's likelihood of achieving maximal lifespan. Poor husbandry and incorrect nutrition are known to cause reduced lifespan, compared to the wild in some species, e.g. the giraffe (Potter & Clauss, 2005). Efforts to rectify any lack of understanding of nutritional needs and provide information on procurement of specific ingredients for zoo diets are required to support animals across all stages of their lives.
- Research. Zoos engage in scientific research that provides vital evidence to the evolution of housing, husbandry, and management. Research can only be valid if animals are healthy and behaving normally. Animals can only be healthy and behave normally if their diets have been researched and prepared correctly. Zoo scientists and animal care staff need to collaborate, across institutions, to ensure that the husbandry inputs support research outputs and vice versa.
- Education and awareness. Zoos serve as educational institutions; they play a vital role in raising awareness about wildlife and conservation actions and illustrate the sheer variety of life on Earth. Presenting accurate information on diets to visitors, and explaining how animals

- Is the total, current diet based on reliable, empirical evidence?
- How much food is being fed out and when is this food being provided?
- Is the feeding schedule in line with the species' time-activity pattern and circadian rhythm?
- Has the diet been nutritionally analysed? E.g., using a bespoke software, such as "Zootrition"?
- Is each animal's body condition and faecal consistency regularly scored?
- Is the mass of each animal weighed or estimated regularly?
- Is the diet provided in an enriching and biologically relevant manner?
- Can all dietary ingredients be weighed and provided on an individual animal basis?
- Are any behavioural disturbances noted before, during or after feeding?
- Would behavioural disturbances be alleviated with a change in diet presentation or feeding schedule?
- Do all animal care staff understand the species' dietary regime (of what, when and how to feed)?
- Is training provided to reduce "keeper drift" from best practice approaches to a species' in-zoo nutrition?
- Is nutritional husbandry regularly reviewed and updated and is research used to fill gaps in knowledge?
- Is the animals' diet explained accurately to zoo visitors?

Figure 8.7 A checklist for evaluation and review of dietary suitability that could be utilised by animal care staff to ensure that diets for all species in their organisation are fit for the purpose of upholding positive welfare.

are fed, and why they are fed in a specific way, promotes these educational and public awareness aims of the modern zoo.

- Zoo animal nutrition is an essential aspect of the daily, correct-for-that-species, husbandry in all zoos. Providing species-appropriate nutrition is fundamental for maintaining the health and well-being of zoo animals, promoting natural behaviours, supporting breeding efforts, and supporting public education by the presentation of a healthy, behaviourally normal living collection of animals. Animal care staff could employ a checklist, such as that detailed in Figure 8.7, to evaluate and examine the diets of all species in their care and thus make informed improvements where needed.

8.5 NUTRITION WORKING GROUPS AND PROMOTING ANIMAL WELFARE

Many of the large zoo organisations have specific nutrition committees or working groups that aim to promote the importance of correct diet to health and wellbeing, support zoos with nutritional analysis of diets, and provide expertise and advice on diet formulation and presentation. As an example, the European Association of Zoos and Aquaria (EAZA) Nutrition Group (ENG) is a specialised working group that focuses specifically on the science, practice, and advancement of animal nutrition for EAZA member zoos and aquariums (EAZA, 2023a). The ENG's primary goal is to promote the health of captive animals through the development of appropriate and species-specific diets. The group fosters research on zoo nutrition by facilitating collaborations amongst stakeholders across, and beyond, the EAZA community. By promoting the exchange of knowledge through seminars, workshops, and conferences (including the long-running European Zoo Nutrition Conferences), and publications (such as nutrition-focused special editions of EAZA's "Zooquaria" magazine), the ENG attempts to make the latest scientific findings in nutrition accessible to those that need to use them (EAZA, 2023a). Encouraging participation in continuing professional development activities helps develop skills in nutritional analysis to promote good practice across zoos, thus improving care and welfare for the maximum number of animals.

As EAZA moves away from husbandry manuals to best practice guidelines for the species within its member zoos (EAZA, 2023b), evidence for captive diets is essential. Therefore, the ENG develops and updates nutritional guidelines for different zoo-housed species. These guidelines provide crucial information to zoos for the planning and preparation of diets that meet a species' nutritional requirements in a manner that is palatable and accessible.

Work by researchers affiliated with or members of the ENG has been key to changes in attitudes towards zoo animal care. For example, obligate browsing species such as the giraffe used to be provided with cut tree branches ("browse") as a luxury (i.e. an optional extra). Work has shown the importance of browse to the functioning of the giraffe's gastrointestinal system and its digestive physiology (Clauss et al., 2002; Hatt et al., 2005) and now browse is not seen as optional, but as an essential element of optimal giraffe husbandry and care (Rose, 2023) that promotes multiple behavioural benefits (Newnham & Rose, 2025).

The ENG assists member institutions in formulating appropriate diets for the animals in their care. This involves analysing the nutritional composition of dietary ingredients, ensuring that they meet the specific requirements of different species, as well as complete diets to provide as accurate as possible information on dietary suitability and quality. Building relationships with commercial animal nutrition companies has catalysed the development of specialised diets and nutritional supplements that meet the unique nutrient requirements of non-domestic species. However, the development and use of such commercial diets, in the form of concentrate or pelleted rations, has resulted in further work for these zoo-focused nutrition groups to suggest ways of presenting and feeding these pelleted diets in a manner that is interesting and stimulating to the animal.

EAZA member institutions have access to a network of experts in zoo animal nutrition, allowing them to seek guidance on creating appropriate diets for their animals. This expertise is especially valuable when dealing with species whose dietary needs in captivity can be challenging, and therefore the sharing of good practice and innovations helps advance husbandry. By synthesising scientific research and translating it into practical dietary recommendations, the ENG makes research outputs more accessible and applicable to those working in zoos and, by association, to the animals themselves.

Several books have been published after European Zoo Nutrition Conferences (Clauss et al., 2009; Fidgett et al., 2003, 2006; Nijboer et al., 2000) and these publications are valuable pieces of evidence that others can use when seeking empirical and reliable information on the nutritional requirements of captive wild animals. Such outputs transcend one zoo organisation (e.g. in this case EAZA) by making scientific evidence and research results accessible to all those attempting to develop and improve the diets for their zoo-housed species. Therefore, the outputs of one zoo organisation's nutrition group bolster and uphold positive welfare for many different zoo animals around the globe.

8.6 CONCLUSIONS

The correct feeding of zoo-housed species needs to be based on scientific evidence, but the process itself is more than just a scientific one. Numerous stakeholders are involved and there are practical and logistical challenges to be overcome to ensure that captive diets can be the best substitutes possible for what is selected and consumed by free-living individuals of a species. However, individual animals (and their needs) will differ. Zoos need to work alongside available best practice guidelines to support welfare-focused husbandry that centres around nutrition and diet formulation. The knowledge and experience of animal care staff who know their individual charges, as well as inputs from veterinary surgeons, animal nutritionists and curatorial staff, should be layered on top of species-wide best practice information to ensure that such correct diets are relevant, accessible and palatable for an individual animal at a specific life stage (e.g. elderly and infirm animals) or when in a specific type of care regime (e.g. during quarantine, medical treatment plan). Continual improvement and review of welfare assessment protocols, a focus on biologically relevant inputs (housing, husbandry, social group, diet) to improve welfare positive outputs, will create a holistic approach to the evaluation and review of animal welfare states. Dietary analysis (a combination of laboratory protocols and via specialised computer software) to ascertain the accuracy of what is being fed supports the delivery of these welfare-focussed inputs and provides behavioural, psychological, and physical evidence of dietary correctness. Sharing such information, working collaboratively across zoological institutions, and remaining current regarding the latest nutritional scientific outputs, technologies and approaches ultimately is the best way for zoo personnel to be as confident as possible in the relevance of the captive diets they are providing to a species' physical, psychological, and behavioural health.

REFERENCES

Bloomsmith, M. A., Brent, L. Y. & Schapiro, S. J. (1991). Guidelines for developing and managing an environmental enrichment program for nonhuman primates. *Laboratory Animal Science*, *41*(4), 372–377.

Broom, D. M. (1986). Indicators of poor welfare. *British Veterinary Journal*, *142*(6), 524–526.

Clauss, M. & Dierenfeld, E. S. (2008). The nutrition of" browsers". In M. E. Fowler & R. E. Miller (Eds.), *Zoo and wild animal medicine: Current therapy* (Vol. 6, pp. 444–454). St Louis, MO: Elsevier.

Clauss, M., Fidgett, A., Janssens, G., Hatt, J.-M., Huisman, T., Hummel, J., Nijboer, J. & Plowman, A. (Eds.). (2009). *Zoo animal nutrition IV.* Fürth, Germany: Filander Verlag.

Clauss, M., Lechner-Doll, M., Flach, E. J., Wisser, J. & Hatt, J.-M. (2002). Digestive tract pathology of captive giraffe. A unifying hypothesis. *Proceedings of the European Association of Zoo and Wildlife Veterinarians*, *4*, 99–107.

Clauss, M., Lechner-Doll, M., Flach, E. J., Tack, C. & Hatt, J. M. (2001). Comparative use of four different marker systems for the estimation of digestibility and low food intake in a group of captive giraffes (Giraffa camelopardalis). *Zoo Biology*, *20*(4), 315–329.

Cox, L., O'Regan, S., Harrol, D. & Plowman, A. (2019). *An investigation into the effects of dietary alterations on the behaviour and daily food intake of giraffes at Paignton Zoo Environmental Park.* Paper presented at the 10th European Zoo Nutrition Conference, Marwell Zoo, Winchester, UK.

Crissey, S. D. (2001). The history of zoo nutrition. In C. Bell (Ed.), *Encyclopedia of the world's zoos* (pp. 45–50). Chicago, IL: Fitzroy Dearborn Publishers.

Davies, H., Bond, J. & Hunt, K. (2019). *Nutritional analysis on captive aardvarks and separation solution of Termant using emulsifiers.* Paper presented at the 10th European Zoo Nutrition Conference, Marwell Zoo, Winchester, UK.

DeMello, M. (2021). *Animals and society: An introduction to human-animal studies.* New York, NY: Columbia University Press.

Dierenfeld, E. S. (1997). Captive wild animal nutrition: A historical perspective. *Proceedings of the Nutrition Society*, *56*(3), 989–999.

Donoghue, S. (2006). Nutrition. In D. R. Mader (Ed.), *Reptile medicine and surgery* (2nd ed., pp. 251–298). Saint Louis, MO: Saunders.

EAZA. (2006). *EAZA husbandry and management guidelines for Giraffa camelopardalis.* Burger's Zoo, Arnhem, The Netherlands: EAZA Giraffe EEP.

EAZA. (2023a). EAZA Nutrition Group. European Association of Zoos and Aquaria. https://www.eaza.net/about-us/areas-of-activity/eaza-nutrition-group/.

EAZA. (2023b). Programmes. European Association of Zoos and Aquaria. https://www.eaza.net/conservation/programmes/. (Access Date 22/05/2023).

Egerton, S., Culloty, S., Whooley, J., Stanton, C. & Ross, R. P. (2018). The gut microbiota of marine fish. *Frontiers in Microbiology*, *9*, 873.

Fidgett, A., Clauss, M., Eulenberger, K., Hatt, J.-M., Hume, I., Janssens, G. & Nigboer, J. (Eds.). (2006). *Zoo animal nutrition III.* Fürth, Germany: Filander Verlag.

Fidgett, A., Clauss, M., Gansloßer, U., Hatt, J.-M. & J., N. (Eds.). (2003). *Zoo animal nutrition II.* Fürth, Germany: Filander Verlag.

Föllmi, J., Steiger, A., Walzer, C., Robert, N., Geissbühler, U., Doherr, M. G. & Wenker, C. (2007). A scoring system to evaluate physical condition and quality of life in geriatric zoo mammals. *Animal Welfare*, *16*(3), 309–318.

Frankel, J. S., Mallott, E. K., Hopper, L. M., Ross, S. R. & Amato, K. R. (2019). The effect of captivity on the primate gut microbiome varies with host dietary niche. *American Journal of Primatology*, *81*(12), e23061.

Gray, R. F. (1960). Hippopotamus milk analysis. *International Zoo Yearbook*, *1*(1), 46.

Hadfield, C. A., Clayton, L. A. & Barnett, S. L. (2006). Nutritional support of amphibians. *Journal of Exotic Pet Medicine*, *15*(4), 255–263.

Hatt, J. M., Schaub, D., Wanner, M., Wettstein, H. R., Flach, E. J., Tack, C., Hässig, M., Ortmann, S., Hummel, J. & Clauss, M. (2005). Energy and fibre intake in a group of captive giraffe

(Giraffa camelopardalis) offered increasing amounts of browse. *Journal of Veterinary Medicine Series A*, *52*(10), 485–490.
Hill, S. P. (2018). 'Regurgitation and Reingestion' (R/R) in great apes: A review of current knowledge. *International Zoo Yearbook*, *52*(1), 62–78.
Hofmann, R. R. & Matern, B. (1988). Changes in gastrointestinal morphology related to nutrition in giraffes Giraffa camelopardalis: A comparison of wild and zoo specimens. *International Zoo Yearbook*, *27*(1), 168–176.
Hummel, J., Zimmermann, W., Langenhorst, T., Schleussner, G., Damen, M. & Clauss, M. (2006). *Giraffe husbandry and feeding practices in Europe. Results of an EEP survey.* Paper presented at the 6th Congress of the European Association of Zoo and Wildlife Veterinarians, Budapest, Hungary.
Keady, M. M., Prado, N., Lim, H. C., Brown, J., Paris, S. & Muletz-Wolz, C. R. (2021). Clinical health issues, reproductive hormones, and metabolic hormones associated with gut microbiome structure in African and Asian elephants. *Animal Microbiome*, *3*(1), 1–19.
Kirschner, A. C., Putnam, L. A., Calvin, A. A. & Irlbeck, N. A. (1999). *Browse species preference and palatability of Colobus guereza kukuyuensis at the Denver Zoological Gardens.* Paper presented at the Proceedings of the Nutritional Advisory Group Conference.
Leus, K. & Macdonald, A. A. (1997). From babirusa (Babyrousa babyrussa) to domestic pig: The nutrition of swine. *Proceedings of the Nutrition Society*, *56*(3), 1001–1012.
McDonald, P., Edwards, R. A., Greenhalgh, J. F. D., Morgan, C. A., Sinclair, L. A. & Wilkinson, R. G. (2022). *Animal nutrition* (8th ed.). Harlow, UK: Pearson.
Meehan, C. L., Millam, J. R. & Mench, J. A. (2003). Foraging opportunity and increased physical complexity both prevent and reduce psychogenic feather picking by young Amazon parrots. *Applied Animal Behaviour Science*, *80*(1), 71–85.
Mellor, E., Brilot, B. & Collins, S. (2018). Abnormal repetitive behaviours in captive birds: A Tinbergian review. *Applied Animal Behaviour Science*, *198*, 109–120.
Mellor, E. L., McDonald Kinkaid, H. K., Mendl, M. T., Cuthill, I. C., van Zeeland, Y. R. A. & Mason, G. J. (2021). Nature calls: Intelligence and natural foraging style predict poor welfare in captive parrots. *Proceedings of the Royal Society B: Biological Sciences*, *288*(1960), 20211952.
National Zoological Park. (1960). Horsemeat substitute at the National Zoological Park, Washington. *International Zoo Yearbook*, *1*(1), 44.
Newnham, Z. & Rose, P. (2025). Quantifying the influence of cut tree branches on zoo giraffe behaviour during and post browsing: A case study on five animals and four tree species. *Journal of Veterinary Behavior*, *80*, 10–19. https://doi.org/10.1016/j.jveb.2025.05.005
Nijboer, J., Hatt, J.-M., Kaumanns, W., Beijnen, A. & Gansloßer, U. (Eds.). (2000). *Zoo animal nutrition I*. Fürth, Germany: Filander Verlag.
Peaker, M. (1990). Nutritional requirements and diets for hummingbirds and sunbirds. *International Zoo Yearbook*, *29*(1), 109–118.
Perry, T., West, E., Eisenhofer, R., Stenhouse, A., Wilson, I., Laming, B., Rismiller, P., Shaw, M. & Grützner, F. (2022). Characterising the gut microbiomes in wild and captive short-beaked echidnas reveals diet-associated changes. *Frontiers in Microbiology*, *13*, 687115.
Potter, J. S. & Clauss, M. (2005). Mortality of captive giraffe (Giraffa camelopardalis) associated with serous fat atrophy: a review of five cases at Auckland Zoo. *Journal of Zoo and Wildlife Medicine*, *36*(2), 301–307.
Rabb, G. B. (2004). The evolution of zoos from menageries to centers of conservation and caring. *Curator: The Museum Journal*, *47*(3), 237–246.
Radwinska, J. & Zarczynska, K. (2014). Effects of mineral deficiency on the health of young ruminants. *Journal of Elementology*, *19*(3), 915–928.

Rose, P. E. (2023). Identifying essential elements of good giraffe welfare - can we use knowledge of a species' fundamental needs to develop welfare-focussed husbandry? *Journal of Zoological and Botanical Gardens, 4*(3), 549–566.

Rose, P. E. & Riley, L. M. (2022). Expanding the role of the future zoo: Wellbeing should become the fifth aim for modern zoos. *Frontiers in Psychology, 13, 6649.*

Rose, P. E. & Soole, L. (2020). What influences aggression and foraging activity in social birds? Measuring individual, group and environmental characteristics. *Ethology, 126*(9), 900–913.

University of Cambridge. (2023). Behavioural Ecology. University of Cambridge. https://www.zoo.cam.ac.uk/research/behavioural-ecology. (Access Date 22/08/2023).

Valdes, E. V. & Schlegel, M. (2014). Advances in giraffe nutrition. In E. R. Miller & M. E. Fowler (Eds.), *Fowler's zoo and wild animal medicine* (Vol. 7, pp. 612–618). St Loius, MO: Elsevier Health Sciences.

van Zeeland, Y. R. & Schoemaker, N. J. (2014). Plumage disorders in psittacine birds-part 1: Feather abnormalities. *European Journal of Companion Animal Practice, 24*(1), 34–47.

Venesky, M. D., Wilcoxen, T. E., Rensel, M. A., Rollins-Smith, L., Kerby, J. L. & Parris, M. J. (2012). Dietary protein restriction impairs growth, immunity, and disease resistance in southern leopard frog tadpoles. *Oecologia, 169*, 23–31.

Voss, G. (1960). Banana diet for American moose. *International Zoo Yearbook, 1*(1), 44.

Waterman, J. O., McNally, R., Harrold, D., Cook, M., Garcia, G., Fidgett, A. L. & Holmes, L. (2021). Evaluating environmental enrichment methods in three zoo-housed Varanidae lizard species. *Journal of Zoological and Botanical Gardens, 2*(4), 716–727.

Yealland, J. J. (1960). Pellet food for birds at the London Zoo. *International Zoo Yearbook, 1*(1), 44.

Young, R. J. (1997). The importance of food presentation for animal welfare and conservation. *Proceedings of the Nutrition Society, 56*(3), 1095–1104.

Zootrition. (2021). Diet balancing tool for zoo and wildlife professionals Zootrition Software. http://www.zootrition.org/. (Access Date 16/02/2025)

9

People and animal welfare: Impacts of human presence in the zoo

PAUL ROSE AND IAN HICKEY

9.1 INTRODUCTION

Human presence is ubiquitous in zoos and aquariums and has the potential to be a source of both stress and enrichment to the zoo's occupants. Challenges emerge in the identification of which species are stressed or enriched by human presence, and in what aspect of human presence they find to be stressful or enriching. To ameliorate negative welfare and to enhance positive human-animal interactions in zoos, it is important that individual animal responses to humans in the zoo be investigated. Most research on human-zoo animal interactions focuses on the "visitor effect", which can be defined as the behavioural impact of visitor presence on zoo animals (Hosey, 2000). The idea that the zoo's inhabitants will be responsive to those humans in and around their managed environment is not new; however, being documented by Hediger (1965) in the middle of the 20th century. Different forms of human-animal interaction exist within zoos, such as the relationship between animals and their keepers (Melfi et al., 2022), which may bring with them different welfare connotations. Regular keeper-focused interactions may allow both the animal and keeper to predict the outcome of future interactions, for example, developing a positive human-animal relationship (Hosey, 2008) resulting in a potential improvement to the animal's overall quality of life.

Any change in behaviour that results from human presence is the result of the animal processing a stimulus from the external environment and thus directing (what the animal considers to be) the most appropriate response. Therefore, to better understand any impact of keepers, visitors, and vets, we need to understand how the animals perceive these people (as a stimulus), how they process such information, how they react to human presence, and what their reactions mean in the context of their welfare state (Figure 9.1). For many species, we may not be aware of the perceived impact of humans on animal welfare states as we are unable to determine, measure, or comprehend how a species is perceiving their environment and responding accordingly. Thus, Figure 9.1 encourages us to consider what we know about species' responses to their environment, how they filter and process

DOI: 10.1201/9781003429425-9

information, and ultimately how they direct their behaviour when faced with differing human presence in the zoo.

9.2 THE ZOO'S HUMANS

Although research in this area has largely focused on visitor effects, many other (potential or realised) effects of human presence are found at the zoo. Animals are likely to encounter their keepers daily, and other zoo staff may also be present at various levels, including educational presenters, maintenance staff, horticulturists, veterinary and animal health personnel, or office-based staff (e.g. marketing and public relations). The differing job roles of these people result in a variety of interactions with the zoo's animals and, as a result, there may be different welfare impacts associated with their presence (Figure 9.2). The presence of different people at an enclosure and what this could mean for the animal is explored in Chapter 1 (Figure 1.3).

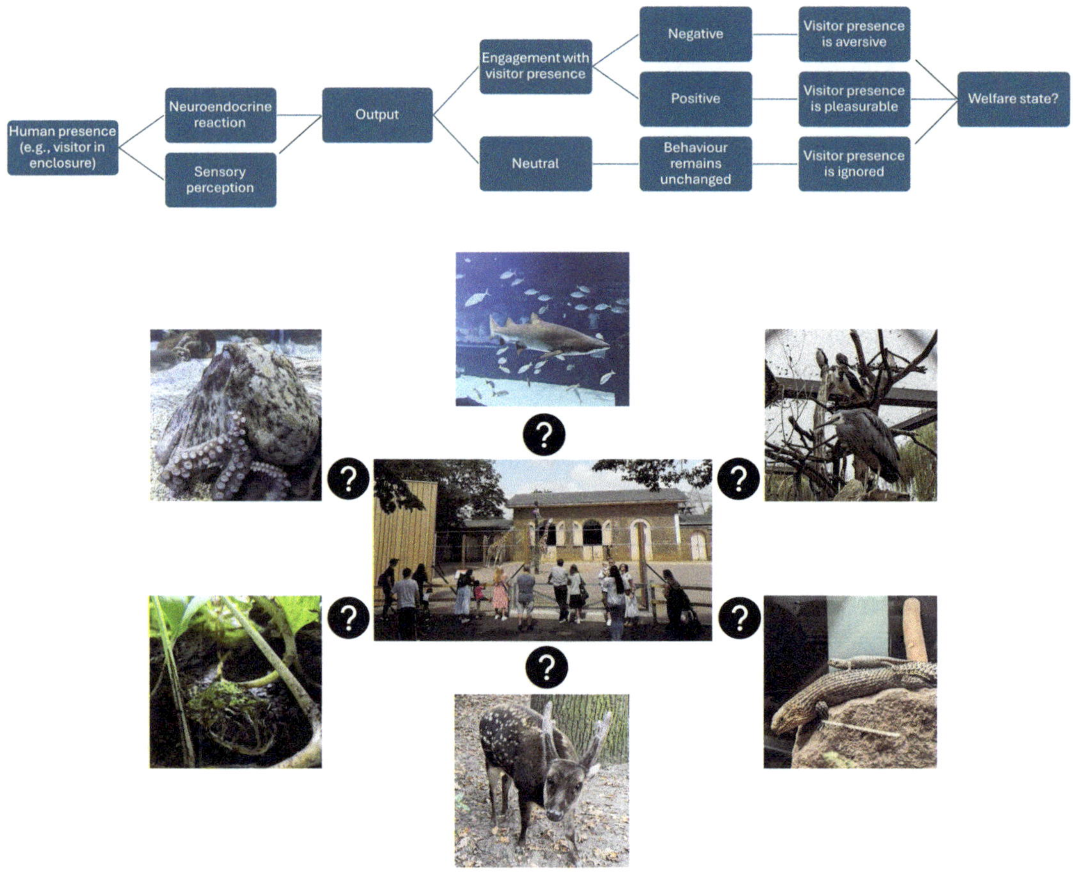

Figure 9.1 Stimulus-response reaction showing the three main outcomes that zoo animals provide to human presence. But how do zoo animals across different taxa perceive, and then respond to, human presence? Behaviour is the result of a neuroendocrine response (a link between the animal's nervous system and a resulting release of hormones). How the animal perceives the stimulus via species-specific sensory modalities will influence the ultimate response that we can see (and measure). The presence of a human may be engaged with (both favourably or not) as well as ignored, and the individual's behaviour remains unchanged. Determining what such behavioural outcomes mean for welfare (even if "ignore" is the outcome) is important here.

The likelihood of an animal encountering different people in different roles at the zoo varies temporally (e.g. peak times for visitation of an enclosure, daily health checks by veterinary staff) and, to some extent, seasonally (e.g. horticulture staff). The endpoint of any human-created encounter will also differ based on both the type of interactions and the species. Animal care staff, for example, are likely to provide food or positive interaction opportunities, and their presence is often the most valuable for the animal (Learmonth et al., 2021). However, such engagement can depend on the scale of visitor presence at an enclosure. For example, research on black casqued hornbills (*Ceratogymna atrata*) showed that bird interest in their keepers was stronger than that directed at visitors, except when the number of visitors was especially high (Rose et al., 2020). This "environmental context" of keeper presence, whereby animals change their activity when keepers are present (Pacheco et al., 2024) should be researched in more detail across a wide range of taxa to help identify the positive impacts of the keeper's presence on animal welfare states. If we can integrate welfare measures into human-animal interactions and understand more about a species' sensory modalities that enable them to process information from the humans around them, we may be able to determine the benefits of working closely that both parties (animal and human) will experience (Figure 9.3).

Encounters with education staff and presenters may be infrequent for many zoo animals but, for select species (such as those that are the focus of a presentation), these encounters may occur multiple times per week (Powell et al., 2020). Similarly, where different species of animals are included in an enclosure that deliberately encourages visitor-animal interaction (e.g. petting zoos), consideration must be given to how each species

Figure 9.2 The types of people that a zoo animal might encounter and the interactions they may receive from these people are varied. The black icons indicate a highly likely interaction, and the grey icons indicate a less likely interaction. indicates feeding, indicates entry into exhibit, indicates a medical procedure such as injection, indicates sound and indicates physical contact such as handling or restraint.

Figure 9.3 The hypothetical conservation between an animal and its caregivers. Validating measures of how animals respond so that the context of the response can be objectively judged and evaluated would be helpful in quantifying the value of animal-keeper bonds in the zoo, and the impact of these bonds on husbandry, population management, visitor engagement, and welfare states. Animal concepts are shown in thought clouds; keeper concepts in speech bubbles.

will respond to levels of visitor presence that will fluctuate greatly within the animals' direct living space (Farrand et al., 2014). Interpretation and explanation of the animals' behaviours and their needs, as well as direct management from the zoo to regulate visitor presence around direct contact/petting animals, is needed to protect good welfare states (Ramont et al., 2021).

Presenters are also a source of stimuli, both positive (e.g. feeding opportunities, training) and negative (e.g. large crowds, increased sound levels), and so they are likely to be anticipated by animals, especially if talks occur at a set time. The occurrence of anticipatory behaviour is thought by some to be a useful indicator of an individual's current welfare state, especially when humans are involved in specific aspects of an animal's daily routine (Podturkin et al., 2023). Measures of such behaviour should therefore be taken at different levels (and contexts) of human presence to validate this.

9.2.1 DIFFERENT CATEGORIES OF ZOO HUMAN INTERACTIONS

Although visitor effect research is commonplace, there remain biases in the species under study (i.e. mainly mammals and, often, mainly primates), and discrepancies in how animals respond to visitors. For this latter point, research shows that responses to visitors can differ to the extent that the same species housed in two different facilities will show variation in their responses to visitor presence. It is believed that an animal's social environment and exhibit complexity can mediate some negative visitor effects (Sherwen & Hemsworth, 2019), though much of the impact may be reflective of individual animal preferences and personality.

It is important, therefore, that animals (individuals or species) that are sensitive to the visitor effect are rapidly identified, and that measures are put in place to mitigate any negative impacts on welfare (Figure 9.4).

While there are clearly confounding factors as described above, it has been suggested that species that are nocturnal and small-bodied species from closed habitats (i.e., tropical rainforests with poor long-distance visibility) are at greatest risk of disturbance from sudden or prolonged human presence. A review of primate responses to visitors identified that smaller, arboreal species, such as tamarins (Callitrichidae), were more likely to respond negatively compared to larger, terrestrial species such as baboons, *Papio* sp. (Hosey et al.,

Figure 9.4 Examples of good practice in managing visitor behaviour around species that are sensitive to human presence include altering public viewing opportunities based on welfare assessment data and using facts about the animals' own behaviour patterns to encourage understanding of their needs. It is important to explain to visitors that some species are sensitive to disturbance and, thus, they (the visitors) are being managed in a particular way.

2023). It has been suggested that smaller primates are less able to use their wild-type avoidance strategies, such as climbing high into trees, when held in enclosures that constrain arboreal activity. Across all species held in the zoo, this example shows the importance of embedding natural history information in enclosure design to reduce the negative impacts of visitors and to ensure that animals can perform naturally evolved coping strategies to mitigate or avoid what they perceive as potential threats. This concept is explored further in Section 9.4.

When visitors are encouraged to directly engage with animals, for example in petting zoos or children's farms, the management of such interactions is essential to ensure that animal welfare is not compromised by direct visitor contact. The potential of such direct encounters to enhance human well-being and connection to nature is huge, but the animals that are central to this experience must be allowed space away from visitors and have access to retreats that only they can use. Zoo associations, such as BIAZA, have policies on close contact that guide and advise zoos on key elements of good practice concerning direct animal-visitor interaction

9.2.2 ZOO STAFF

For the majority of individuals, zoo animals are more likely to directly interact with their caregivers than with other humans in the zoo. Keepers provide positive stimuli for animals in the form of food, training, and enrichment, but can also be disruptive during their husbandry and health check routines. Although keepers are intrinsically important in keeping animals alive and thriving, the animal-keeper interaction is studied less frequently compared to questions pertaining to visitor effects. The available evidence demonstrates that animals will readily interact with their keepers (Cole & Fraser, 2018) and that some animals will enjoy such interactions. Aldabra tortoises (*Aldabrachelys gigantea*), for example, consistently chose to spend time interacting with their keepers when their reactions to different forms of enrichment were tested (Learmonth et al., 2021).

While rarely researched using a behavioural methodology, animal-keeper interactions have nonetheless been studied in a social science setting. Research has aimed to quantify the human-animal relationship using self-reporting techniques by keepers (Hosey & Melfi, 2012) and such research demonstrates that positive relationships between keeper and animal can make the animal easier to work with and increase the keeper's sense of job satisfaction (Carlstead et al., 2019). A self-reported methodology does, however, make identifying any benefits from the animal's perspective more challenging; therefore, incorporation of keeper presence into quantitative (e.g. time budget) and qualitative (e.g. description of behavioural expression) research protocols is recommended to objectively judge an animal's perception of their care staff.

The role of veterinary-focused staff in the protection of animal welfare in the zoo is clear (see Chapter 15), and their presence at an animal's enclosure will usually be circumstance-specific. Healthcare staff may become associated with negatively valenced experiences by the animal, such as treatment and medication procedures that can involve unusual and potentially stressful handling and removal from home enclosures. However, as husbandry training for minor veterinary procedures (e.g. blood draw, vaccinations, and providing medication) can support a positive working relationship between animals and their caregivers/animal health team (Newnham et al., 2024), the presence of veterinary personnel may not always end in a negative outcome. Once again, we must return to the animal's perception of the interaction and how information is being provided, processed, and responded to.

9.3 STIMULUS DETECTION AND PERCEPTION

Humans are visual beings, relying less on our other senses and, as such, it is easy to take an anthropocentric view of sensory perception when investigating animal welfare. Although we may be familiar with differences in sensory perception across species (e.g. electroreception in sharks and infrasound in elephants), we have a tendency to assess welfare using our own preferred sensory modalities. We rely heavily on sight in animal welfare assessments, where behavioural observations, based primarily on *observable* behaviours, are used. Even the word behaviour relates to how observable a response that an animal gives to stimuli is (Barnard, 1983). When studying the visitor effect, we quantify visitor presence visually (i.e., as a count of individuals) but we do not readily consider the other ways in which being present can be detected by animals. Some research has attempted to document "noise", but this again is a human creation.

There is no such thing as "noise" to a non-human animal; simply a change in the sound environment around the individual that the animal may or may not perceive. Removing the word "noise" from visitor effect/human impact research in zoos would be a welcome first step in changing the context of such research, away from how we describe some sounds towards a focus on how the animal may perceive them. Other senses, such as olfaction and gustation, are more challenging to quantify but may be essential for advancing our understanding of human presence impacts. Determining the primary sensory modalities of a species is a first step in understanding how humans at the zoo are perceived by them. Naked mole rats (*Heterocephalus glaber*), for example, are likely able to hear visitors and feel their vibrations but not see them. Use of a radio to habituate captive mole rats to human speech (Amari et al., 2024) could improve the animal's ability to cope with changes to the sound environment around their colony and thus enable positive welfare to be experienced.

Future researchers should consider how animals perceive human presence to assess welfare from a "how animals view the world" perspective. Such questions should consider stimuli-response via complicated and highly specialised senses such as electroreception and barorecption, as well as sight, sound, gustation, touch and olfaction. Large-scale research projects, that have involved multidisciplinary approaches have attempted to unpick how animals perceive specific information from their environment (i.e., changes to the sounds around them) and evaluate whether this translates into impacts on welfare (Box 9.1). Reviewing and evolving the methods from such projects to different sensory modalities in different species could generate a wealth of impactful information to be built into zoo husbandry and management.

BOX 9.1: Listening to the zoo

As an example of research that has looked at how animals perceive humans and how humans perceive animals, the Listening to the Zoo (LTZ) project was run at the University of Exeter from 2018 to 2021. https://www.exeter.ac.uk/research/centres/ease/research/listeningtothezoo/

LTZ was funded by one of the largest UK funding agencies, and it investigated how animals respond to sound at the zoo. The project involved anthropologists, ethicists, and animal behaviour scientists to recreate a multidisciplinary view of how animals (and people) hear the world around them. Researchers surveyed zoo visitors to understand their experiences of sound at the zoo and how such experiences modulated pleasurable or aversive elements of their visit. Important animal welfare applications from this project are:

- Identifying the need to validate measures of sensory perception when attempting to link to welfare outputs (i.e., what are animals hearing, and can this be measured?).
- Considering differences in how populations respond in different environments. Some sound metrics may predict welfare state in some zoos for some species but not for others.
- Effects of enclosure size and shape, as well as weather and climate, also influence how zoo birds respond to visitors.

Key outputs from this project can be found at these references:

HUMAN ELEMENTS

Badman-King, A., Rice, T., Hurn, S., Rose, P. E. & Reed, A. (2024). Why we should keep quiet at the zoo. *Ethics in Science and Environmental Politics*, *24*, 77–88.

Rice, T., Badman-King, A., Hurn, S., Rose, P. E. & Reed, A. (2021a). Listening after the animals: Sound and pastoral care in the zoo. *Journal of the Royal Anthropological Institute*, *27*(4), 850–869.

Rice, T., Reed, A., Badman-King, A., Hurn, S. & Rose, P. E. (2021b). Listening to the zoo: Challenging zoo visiting conventions. *Ethnos*, *89*(1), 1–20.

Rice, T., Reed, A., Badman-King, A., Hurn, S. & Rose, P. E. (2023). A desirable privation: Exploring silence as a mode of zoo visiting. *TRACE: Journal for Human-Animal Studies*, *9*, 6–32.

ANIMAL ELEMENTS

Rose, P. E., Badman-King, A., Hurn, S. & Rice, T. (2021). Visitor presence and a changing soundscape, alongside environmental parameters, can predict enclosure usage in captive flamingos. *Zoo Biology*, *40*(5), 363–375.

McConnell, H., Brereton, J., Rice, T. & Rose, P. E. (2022). Do birds of a feather always flock together? Assessing differences in group and individual zoo enclosure usage by comparing commonly available methods. *Journal of Zoological and Botanical Gardens*, *3*(1), 71–88.

Rose, P. E., Reed, A., Hurn, S., Badman-King, A. & Rice, T. (2022). Does the sound environment influence the behaviour of zoo-housed birds? A preliminary investigation of ten species across two zoos. *Behavioural Processes*, *203*, 104763.

While mitigating any direct effects of human presence, we must also consider any indirect effects on animal welfare from human decisions and actions in the zoo (e.g. through enclosure design features). Waterfalls, for example, are often used within enclosures to create a more realistic representation of a species' native habitat and, as such, make an exhibit more attractive to visitors. While these can provide valuable environmental variation for the enclosure's inhabitants, they are likely to produce a constant sound within the animal's environment, the volume of which will inherently get louder with increased size and flow. Therefore, when catering for all those who have a vested interest in an enclosure, exhibit designers must think of the indirect ways in which such features can impact animal welfare. These features should be located in such a way that an individual animal can choose whether to spend time near this sound or move away to a quieter part of its enclosure, without removing the choice to use other enclosure features. This concept of choice within enclosure usage is explored further in Chapter 7. While this waterfall example is one of indirect human influences relating to change to the sound environment, the majority of visitor impacts will be through their direct presence within and around an animal's enclosure.

9.4 CONSIDERING PEOPLE IN ENCLOSURE DESIGN

Across exhibit styles within the zoo, any impact of people will be experienced in different ways by the animals within those exhibits. Good enclosure design is central to ensuring positive animal welfare, and a strong understanding of how an animal uses its environment can be employed to develop enclosure designs that both mitigate negative visitor effects and scaffold opportunities for positive human-animal interaction. There is, however, an inherent conflict of interest which occurs here, particularly for zoos that are open to visitors.

One of the central pillars of captive animal welfare is the provision of choice, particularly surrounding the animal's use of their environment. When providing animals with the choice of where to spend time and how to use their enclosure, individuals may naturally choose to occupy spaces away from visitors. This does not directly align with the common, and understandable, expectation that zoo visitors have – to view the animals that they have likely paid entry to see. Moreover, not only can the potential for animals being out-of-view compromise visitor experience, it may also dilute educational opportunities for the zoo to educate on species-relevant information and share broader conservation messaging which is now a cornerstone of the modern zoo's mission. While these two, often conflicting, factors (animal welfare and visitor experience) will usually require compromise, this compromise can be limited with thoughtful enclosure design. One common example where we see this conflict is in walkthrough animal exhibits.

The concept of a walkthrough exhibit allows visitors and animals to share the same space, creating a greater feeling of immersion. Not only can this provide an elevated viewing experience, compared to traditional viewing through a barrier, but it can also leave visitors with heightened empathy for the natural environment (Pavitt and Moss, 2019). However, by taking visitors through an enclosure, we inherently impact upon the animals' use of that space. An example of this can be seen in Figure 9.5, which details such an impact for flamingos (Phoenicopteriformes) and a simple solution for how this can be overcome without compromising visitor experience.

As can be seen from Figure 9.5, there are ways of managing potential visitor impacts on animal welfare in enclosures. However, for complex species such as flamingos, individual flock differences need to be considered because some groups are known to respond to different environmental changes (caused by visitors) compared to others (Rose et al., 2021). We can use species-specific outputs to measure comfort in an enclosure and how secure flamingos feel in their exhibit to generalise the interpretation of welfare across flocks when people are present. For example, measurement of space use where individuals are not all consistently grouped in one small enclosure area (Rose et al., 2014; Anderson & Rodman, 2025).

While walkthrough exhibits take visitors directly through an animal's space, the majority of zoo enclosure viewing is through a physical barrier. While generally less invasive, this viewing can still negatively impact individuals if not designed in the right way. Cross-viewing (whereby visitors can see other visitors across another part of an exhibit) is generally avoided in modern zoo design as it can remove the feeling of "immersion" (Coe, 1994) but such viewing can also greatly impact animal welfare. Research has shown that increased visitor presence can have negative impacts for certain species (Sherwen & Hemsworth, 2019) and we must therefore consider this when designing viewing points into an exhibit. Limiting viewing opportunities to one or two sides of an enclosure,

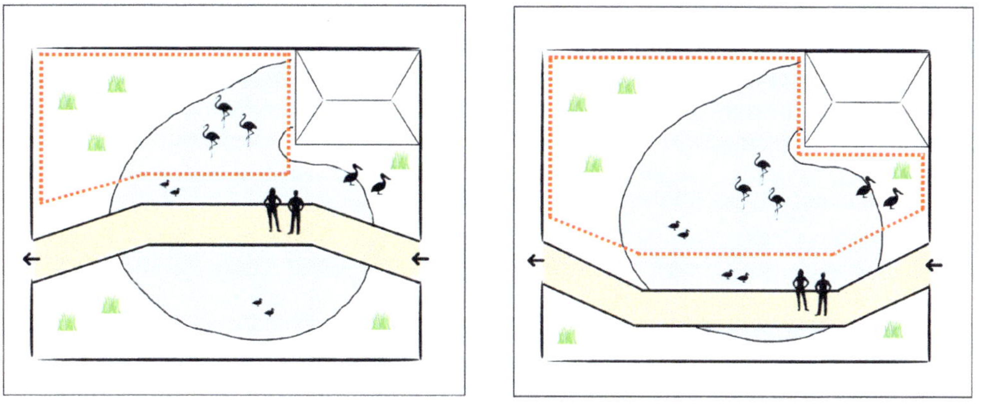

Figure 9.5 Two hypothetical enclosure designs for a mixed-species aviary focusing on flamingos, showing how the placement of a visitor boardwalk through the middle of an exhibit can impact the available space for different species (left), and a simple way that this can be improved (right). The red dotted line indicates the available space for flamingos, taking into account an indicative distance that they will likely maintain from visitors. Other waterbirds (such as pelicans) may feel more comfortable in closer proximity to visitors, while smaller species like ducks may feel comfortable swimming under, or flying over, a boardwalk, resulting in half of the water body only being accessible to these smaller species.

and to specific points along those sides, is essential to ensure that animals have space to retreat and can choose to spend time away from visitors. Hediger (1950) noted the importance of considering the distance from visitors around an enclosure when encouraging an animal to feel comfortable engaging with resources, thus avoiding animals being in a constant state of "fight or flight". This is particularly important for "prey" species that may be more predisposed to feeling unsafe when surrounded. Figure 9.6 shows such design in practice in a Humboldt penguin (*Spheniscus humboldti*) exhibit, with visitors able to surround the animals' only water source and lean into the enclosure, removing any area of water where animals can feel safe or secluded.

While we must always ensure that animals have the choice to move away from visitors and design

Figure 9.6 It is essential to create enclosures whereby animals can choose to spend time in spaces away from visitors. For aquatic species like penguins, we should provide areas of water where visitors cannot follow. Despite attempting to limit viewing into the water through three windows, the poor design of the barriers at this Humboldt penguin exhibit allows visitors to stand on ledges and lean over both walls and glass, resulting in the animals' only water body being surrounded by visitors and facilitating potentially unwelcome hands-on contact.

such refuge spaces within an enclosure, designers should also consider species differences when it comes to visitor effects. For example, Holmes (2010) demonstrated differences in behavioural responses between wild king (*Aptenodytes patagonicus*), gentoo (*Pygoscelis papua*), and royal (*Eudyptes schlegeli*) penguins to on-ground human visitors. Furthermore, Ozella et al. (2015) observed a decrease in water use by captive African penguins (*Spheniscus demersus*) with increased visitor presence, while noting that previous research showed that the presence of visitors had a positive effect on the diving behaviour of Humboldt penguins. This highlights the importance of tailoring how visitor viewing is achieved at an enclosure to each species housed within. However, it is recommended to always allow animals opportunities to access spaces away from visitors and their presence, as behavioural differences are individual to the animal and not just species-specific.

9.5 WHAT ARE ANIMALS RESPONDING TO?

It is essential that we consider how animals process the entire presence of humans around them. The sight, sound, and smell of a visitor could all elicit a response from the individual animal, and some of these stimuli might be heightened according to the animal's sensory perception compared to how we, as fellow humans, perceive them. Stressful stimuli activate the sympathetic nervous system, resulting in the "fight or flight" response and disrupting homeostasis away from normal "rest and digest". Measures of tolerance and retreat proposed by Leroy et al. (2024) maybe the 21st century way of understanding how comfortable animals are within a managed space. Positive stimuli that cause excitement yield a change to homeostasis in a similar way to that seen in a stress response, and so context around behavioural and physiological measures of stress is required to ensure that eustress (Villalba & Manteca, 2019) – i.e., a positive stressor – is not incorrectly judged as an aversive situation.

Opportunities to engage in play behaviour may activate the sympathetic nervous system and thus excite the animal but are not considered stressful from the animal's perspective.

Animals can become sensitised to stimuli that would not carry any biological relevance in the wild (e.g. the sound of keys, the presence of coloured clothing, the tone and intonation of an individual's voice). Such classical conditioning i.e., a behaviour being elicited by a previously neutral stimulus that is paired with a specific event or outcome, may enable animals to be prepared for particular moments or happenings during the working day of the zoo. Habituation (where response to a stimulus wanes and extinguishes over time) and learned helplessness (where an animal loses the capacity to respond because the responses they have been given do not result in a change to the animal's situation) are also possible when zoo-housed species have prolonged exposure to different people. Habituation to visitors may have welfare benefits, as animals filter out the presence of people around them and continue with more important (or functional) behaviours. However, learned helplessness (which could manifest as lethargy, inactivity, or stereotypy) during the presence of humans at, or in, an enclosure is more concerning and the root cause of this (lack of) response should be investigated.

9.6 CONSIDERING ANIMAL RESPONSES

Several zoo organisations, as well as government ministries pertaining to zoos, have put steps in place to protect animal welfare during periods of human contact, whether these are direct or indirect. In the UK, the SSSMZP provides specific guidelines for zoos managing close-contact areas (Appendix 6 of the 2012 version current at the time of writing) to protect both animal welfare and human health. BIAZA has published a "Close Contact Policy" (BIAZA, 2024) that requires zoos to follow national policy and health and safety requirements when animals and visitors are within the same area or have opportunities for contact. Best practice guidelines for species, such as those published by EAZA, also provide instruction on how to design enclosures and provide husbandry that considers the impacts of different human presence around animals and thus protects welfare.

When measuring welfare impacts from humans, care must be taken to ensure that behaviour change is truly the result of human presence. For example, other variables (such as weather) can correlate with both visitor presence and animal behaviour. Research on hornbills, for example, showed temperature to have a much greater impact on behaviour than visitors, and similar findings were identified for ring-tailed lemurs, *Lemur catta* (Goodenough et al., 2019; Rose et al., 2020). Different populations of similar animals in zoos not too far apart can also respond differently to weather and to visitors (Rose et al., 2021), thus complicating how we draw conclusions. Ultimately, researchers studying behaviour and human presence need to record all variables that animals may be responding to, alongside behavioural outcomes, if they are to generate valid and relevant applications from their work. Even when physiological measures are taken, one must remember that season, time of day, physiological state (e.g. growth, reproduction, moulting, etc.), health status, presence of conspecifics and previous experiences will all impact on any measure of glucocorticoids and thus the interpretation of these outputs. Such considerations are illustrated by research on zoo-housed meerkats (*Suricata suricatta*), which determined that group size, enclosure size and visitor presence all impact on circulating glucocorticoid concentrations (Scott et al., 2017). Analysis of both behaviour patterns and space use can provide a useful evaluation of the impact of people, especially when such methods are coupled with ways of assessing internal emotional states (e.g. Qualitative Behavioural Assessment, see Chapter 1). These mixed-methods approaches could help zoo researchers unravel how different degrees of disruption, from human activities, impact on individual and population-level welfare states (Figure 9.7).

9.7 CONCLUSIONS

This chapter has explored the potential impacts of human presence on the welfare of zoo animals. Ensuring that animals in zoos can experience positive welfare is essential, and human presence need not be detrimental to this goal. Keepers play a key role in fostering positive interactions with the animals in their care, which can enhance welfare outcomes and support effective animal husbandry. With thoughtful approaches to enclosure design and visitor interaction programmes, positive visitor interactions can be encouraged, and negative effects minimised. Future research efforts should focus on advancing methods to assess the impact of human presence on animal welfare, including

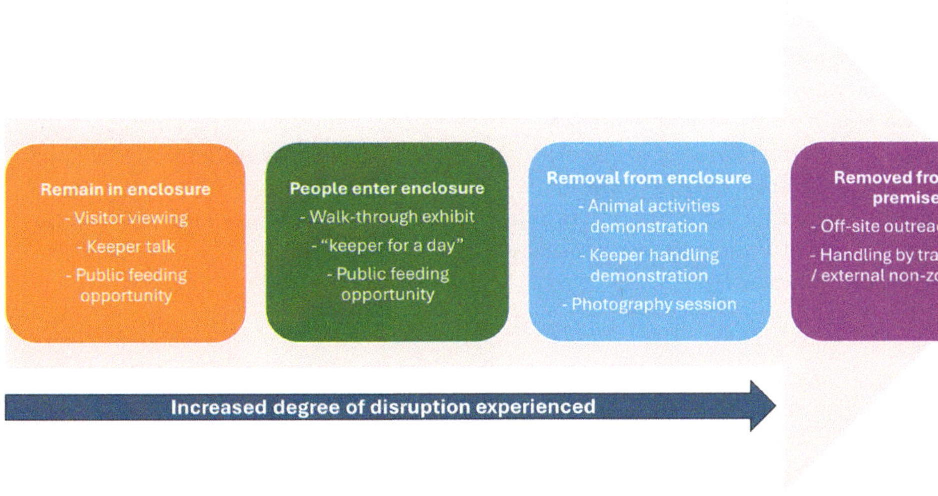

Figure 9.7 The degree of disruption experienced by animals during different interactions with different humans in different roles at the zoo needs to be the subject of future applied animal welfare research to understand the long-term impacts on health and wellbeing of all kinds of human-animal interactions.

the development of mixed methods approaches. This includes improving how we measure human activity and understanding animals' responses from their perspective, while considering how they perceive their surroundings. Furthermore, all forms of human interaction, not just visitor presence, should be evaluated. Exploring how animals respond differently to various types of human interactions could reveal valuable insights into what they find rewarding or aversive in their environment. Box 9.2 provides examples of questions that relate to impacts from people on animal welfare in the zoo which we could consider trying to answer to add to our knowledge of the relationship between the human and the non-human animal within a managed environment.

9.8 ACKNOWLEDGEMENT

We would like to thank Mr James Brereton of University Centre Sparsholt for assistance with developing some of the original concepts for this chapter.

BOX 9.2: Future questions to be asked to develop human-animal interaction studies

- Do animals respond differently to keepers, vets and visitors? If so, which species are able to discriminate between these different groups of people, and what cues do they use to identify each in turn?
- Should keepers work 'hands-on' with their animals? If so, under what scenarios is hands-on work appropriate? A majority of zoo animals will not be released to the wild, and physical contact could aid in the development of the human-animal relationship.
- What is the impact of an invisible public? The fitting of one-way mirrored windows, or similar, is becoming more commonplace and therefore the perception of visitors from within the enclosure should be considered (any impacts of disembodied noise, for example).
- Are captive animals stressed by the scent of the public? Is there any welfare impact of building exhibits downwind of facilities that may be particularly odorous, e.g. cafes, restaurants, waste disposal areas?
- Is sound a better predictor of visitor impact on zoo animal welfare than sight for certain species?
- What is the impact of a preferred keeper on animal behaviour, welfare and comfort? Is this differentiation between humans based on specific attributes (e.g. voice, stature, posture, etc.)?
- How can we assess whether neutral responses to zoo visitors have no welfare impact? Is there a latent effect from the presence of visitors that occurs later, and therefore the animal may respond but at a different point in time?
- How long does it take an individual or a group to recover from any visitor effect? When do behavioural states return to normal or baseline and is this impacted by different numbers of pre-experienced people?

REFERENCES

Amari, M., Mary, A., Zablocki-Thomas, P., Bourgeois, A. & Pouydebat, E. (2024). Positive effect of a diggable substrate on the behaviour of a captive naked mole rat colony. *Scientific Reports, 14*(1), 20138.

Anderson, M. J. & Rodman, E. C. (2025). Examining the impact of the 'visitor effect' and weather variables on the behavior of Chilean flamingos (*Phoenicopterus chilensis*) via webcam. *Avian Biology Research*. https://doi.org/10.1177/17581559251320325

Barnard, C. J. (1983). *Animal behaviour.* Boston, MA: Springer.

BIAZA. (2024). *BIAZA close contact policy.* London, UK: British & Irish Association of Zoos and Aquariums.

Carlstead, K. C., Paris, S. & Brown, J. L. (2019). Good keeper-elephant relationships in North American zoos are mutually beneficial to welfare. *Applied Animal Behaviour Science, 211*, 103–111.

Coe, J. C. (1994) Lanscape Immersion - Origins and Concepts. In: Landscape Immersion Exhibits: How Are They Proving as Educational Settings? (Panel Subject). *1994 AZA Convention Proceedings*. American Zoo and Aquarium Association, Bethesda, MD.

Cole, J. & Fraser, D. (2018). Zoo animal welfare: The human dimension. *Journal of Applied Animal Welfare Science, 21*(sup1), 49–58.

Farrand, A., Hosey, G. & Buchanan-Smith, H. M. (2014). The visitor effect in petting zoo-housed animals: Aversive or enriching? *Applied Animal Behaviour Science, 151*, 117–127.

Goodenough, A. E., McDonald, K., Moody, K. & Wheeler, C. (2019). Are "visitor effects" overestimated? Behaviour in captive lemurs is mainly driven by co-variation with time and weather. *Journal of Zoo and Aquarium Research, 7*(2), 59–66.

Hediger, H. (1950). *Wild animals in captivity.* London, UK: Butterworths Scientific Publications.

Hediger, H. (1965). Man as a social partner of animals, and vice versa. *Symposia of the Zoological Society of London, 14*, 291–300.

Holmes, N. D. (2010). Comparing King, Gentoo, and Royal Penguin responses to pedestrian visitation. *The Journal of Wildlife Management, 71*(8), 2575–2582.

Hosey, G. (2008). A preliminary model of human–animal relationships in the zoo. *Applied Animal Behaviour Science, 109*(2–4), 105–127.

Hosey, G. & Melfi, V. A. (2012). Human–animal bonds between zoo professionals and the animals in their care. *Zoo Biology, 31*(1), 13–26.

Hosey, G., Ward, S. & Melfi, V. (2023). The effect of visitors on the behaviour of zoo-housed primates: A test of four hypotheses. *Applied Animal Behaviour Science*, 263, 105938.

Hosey, G. R. (2000). Zoo animals and their human audiences: What is the visitor effect? *Animal Welfare, 9*(4), 343–357.

Learmonth, M. J., Sherwen, S. L. & Hemsworth, P. H. (2021). Assessing preferences of two zoo-housed Aldabran giant tortoises (Aldabrachelys gigantea) for three stimuli using a novel preference test. *Zoo Biology, 40*(2), 98–106.

Leroy, H., Montedoro, C., Guery, J.-P. & Hance, T. (2024). Visitor impacts on social dynamics and space use in captive Western lowland gorillas (Gorilla gorilla gorilla) in a naturalistic zoo exhibit: A new methodological approach. *Applied Animal Behaviour Science, 281*(2), 106433. https://doi.org/10.1016/j.applanim.2024.106433.

Melfi, V., Skyner, L., Birke, L., Ward, S. J., Shaw, W. S. & Hosey, G. (2022). Furred and feathered friends: How attached are zookeepers to the animals in their care? *Zoo Biology, 41*(2), 122–129.

Newnham, Z., Keywood, L., Reynard, J. & Constable-Dakeyne, G. (2024). Blood draw training in okapi (Okapia johnstoni). *CABI Animal Behaviour and Welfare Cases*. https://doi.org/10.1079/abwcases.2024.0015.

Ozella, L., Favaro, L., Carnovale, I., & Pessani, D. (2015). Pond use by captive African penguins (Spheniscus demersus) in an immersive exhibit adjacent to human bathers. *Journal of Applied Animal Welfare Science, 18*(3), 303–309.

Pacheco, E., Krebs, B. L. & Watters, J. V. (2024). Keeper effect: Animals are more active in the presence of their caretakers. *Zoo Biology, 43*(4), 306–314.

Pavitt, B. & Moss, A. (2019). Assessing the effect of zoo exhibit design on visitor engagement and attitudes towards conservation. *Journal of Zoo and Aquarium Research, 7*(4), 186–194.

Podturkin, A. A., Krebs, B. L. & Watters, J. V. (2023). A quantitative approach for using anticipatory behavior as a graded welfare assessment. *Journal of Applied Animal Welfare Science, 26*(3), 463–477.

Powell, D. M., Kozlowski, C. P., Clark, J. A., Seyfried, A., Baskir, E. & Franklin, A. D. (2020). Physical and physiological indicators of welfare in guinea pigs (Cavia porcellus) serving as ambassador animals. *Animals, 10*(5), 815.

Ramont, M., Leahy, M. & Cronin, K. A. (2021). The welfare of domestic goats (Capra hircus) in a zoo-based animal-visitor interaction program. *Animal Behavior and Cognition, 8*(4), 493–506.

Rose, P. E., Badman-King, A., Hurn, S. & Rice, T. (2021). Visitor presence and a changing soundscape, alongside environmental parameters, can predict enclosure usage in captive flamingos. *Zoo Biology, 40*(5), 363–375.

Rose, P. E., Croft, D. P., & Lee, R. (2014). A review of captive flamingo (Phoenicopteridae) welfare: A synthesis of current knowledge and future directions. *International Zoo Yearbook, 48*(1), 139–155.

Rose, P. E., Scales, J. S. & Brereton, J. E. (2020). Why the "visitor effect" is complicated. Unraveling individual animal, visitor number, and climatic influences on behavior, space use and interactions with keepers- a case study on captive hornbills. *Frontiers in Veterinary Science, 7*, 236.

Scott, K., Heistermann, M., Cant, M. A. & Vitikainen, E. I. K. (2017). Group size and visitor numbers predict faecal glucocorticoid concentrations in zoo meerkats. *Royal Society Open Science*, 4(4), 161017.

Sherwen, S. L. & Hemsworth, P. H. (2019). The visitor effect on zoo animals: Implications and opportunities for zoo animal welfare. *Animals, 9*(6), 366.

Villalba, J. J. & Manteca, X. (2019). A case for eustress in grazing animals. *Frontiers in Veterinary Science, 6*, 303.

10

Conservation welfare: Animal well-being, behavioural ecology, and the One Plan Approach to conservation

PAUL ROSE

10.1 INTRODUCTION

Conservation has been one of the main aims of modern zoos, aquariums, safari parks, and other such zoological organisations for 40 years (Kleiman, 1985) and the idea that zoos should consider the preservation of threatened species was championed by Gerald Durrell in his text "The Stationary Ark" published in 1976. As the aims of the modern zoo evolve to include animal welfare (Rose & Riley, 2022), implementing what can be termed compassion conservation (Bekoff, 2019) by considering animal welfare can support the long-term success of conservation action. Conservation performed in the zoo is an integral part of the IUCN's "One Plan Approach" to conservation (CPSG, 2022), whereby zoo populations are given equal weight to those free-living populations regarding the actions and objectives of species-specific conservation planning. Such a metapopulation approach, linking all populations of a species wherever they are found to be useful for conservation work (Conde et al., 2013), provides zoos with a unique opportunity to actively use their animals to support wider global conservation outcomes.

Species-focussed conservation has traditionally been separated into in situ and ex situ actions. In situ efforts, where conservation of species occurs in their habitats within their natural, free-living environment, focus on protecting biodiversity and ensuring its survival where it naturally occurs. In situ conservation is predicated on three key mechanisms (Figure 10.1). In situ conservation strategies aim to maintain the ecological balance and wider functioning of ecosystems; they feed into wider, holistic aims of conservation to protect and promote ecosystem services that are beneficial to other species and to humans. An in situ conservation approach typically involves actions such as the establishment of protected areas (e.g. national

DOI: 10.1201/9781003429425-10

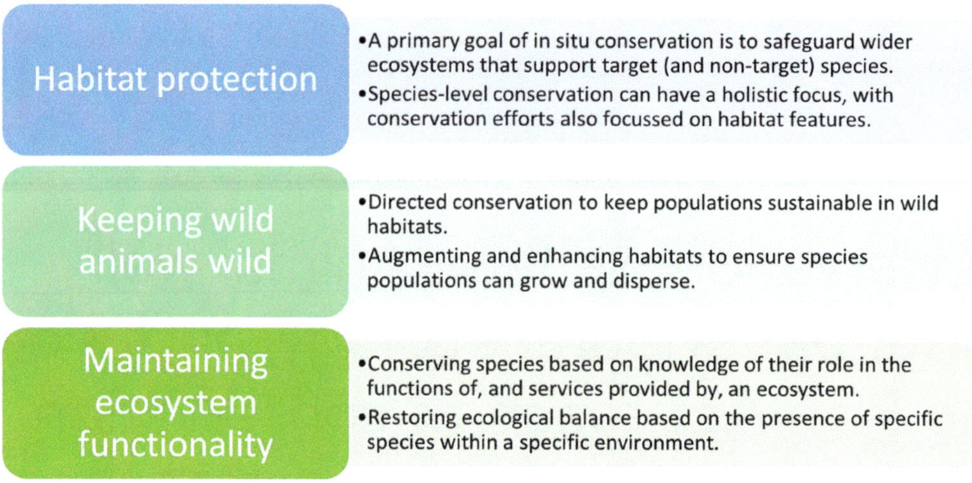

Figure 10.1 The roles of in situ conservation focus on a species' role in its habitat and therefore its wider impact on the persistence of other species around it, as well as support given to key ecosystem services.

parks, wildlife reserves, and nature conservancies) alongside habitat restoration and renewal, augmentation or translocation of populations between different areas of their natural range, enhanced legal protection, and implementation of multilateral policies to ensure species survival, and the management of human-wildlife conflicts to reduce (and eventually eradicate) such threats to free-living populations.

Ex situ conservation, on the other hand, involves the conservation of species outside their natural habitat, often in the controlled environments of zoos, aquariums, and other captive breeding facilities. Ex situ conservation can centre around three key approaches outlined in Figure 10.2. This approach is employed when a species' survival in the wild is at risk due to factors considered "the big five" threats (pollution, overexploitation, habitat loss, invasive species, climate change) to biodiversity (Brook et al., 2008). Ex situ conservation serves as a safety net, maintaining populations of endangered species and their genetic diversity for potential reintroduction, translocation, or supplementation and restocking into the wild. Ex situ conservation and the maintenance of these populations is often termed the "Ark Paradigm" (Harley, 2022), i.e. a concept that describes the role of the zoo as a safe location (in an Ark) that ensures a species will survive by physically protecting it from any threats faced in the wild, and to one day provide a population of sufficient size and diversity that reintroduction could be viable (Bowkett, 2009). Understanding animal behaviour, its observation and measurement, has been key to the successful sailing of this Ark for species on the brink of extinction in the wild, such as the Arabian oryx (*Oryx leucoryx*), Hawaiian goose (*Branta sandvicensis*), pink pigeon (*Nesoenas mayeri*), and many others.

The terms in situ and ex situ are now considered unhelpful and even controversial to how we view conservation, its aims, ideals, and objectives (Braverman, 2014). The terms "in situ and ex situ" literally translate as "in and out of place" and therefore suggest they are contradictory and conflicting to what is hoped to be achieved, rather than different means to the same ultimate goal – the protection of species from extinction and the reestablishment of sustainable populations of wild animals in functioning ecosystems. It is clear that not all species thrive in captivity and that attempting to maintain ex situ populations for all threatened species would be counterintuitive to keeping them alive and thriving. Therefore, some species (such as the mountain gorilla, *Gorilla beringei beringei*) must be conserved in situ, whilst others, like the mountain chicken frog (*Leptodactylus fallax*), are capable of breeding and sustaining a viable population in zoo environments. Therefore, the IUCN Species Survival Commission Conservation Planning Specialist Group (CPSG) launched the "One Plan Approach" to conservation in 2011 as a

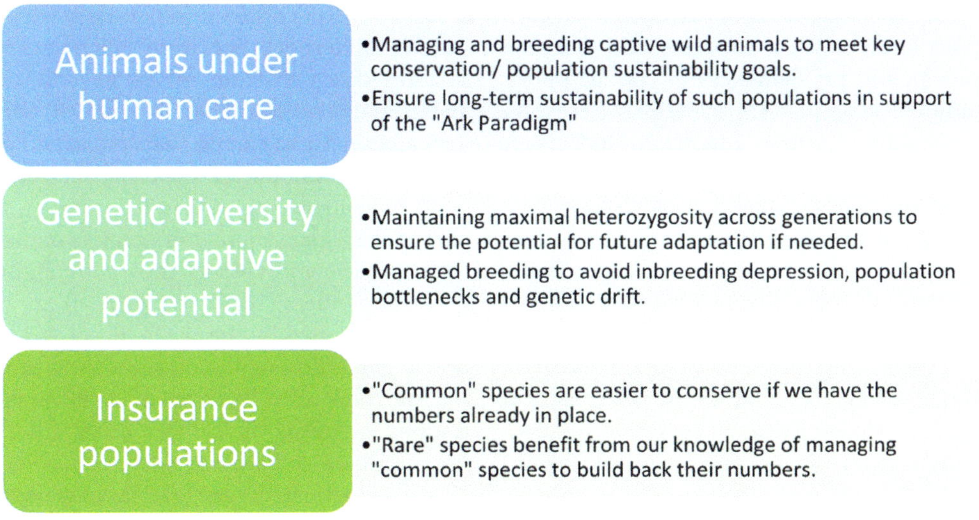

Figure 10.2 The key aims of ex situ conservation and how captive populations support wider species recovery and population sustainability initiatives.

way of better describing the degree of management need and focus of conservation efforts on a species-specific basis (Byers et al., 2013). As the One Plan Approach (OPA) puts more emphasis on the roles of zoo populations in species conservation action planning and conservation implementation, for example, by aligning regional collection plans with conservation aims and objectives (Traylor-Holzer et al., 2019), animal welfare considerations become even more important to such species recovery and survival plans. With more emphasis on ex situ conservation, placing more animals under human care, we have a legal and moral obligation to promote and provide for positive welfare in all such interventions and activities. Ultimately, OPA's goal is to support species conservation in the wild using integrated population management approaches that include the work of zoos and aquariums (Traylor-Holzer et al., 2018).

10.1.1 WHAT IS THE ZOO'S CONSERVATION PLAN?

In the zoo, species survival and persistence plans can be broadly termed as captive breeding or conservation breeding programmes, dependent on the outcome of the programme. Captive breeding may refer to the controlled breeding of animals in captivity with the primary goal of producing offspring and the maintenance of a viable, sustainable population in the zoo (i.e. one that produces young that themselves go on to breed to maintain a population size conducive to long-term survival). Captive breeding programmes can be initiated to increase the numbers of a species held in zoos, especially if this species faces a risk of extinction in the wild. However, non-threatened species will also be the focus of captive breeding programmes that have their own defined long-term population sustainability goals. If we consider that it is easier to conserve something when it is already common, then captive breeding has wide-scale conservation benefits for that species – zoos working together to breed animals within their facilities to ensure that populations remain large enough to be stable, viable, and representative of that species' adaptive traits and ecology. There are clear animal welfare considerations for captive breeding programmes that will see individuals live for their whole lives under human care, concerning housing, husbandry, and population management activities that are suitable for an individual at each stage of its life.

Conservation breeding programmes are commonly implemented for species with specific conservation needs, for example, an end goal of actual reintroduction or the need to produce a specific number of individuals of a species to augment a wild population. Conservation breeding encompasses not only the breeding of animals in captivity but also other conservation actions aimed at conserving a species, such as mitigation of threats in the wild, working with local people to gain indigenous knowledge on a species, and lobbying governments and policymakers to change policy and promote widespread protection. Research, habitat restoration, public education, and direct population enhancement actions (translocation, reintroduction, and release schemes) all form part of conservation breeding actions. Again, animal welfare is a consideration for this work, specifically when it comes to preparing captive-bred animals for a future life in the wild.

In summary, in situ conservation involves protecting species and habitats in their natural environment, while ex situ conservation involves conserving species outside their natural habitat, often in captivity. Discussion around the unhelpful nature of these terms has resulted in a sliding scale of conservation management under OPA. Captive breeding focuses on the reproduction of animals in captivity, while conservation breeding takes a more holistic approach, considering various strategies for preserving species in addition to breeding. Regardless of the name or definition of what conservation is, species within these programmes need to be managed in a way that promotes key adaptive behaviours, enables choice and control within their environment, and is managed in a way that facilitates positive welfare outcomes.

10.2 THE NEED FOR WELFARE IN CONSERVATION

Zoos and their conservation action plans, both in native range states and under human care in controlled conditions, play an increasingly vital role in safeguarding endangered species and restoring ecosystem functionality and health. However, these important activities are not without their ethical and practical complexities. A critical, yet often underappreciated, aspect of conservation planning in all settings where it occurs is the welfare of the animals that are directly involved. Knowledge of the welfare states of captive species, and how these welfare states are expressed, may be paramount for supporting the eventual success of conservation action plans and for upholding ethical principles that are central to how conservation work is executed and evaluated. Figure 10.3 illustrates some of the key concepts and ideals of what adding welfare-centric data (i.e. from welfare assessments of individuals or from reviewing and changing an animal's environment to make it more ecologically relevant) can do to improve and enhance conservation action for a specific species, specifically, by considering individual animal's needs as part of the wider population's captive management strategy. The individuals shown in Figure 10.3, spectacled eiders (*Somateria fischeri*) are listed as Near Threatened by the IUCN (as of 2023) with a decreasing wild population trend, and are part of the Sea Duck Joint Venture (an initiative of organisations including the US Fish & Wildlife Service, Birds Canada and others under the auspices of the North American Waterfowl Management Plan (NAWMP, 2023)) because they have been federally listed as "threatened" since 1997 (SDJV, 2023; USFWS, 2023). The popularity of this species in avicultural collections means there is a strong chance that consistent, viable and well-managed breeding can be a useful tool for future conservation action, if needed. As a specialised sea duck with specific captive care requirements, knowing how birds (at the group and individual levels) respond to their management and what may impact their probability of successful breeding would be essential to the development and implementation of an integrated, multi-party conservation strategy. Research by the USFWS sea duck programme into duck habits and habitats provides essential information for assessing the relevance of captive husbandry and housing to ensure adaptive traits and essential behavioural skills are maintained in flocks under human care.

If welfare is an individual's state during attempts to cope with its environment (Broom, 1986) and is a combination of inputs and outputs that can cause welfare states to fluctuate between positive and negative over time, then measuring welfare states of animals in conservation programmes can illuminate impacts on their feelings and fitness. Biological fitness (e.g. attaining natural longevity and increased reproductive viability) and opportunities to feel positive emotions could be enhanced when animal welfare is embedded into conservation plans, particularly if animals experience eustress (positive stressors) that enable behavioural outputs required for survival and reproduction.

Positive arousal requires positive valence (e.g. experiencing positive emotions such as excitement)

Figure 10.3 What does welfare add to conservation? An example of how considering, assessing and understanding animal welfare states and how they are expressed can benefit conservation in the zoo and assist with species' population management more widely.

and is therefore rewarding (Mendl et al., 2010). Such positive experiences are important to conservation work as providing a biologically relevant challenge (e.g. enrichment, structurally complex habitat replication) that the animal can solve by applying key natural behaviours also develops memories of how to overcome the challenge. This, in turn, creates behavioural resilience that improves welfare overall by giving the individual animal more opportunities for control and choice over their environment.

Ensuring the challenge is beneficial and biologically relevant is key to promoting such opportunities for learning and development. The animal's perspective should be used when perceiving what welfare is, rather than from a human-centric viewpoint (Wolfensohn et al., 2018), i.e. we need to look "through the animal's eyes" as best as we can to understand and identify challenges, threats, aversive stimuli, and other potential impacts on welfare states that the animal may be perceiving from the environment that it is in. Utilising existing welfare assessment approaches (behaviour, physiology, psychology) holistically, we will gain data on what stimuli elicit what responses and why, using such information to improve the design of conservation actions and animal management regimes. Within conservation work, just because something looks right from a human perspective may not necessarily mean it will be comfortable, relevant, or useful to the animals within the programme. Unpicking such subjectivity is a key future research area for conservation welfare.

Therefore, welfare considerations for conservation can be summarised as:

- Arguments concerning animal sentience

  At the heart of ethical considerations for species under human care lies the recognition of the sentience of animals (Proctor, 2012; Proctor et al., 2013). Removed from their natural habitats and living in an artificial environment, animals will continue to experience the same range of physical and emotional states as their wild counterparts. Acknowledging their sentience also calls for an ethical responsibility to ensure their welfare in all aspects of their management and care. This is all underpinned by legal and moral obligations to ensure animal care is species-appropriate.

- Our moral obligations

  All stakeholders within conservation projects and with animal management responsibilities share a moral obligation, and a legal obligation too, to ensure that animals that are being managed are treated with compassion and respect. Ethical review of all interventions, especially those relating to practices that may cause suffering (such as opportunities for predators to catch and dispatch prey), is essential for compassionate conservation. The welfare of captive wild animals is not just an ethical nicety but a fundamental duty that aligns with a commitment to biodiversity conservation wherever it occurs.

- The role of zoo animal management

  As zoos serve as arks for endangered species, housing them and breeding them under controlled conditions, these conservation-focused populations act as a vital insurance policy against the potential extinction of species in the wild. It is essential that zoo animal management is species-specific, evidence-based, and relevant to the conservation aims of the species being managed, increasing opportunities for species to experience control and choice over their environment.

- Protection of wild ecosystems and mitigation of threats

  Field-based conservation focuses on the management and restoration of natural habitats and ecosystems to ensure they remain functional, viable, and resilient. This approach is critical to the maintenance of ecological processes that support the survival of all species that are connected to the ecosystem, including humans. In some cases, field-based conservation may involve managing human-wildlife conflicts to protect both animals and people. Finding ethical solutions that protect animal welfare while addressing human concerns is a challenging but necessary aspect of this work. Zoos can assist with such work by fostering a sense of connection to nature (Rose & Riley, 2023), thereby encouraging an appreciation of biodiversity and promoting a "One Health" approach to human, animal, and ecosystem wellbeing (Greenwell et al., 2023; Rose & Riley, 2022).

- Evolving standards and expectations

  Not all sections of society agree that zoos have a role or are a force for good animal care (Malamud, 2015; Wickins-Dražilová, 2006). Welfare standards have evolved over time, and public expectations for "good" animal care have also increased (De la Fuente et al., 2017). The pressure on zoos to provide the most appropriate care for all their animals, irrespective of what programme the species is part of, is growing. This challenge can only be met by directed research into how animals respond to their care, and how improvements to care manifest in enhanced fitness, viability, and evidence of positive affective states experienced by the animals in these programmes.

- Ongoing monitoring

  Consistent and ongoing monitoring of animal welfare is crucial in conservation programmes to protect and promote biologically relevant welfare states (e.g. an animal's abilities to experience control and choice) at all stages of the process. Welfare monitoring allows institutions to adapt and improve conditions for their animals based on evidence from the animals themselves. This approach fosters a culture of continuous improvement to animal management practices that can translate into field-based conservation work. For example, where long term post-release monitoring of a species is required to ensure the success of the conservation work, data collected on the animals during such monitoring can be used to inform changes to approaches or actions to benefit other species in similar programmes.

- Enhancing a species' behavioural capacities

  Applying animal welfare principles to captive and conservation breeding programmes can improve the success rates of such efforts. Minimising unhelpful stressors (e.g. public presence or opportunities for habituation), ensuring the correct nutrition for a species, and fostering the performance of natural behaviours can enhance the health and reproductive outputs of the animals within a programme (including behaviours that relate to the presence of predators or how to cope with competition). Reintroduction initiatives are the bridge between captivity and the wild, and therefore animals need to be thoroughly prepared (e.g. ensuring they possess the correct suite of behaviours for survival in a natural system) before release. These programmes should prioritise the welfare of animals during the release process, ensuring that animals have the best chance of survival and successful integration into their natural habitats by use of appropriate enclosure design that mirrors important aspects of the wild habitat, training procedures that develop an animal's physical and psychological capacities, environmental enrichment to stretch and challenge problem solving skills and allow animals to cope with unpredictable occurrence of resources, and management of human-animal interactions to ensure individuals are not orientated towards people upon release.

- Considering the challenges of the wild

Species destined for reintroduction need to experience hardship (e.g. competition, limited resources, and habitat variability) if they are able to succeed upon release. Disease and predation are also factors that wild animals must cope with, and therefore conservation biologists, veterinary experts and animal welfare scientists should consider how to prepare animals for such stressors too. Ensuring that animals in conservation reintroduction programmes are not always "pampered" is essential to the population's welfare. If individuals are released without the behavioural or cognitive capacities to learn, adapt and ultimately survive, their welfare will become poor. Likewise, physiologically weak or immunologically deficient animals will have a reduced chance of thriving and surviving post-release. These factors combined could see the aims of the conservation reintroduction fail to be met.

10.3 ONE PLAN FOR CONSERVATION

The International Union for Conservation of Nature (IUCN), founded in 1948, and its Conservation Planning Specialist Group (CPSG) play pivotal roles in the implementation of OPA to conservation action for specific species. This global network of representatives from biological and ecological organisations, government bodies, and other interested parties is dedicated to conserving nature and promoting sustainable development. Conservation action is directed via the Red List of Threatened Species, a comprehensive database that assesses the conservation status of species worldwide, identifies threats and publishes species conservation goals. The resources provided by the IUCN, its structures, and policies provide a unifying framework for conservation efforts and act as a platform for collaboration and knowledge exchange among various stakeholders.

The IUCN's mission statement on the role of zoos, aquariums and botanical gardens in the conservation of nature, published in October 2023, emphasises the value of such ex situ organisations to long-term and sustainable species conservation work (IUCN SSC, 2023). Therefore, providing support for those working in zoos on conservation work and actively advocating for the role of zoos in current and future OPA to species conservation programmes are essential. The CPSG employs a systematic, evidence-based approach to create conservation plans tailored to each species. These plans encompass field-based and captive-based strategies and prioritise actions necessary for species recovery. OPA is integrated into these strategies, ensuring that all aspects of conservation are considered, and form part of WAZA's world zoo and aquarium conservation strategy (Barongi et al., 2015). Both the IUCN and the CPSG promote collaboration among governments, NGOs, zoos, and local communities. This collaborative effort is essential for the successful implementation of OPA, as it brings together the expertise, resources, and support necessary for effective conservation. The CPSG focusses on capacity building by providing training and technical assistance to conservation practitioners and conservation partners. OPA assists with capacity building by extending the ways in which information and expertise can flow between field-based and zoo-based conservation and species experts. The aims and objectives, and overall premise of OPA are summarised by Figure 10.4.

10.4 A CASE STUDY ON INTEGRATED CONSERVATION

Zoos play a crucial and multifaceted role within OPA, which aims to integrate in situ and ex situ efforts into species-specific population management. The expertise of those working in zoos, their knowledge of animal husbandry, population management, record keeping, and providing the right conditions for species to reproduce, are essential to the implementation of integrated conservation practices. The utilisation of proxies in zoo animal husbandry has been a longstanding practice (Rose, 2021). This approach revolves around formulating husbandry strategies for species that are rarer, more challenging to care for, or are in urgent need of conservation attention (Kerr, 2021). Drawing on our experiences with the management of more common or familiar zoo species, this concept aims to enhance the well-being and conservation efforts for those requiring specialised care to which we are less accustomed.

One example of this is the work by WWT and its partners at Durrell Wildlife Conservation Trust (DWCT) in rescuing the "world's rarest duck", the Madagascar pochard (*Aythya innotata*), from extinction. Madagascar pochards were thought to be extinct after the death of the perceived last known individual in 1991. However, a small flock was rediscovered in 2006 in poor quality habitat, and so a captive breeding and reintroduction project was initiated in 2009 (WWT, 2023). Research identified that steep-sided lakes, which these remaining pochards were forced to live on, provided little food for growing ducklings. The lakes were too deep for ducklings to dive down for invertebrate food and consequently, whilst nesting and hatching rates were comparable to other *Aythya* pochard species, the successful fledging of ducklings was only 4% (Bamford et al., 2015). Ducklings were starving due to unsuitable habitat and a lack of suitable diet.

To ensure the long-term well-being of the wild pochards, a key aspect of the conservation strategy involved habitat restoration. Efforts were made to replant native vegetation, combat invasive species, and raise awareness about sustainable land use practices. Known as "Fotsimaso" in the local dialect, community engagement played a crucial role in the project's success, as local people became

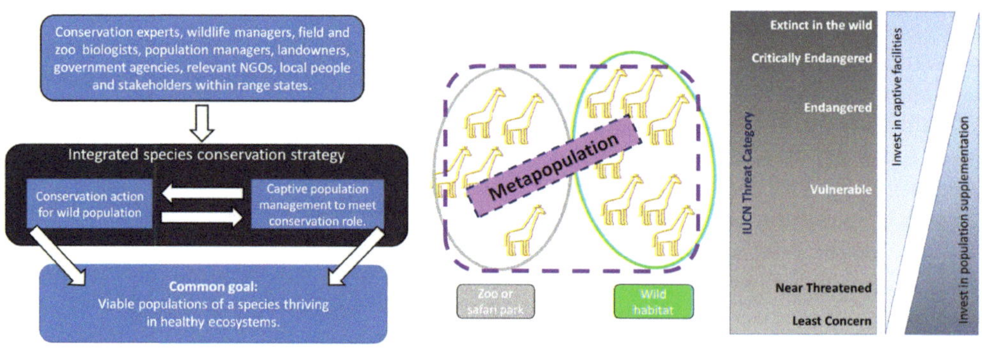

Figure 10.4 The One Plan Approach relies on an integrated species conservation strategy that brings together wild and captive conservation actions to fulfil a common goal of species thriving in healthy ecosystems. To do this, species within an OPA are managed on a sliding scale of intensity, between captive and wild settings, based on their needs, and all individuals (wherever they are) are included in this metapopulation. The categories from the IUCN Red List provide the foundation for the strength, urgency, and timeframe of conservation action (and what form of conservation action is best) based on the species' current threat level, current population trend, and future predictions.

stewards of their natural resources and invested partners in the duck's conservation process. WWT and DWCT collaborated with local communities and the Malagasy government to address the root causes of the pochard's decline – to promote and protect the health of the duck's wetland ecosystems to ensure the health and vitality of the breeding pochards and the ultimate survival of their ducklings. It would have been unethical and detrimental to the pochard population's overall welfare had reintroduction effects been made without having habitat protection, habitat management, and threat mitigation strategies that were causing high duckling mortality and consequential population declines.

To prevent the immediate extinction of the Madagascar pochard, a biosecure breeding centre was established in Antsohihy to house a population of this species under human care, but still within its native range (Cranswick, 2020). Avicultural knowledge from the care of other *Aythya* species would be key to the successful long-term management of these captive pochards. Trialling rearing procedures on tufted ducks (*A. fuligula*) and Baer's pochards (*A. baeri*) and using these captive pochards to test out ways of encouraging ducklings to feed in a specific area where food quality is good has also directly benefited the conservation of reintroduced Madagascar pochards. Captive-reared tufted ducks were given access to a large salmon farming cage, a floating aviary as it were, that enabled birds to practice diving and foraging underwater (Cranswick, 2020; WWT, 2023). Exposing ducklings to deeper water as they grew gave them training in how to dive deeply and forage more successfully. Birds also learned that the floating aviaries provided a safe refuge for loafing and resting.

Young tufted ducks were housed in the floating aviary for a period of 10 days and then allowed to come and go as they pleased. Transmitters on the birds allowed scientists to track their movements, and by feeding them in the floating aviary, conservation staff could ensure the birds were able to access feed and forage naturally in a safe location. This practice worked so well that it was eventually employed on Lake Sofia in Madagascar, where newly hatched Madagascar pochards were housed in these floating aviaries. Ducklings learned to use the artificial feeding stations and, when given access to the lake overall, returned to their floating housing to feed and rest. The habitat quality of Lake Sofia, although degraded, offer the best possible chance for Madagascar pochards to re-establish themselves in the wild. Therefore, it was essential that reintroduced birds did not leave this lake. Rearing a sufficient number of pochards to be an "anchor" to future releases – to keep new birds on Lake Sofia so they can learn from the original pioneers of the conservation release scheme – was a key aim of this work (Cranswick, 2020). As of 2023, the Madagascar pochard conservation story shows many signs of success, with 83 birds being released and a wild population estimation of 119 ducks in total (Durrell, 2023). Pochards have been seen moving between Malagasy wetlands, with the first birds sighted at Lake Alaotra (the ancestral home of this species) for the first time in over 30 years. Ensuring the well-being of the pioneers of this reintroduction and managing the initial population at the safety of Lake Sofia have certainly contributed to its success.

The Madagascar pochard is a prime example of a "silent extinction", a species deemed locally common but actually undergoing a fairly rapid decline that went on unnoticed until the bird was all but vanished (Bamford et al., 2015). Working with field biologists and avicultural experts, animal husbandry, population management and reintroduction efforts were devised that both protected the welfare of pochards at all stages of the conservation work and that enabled the successful augmentation and growth of wild populations using captive-reared birds (Figure 10.5). This project is an excellent example of how animal husbandry and management knowledge can be integrated into conservation work to provide the best possible opportunities for captive individuals to learn and develop skills essential for long-term survival in the wild. Whilst welfare may be challenged when pochards are learning to dive and locate food, ultimately, their overall welfare will be improved by being housed in a way that encourages development and refinement of such important behaviours.

Figure 10.5 Interpretation of the Madagascar pochard conservation programme in the "Mission Possible" exhibit at WWT Slimbridge. 1) Enclosure themed like a Malagasy lake where the pochard would be released. 2) Baer's pochard on display in the enclosure representing the Madagascar pochard. 3) Avicultural creativity – rearing the first batch of pochard ducklings in a hotel bathroom. 4) Equipment used for a successful release of birds onto Lake Sofia. 5) The floating aviaries, in situ on Lake Sofia that were used to train pochards for a life back in the wild.

10.5 POPULATION SUSTAINABILITY, CONSERVATION SUCCESS AND ANIMAL WELFARE SCIENCE

The ultimate goal of conservation is to ensure species' survival and persistence in the wild. To survive for the long term, such a population needs to be sustainable and balance birth rates, death rates, immigration, and emigration. Population sustainability is defined by Lees and Wilcken (2009) as "a population that is able to persist, indefinitely, with the resources available to it." Conservation action takes into account the health and vitality of the habitat to ensure a carrying capacity is present that promotes viability. In the zoo, collaboration between institutions can support the carrying capacity of all institutions involved. And whilst beyond the scope of this chapter, reintroductions and the augmentation, translocation or direct manipulation of wild populations come with welfare-associated risks for other species living in the area of conservation work. There should be a focus on the potential welfare impacts to non-target species when an integrated conservation approach reaches its aims to release a target species back into the wild. Beyond the metapopulation of the target species and looking at other species in the same habitat or ecosystem that may be impacted by conservation activities.

Similarly, population sustainability may only be achieved when key survival behaviours are trained for in captive individuals (Roberts and Luther, 2024). For example, ensuring that prey can respond to predator cues but also that carnivorous species can successfully stalk, catch and dispatch prey. The welfare of all species involved in such a venture should be monitored, managed, assessed and all protocols ethically reviewed. Both the species used as prey and the animal due for release into the wild would suffer unduly if such a process were not given due consideration and review to uphold the best possible implementation.

Interactions with a species' environment can also take the form of social flexibility and changes to social structure. Social systems in the zoo that are not aligned with wild systems impact on animal welfare states (Morgan & Tromborg, 2007) and such an unnatural social environment can impact reproductive outputs and therefore a population's viability. Whilst short-term, acute stressors that promote behavioural diversity due to positive challenge can be beneficial (Meehan & Mench, 2007; Meehan et al., 2003), chronic stress that is beyond the animal's coping ability is detrimental to animal welfare and an individual's chances of thriving in their current situation (Broom & Johnson, 2019). Animals in zoo-based conservation programmes may need to experience different social environments as they develop to provide experiences of a wide variety of interactions with conspecifics. Static social structures and unchallenging captive habitats can alter a species' behavioural repertoire in ways that do not promote social cohesion (Kelley et al., 2006) and these types of behavioural deviation may hamper long-term population survival.

Regardless of the space and complexity of an environment created in the zoo, individuals within zoo-based conservation programmes may be subject to spatial and behavioural restrictions, which will impact their attainment of positive welfare states. Knowledge of behaviours that are highly motivated in their performance, and whose performance is rewarding irrespective of the animal reaching the end goal of the behaviour, i.e. the behaviour's consummatory act (Hughes & Duncan, 1988) is therefore essential to "conservation welfare" – a way of integrating behavioural ecology, conservation biology and animal welfare science. Planning for animal welfare in the management of populations, their enclosure design, training programmes and overall husbandry regimes embeds good welfare firmly into conservation objectives.

To promote the relevance of animal behaviour study to the development of conservation actions, Berger-Tal et al. (2011) designed a conceptual framework to encourage wider integration of the fields of animal behaviour and conservation. The original framework links together three key processes: i) Anthropogenic impacts on animal behaviour, ii) Behaviour-based management, and iii) Behavioural indicators around key behavioural domains of movement and space use, foraging and vigilance, social organisation and reproductive behaviours. These behavioural domains are selected because they are essential for the attainment of individual peak fitness (survival and reproductive potential), influence recruitment into a population, and therefore are integral to conservation outcomes (Berger-Tal et al., 2011). This model could be adapted to replace the behavioural domains with key welfare concepts to illustrate the links between management, human impact and outputs observed in the animal that are all important ways of strengthening conservation work (Figure 10.6).

Control and choice – i.e. ensuring that individuals of a species can experience autonomy in their decision making and actions within their current environment (Ross, 2006) – as well as promoting the performance of appetitive and consummatory behaviours that have adaptive benefits, and ensuring individuals can express positive moods and emotions that indicate environmental suitability are all important aspects of welfare measurement that can be used to enhance conservation projects. Performing behaviours that convey adaptive benefits has welfare benefits too; individual animals are more likely to be equipped with a behavioural repertoire that enables them to meet the challenges of daily life and be satiated from their actions. Appetitive behaviours (activities that take place when the animal is searching for stimuli to interact with) and consummatory acts (the end goal of behaviour that leads to satiation) are important mechanical activities to promote in managed systems because they are likely to be highly motivated to engage in such behaviours (Hinde, 1953) and can gain adaptive benefits from their performance (Arndt et al., 2022). Appetitive behaviours are in themselves rewarding (Spruijt et al., 2001) and therefore environments should promote opportunities for their performance; such self-fulfilling performance of appetitive behaviours without the consummatory reward explains the phenomenon of contrafreeloading. Therefore, impacts from the environment that can also cause behavioural restrictions will impact on the animal's ability to experience good welfare.

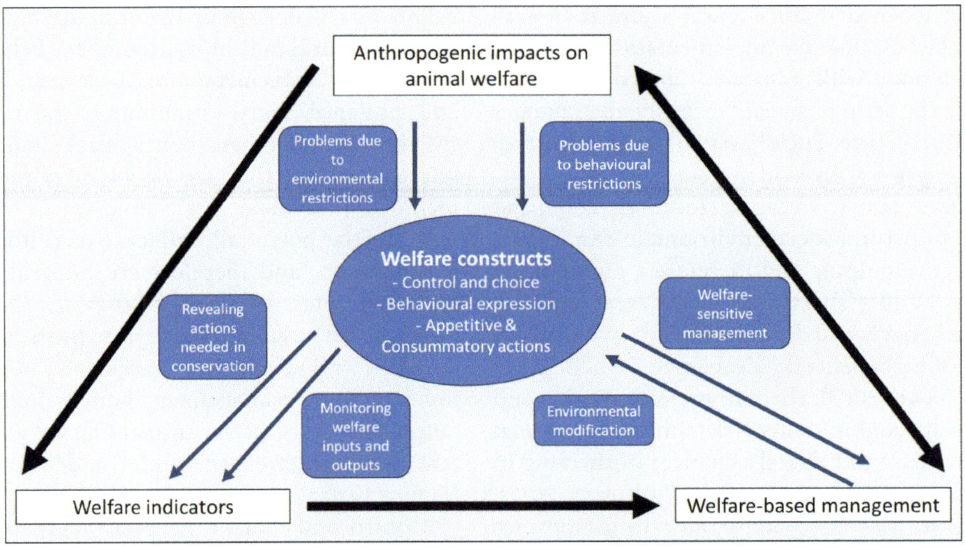

Figure 10.6 An adapted conceptual map of welfare science integrated into conservation biology, taking the three-pronged approach of Berger-Tal et al. (2011) and replacing "behaviour" with welfare. This conservation welfare framework is composed of three fundamental interrelated conservation themes: 1) Anthropogenic impacts on animal welfare; 2) Welfare-based management; and 3) Welfare indicators. Black arrows represent interactions between the conservation themes. Blue arrows show the connections between each theme and specific welfare constructs.

10.5.1 WHAT IS THE ZOO'S CONSERVATION PLAN?

To encourage further engagement with animal welfare science and conservation biology, this chapter strongly advocates for the idea of "conservation welfare." Several other publications have expanded on the concept of conservation welfare to how we treat animals in conservation projects, how conservation can become more compassionate, and how welfare can enhance conservation breeding programmes (Beausoleil, 2020; Beausoleil et al., 2018; Greggor et al., 2018). Using methods and approaches from different areas of behavioural science could enable welfare to be better assessed for individuals across all stages of a conservation project (e.g. when under human care, when being prepared for release, when living post-release in the wild). An example could be borrowing the skills and techniques from those working in sensory ecology to enable us to better understand how animals perceive and interact with the world around them (Swaisgood, 2010). Use of sensory ecology information would provide insight into how the animal is gathering and responding to information and stimuli around it from the animal's perspective, rather than from our own perspective, and therefore we would gain better inferences of the welfare state that the animal may be experiencing.

This chapter presents the term conservation welfare as the following concept:

- Integration of the fields, expertise and methods from conservation biology, behavioural ecology, developmental psychology, veterinary science, and applied animal welfare science.
- A way of considering the relevance of inputs provided to animals within a conservation programme as well as measurement and assessment of the outputs from the animals themselves.
- Providing opportunities for behavioural diversity and the development of appetitive behaviours and adaptive behaviours that convey fitness benefits and welfare benefits too.
- Considering the welfare of the animal within the conservation programme, as well as afterwards. Ultimately, ensuring that animals are well prepared for the desired outcomes of the conservation action (e.g. a life back living in a wild habitat).

We can further develop and enhance the concept and application of conservation welfare by testing and answering the questions posed in Box 10.1.

BOX 10.1: Future questions to be asked to further develop conservation welfare.

- Should we ensure individuals in conservation programmes are fully prepared for reintroduction into the wild by providing experiences of aversive stimuli? Ultimately, this would improve population (and individual) welfare post release.
- Should we maintain more complexity and environmental upheaval (e.g. experiences of unpredictability, sudden and dramatic changes to resources and social groups) within captive populations of animals within conservation programmes to enable individuals to build resilience that, in the long term, would have a positive impact on welfare?
- Should we ensure that environmental enrichment strategies used for captive populations are always fully evaluated and assessed for their usefulness in promoting adaptive behaviours and other actions with a survival function?
- Should we further incorporate mate choice and key elements of a species' sexually selected traits and characteristics into conservation breeding programmes to ensure that mating systems are not overlooked, and individuals get an active choice in who they wish to breed with.
- Should we be working hard to conserve behaviour and the culture of different populations to ensure that important behaviours relevant to how a species learns, develops and adapts to its surroundings are not lost after multiple generations in captivity?
- Should we consider increasing the size, shape and structure of enclosures created for captive species, to make them more dynamic, changeable, and varied on a day-to-day basis?
- Should we remove more of the predictability associated with husbandry and daily captive care?
- Should we develop bespoke animal welfare auditing and assessment tools for individuals within active conservation programmes to ensure that we have some way of estimating resilience and how species will cope in the future, based on what they have currently experienced and how they are developing coping strategies?
- Should we build animal welfare auditing and assessment protocols into release programmes, whereby animals that are part of reintroduction schemes have their welfare monitored for a period of time once they are returned to the wild?
- Defining when animals stop being in a conservation programme and therefore how long do we have a responsibility for them? This means, for how long post-release is an individual monitored until it is no longer part of a reintroduction or translocation scheme, but simply another individual within a population out in the field?

10.6 CONCLUSIONS

This chapter has focussed on the relevance of measuring and upholding good animal welfare to the implementation and success of zoo and aquarium-centred conservation work. Animals in conservation programmes need to experience positive challenges – situations that are uncomfortable yet can be rectified by the animal's responses and that provide an opportunity for learning and behavioural development. To build behavioural resilience, individuals must be exposed to stimuli that elicit a diverse array of behavioural responses that, ultimately, enable an individual to meet said challenge and learn from the experience. Animal welfare considerations in conservation also extend to other species that are directly or indirectly involved. For example, animals used as live prey for training key survival behaviours (e.g. stalking, hunting and prey dispatch) or species that may experience extra competition or displacement due to introductions of formally extirpated species that increase the population density of a given area.

To ensure the welfare of animals in captive- or conservation-breeding programmes, we must continue to develop robust welfare assessment tools that will allow us to ensure "welfare-focussed husbandry" for these important populations. Such assessment tools should encompass measurement of physical health alongside inferences of psychological wellbeing, such as stress responses, behavioural expression, and social choice. Conservation Welfare can be embedded into the One Plan Approach by evaluating the responses of animals to different experiences that build survival skills or conserve adaptive traits or enhance resilience for individuals that need survive in a wild system. Experiencing short-term welfare challenges (fear, competition, alarm) to enable appropriate responses to be developed and used effectively can maximise welfare in the longer term. Ultimately, knowing and understanding the needs of the individual animal (from their perspective of the world around them) will enable conservation success, particularly if animals can be better trained, prepared and readied for a life back out in the unpredictability of the wild.

For further reading and information on the key areas of conservation practice documented in this chapter, please see:

The IUCN Red List https://www.iucnredlist.org/

The IUCN Species Survival Commission https://www.iucn.org/our-union/commissions/species-survival-commission

The IUCN's conservation tools resources page https://www.iucn.org/resources#resource_types-resource_type_conservation_tool

The IUCN SSC CBSG One Plan Approach information page https://www.cpsg.org/our-approach/one-plan-approach-conservation

The World Association of Zoos and Aquariums conservation strategies page https://www.waza.org/priorities/conservation/conservation-strategies/

ACKNOWLEDGEMENTS

With grateful thanks to Dr Mark Whiteside for his comments, edits and ideas to further develop and clarify this chapter.

REFERENCES

Arndt, S. S., Goerlich, V. C. & van der Staay, F. J. (2022). A dynamic concept of animal welfare: The role of appetitive and adverse internal and external factors and the animal's ability to adapt to them. *Frontiers in Animal Science*, *3*, 908513.

Bamford, A. J., Sam, T. S., Razafindrajao, F., Robson, H., Woolaver, L. G. & RenÉ de Roland, L. A. (2015). The status and ecology of the last wild population of Madagascar Pochard Aythya innotata. *Bird Conservation International*, *25*(1), 97–110.

Barongi, R., Fisken, F. A., Parker, M. & Gusset, M. (2015). *Committing to conservation: The world zoo and aquarium conservation strategy.* Gland, Switzerland: WAZA.

Beausoleil, N. J. (2020). I am a compassionate conservation welfare scientist: Considering the theoretical and practical differences between compassionate conservation and conservation welfare. *Animals*, *10*(2), 257.

Beausoleil, N. J., Mellor, D. J., Baker, L., Baker, S. E., Bellio, M., Clarke, A. S., Dale, A., Garlick, S., Jones, B. & Harvey, A. (2018). "Feelings and fitness" not "feelings or fitness"–the raison d'être of conservation welfare, which aligns conservation and animal welfare objectives. *Frontiers in Veterinary Science*, *5*, 296.

Bekoff, M. (2019). *Ignoring nature no more: The case for compassionate conservation.* Chicago, IL: University of Chicago Press.

Berger-Tal, O., Polak, T., Oron, A., Lubin, Y., Kotler, B. P. & Saltz, D. (2011). Integrating animal behavior and conservation biology: A conceptual framework. *Behavioral Ecology*, *22*(2), 236–239.

Bowkett, A. E. (2009). Recent captive-breeding proposals and the return of the ark concept to global species conservation. *Conservation Biology*, *23*(3), 773–776.

Braverman, I. (2014). Conservation without nature: The trouble with in situ versus ex situ conservation. *Geoforum*, *51*, 47–57.

Brook, B. W., Sodhi, N. S. & Bradshaw, C. J. A. (2008). Synergies among extinction drivers under global change. *Trends in Ecology & Evolution, 23*(8), 453–460.

Broom, D. M. (1986). Indicators of poor welfare. *British Veterinary Journal, 142*(6), 524–526.

Broom, D. M. & Johnson, K. G. (2019). *Stress and animal welfare* (2nd ed.). Cham, Switzerland: Springer.

Byers, O., Lees, C., Wilcken, J. & Schwitzer, C. (2013). The one plan approach: The philosophy and implementation of CBSG's approach to integrated species conservation planning. *WAZA Magazine, 14*, 2–5.

Conde, D. A., Colchero, F., Gusset, M., Pearce-Kelly, P., Byers, O., Flesness, N., Browne, R. K. & Jones, O. R. (2013). Zoos through the lens of the IUCN red list: A global metapopulation approach to support conservation breeding programs. *PLoS One, 8*(12), e80311.

CPSG. (2022). The one plan approach to conservation. IUCN SSC Conservation Planning Specialist Group. https://www.cpsg.org/our-approach/one-plan-approach-conservation. (Access Date 24/10/2022).

Cranswick, P. (2020). Madagascar pochard (Aythya innotata). Mohamed bin Zayed Species Conservation Fund. https://www.speciesconservation.org/case-studies-projects/madagascar-pochard/18663. (Access Date 13/11/2023).

De la Fuente, M. F., Souto, A., Caselli, C. & Schiel, N. (2017). People's perception on animal welfare: Why does it matter? *Ethnobiology and Conservation, 6*(18), 1–7.

Durrell. (2023). World's rarest duck returns home. Durrell Wildlife Conservation Trust, https://www.durrell.org/news/world-s-rarest-duck-returns-home/. (Access Date 14/11/2023).

Greenwell, P. J., Riley, L. M., Lemos de Figueiredo, R., Brereton, J. E., Mooney, A. & Rose, P. E. (2023). The societal value of the modern zoo: A commentary on how zoos can positively impact on human populations locally and globally. *Journal of Zoological and Botanical Gardens, 4*(1), 53–69.

Greggor, A. L., Vicino, G. A., Swaisgood, R. R., Fidgett, A., Brenner, D., Kinney, M. E., Farabaugh, S., Masuda, B. & Lamberski, N. (2018). Animal welfare in conservation breeding: Applications and challenges. *Frontiers in Veterinary Science, 5*, 323.

Harley, J. (2022). Behavioural biology in animal collection planning and conservation. In P. E. Rose (Ed.), *The behavioural biology of zoo animals* (pp. 11–19). Boca Raton, FL: CRC Press.

Hinde, R. A. (1953). Appetitive behaviour, consummatory act, and the hierarchical organisation of behaviour: With special reference to the great tit (Parus major). *Behaviour, 5*, 189–224.

Hughes, B. O. & Duncan, I. J. H. (1988). The notion of ethological 'need', models of motivation and animal welfare. *Animal Behaviour, 36*(6), 1696–1707.

IUCN SSC. (2023). *Position Statement on the role of botanic gardens, aquariums, and zoos in species conservation*. Gland, Switzerland: IUCN Species Survival Commission.

Kelley, J. L., Magurran, A. E. & García, C. M. (2006). Captive breeding promotes aggression in an endangered Mexican fish. *Biological Conservation, 133*(2), 169–177.

Kerr, K. C. R. (2021). Zoo animals as "proxy species" for threatened sister taxa: Defining a novel form of species surrogacy. *Zoo Biology, 40*(1), 65–75.

Kleiman, D. G. (1985). Criteria for the evaluation of zoo research projects. *Zoo Biology, 4*(2), 93–98.

Lees, C. M. & Wilcken, J. (2009). Sustaining the Ark: The challenges faced by zoos in maintaining viable populations. *International Zoo Yearbook, 43*(1), 6–18.

Malamud, R. (2015). The problem with zoos. In L. Kalof (Ed.), *The Oxford hanbook of animal studies* (Vol. 6, pp. 397–410). Oxford, UK: Oxford University Press.

Meehan, C. L. & Mench, J. A. (2007). The challenge of challenge: Can problem solving opportunities enhance animal welfare? *Applied Animal Behaviour Science, 102*(3–4), 246–261.

Meehan, C. L., Millam, J. R. & Mench, J. A. (2003). Foraging opportunity and increased physical complexity both prevent and reduce psychogenic feather picking by young Amazon parrots. *Applied Animal Behaviour Science, 80*(1), 71–85.

Mendl, M., Burman, O. H. P. & Paul, E. S. (2010). An integrative and functional framework for the study of animal emotion and mood. *Proceedings of the Royal Society B: Biological Sciences, 277*(1696), 2895–2904.

Morgan, K. N. & Tromborg, C. T. (2007). Sources of stress in captivity. *Applied Animal Behaviour Science, 102*(3–4), 262–302.

NAWMP. (2023). About. North American Waterfowl Management Plan. https://nawmp.org/content/north-american-waterfowl-management-plan. (Access Date 27/11/2023).

Proctor, H. (2012). Animal sentience: Where are we and where are we heading? *Animals, 2*(4), 628–639.

Proctor, H. S., Carder, G. & Cornish, A. R. (2013). Searching for animal sentience: A systematic review of the scientific literature. *Animals, 3*(3), 882–906.

Roberts, J. L. & Luther, D. (2024). Captive birds exhibit greater foraging efficiency and vigilance after anti-predator training. *Behavioral Ecology, 35*(5), arae051.

Rose, P. E. (2021). Evidence for aviculture: Identifying research needs to advance the role of ex situ bird populations in conservation initiatives and collection planning. *Birds, 2*(1), 77–95.

Rose, P. E. & Riley, L. M. (2022). Expanding the role of the future zoo: Wellbeing should become the fifth aim for modern zoos. *Frontiers in Psychology, 13*, 6649.

Rose, P. E. & Riley, L. M. (2023). Five ways to wellbeing at the zoo: improving human health and connection to nature. *Frontiers in Psychology, 13*. https://doi.org/10.3389/fpsyg.2023.1018722.

Ross, S. R. (2006). Issues of choice and control in the behaviour of a pair of captive polar bears (Ursus maritimus). *Behavioural Processes, 73*(1), 117–120.

SDJV. (2023). What is the Sea Duck Joint Venture? Sea Duck Joint Venture. https://seaduckjv.org/about-the-partnership/. (Access Date 27/11/2023).

Spruijt, B. M., Van den Bos, R. & Pijlman, F. T. A. (2001). A concept of welfare based on reward evaluating mechanisms in the brain: Anticipatory behaviour as an indicator for the state of reward systems. *Applied Animal Behaviour Science, 72*(2), 145–171.

Swaisgood, R. R. (2010). The conservation-welfare nexus in reintroduction programmes: A role for sensory ecology. *Animal Welfare, 19*(2), 125–137.

Traylor-Holzer, K., Leus, K. & Bauman, K. (2019). Integrated collection assessment and planning (ICAP) workshop: helping zoos move toward the one plan approach. *Zoo Biology, 38*(1), 95–105.

Traylor-Holzer, K., Leus, K. & Byers, O. (2018). Integrating ex situ management options as part of a one plan approach to species conservation. In B. A. Minteer, J. Maienschein, & J. P. Collins (Eds.), *The ark and beyond: The evolution of zoo and aquarium conservation* (pp. 129–141). Chicago, IL: University of Chicago Press.

USFWS. (2023). Sea Duck Joint Venture. US Fish & Wildlife Service. https://www.fws.gov/project/sea-duck-joint-venture. (Access Date 27/11/2023).

Wickins-Dražilová, D. (2006). Zoo animal welfare. *Journal of Agricultural and Environmental Ethics, 19*(1), 27–36.

Wolfensohn, S., Shotton, J., Bowley, H., Davies, S., Thompson, S. & Justice, W. S. M. (2018). Assessment of welfare in zoo animals: Towards optimum quality of life. *Animals, 8*(7), 110.

WWT. (2023). Saving the Madagascar pochard. Wildfowl & Wetlands Trust. https://www.wwt.org.uk/our-work/projects/saving-the-madagascar-pochard/. (Access Date 14/11/2023).

11

New technology – challenges and opportunities for zoo animal welfare science

LISA K. YON

11.1 INTRODUCTION

There have been numerous developments in equipment and in scientific approaches over the past 10–20 years that may be particularly useful in evaluating and/or improving the welfare of wildlife under human care. Methods that allow facilities to collect data when staff and visitors are not present can eliminate both potential observer effects and the impact of visitors on behaviours exhibited. Technology could also be used to automate the collection of welfare-related data, thus reducing the staff time required to monitor animal well-being. For those behaviours that are expressed less frequently, continuous behavioural monitoring facilitated by new automated technology ensures that these rarer behaviours will still be captured. Such technology can also remove the known risks of observer fatigue that result from people needing to spend extended periods of time observing and recording behaviour. Moreover, it can provide new ways to capture aspects of an animal's welfare, including new physical and physiological measures. Those hoping to utilise technology to support welfare assessment in zoo and aquarium populations could review the literature from the agricultural sciences to aid their own developments (e.g., Kovács et al., 2014), especially when considering integrated concepts of animal welfare (Reimert et al., 2023). Examples of these new methods and approaches are discussed in this chapter.

11.2 SENSOR TECHNOLOGIES

11.2.1 ACCELEROMETERS

Accelerometers have been used since the 1990s in numerous terrestrial, aquatic, and avian species (including domestic chickens (*Gallus gallus*

DOI: 10.1201/9781003429425-11

domesticus), goats (*Capra hircus*), cattle (Bos taurus), and horses (*Equus ferus caballus*), as well as red deer (*Cervus elaphus*), sea turtles (Cheloniidae), penguins (Spheniscidae), lemurs (Lemuridae), koalas (Phascolarctidae), and elephants (Elephantidae)) – resulting in recorded data for over 120 species (Whitham et al., 2016). These data can enable analysis of activity budgets, sleep patterns, enclosure usage, and positive and negative social interactions. Triaxial accelerometers, specifically, record bodily motion in three different planes or axes: lateral (x axis, left-right), anterior-posterior (y axis, front-back), and dorso-ventral (z axis, up-down). In combination, these axes can identify movements involving changes in orientation within the same location (static movements) and movement between different locations (dynamic movements). Additionally, periodic movement or oscillation may be categorised (e.g. walking or repetitive swaying) (Soltis et al., 2016). Once these data have been collected, supervised machine learning algorithms can be applied to infer behaviour. Training the algorithm involves coupling acceleration data with direct behavioural observations to confirm accuracy. This methodology is discussed in greater detail in section 11.5.

While there are many possible uses of accelerometer data, validation for the species of interest is essential. This should include sufficient sample size, of both sexes, a wide age range, and a greater range of behaviours than previously considered. Validation with additional indicators would also provide greater confidence in the resulting data. Such data have been used to successfully distinguish between resting, feeding, bathing, walking, rapid walking, and lying down in elephants (Soltis et al., 2016). By attaching two accelerometers to different legs, Soltis et al. were able to evaluate gait and detect asymmetric stepping or other gait abnormalities. Data such as this can also determine if interventions have positively affected behaviour as intended.

Before recording data, the location and method of attachment of accelerometers must be decided, while confirming the equipment itself doesn't cause behavioural changes (Whitham et al., 2016). It's important to confirm that neither the subject nor other animals can remove or damage the equipment. Additionally, time is required to habituate the animal prior to data collection. While triaxial accelerometers might, in principle, distinguish between different affective states, e.g. in elephants (Soltis et al., 2016), the evidence is insufficiently conclusive. Increasing the sensitivity of this equipment may distinguish between subtle differences in posture or movement, but more comprehensive, scientifically robust studies are required to determine if this methodology could reliably discriminate between different affective states.

11.2.2 GLOBAL POSITIONING SYSTEMS (GPS)

Global Position Systems (GPS) use satellite telemetry to triangulate an animal's location. These can be attached to an animal's neck (via a collar) or attached with a strap around a limb (Jukan et al., 2017). This can be particularly useful for cryptic species, those in an environment where they are difficult to observe, and those roaming over distances of hundreds or thousands of miles (Ossi et al., 2016). For captive wildlife, GPS sensors can evaluate activity (movement), distance travelled, and interaction with resources or conspecifics (based on distances between them) (Hacker et al., 2015). These sensors have been used in numerous species including green turtles (*Chelonia mydas*), king penguins (*Aptenodytes patagonicus*), and African elephants (*Loxodonta africana*).

There are, however, numerous challenges and limitations when using GPS to track behaviour and movement. Project design must consider appropriate equipment specifications for the study species, including device size and position. It is worth noting that devices worn as anklets in quadrupedal species may not be as accurate, as the animal's body can block location attempts. It's also important to test that the size and position of the device do not influence animal behaviour, and to train the animal to accept the placement and removal of the device (Ossi et al., 2016).

Environmental features can also influence location accuracy, including the influence of terrestrial atmosphere, habitat, topography, and weather. For example, gaining accurate location measurement indoors can be difficult as satellite signals can be poorly received by the GPS device. Outdoor locational accuracy in different GPS systems ranges from 1–10 metres; depending on habitat size and the study species; this could result in behaviourally meaningful differences. It is therefore crucial to test the accuracy of the GPS system for that species and habitat, prior to use in a study. Device locational accuracy should be tested comparing results with a hand-held GPS unit, placed at pre-measured distances. When GPS data will be used to predict behaviours, validation (using behavioural observations) is required, for the species and for individuals, as behavioural idiosyncrasies may be difficult to interpret from GPS data alone.

11.2.3 RADIO FREQUENCY IDENTIFICATION SYSTEMS (RFID)

Radio frequency identification systems (RFID) have been used in the livestock industry for decades (Kalafut et al., 2020). These systems consist of RFID tags worn by the animals, antennae which detect proximity of a tag, and a reader/receiver that receives and stores data. Whenever a tag is within range of an antenna, electromagnetic induction activates it, and the tag's unique number is sent to the reader, which records the date and time of the tag's detection and location of the antenna that detected it; all of these data are stored by the reader. The tags can be purely passive so can be very lightweight, enough that they can be attached to insects such as bees. Such passive tags are also inexpensive and have an essentially unlimited operational life. Alternatively, tags can be active, with a battery and internal transmitter, actively transmitting their data. Active tags can be detected up to 150 m away, so are appropriate for larger distances between the device/animal and the receiver. The trade-off for such tags is usually a larger size, greater weight, and the need to charge or change out the energy source for such tags. Unlike GPS, RFID tags work in more settings, including inside buildings and in burrows or nests, using local readers rather than satellites. Multiple tags can be read simultaneously, and a line of sight between tag and reader isn't required. Trilateration can be used to pinpoint location, and this system can investigate daily or seasonal space use and environmental interaction. This can include interactions with other animals, if they are also wearing tags.

With RFID tags, researchers have investigated animals' interactions with both their social and physical environments, including individual preferences in resource use, feeding rates, and social interactions. RFID systems have provided insights into social dynamics, encounter mapping, and social networks in various species, including ants, badgers and bumblebees. They have also enabled opportunities for choice by zoo animals, such as turning lights on or off, accessing specific areas of their exhibit, or accessing feeders (keyed to individual RFID tags) (Coe et al., 2020). Providing more opportunities for animals to make choices and have more control over their environment is held to be fundamental in providing positive welfare to any animals in captivity (Englund et al., 2023), and methods are increasingly being sought to provide more such opportunities.

There are, though, several considerations or cautions that should be taken into account when seeking to use RFID systems in a zoo setting. As with any such technology, careful planning is essential. Careful consideration must be given to tag placement on all the relevant resources and antennae at appropriate locations. There can be interference from adjacent readers, and electromagnetic interference from certain environmental features. It's possible that a tag won't be detected due to its orientation relative to the antenna, or it may be detected multiple times if the animal moves slowly. The devices (tags, antennae, readers) must also be durable, weather-proof, and safe for the animals and members of the public, while being conveniently accessible. Data analysis can be time-consuming, and raw data must be reviewed initially to create 'rules' for interpreting the generated datasets. When setting up the system, it's important to consider frequency of equipment servicing, but also time and planning requirements to set up and install the system.

11.2.4 BIOACOUSTICS

We are becoming increasingly aware of the potential impacts (both positive and negative) that the sound environment can have on captive animals (Rose et al., 2022). Devices placed on animals or within the animals' enclosure can be used to record sounds produced by the individual, or can monitor for noise pollution within and around the enclosure. Such devices can be fixed to collars or other types of mountings near the animal's head, to record vocalisations that they produce. This is most feasible on larger animals that can tolerate equipment weight from the sound detector and associated power source (Orban et al., 2017). Alternatively, devices can be placed in a fixed location to capture animals' vocalisations and environmental sounds. This is useful for smaller animals for whom collar-mounted devices are unfeasible due to size and/or weight. Where individuals can be identified from acoustic features of their vocalisations, then individual animals can be monitored using an environmental hydrophone. The advantage of having an environmental microphone is that it can be attached to mains power, or to larger batteries, which can power the microphone for longer time periods; these batteries can be readily switched out as required. Such devices can continuously monitor these sounds in real time.

Vocalisations produced by an animal can reflect physiological and psychological aspects of welfare, provided the vocalisations of that species are well understood. Vocalisations can be associated with activation of the hypothalamic-pituitary-adrenal axis during stress or excitement (physiological arousal), and vocalisation studies have typically focussed on negative welfare indicators (stress or pain). However, recent studies have explored vocalisations associated with positive welfare states, such as 'all's well' sounds including chirps by rats (*Rattus norvegicus domestica*), laughter-type sounds by great apes, and 'let's go' rumbles by African elephants (*Loxodonta africana*) (Orban et al., 2017). If individuals can be identified by their individual acoustic 'fingerprints', then it's possible to monitor individual responses to the environment and to conspecifics.

There are, however, some challenges that must be surmounted or considered when interpreting vocalisation data. Foremost among these, for many species, is if their 'language' is poorly understood. This would make it difficult to interpret vocalisations, or to identify with certainty the vocalisations' association with positive or negative welfare. To more confidently distinguish between positive and negative welfare states, validation is necessary for the focal species. This involves collecting data on vocalisations alongside other types of welfare indicators, such as a combination of behavioural, physiological or animal health parameters. It is important to ensure that sufficient numbers of animals of both sexes, and across age classes, are included in preliminary validation, alongside eliciting and spectrographically analysing the calls produced. It is crucial to select recording equipment appropriate to the species, which can detect sounds in the relevant frequencies (possibly including infra- or ultra-sonic ranges). Selecting unidirectional or omnidirectional microphones, and appropriate sampling rates and bit depth, is also important (Blumstein et al., 2011).

If the species is group-living and too small to wear individual microphones, it might not be possible to evaluate vocalisations from individual animals, unless each individual has a unique soundprint. Recording vocalisations is particularly challenging in an aquatic environment, and will require numerous additional equipment specifications. Furthermore, in any environment, there will be inconsistency of vocal responses to specific situations, even for the same animal at different times. It's also possible, or even likely, that vocalisations won't reflect longer-term welfare states, as they are often made as an immediate response to particular circumstances.

11.2.5 INFRARED THERMOGRAPHY

Infrared thermography (IRT) involves using specialised equipment to detect changes in the electromagnetic radiation emitted from the surfaces of regions of the body. IRT has been used in numerous species, including insects, reptiles, birds and mammals, such as macaques (*Macaca mulata*), horses (*Equus ferus caballus*) and cattle (*Bos taurus*) (Travain et al., 2021). IRT can non-invasively evaluate temperature in various regions of the body, while recording the thermal image using photographs or captured video footage. The equipment ranges from small handheld units to large free-standing scanning stations, and can be used at very short range (less than a metre) and up to large distances of over 1,000 metres.

Measurement of heat radiated from body surfaces can indicate a variety of conditions. Starvation results in decreased heat-producing metabolic processes in order to conserve energy, thereby resulting in lower body surface temperatures. IRT can detect an infectious disease, since the body's response to infectious organisms is to release mediators which increase body temperature. This was identified in domestic cattle infected with Bovine Viral Diarrhoea, wild raccoons (*Procyon lotor*) with rabies, and mule deer (*Odocoileus hemionus*) with foot and mouth disease (Travain et al., 2021). Inflammation results in local vasodilation, increasing temperature in areas of the body with inflammation. IRT has been used to detect inflammation in the feet and legs of Asian elephants (*Elephas maximus*), black rhinoceros (*Diceros bicornis*), giraffes (*Giraffa camelopardalis reticulata*) and hippopotami (*Hippopotamus amphibius*).

IRT has also been used to detect changes in temperature in various body regions in association with negative emotional states (Wongsaengchan et al., 2023). Stress stimulates increased core body temperature through increased heart rate and heat production by brown adipose tissue. Stress also triggers catecholamine release from the adrenal medulla, causing peripheral vasoconstriction and decreased surface temperature on numerous body regions (Ghezzi et. al., 2004). Thus, there is a complex balance between increased and decreased temperature on various body surfaces, which depends on the species and the presence of fur or feathers, which have insulative properties. Changes in surface temperatures, associated with a presumed negative emotional state, have been detected in horses, laboratory rodents, and non-human primates such as macaques, chimpanzees (*Pan troglodytes*) and gorillas (*Gorilla gorilla*) (Travain et al., 2021).

A commonly used site for monitoring thermal changes is the eye as it is unhaired and accessible, regardless of the presence/colour/thickness of the animal's haircoat (or feathers), with extensive capillary beds that respond to sympathetic and parasympathetic stimulation. Additionally, the eye most consistently exhibits subtle temperature changes during stress or high physiological arousal. Other body areas used to monitor thermal changes include the nose, other regions of the face, ears, the axillary region and, more generally, body surface temperatures from the dorsum or thoracoabdominal regions. In great apes, a negative association has been identified between nasal temperature and social stress (Travain and Valsecchi, 2021). This was interpreted as resulting from vasoconstriction in vessels supplying the nasal region, due to increased physiological arousal and release of vasoconstrictors such as catecholamines from the adrenal medulla. IRT can also indicate an animal's thermal state or comfort, vasoconstricting in cooler temperatures to conserve heat, resulting in lower body surface temperatures, or vasodilating to shed heat, resulting in higher body surface temperatures. This can highlight conditions when the animal is unable to either stay sufficiently warm (and needs access to heated areas), or is insufficiently able to shed heat, thus requiring shaded areas or a way to cool off through access to water or a mud wallow, for example (Mota-Rojas et al., 2022).

As with other sensor methods, there are numerous challenges in using IRT to monitor well-being in captive wildlife. Firstly, there are numerous potentially confounding factors, including handling the animal, and variations in environmental temperatures, sunlight, shade and wind can all affect body surface temperatures. Another challenge involves selecting an appropriate site on the body to monitor, since fur or feathers can make it difficult to detect changing body surface temperature. Additionally, whether temperature increases or decreases under different conditions varies for different areas of the body, making the selection of the body region to monitor an essential part of study design. Moreover, the hair or feather colour, whether they are wet or dry, and whether or not they are dirty, also affect surface temperature. It's crucial that the IRT device is perpendicular to the surface being measured, otherwise it will read the surface as being colder than it actually is.

It's important too, that the selected body region (e.g. eye, muzzle, dorsal surface, joints, feet) is appropriate for that species (often hairless areas in mammals e.g. face, muzzle or eye, or featherless areas in avian species e.g. eyes or legs/feet). Crucially, a rigorous process of experimental validation is required to confirm whether the selected body part reflects body temperature changes in response to physical, thermal or emotional stress. However, for any given stress or species, the literature is inconsistent over whether temperatures increase or decrease in response, and in which body part temperature changes, making it difficult to interpret thermal data in isolation. It is therefore necessary to analyse this parameter alongside other welfare measures (e.g. behavioural, endocrine, oxidative stress) to facilitate data interpretation.

11.2.6 HEART RATE AND HEART RATE VARIABILITY

For decades, it has been suggested that heart rate can indicate a welfare state. Heart rate is determined by the balance of the sympathetic and parasympathetic nervous systems. The parasympathetic nervous system (PNS), is most active during rest and relaxation. When the parasympathetic system predominates, heart rate will tend to be slower. At times of stress or excitement, the sympathetic nervous system (SNS) predominates, and a higher heart rate results. Sensors to measure heart rate usually involve a monitor, which is secured onto the animal's body; this is frequently a Holter monitor worn in a harness. While such monitors are typically placed on the chest directly over the heart, some may be positioned over a peripheral blood vessel, e.g. on a limb. The location and form of attachment must be validated for the species in question. Alternatively, heart rate monitors have recently been incorporated into an ingestible pill which the animals swallow; heart rate is monitored as it moves through the digestive tract, and data are downloaded once the device has finished its transit through the animal (Von Borell et al., 2007). More recently, heart rate variability (HRV), which is influenced by the balance of parasympathetic and sympathetic activity (Wascher, 2021), has been suggested to indicate a welfare state. At rest, when the PNS predominates, heart rate is inherently variable. However, during stress, when the SNS predominates, the heart rate settles into a more fixed pattern, and HRV decreases.

While there are a number of ways to measure HRV, they all ultimately evaluate variations in interbeat intervals (IBIs) (Von Borell et al., 2007). The most commonly used approach evaluates variations in IBI over 5-minute intervals. While Holter monitors are usually used, other types of equipment have also been successful, with some studies suggesting that it may be possible to evaluate both heart rate and heart rate variability using photos or video footage (Al-Naji et al., 2019). However, caution is warranted on the validity of this approach, given the limitations mentioned below, even if using more standardised approaches to measurement. Though heart rate can increase during stressful situations which negatively impact the welfare of that animal, it can also increase during times of physical activity or social excitement. Therefore, heart rate is more appropriately considered an indicator of physiological arousal which may reflect either a decrement or an improvement in welfare – thus heart rate doesn't provide any information about the valence, and other data (such as measures of behaviour before, during and after any changes in heart rate) are needed in order to accurately interpret what changes in heart rate may mean. In addition, a period of training and of habituation is required prior to any study involving heart rate measurement. Equipment used may be costly, and the resulting data time-consuming to interpret.

HRV is described as a measure of vago-sympathetic balance; with good health (physical or mental) equating to parasympathetic dominance in the resting state. However, even within a single species, there is tremendous variability in this measure and, correspondingly, less certainty in the use of this as a measure of welfare across other species. As with all new welfare measures, it is sensible to first develop experiments to systematically and methodically evaluate validity for the species of interest, and to cross-validate using a suite of more established measures.

11.3 ENRICHMENT TECHNOLOGIES

Further to improving the collection of welfare-driven research data, advancements in technology have also created numerous opportunities to create interesting, cognitively stimulating environments, providing opportunities for captive animals to express natural behaviours and to make choices with a sense of personal agency. There are new interactive devices that have been used with numerous primate species, and have shown to be intrinsically rewarding (Coe et al., 2020). Other devices used levers, switches or interactive touchscreens, enabling animals to exert control over their environment. For example, orangutans (*Pongo* spp.) changed lighting or heating/cooling in their exhibit (Carter et al., 2021) and automated showers allowed elephants to control water provision using levers (Coe et al., 2020).

RFID chips (either implanted in the skin or on collars worn by animals) can complement enrichment strategies, allowing individual animals to control their own access to food and other resources, or use motion detection to open doors to access different areas of their exhibit (see Coe and Hoy, 2020). In other cases, randomising timing for access to food and other resources has been shown to decrease stereotypies and encourage exhibit use. Coe and Hoy (2020) reported increased elephant locomotion and enclosure use when feeder access occurred at random times, determining that the elephants walked daily distances comparable to those travelled by their wild conspecifics.

Virtual reality (VR) technology may also present tremendous opportunities for enrichment, creating a virtual environment which could incorporate numerous sights and sounds similar to a species' natural ecosystem. This could provide a sense of living in a larger, and seemingly more complex, space (similar to their natural environment) than is actually the case. Preliminary studies in chimpanzees and bonobos (*Pan paniscus*) indicated VR technology was successfully used to engage these species (Coe and Hoy, 2020). In future, such technology may be beneficial in providing more opportunities for, and encouraging, the expression of natural behaviours.

There are a number of issues, however, which must be considered when planning to utilise such equipment. To start with, this equipment can be quite costly to purchase and maintain. An extensive period of behaviour training may be required, not only for interactive systems, but also for animals to accept collars or other equipment housing sensors. Staff training will also be required to monitor and control equipment access, and to train the animals to use the equipment effectively and safely. Safety considerations, more generally for all the animals in the enclosure, for keepers and for members of the public, must be adhered to. For example, if microchips are used, they can migrate under the skin. Staff may also lack the time required to train, monitor and maintain use of the equipment.

More studies are needed to better understand individual differences in response to these technologies, and longer-term impacts of these devices on the well-being of the animals involved. The equipment or technology must be tailored to the species of concern, to ensure it is safe and usable, but also validated for use in that species. It's important that the institution has a system for evaluating device efficacy and impact, both in the immediate and longer term. It is quite possible that, over time, habituation to the devices will occur, the level of interaction will lessen, and positive impacts from the device will decrease. Ultimately, as with any form of enrichment, it is essential that there are clear and measurable behavioural goals, that these are systematically and routinely evaluated, and adjustments or changes are made based on the results of those evaluations.

11.4 SOFTWARE

There are a number of programmes that have been developed to collect and analyse data for measuring and monitoring welfare (see van der Marel et al., 2021; Galante et al., 2022). Some software are intended for a wide range of species, while others may be species-specific. The former includes software such as 'ZooMonitor', 'BORIS', 'Animal Observer', 'Animal Behaviour Pro', 'Prim8' and 'Welfare Trak'; and all programmes that help to capture and collate behavioural data. Some of these are free of charge to users (e.g. 'Animal Observer', 'BORIS', and 'Prim8') while others have a fee for use ('Welfare Trak', 'ZooMonitor', 'Animal Behaviour Pro') and may include additional features. For example, 'ZooMonitor' permits the user to capture accompanying metadata, including the location of the animal within the enclosure (Wark et al., 2019). Some species-specific software have been scientifically developed and validated for use, such as the 'Elephant Welfare App' (see Yon et al., 2019). These software systems can save time and effort in collecting (and potentially help in analysing) behavioural data (Figure 11.1). It seems likely that these will continue to be developed in the future, given the increasing drive to evaluate and improve welfare in captive wildlife held in zoos around the globe.

While such software may represent a tremendous opportunity for collecting and analysing welfare-related data, it is essential that background work is first undertaken to identify appropriately validated behavioural welfare measures for the species in question, thus utilising or developing a validated behavioural sampling protocol that is an accurate representation of the animal's behaviour. As with any such measures, robust scientific validation of the methods is vital in order to be confident that the results accurately reflect that aspect of the animal's welfare state; if not, then these data gathered, using what might seem to be a scientific and quantitative methodology, must at best

Figure 11.1 Zookeeper using the 'Elephant Welfare App' to collect Qualitative Behavioural Assessment (QBA) data for captive elephants. Photo credit: Alex Wilkinson Media/Elephant Welfare International.

be viewed as tentative and may not represent the animal's true welfare state. Utilising programmes and apps which have already undergone rigorous scientific validation is therefore a cost-effective and time-efficient option for facilities. It would seem prudent to explore collaboration with external colleagues working in the fields of animal behaviour, health or welfare, or in mathematics or computer science, who may bring a fresh perspective and new methodologies to approach the continuing challenges of providing and monitoring welfare in captive wildlife.

11.5 FUTURE POSSIBILITIES: MACHINE LEARNING AND ARTIFICIAL INTELLIGENCE (AI)

As is highlighted and discussed throughout this book, the measurement of animal welfare requires a comprehensive, integrated approach, utilising measures across numerous disciplines to develop a holistic understanding of an animal's well-being. Many of the new technological developments mentioned in this chapter will, however, generate enormous quantities of data, with both the volume of data generated and the multiplicity of data types being time-consuming and complex to analyse. Machine Learning can be used to overcome this challenge, using rich datasets to train models and develop algorithms able to interpret and integrate extensive and complex data (Zuerl et al., 2022). Once developed, these algorithms could be utilised to rapidly process and analyse data, enabling real-time analysis. This would enable the development of early warning systems for detecting problems quickly and would also enable ongoing monitoring for a more comprehensive approach to evaluating welfare in captive wildlife. Such algorithms have already been developed and are currently utilised by many livestock farmers who are using 'precision livestock farming' to integrate data from multiple sources, including sensors both on the animal and in the environment to monitor health and welfare parameters in real time. This results in improved animal welfare outcomes and increased farming efficiency (Kleen et al., 2023).

Although there is great promise in the utilisation of machine learning and artificial intelligence to process and integrate complex datasets rapidly, the training of the algorithms with appropriate and extensive datasets is essential; such data may be cumbersome, time-consuming and potentially costly to obtain. Each system developed will also only be applicable to the animals and environments on which it has been trained. Thus, it is imperative to garner the largest possible datasets, including animals of both sexes and across all age classes. The trained algorithm will be specific to a particular species, and this same process of data gathering and algorithm training will be required for each species of interest. It may be hoped that, if the captive wildlife community came together, they could share out the process of developing and training algorithms for different species, thus making the process more efficient.

Table 11.1 references each technology as detailed in this chapter, highlighting their current uses, potential challenges and suggestions for future directions where each could be taken. Table 11.1 outlines a selection of technologies an their applications to animal welfare science in the zoo, but this is a broad subject area with more applications than have been possible to discuss in this chapter. Box 11.1 lists a selection of texts for further recommended reading on the subject.

BOX 11.1: Suggested further reading

With the application of new technologies being such an evolving area of zoo animal welfare science, additional articles of suggested reading are listed below which may be used to further expand comprehension of this topic:

Clay, A. W., Perdue, B. M., Gaalema, D. E., Dolins, F. L., & Bloomsmith, M. A. (2011). The use of technology to enhance zoological parks. *Zoo Biology, 30*(5), 487–497.

Diana, A., Salas, M., Pereboom, Z., Mendl, M., & Norton, T. (2021). A systematic review of the use of technology to monitor welfare in zoo animals: Is there space for improvement? *Animals, 11*(11), 3048.

Grünewälder, S., Broekhuis, F., Macdonald, D. W., Wilson, A. M., McNutt, J. W.,& Shawe-Taylor, J. (2012). Movement activity-based classification of animal behavior with an application to data from cheetah (*Acinonyx jubatus*). *PLOS One, 7*(11), e49120.

Leoni, J., Tanelli, M., Strada, S. C., & Berger-Wolf, T. (2020). Data-driven collaborative intelligent system for automatic activities monitoring of wild animals. *Proceedings of the 2020 IEEE International Conference on Human-Machine Systems (ICHMS)*, 1–6.

Menaker, T., Zamansky, A., van der Linden, D., Kaplun, D., Sinitica, A., Karl, S., & Huber, L. (2020). Towards a methodology for data-driven automatic analysis of animal behavioral patterns. *Proceedings of the Seventh International Conference on Animal-Computer Interaction, 12*, 1–6.

Picco, G. P., Molteni, D., Murphy, A. L., Ossi, F., Cagnacci, F., Corrà, M., & Nicoloso, S. (2015). Geo-referenced proximity detection of wildlife with WildScope: Design and characterization. *Proceedings of the 14th International Conference on Information Processing in Sensor Networks*, 238–249.

Table 11.1 Uses, opportunities and challenges of new technologies in zoo animal welfare science

Technology/new measures	Details	Uses	Caveats/costs	Future directions
GPS	Uses satellite telemetry to triangulate an animal's location.	Activity, infer resource use and social interactions based on time spent in proximity to them.	Validation, how and where to attach, animal training, costs. Weather, topography can interfere. Poor measurement indoors. Extensive data analysis.	Combine with accelerometers, Machine learning, integrate data, real-time monitoring of movement and behaviour.
Radio frequency identification systems (RFID) tags	Antennae detect tag proximity to resources or conspecifics.	Monitor use of space & resources, social interactions. Can be used to gain access to resources.	Validation, how and where to attach, location of tags and antennae. Interference from adjacent readers, environmental features. Extensive data analysis.	Integration with other devices/ measures. Machine learning for rapid data analysis.
Triaxial accelerometers	Detects changes in body orientation or dynamic movements.	Identifies rest, locomotion, interaction with physical and social environment, and gait.	Validation, how and where to attach, sturdy and safe equipment, training needed, equipment cost, extensive data analysis.	More sensitive equipment, detects subtle changes in posture and movement, may detect different affective states.
Bioacoustics	Device to detect sound worn on the animal, or placed in the enclosure.	Monitor noise pollution, or sound from animals, may link with affective state.	Validation, individual variation and inconsistency; device weight, size, durability, power and frequency range.	Individual acoustic 'fingerprints', can track in environment, interactions with conspecifics.

(Continued)

Table 11.1 (Continued)

Technology/new measures	Details	Uses	Caveats/costs	Future directions
Infrared thermography	Detects electromagnetic radiation emitted from body surfaces.	Detects starvation, inflammation, infectious diseases, thermal balance, and potentially emotional states.	Validation, affected by hair or feathers (colour, thickness), environmental conditions, variation in direction of temperature change.	More sensitive instruments: longer distances (field conditions); better accuracy, detect subtler changes.
Heart rate variability	Sensor(s) to monitor heart rate (Holter monitor and harness).	Heart rate is more variable when relaxed; less variable when stressed.	Validation, highly variable, even within species.	Monitor using photos or videos. Monitoring device contained in ingestible pill.
Enrichment with tech	Interactive electronic devices or RFID tags.	Cognitively stimulating environment, elicits natural behaviours, provide agency.	Validation, costly equipment and maintenance, training (staff and animals), habituation, sturdy safe equipment.	Virtual environments, naturalistic, increased sense of space or complexity.
Software	Electronically collect welfare-related data, helps organise data collection and storage.	May analyse collected data, summarise results. Faster, easier welfare monitoring.	Validation for species is crucial, caution extrapolating from collected data.	Multiple welfare measures in one App, machine learning for complex data analysis.
Machine learning	Train models, develop algorithms to interpret and integrate extensive and complex data.	Rapid data analysis in real time. Detect problems early, comprehensive welfare evaluation.	Time consuming to train algorithms. Only applicable to animals and environments algorithm was trained on. Species specificity.	Integrate data from multiple sources in real time, tailor care to each individual.

11.6 CONCLUSIONS

With so many new devices and developing technologies, it is important to ensure that there has been a rigorous process of scientific development and validation of the tools and equipment for the species in question, to ensure that what is measured is definitively linked to welfare and that it is being captured in an accurate and representative way. Given the potential complexity and costs involved for many of these technologies, it is important that a facility spends time brainstorming ideas with zoo staff prior to the selection and implementation of a particular tool, to ensure that the method is both feasible and practical. Facilities should undertake pilot testing of a device or system prior to implementation to ensure that it is working correctly, whilst also ensuring that outputs from any devices are regularly and routinely evaluated to ensure that the device or system is functioning correctly. It is also important to regularly check for any updates to the technology being used, to ensure it stays up to date.

There is work underway to develop devices utilising multiple sensors (e.g. temperature, respiration and heart rate monitoring), enabling the integrated monitoring of multiple physiological systems. Other developments focus on animal behaviour and movement technology, for example integrating a microphone, GPS locator and accelerometer to gather activity, location and vocalisation data. Drones with video cameras and microphones are also increasingly being deployed to enable remote monitoring of wild animal behaviour. As all these technologies become less costly, they can be more readily trialled and adopted by facilities in the future.

REFERENCES

Al-Naji, A., Tao, Y., Smith, I., & Chahl, J. (2019). A pilot study for estimating the cardiopulmonary signals of diverse exotic animals using a digital camera. *Sensors, 19*(24), 5445.

Blumstein, D. T., Mennill, D. J., Clemins, P., Girod, L., Yao, K., & Patricelli, G. (2011). Acoustic monitoring in terrestrial environments using microphone arrays: Applications, technological considerations and prospectus. *Journal of Applied Ecology, 48*(3), 758–767.

Carter, M., Sherwen, S., & Webber, S. (2021). An evaluation of interactive projections as digital enrichment for orangutans. *Zoo Biology, 40*(2), 107–114.

Coe, J., & Hoy, J. (2020). Choice, control and computers: Empowering wildlife in human care. *Multimodal Technologies and Interaction, 4*(4), 92.

Englund, M. D., & Cronin, K. A. (2023). Choice, control, and animal welfare: Definitions and essential inquiries to advance animal welfare science. *Frontiers in Veterinary Science, 10,* 1250251.

Galante, J., & Margulis, S. W. (2022). Comparing data collection tools for zoo management decision-making: A case study examining behavioral measures of Humboldt penguin bond strength. *Animals, 12*(21), 3031.

Ghezzi, M. D., Ceriani, M. C., Domínguez-Oliva, A., Lendez, P. A., Olmos-Hernández, A., & Casas-Alvarado, A. (2024). Use of infrared thermography and heart rate variability to evaluate autonomic activity in domestic animals. *Animals, 14*(9), 1366.

Hacker, C. E., Horback, K. M., & Miller, L. J. (2015). GPS technology as a proxy tool for determining relationships in social animals: An example with African elephants. *Applied Animal Behaviour Science, 163*, 175–182.

Jukan, A., Masip-Bruin, X., & Amla, N. (2017). Smart computing and sensing technologies for animal welfare: A systematic review. *ACM Computing Surveys, 50*(1), 1–27.

Kalafut, K., & Kinley, R. (2020). Using radio frequency identification for behavioral monitoring in little blue penguins. *Journal of Applied Animal Welfare Science, 23*(1), 62–73.

Kleen, J. L., & Guatteo, R. (2023). Precision livestock farming: What does it contain and what are the perspectives? *Animals, 13*(5), 779.

Kovács, L., Jurkovich, V., Bakony, M., Szenci, O., Póti, P., & Tőzsér, J. (2014). Welfare implication of measuring heart rate and heart rate variability in dairy cattle: Literature review and conclusions for future research. *animal, 8*(2), 316–330.

Mota-Rojas, D., Pereira, A. M., Martínez-Burnes, J., Domínguez-Oliva, A., Mora-Medina, P., Casas-Alvarado, A., Rios-Sandoval, J., de Mira Geraldo, A., & Wang, D. (2022). Thermal imaging to assess the health status in wildlife animals under human care: Limitations and perspectives. *Animals*, *12*(24), 3558.

Orban, D. A., Soltis, J., Perkins, L., & Mellen, J. D. (2017). Sound at the zoo: Using animal monitoring, sound measurement, and noise reduction in zoo animal management. *Zoo Biology*, *36*(3), 231–236.

Ossi, F., Focardi, S., Picco, G. P., Murphy, A., Molteni, D., Tolhurst, B., Giannini, N., Gaillard, J-M., & Cagnacci, F. (2016). Understanding and geo-referencing animal contacts: Proximity sensor networks integrated with GPS-based telemetry. *Animal Biotelemetry*, *4*(21), 1–14.

Reimert, I., Webb, L. E., van Marwijk, M. A., & Bolhuis, J. E. (2023). Review: Towards an integrated concept of animal welfare. *animal*, *17*(4), 100838.

Rose, P. E., Reed, A., Hurn, S., Badman-King, A., & Rice, T. (2022). Does the sound environment influence the behaviour of zoo-housed birds? A preliminary investigation of ten species across two zoos. *Behavioural Processes*, *203*, 104763.

Soltis, J., King, L., Vollrath, F., & Douglas-Hamilton, I. (2016). Accelerometers and simple algorithms identify activity budgets and body orientation in African elephants *Loxodonta africana*. *Endangered Species Research*, *31*, 1–12.

Travain, T., & Valsecchi, P. (2021). Infrared thermography in the study of animals' emotional responses: A critical review. *Animals*, *11*(9), 2510.

van der Marel, A., O'Connell, C. L., Prasher, S., Carminito, C., Francis, X., & Hobson, E. A. (2021). A comparison of low-cost behavioral observation software applications and recommendations for use. *arXiv*, 2103.07576.

Von Borell, E., Langbein, J., Després, G., Hansen, S., Leterrier, C., & Marchant, J. (2007). Heart rate variability as a measure of autonomic regulation of cardiac activity for assessing stress and welfare in farm animals—A review. *Physiology & Behavior*, *92*(3), 293–316.

Wark, J. D., Cronin, K. A., Niemann, T., Shender, M. A., Horrigan, A., & Kao, A. (2019). Monitoring the behavior and habitat use of animals to enhance welfare using the ZooMonitor app. *Animal Behavior and Cognition*, *6*(3), 158–167.

Wascher, C. A. (2021). Heart rate as a measure of emotional arousal in evolutionary biology. *Philosophical Transactions of the Royal Society B*, *376*(1831), 20200479.

Whitham, J. C., & Miller, L. J. (2016). Using technology to monitor and improve zoo animal welfare. *Animal Welfare*, *25*(4), 395–409.

Wongsaengchan, C., McCafferty, D. J., Lennox, K., Nager, R. G., & McKeegan, D. E. (2023). Non-invasive assessment of positive affective state using infra-red thermography in rats. *Animal Welfare*, *32*(e66), 1–15.

Yon, L., Williams, E., Harvey, N. D., & Asher, L. (2019). Development of a behavioural welfare assessment tool for routine use with captive elephants. *PLoS One*, *14*(2), e0210783.

Zuerl, M., Stoll, P., Brehm, I., Raab, R., Zanca, D., & Kabri, S. (2022). Automated video-based analysis framework for behavior monitoring of individual animals in zoos using deep learning—A study on polar bears. *Animals*, *12*(6), 748.

12

Animal welfare and zoo conservation outputs

JONATHAN BEILBY, WILLIAM COSTA AND CHRISTINA R. STANLEY

12.1 INTRODUCTION

The role of modern zoos has evolved considerably from their traditional function of a living museum. Conservation outputs are no longer an aspiration or an added extra; they are both the drivers and required outcomes of all zoological collections. Although not all species housed in zoos are highly endangered in the wild, each can contribute to conservation in a variety of ways, including via education, research, and fundraising. For those species whose wild populations are in serious threat of decline, ex situ populations are invaluable as insurance populations and reserves of genetic diversity. Whilst some individuals will live their entire lives in captivity, others might be reintroduced to their wild range to boost wild population numbers or genetic pools, playing an important part in key conservation outputs. Considerations regarding the welfare of these individuals can vary according to the context of their current environment.

Since conservation is the intended outcome of zoos, there can be a difficult balance to strike between the welfare of individuals and the survival of a species. Captive breeding is commonly governed by genetics; studbooks are maintained so that individuals can be translocated between captive groups when required to safeguard genetic diversity and maximise population-level reproductive success. However, in the case of social species, moving animals to new groups can have negative impacts on both the welfare of the translocated animals, for example when joining a new group, and on the welfare of those individuals in the groups the animals are both leaving and joining. For solitary animals, the stress involved during introduction to a potential mate can also be considerable; despite careful planning and safeguards, it is hard to predict with certainty the outcome of such introductions. Breeding animals for potential future reintroduction introduces even more layers of complexity in terms of welfare. Assuring the welfare of captive zoo animals is a responsibility that zoo staff take extremely seriously, but the welfare of those animals once released into the wild can often, by necessity, become compromised. In their natural environment, wild animals commonly experience competition, predation, disease, parasites and many other hardships from which their captive counterparts are to some extent protected. Whilst zoos can manage species in a way

DOI: 10.1201/9781003429425-12

that maximises future survival prospects, this can sometimes incur, by necessity, challenges to individual welfare.

In this chapter, we discuss how to balance individual welfare and successful conservation outputs. First, we discuss the inherent welfare challenges of captive breeding. We then consider how welfare compromises can be an integral part of preparing individuals for future reintroduction, contextualising these by discussing the inherent differences in welfare between animals in captivity and those in the wild. Finally, we put this into context using a case study of a 'headstarting' programme for the black-tailed godwit (*Limosa limosa limosa)*, considering welfare implications for both adult and juvenile birds.

12.2 CAPTIVE BREEDING

Ensuring the best possible welfare of animals involved in captive breeding programmes is essential as compromised welfare can lead to reduced reproductive output. A vast range of species-specific behaviours and physiological mechanisms have evolved that are associated with mating and reproduction. These are frequently adaptations to address the challenges posed by wild populations' environments, both physical and social. When animals are brought into captivity, their environments change drastically; the most obvious difference is usually in terms of restricted space, particularly for animals with large natural ranges, as even with the most naturalistic exhibit, animals' movements are controlled. Their choice of social and sexual partners is also inherently restricted in a zoo; it is not always possible to replicate the group sizes and structures experienced in the wild in captivity. It is therefore the responsibility of the zoo to ensure that the breeding animals' needs are met, especially in terms of their social choices.

Modern zoos overcome the barriers imposed by captivity by moving animals between institutions to replicate natural dispersal events and encounters with genetically dissimilar potential mates. Studbooks are maintained to record pedigree and genetic make-up of captive populations. The species coordinator uses this information to make recommendations for pairings of particular individuals for breeding to maximise genetic diversity within a species. This broad overview of the genetic pool enables the species coordinator to make decisions in terms of which pairings will maximise overall genetic diversity; however, welfare challenges can be presented at various stages as depicted in Figure 12.1. For solitary species, the introduction of a potential mate can also be a risky business, potentially compromising the welfare of both individuals; genetic compatibility on paper does not always translate to a harmonious coupling when two individuals are introduced. These factors must be considered prior to mixing and mitigation strategies put in place.

It is not only the process of translocation that requires careful consideration, but also the choice of which individuals to move in the context of the group's social environment. Most animals show some level of sociality, and it is well known that the social environment can have a significant impact on health and survival (Snyder-Mackler et al., 2020). For many species, the character of social bonds experienced by individuals can have a significant impact upon reproductive success. For example, in free-ranging baboons (*Papio cynocephalus ursinus)*, offspring whose mothers were strongly bonded to other females lived significantly longer than those whose mothers experienced weaker social bonds, independent of their mother's dominance rank or number of close relatives in the group (Silk et al., 2009) whilst in feral horses (*Equus caballus)*, more socially integrated females have both higher birth rates and offspring survival due to these females experiencing lower rates of male harassment (Cameron, 2009). To maximise the welfare (and reproductive success) of strongly bonded individuals, it can be pertinent to move them to a new group as a pair; animals experiencing positive social relationships are known to be able to manage the stress response more easily, a phenomenon known as social buffering (Hennessy et al., 2009), meaning they will be better able to cope with the potential stressors of their new environment if moved together. Social network analysis can be extremely useful to quantify the strength and character of individual social bonds, as has been applied to the captive population of Livingstone's fruit bats (*Pteropus livingstonii)* (Figure 12.2). Social buffering is known to occur in this species (Edwards et al., 2022); therefore, strongly bonded individuals should be moved together to new groups, and individuals that are

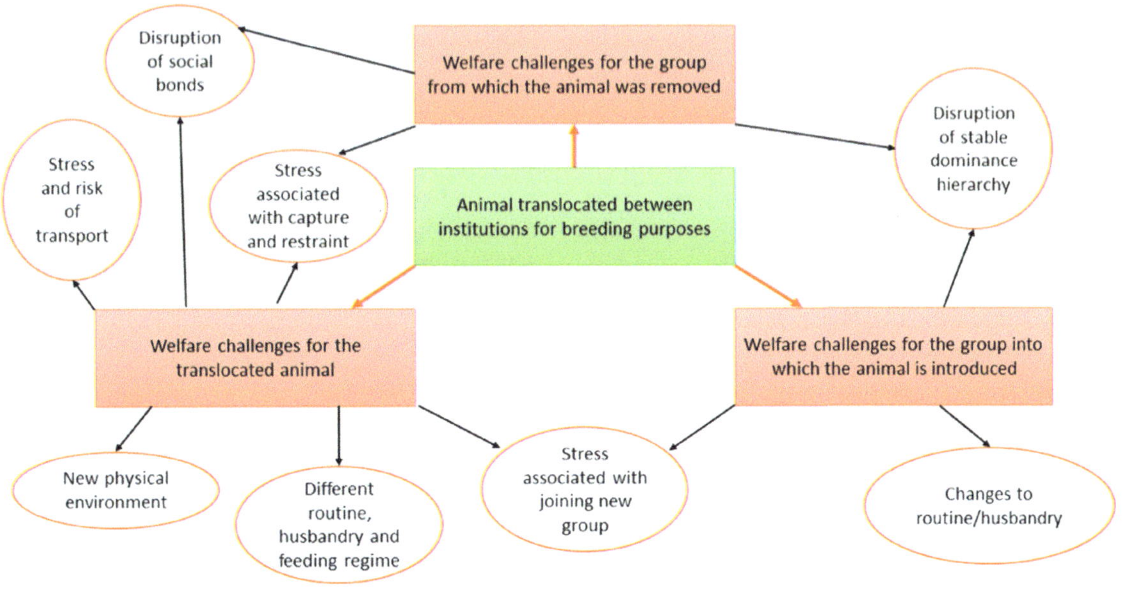

Figure 12.1 Animals are frequently moved between captive groups for the ultimate purpose of maximising genetic diversity in offspring and preventing inbreeding. This practice can pose potential welfare challenges. This diagram illustrates how the initial decision to translocate an animal can have welfare implications for the animal itself, the group from which it was removed, and the group into which it is introduced. Various potential challenges to welfare (indicated by ovals) must be considered when planning a translocation for breeding purposes, and mitigation strategies must be put in place. For example, the choice of method of introduction of the individual to the new group will need to consider the personality of the individual being introduced, the social structure and personalities of the group it will join, the enclosure's capacity to allow animals to avoid aggressors, and the species' behavioural ecology.

Figure 12.2 Livingstone's fruit bats show clear social preferences, with individuals of similar age and dominance rank spending more time in close proximity. Positive social relationships have been found to correlate with reduced stress levels in this species, as measured by cortisol levels in faeces. The photograph was taken at Jersey Zoo by Christina R. Stanley.

not yet well integrated within the social group can be identified as candidates for translocation (Welch et al., 2020). Consideration of individual social roles is also essential; the removal of this individual could have a catastrophic effect on the social structure of the group they leave. When certain individuals (termed "policers") were experimentally removed from captive groups of pigtailed macaques (*Macaca nemestrina*), the remaining groups became fragmented, with higher levels of aggression and less integrated grooming and play networks (Flack et al., 2006). The decision of which individuals to translocate for breeding purposes should therefore never be purely based on genetics; information about individuals' personalities and social relationships needs to be available to maximise the welfare of captive breeding groups.

The change in social grouping of animals, whilst often natural, can be a stressful experience. Some species that rely on visual signals may show obvious signs (e.g. feather plucking or over-grooming), whilst for other species, hormonal changes are the only available indicators of stress. There are many methods by which stress can be quantified; these are well summarised by Beaulieu (2024).

Hormonal studies not only allow the measurement of stress-related hormones such as cortisone and corticosterone, but can also provide precise timings for mixing animals, as in the example of the Eastern black rhinoceros (*Diceros bicornis michaeli*). Black rhinos are critically endangered, being heavily persecuted for their keratinous horn, which is used in traditional Chinese medicine. They are a perfect candidate for a captive breeding programme, providing an insurance population with the eventual goal of reintroduction. For the ex-situ population to maintain its role, it must be self-sustaining, but in the case of the black rhino, productivity has historically been low. As a solitary species, wild rhinos only come together to breed, where males follow olfactory signals from reproductively active females. Many captive pairs of rhinos were deemed incompatible as mixes resulted in aggressive behaviour towards one another and caused potential injury. This short-term stress could have led to long-term benefits for the population if copulation was successful; in Europe, the population increase was not at the ideal target of 5% per annum. Owing to the urgent need for improved success with this species, a population-wide endocrine study was carried out (Edwards, 2013).

Keepers primarily rely on changes in the behaviour or appearance of their animals to indicate optimal mixing times. For black rhino females, these cues include increased vocalisation, restlessness, and cocking of the tail. Edwards (2013) was able to show that some females (particularly those that had not previously bred) would not show these behavioural signs of coming into oestrus and therefore would not be introduced to males at an appropriate time. The routine measurement of hormonal levels was therefore recommended to improve breeding success. Monitoring of hormonal levels has several benefits for animals' welfare; samples can be taken from faeces non-invasively; cycles can be monitored to optimise mixing times (thereby reducing stress and aggression), and animals that do not exhibit behavioural cues of oestrus can be identified and monitored.

12.3 PREPARING FOR REINTRODUCTION

Reintroduction is defined by the IUCN (The International Union for Conservation of Nature) as "an attempt to establish a species in an area which was once part of its historical range, but from which it has been extirpated or become extinct." Reintroductions can involve high stress levels for animals during capture, transport, the release itself, and subsequent exploration of a new environment. In these situations, costs and benefits must be clearly evaluated; sometimes, individual welfare may need to be compromised for the benefit of the species.

One such example is the Javan warty pig (*Sus verrucosus*), a suid endemic to the island of Java. Owing to a huge human population increase, persecution, and the release of domestic dogs (*Canis lupus familiaris*) and pigs (*Sus domesticus*), Javan warty pigs have found it difficult to adapt to their new environment. In 2007, a captive breeding programme was established at the Cikananga Conservation Breeding Centre to breed animals with the view to release them in the future. Given the lack of suitable habitat and the prevalence of domestic pigs on the island, along with the native subspecies of the wild boar (*Sus scrofa vittatus*), hybridisation often occurs (Drygala et al., 2020). The first step in a reintroduction is to confirm the absence of a population. In Baluran National Park, East Java, camera traps were used to declare the species locally extinct. The next stage was to establish the purity of the selected founder animals. Owing to the aforementioned hybridisation of pigs on the island of Java, it was essential to reintroduce genetically pure animals; hybrid vigour can allow hybrids to outcompete 'pure' individuals, or conversely, hybridisation could lead to genetic disorders that could reduce the fitness of the population. In both instances, this leads to decreased survival of pure Javan warty pigs and reduced individual welfare, as hybrid animals can be rejected from groups or may not survive (Simberloff, 1996).

A small piece of tissue is required from each potential founder individual for DNA analysis to confirm genetic purity. However, as the Javan warty pigs were kept in semi-wild conditions, manual restraint was not an option as individuals could suffer from potentially fatal capture myopathy. Anaesthesia and sedation were also considered and rejected due to cost to the animals' welfare (high risk of injury or potential death) outweighing the benefits of obtaining a genetic sample. Instead, the pigs were offered fish filled with glitter, with individual pigs receiving distinct colours, allowing their faeces to be individually identifiable. Despite faecal DNA occurring in lower concentration than in tissue samples and being more difficult to extract, this non-invasive DNA sampling method is used where possible to avoid the negative welfare impacts of handling and restraint.

Individual identification of captive animals is important to monitor individual health and welfare; in a reintroduction project, this also allows monitoring of individuals' survival, thereby enabling evaluation of the success of the project. The best form of identification will vary from species to species, and even within a species, this will differ depending on the project's aims. For the Javan warty pig, coloured or numbered ear tags were initially used to identify captive pigs, but injuries resulted from conspecifics ripping these from others' ears. Ear notching was then trialled as an alternative method of individual identification. Notching involves cutting small triangular areas of the ears of piglets, in a particular pattern that translates to an individual identification number for each pig, see Figure 12.3. Piglets can be physically restrained by a single person, minimising stress and avoiding the need for anaesthesia. The process causes some pain, and there is an infection risk, but any concerns are addressed during handling as opposed to retrospectively treating an adult pig with an ear tag-related injury. There is also the concurrent benefit that these 'notches' provide a genetic sample. As the pigs grow, the

Figure 12.3 Top left: a Javan warty pig which has had the ear tag ripped out by a conspecific. Bottom left: an example of a notching pattern. When the values of the notches are combined, this can allow for an ID number between 1 and 99. For example, top right is number 77; bottom right is 82. Diagram by Jodie Clements.

notching remains clear, meaning that a single restraint has been sufficient for their whole lives – both for visual identification and genetic sampling. It is, however, imperative to evaluate the advantages and disadvantages of different techniques in terms of welfare impacts when selecting the best method of identification for a particular species.

There are various steps to prepare animals for reintroduction whilst being housed in zoos. Captivity itself imposes phenotypic costs on animals, owing to selection pressures differing significantly from those in the wild (Crates et al., 2023). These differences between captive animals and their wild counterparts can mean they are less likely to survive and successfully reproduce upon reintroduction to their natural habitat; for example, if captive-reared animals do not recognise and respond to potential predators, they are much more likely to be eaten. Management approaches that support wild phenotypes are therefore required. Although such approaches can have negative repercussions on individual welfare, imposing short-term stress can improve survival chances upon future reintroduction, as life in the wild is rarely stress-free. Here we consider some of the management practices employed and their impacts on individual welfare.

12.3.1 ENVIRONMENTAL ENRICHMENT

The aim of enrichment is generally to maximise welfare; by increasing the complexity of an animal's environment, with enrichment design taking into consideration the species' natural history, more natural conditions are simulated with the aim of boredom being relieved. This often goes hand in hand with preparing animals for future reintroduction. The more complex the environment, the greater the capacity for improved physical fitness, increased learning (for example, solving challenges set to gain access to food) and the maintenance of a "natural" behavioural repertoire. Other chapters in this book consider this area in more detail, but it is worth noting here that whilst enrichment is generally seen as a positive approach, it can cause frustration, fear, or other negative states. It is important to take a flexible approach to enrichment, regularly monitoring individuals' responses, and to consider the extent to which it simulates a wild environment and natural behaviour, especially when the animal is to be later reintroduced into that environment.

12.3.2 TRAINING

Animals born into captive environments often lack key behavioural traits required for survival in the wild. Many of these are learned behaviours, either actively taught by parents or other group members to youngsters, e.g. hunting techniques, or acquired by social learning, where individuals copy the behaviour of others, e.g. food selection. In captivity, there might not be opportunities for such learning; active hunting to obtain food is rarely possible for carnivores, and food provided is likely to differ from that in the wild. In these circumstances, many behaviours can be learned prior to reintroduction by employing training techniques.

There are two main areas where training might be beneficial: avoiding predation and finding food. Avoiding predation is key for newly reintroduced animals as naive, disorientated individuals are an easy target for predators. Webb (2020) and Edwards et al. (2021) give examples of how captive individuals across species can be trained to enhance their likelihood of performing key adaptive behaviours when reintroduced.

Techniques employed can vary in the level of stress experienced by the tutee, and perhaps in their efficacy; an evidence-informed choice is required for the individuals in question so that a good balance can be maintained. The efficacy of anti-predator training has been assessed in only a small number of species and is mostly measured in terms of survival rates of released individuals. One exception is the case of the Hawaiian crow, ʻalalā (*Corvus hawaiiensis*). In addition to training individuals to avoid a natural predator, the Hawaiian hawk, ʻio (*Buteo solitaries*), in live predator trials, the team exposed control groups to two other stimuli – a net commonly used to recapture these animals or a live chicken (*Gallus gallus domesticus*) (Greggor et al., 2022). By using this experimental set-up, researchers found that learned responses shown were similar across all groups, meaning predator-specific learning was unlikely to have occurred; birds were not actually learning to respond fearfully towards a specific predator, but to any potentially scary stimulus. This species showed high levels of predator-avoidance behaviour towards all potentially dangerous stimuli, despite generations of captivity. A task that allowed the individuals to learn behaviours that would keep them safer from specific predators would therefore be preferable to justify the short-term stress the birds were exposed to during these trials.

When selecting a training method, consideration of the level of stress that individuals experience in their natural environment could be a useful way to benchmark the level of stress the animals could ethically be exposed to in captivity, since pre-exposure to such stressors could be beneficial in the long term. It is better for individuals to experience stress in an environment where they are being monitored closely; if they do not cope well, either they are not ready for release or they might not be suited to reintroduction due to their life experience or personality. This argument can justify situations that might threaten the lives of the individuals being trained, such as exposure to live predators in captivity. This approach, termed "predator conditioning," has been extremely effective in greater bilbies (*Macrotis lagotis*) destined for reintroduction (Blumstein et al., 2019). Bilbies were maintained in large enclosures that included a small number of domestic cats (*Felis catus*). Cats are known to be a significant threat to this species in its natural range. A control group of bilbies was housed in a predator-free enclosure for the same length of time. When both populations were introduced to a new enclosure that had a higher density of cats, those that had pre-exposure to this predator showed significantly higher survival rates. This approach, that does result in some predation in captivity, might not be suitable for all species. However, this is an example of where the level of stress animals are subjected to during training can be significant, but where this is arguably justifiable due to the evidenced positive outcome.

When training to find food, there might be justified threats to welfare during the process as well-fed animals might not have the motivation to work to obtain their meals, meaning learning is unlikely to take place. Again, the degree of hunger, or frustration when food is not easily available, must be justified in terms of the long-term goal of survival upon reintroduction. There are also welfare implications that must be considered where live prey is used; a death at the paws, beak or talons of a naive predator might not be a quick and "humane" event. It is therefore imperative that the long-term benefits outweigh the costs. To determine the potential benefits of any training, more studies need to quantify these benefits in released animals so that an evidence-informed approach can be taken. One example where this research has been conducted is for the European lynx (*Lynx lynx*), where there is evidence that orphaned lynx that were rehabilitated and trained to hunt using live prey in captivity were significantly more likely to survive in their first year post-release than conspecifics who had not been given this training (Molinari-Jobin et al., 2024). Training within conservation programmes is essential for conserving key wild-type behaviours that ultimately convey an advantage to the individuals that preform them. This reduces the chances of individuals domesticating to the zoo and can enhance welfare by reducing an individual's chance of performing abnormal repetitive behaviours if they are stimulated to display a wild-type time budget (Kelley et al., 2025).

12.3.3 VETERINARY INTERVENTIONS

Whilst animals in zoos are generally protected from disease and parasites, this is not necessarily beneficial for animals destined for reintroduction. Where populations have been bred for generations in captivity, herd immunity is often lost since there is not a sufficiently large population for density-dependent endemic disease levels to be maintained. Similarly, the routine use of anti-parasitic drugs means captive animals can have an abnormally low or absent parasite burden (however, this is not always the case nowadays, due to increased levels of anthelmintic (wormer) resistance in captive populations). Upon exposure to parasites after release, these individuals might experience greater negative health effects due to showing lower immunocompetence than their wild conspecifics. There is therefore an argument that low levels of disease and parasite infection should be permitted in captivity, or even introduced prior to release. Despite negative short-term welfare implications, long-term benefits could be significant.

12.3.4 HANDS-OFF APPROACH

One key benefit of training zoo animals to accept routine medical and husbandry procedures is that the potential risks and stress associated with darting, sedation and restraint can be reduced. However, such training will have negative implications for individuals destined for release. Humans remain the biggest threat to many species, particularly where hunting is a major cause of natural mortality or where wild animals become a nuisance in human settlements. A level of wariness of humans is essential for most species in the wild, meaning that a hands-off approach in captivity, despite potential negative welfare implications, is usually recommended.

12.3.5 COMPETITION

In a captive environment, care is taken to ensure all individuals can access food, shelter, water, mates, and other resources. However, this is unlikely to be the case in the wild. Despite negative welfare implications, a degree of competition in captivity might therefore be necessary. In addition, wild populations benefit from natural selection; those unable to successfully compete with conspecifics will be unable to pass on their genes, meaning advantageous adaptations will become more prevalent in the population. In captivity, natural selection is unlikely as, often for welfare reasons, survival of all individuals is supported. This results in animals that are poorly adapted to the pressures they will face in their new environment upon release. Wild phenotypes in captivity can be supported by ensuring competition persists, but only where environmental pressures mimic those experienced in the wild, for example by allowing animals' exposure to low levels of parasites or disease. This could reduce the short-term welfare of captive animals, but it will have benefits to those individuals that are finally released.

12.4 CONCLUSIONS AND FUTURE CONSIDERATIONS

Life in the wild is not easy. Wild animals are exposed to inclement weather, competition, parasitism, disease, predation, food scarcity and many other challenges from which captive animals are often protected. Assessing the welfare of wild animals is therefore not straightforward using many of the commonly used definitions of welfare and is usually only undertaken to quantify the effects of anthropogenic impacts. One way of defining the welfare of wild animals is to consider whether animals possess the necessary adaptations to cope with a particular stressor (Fraser et al., 1997). If they are being exposed to a level or type of stressor to which they have not evolved physiological or behavioural responses, their welfare can be considered as compromised. It could be useful to take this approach to assessing the welfare of animals that are being bred in captivity for the purposes of reintroduction. This would allow an "accepted" level of welfare to be quantifiable in these populations, which is by necessity likely to be different from that in other captive individuals. Whether or not this is publicly acceptable would depend on the extent to which education is used to justify this approach. Zoos can achieve the delicate balance of safeguarding individual welfare whilst still achieving conservation goals by taking an evidence-informed, evaluative approach, a case study of which is detailed in Box 12.1. Potential costs to welfare must be balanced against the perceived gains in terms of survival of reintroduced populations. Short-term stressors might be justifiable where exposure will reduce the likelihood of long-term stress, or mortality. In many cases, an accurate decision is currently not possible due to a paucity of evidence in the literature to quantify the benefits of different approaches. There is therefore an urgent need for further research in this area, and for long-term monitoring of reintroduced animals, so that the most effective approaches can be taken.

One area that is not covered here is the ethical debate surrounding the concept of individual hardship for the "good of the species". The goal of conservation is often to increase or safeguard the number of individuals in a population; however, this does not necessarily need to compromise the quality of these lives. The field of conservation welfare has a growing voice within this debate (Sekar & Shiller, 2020). Welfare and conservation are both essential in zoos and need not be in opposition. For further reading on this topic, see Box 12.2.

BOX 12.1: Case study – welfare considerations for headstarting black-tailed godwits

Wading birds (Charadriiformes) undertake extended incubation periods followed by a period of neonate care in which the chicks are unable to fly. Reductions in population productivity have led to techniques aimed at circumventing this vulnerable life stage. One technique is 'headstarting,' when a young animal or egg is taken into captivity, reared, and subsequently released. This technique has been used to reinforce the population of black-tailed godwit, hereon referred to as godwit, in the United Kingdom (UK) (Figure 12.4). The release of 206 birds led to a productivity increase of 1900%, boosting the British population of this red-listed species. Temporarily maintaining a prey species vulnerable to stress in captivity whilst managing them for a future release requires careful planning and welfare assessments.

Headstarting begins with the collection of wild eggs. To encourage the parents to lay a replacement clutch and reduce wasted parental investment, eggs are collected as early into incubation as possible. This not only improves welfare for the parents but also allows them to invest in their second brood, thus improving the welfare of future offspring.

Repeated nesting failures impact birds, not only via failed resource investment but also from the stress of inferred predation risk and failed mate choice. The collection of eggs impacts birds on levels comparable to natural nest failures. Repeated breeding failures may be a factor in the emigration of godwits from sites that are deemed "unproductive." Utilising a decision tree to select which nests are collected allows us to centre the process around parental welfare and mitigate impacts where possible.

Predation levels can reduce the chances of wild chick survival to less than 1%. Maintaining chicks during their non-flying period in a predator-proof environment removes predation risk and stress caused by a perceived threat of predation. This improves the welfare of individual chicks; however, maintaining chicks in captivity introduces novel challenges.

Protocols aim to minimise human-induced stress and negative impacts of imprinting. Whilst nidifugous, wading birds rarely experience filial or sexual imprinting on humans; however, they can become desensitised to human presence, which can bring them into conflict once released. Field observations of spoon-billed sandpipers (*Calidris pygmaea*) have shown occasions where headstarted adults approach researchers. As wading birds are vulnerable to hunting along their flyways, this is not a desirable trait.

During the inside rearing stage, a radio playing 'wetland sounds' is used to condition the godwits to auditory stimuli whilst also muffling human sounds. In the first few days of life, godwits are sensitive to loud sounds and will exhibit a predatory avoidance response (dropping to the floor and not moving); the radio helps the chicks to differentiate between harmful noises and ambient noise.

Staff movements when working around the birds are slow and steady. This is particularly important in a flock situation as fear responses quickly spread. Wading birds reared in captivity are susceptible to the development of stress-induced ailments such as aspergillosis, amyloidosis, and feather barring. There is a careful balance between maintaining the birds comfortably in captivity and not producing desensitised birds unsuitable for release.

All godwits released are individually identifiable via four coloured rings attached to their legs for post-release monitoring, as seen in Figure 12.4. Godwits are prone to capture myopathy; thus, all handling sessions are managed to a strict time limit with experienced staff overseeing the operations. Post ringing, the birds are routinely monitored. Behavioural concerns include 'leg flicking' and repeatedly sitting down. If these behaviours continue, the rings are removed; of 206 birds released, only two individuals have had their coloured rings removed.

Geolocators allow migratory routes to be assessed. The devices are glued onto colour rings on the birds' legs. The weight of geo-locators means that the rings can spin around the bird's leg, causing the individual some distress. Larger birds with optimum chick growth curves are selected to carry the devices to minimise the impacts of weight on flight dynamics. Twenty-six per cent of geo-locators were removed pre-release as consistent stress-related behaviours indicated a lack of habituation to the devices. Another consideration is the requirement to catch the bird again post-release to download the data, which can itself have welfare implications. However, eighteen units have been retrieved, and the data have provided invaluable insights into their migration routes. This highlights the delicate balance between animal welfare and project success.

Figure 12.4 Black-tailed godwit headstarting at WWT. Photo credit: WWT

Headstarting projects should only be undertaken following a rigorous cost-benefit assessment, in which all conservation measures are considered, and animal welfare concerns are addressed. The removal of godwit eggs from wild parents and the management of neonates in captivity do have welfare implications, including elements discussed earlier in the chapter, e.g. predator avoidance training and social dynamics. However, without drastic intervention, most godwit eggs laid in the wild would fail to produce fledglings; therefore, a scenario without headstarting can be thought of as having more profound welfare implications for the species.

BOX 12.2: Suggested further reading

The EAZA best practice guidelines provide an in-depth review of captive husbandry from a variety of institutions and can be found here: https://www.eaza.net/BPG/
EAZA also has a wide range of operational guidelines, including those related to conservation, and can be found here: https://www.eaza.net/operational-documents/

REFERENCES

Beaulieu, M. (2024). Capturing wild animal welfare: A physiological perspective. *Biological Reviews*, *99*(1), pp.1–22.

Blumstein, D. T., Letnic, M. & Moseby, K. E. (2019). In situ predator conditioning of naive prey prior to reintroduction. *Philosophical Transactions of the Royal Society B*, *374*(1781), p.20180058.

Cameron, E. Z., Setsaas, T. H. & Linklater, W. L. (2009). Social bonds between unrelated females increase reproductive success in feral horses. *Proceedings of the National Academy of Sciences*, *106*(33), pp.13850–13853.

Crates, R., Stojanovic, D. & Heinsohn, R. (2023). The phenotypic costs of captivity. *Biological Reviews*, *98*(2), pp.434–449.

Drygala, F., Rode-Margono, J., Semiadi, G., & Wirdateti, F. A. C. (2020). Evidence of hybridisation between the common Indonesian banded pig (*Sus scrofa vitattus*) and the endangered Java warty pig (*Sus verrucosus*). *Conservation Genetics*, *21*, pp.1073–1078.

Edwards, K. L. (2013). *Investigating population performance and factors that influence reproductive success in the Eastern black rhinoceros*. Ph.D. Thesis. University of Liverpool. https://core.ac.uk/download/pdf/80771456.pdf (Accessed: 27th July 2024).

Edwards, M. C., Ford, C., Hoy, J. M., FitzGibbon, S. & Murray, P. J. (2021). How to train your wildlife: a review of predator avoidance training. *Applied Animal Behaviour Science*, *234*, 105170.

Edwards, M. J., Stanley, C. R., Hosie, C. A., Richdon, S., Price, E., Wormell, D. & Smith, T. E. (2022). Social roles influence cortisol levels in captive Livingstone's fruit bats (*Pteropus livingstonii*). *Hormones and Behavior*, *144*, p.105228.

Flack, J. C., Girvan, M., De Waal, F. B. & Krakauer, D. C. (2006). Policing stabilizes construction of social niches in primates. *Nature*, *439*(7075), pp.426–429.

Fraser, D., Weary, D. M., Pajor, E. A. & Milligan, B. N. (1997). A scientific conception of animal welfare that reflects ethical concerns. *Animal welfare*, *6*(3), pp.187–205.

Greggor, A. L., Masuda, B., Sabol, A. C. & Swaisgood, R. R. (2022). What do animals learn during anti-predator training? Testing for predator-specific learning in 'alalā (*Corvus hawaiiensis*). *Behavioral Ecology and Sociobiology*, *76*(12), p.165.

Hennessy, M. B., Kaiser, S. & Sachser, N. (2009). Social buffering of the stress response: Diversity, mechanisms, and functions. *Frontiers in neuroendocrinology*, *30*(4), pp.470–482.

Kelley, J. L., Firman, R. C., Garcia, C. M. & Mills, H. R. (2025). Behavioral changes in captivity: Consequences for captive breeding and reintroduction programs. In *Reference Module in Life Sciences*. Elsevier. https://doi.org/10.1016/B978-0-443-29068-8.00105-7

Molinari-Jobin, A., Zimmermann, F., Borel, S., Le Grand, L., Iannino, E., Anders, O. & Vogt, K. (2024). Rehabilitation and release of orphaned Eurasian lynx (*Lynx lynx*) in Europe: Implications for management and conservation. *PLoS One*, *19*(3), e0297789.

Sekar, N. & Shiller, D. (2020). Engage with animal welfare in conservation. *Science*, *369*(6504), pp.629–630.

Silk, J. B., Beehner, J. C., Bergman, T. J., Crockford, C., Engh, A. L., Moscovice, L. R., Wittig, R. M., Seyfarth, R. M. & Cheney, D. L. (2009). The benefits of social capital: Close social bonds among female baboons enhance offspring survival. *Proceedings of the Royal Society B: Biological Sciences*, *276*(1670), pp.3099–3104.

Simberloff, D. (1996). Hybridization between native and introduced wildlife species: Importance for conservation. *Wildlife Biology*, *2*(3), pp.143–150.

Snyder-Mackler, N., Burger, J. R., Gaydosh, L., Belsky, D. W., Noppert, G. A., Campos, F. A., Bartolomucci, A., Yang, Y. C., Aiello, A. E., O'Rand, A. & Harris, K. M. (2020). Social determinants of health and survival in humans and other animals. *Science*, *368*(6493), p.eaax9553.

Webb, J. (2020). Training animals in captivity or the wild, so they can return to the wild. In V. A. Melfi, N. R. Dorey, & S. J. Ward (Eds.), *Zoo Animal Learning and Training* (pp. 289–308). Hoboken, NJ: Wiley-Blackwell.

Welch, M. J., Smith, T., Hosie, C., Wormell, D., Price, E. & Stanley, C. R. (2020). Social experience of captive Livingstone's fruit bats (*Pteropus livingstonii*). *Animals*, *10*(8), p.1321.

13

Designing and implementing animal welfare management assessment programmes

JESSICA J. HARLEY AND LISA M. CLIFFORDE

13.1 INTRODUCTION

Today welfare assessment frameworks and protocols extend beyond the more traditional focus of basic health or resource-based measures to encompass a more comprehensive approach. In other words, animal welfare no longer focuses on the absence of suffering but focuses on ensuring a life worth living. This shift recognises the complexity of animal welfare, moving beyond the provision of adequate resources to consider the animal's overall physical, mental, and emotional state. This holistic approach is demonstrated by frameworks like the Five Domains Model (Mellor et al., 2020) and the Welfare Quality® protocol (Blokhuis et al., 2013). Zoos and aquariums (hereinafter referred to as zoos) are increasingly adopting these comprehensive frameworks to ensure the welfare of the animals in their care. For example, membership and accreditation bodies such as the World Association of Zoos and Aquariums (WAZA) and the British and Irish Association of Zoos and Aquariums (BIAZA) require an animal welfare evaluation process, including regular welfare audits. This requirement reflects the acknowledgement of animal welfare as an underlying operational necessity for zoos. Furthermore, audits must use animal-based indicators, which assess an animal's physical, physiological, and/or behavioural state, to provide insights into its welfare from the animal's perspective. This approach focuses on how the animal is coping with its environment, rather than simply evaluating the resources provided (BIAZA, 2023; WAZA, 2023).

The Five Domains Model has been adopted by the majority of zoo accreditation organisations and emphasises the interrelation of physical health, environmental conditions, behavioural

DOI: 10.1201/9781003429425-13

interactions, and mental state (Mellor & Beausoleil, 2015). By adopting a holistic perspective, zoos can effectively address welfare by ensuring that all aspects of an animal's life are considered. The development of subsequent welfare audit tools based on these frameworks enables a systematic evaluation and monitoring of the welfare of zoo animals. Additionally, technological advancements, such as animal tracking systems, thermal imaging, mobile electrocardiogram (ECG) monitors and bio-logging tags offer novel ways to enhance animal management and welfare monitoring (Diana et al., 2021).

However, despite the considerable progress and resources available, several challenges remain when developing and implementing zoo animal welfare management. One of the primary issues is the vast taxonomic diversity of species housed in zoos, as the specific needs of each species may not be fully understood (Miller & Chinnadurai, 2023; Tallo-Parra et al., 2023). Thus, the challenges and considerations associated with managing animals in captive environments have necessitated the development of tailored approaches to welfare assessment and management practices in zoos. This, in turn has led to the emergence of zoo animal welfare science as a specialised field within the broader discipline of animal welfare (Ward et al., 2018). However, the extensive amount of information and the variety of welfare assessment methods can be difficult to navigate, especially for those who may lack staff specifically trained in welfare. This is further complicated by the fact that each species, and often individual animal, requires tailored assessments, demanding a significant investment of time that smaller or understaffed zoos might struggle to provide. Another challenge faced by zoos is the constraint of limited resources, e.g. financial and spatial, which may limit the ability of zoos to provide ideal environments for all animals. While a comprehensive solution to some of these challenges is beyond the scope of this chapter, we can provide practical strategies and tips for developing audits and maintaining zoo animal welfare management programmes.

In this chapter, we aim to guide readers through the practical considerations of animal welfare management. We will outline a clear process for developing and implementing effective welfare programmes and discuss different methods for assessing animal welfare. We will also provide practical information and strategies to address the specific challenges of integrating welfare programmes into the daily operations of a zoo.

13.2 IMPLEMENTING A WELFARE ASSESSMENT PROGRAMME

The successful implementation of a welfare assessment programme requires an institutional commitment to the process, and more specifically, buy-in from all key stakeholders, such as those responsible for daily operations, strategic planning, staff resources, and financial budgets. This includes the zoo's senior leadership team, animal care teams, veterinarians and welfare officers. Welfare assessment can take many forms but there is commonality in that data collection is required for input and specific actions are required as an output. Thus, every welfare assessment programme will at the very least require individuals with the competency and time to collect information, and a supporting framework to resource and implement resulting action plans. The way in which this is achieved will vary, but many zoos find it helpful to create a welfare policy or a welfare group, to define goals, disseminate information, and review current practices. Some of the crucial steps to implementation are defining the resources that will be available, who the decision makers will be, and how staff will be trained (Kagan et al., 2015).

13.2.1 GOALS FOR A WELFARE ASSESSMENT PROGRAMME

The purpose of a welfare assessment programme is to improve animal welfare. To do this, it is first necessary to ascertain the current welfare state for each animal. Thus, many zoo association bodies require zoos to have a format for routinely assessing their entire animal collection. Historically, the focus has been heavily biased towards mammals and to a lesser extent, birds, but modern programmes should seek to include all taxonomic groups within their remit. A collection-wide welfare assessment enables managers to consider where priority areas for improvement are, and if there are recurrent themes in welfare improvement opportunities across the collection that require consideration of changes to husbandry practices or animal management policies e.g. indoor/outdoor access. See Figure 13.1 for an example welfare assessment process at the Zoological Society of London (ZSL).

The challenges of conducting a collection-wide assessment are, firstly, the considerable effort required to complete the assessments themselves, and secondly, the potential for identifying numerous areas for improvement, each demanding additional resources to address. This can strain existing resources, both in terms of staff time and finances. To facilitate this process, a good welfare assessment programme will consider how to spread the workload over time and personnel and include a prioritisation system for both assessments and action plans. Prioritisation approaches can be based on the severity of the concern, upcoming deadlines for completion (e.g. improve heating before winter), or they may take an alternative approach and consider the number of animals that will benefit, or the ease of completion. This may sound unorthodox, but the concept of prioritising 'quick wins', whereby welfare improvements that can be actioned cheaply or quickly are undertaken first, can generate not only an improvement in the welfare state but also a feeling of productivity and achievement which will benefit the continuation of the welfare programme.

13.2.2 RESOURCING THE WELFARE PROGRAMME

As mentioned earlier, there are two sets of resources required for a welfare assessment programme: competent staff time allowance and financial support for welfare improvements. The allocation of these resources is not something that can be dictated across the industry but must instead be decided within each institution. Many zoo associations now require zoos to have a named member of staff responsible for welfare (e.g. BIAZA, 2023). This person is not responsible for the welfare state of each animal, but rather for the welfare assessment programme, and as such must be educated and up to date on current welfare science assessment practice and zoo association and legislative requirements for welfare assessments. It is the responsibility of this role to ensure that the welfare assessment processes utilised by the zoo are based on current best practice, and that all staff involved in their completion are sufficiently trained to complete their part of the assessment competently. They do not have to deliver this training but should be well informed about the available internal and external resources for training purposes and be able to direct staff towards these.

Zoos may choose to involve a wide breadth of staff in their welfare assessment programmes, and there are no restrictions on who may contribute. Larger zoos may employ welfare specialists to lead

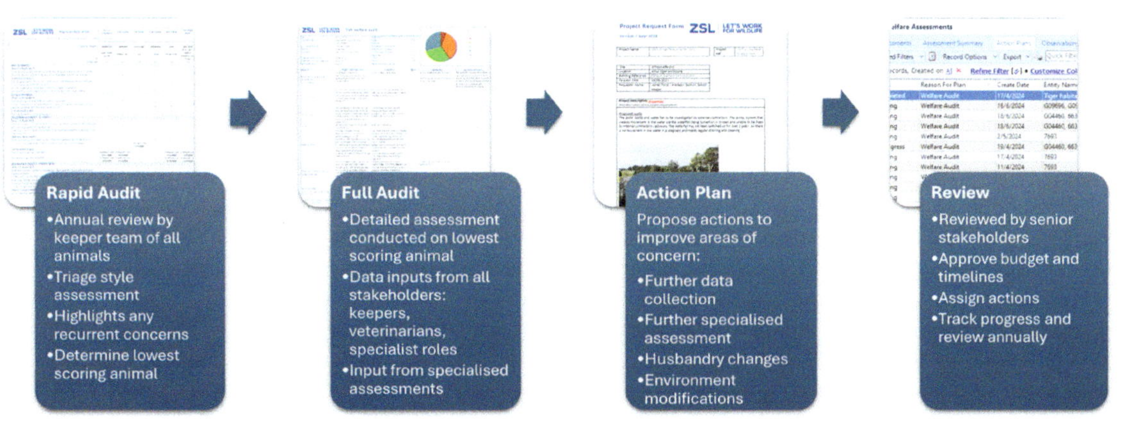

Figure 13.1 ZSL Annual Welfare Assessment Process, an example of how to utilise multiple assessment types within a comprehensive welfare programme.

the programme, but smaller zoos will by necessity have to rely on their animal care staff. Whilst a single person or designated group will be given responsibility for collating information and filling in the assessment, veterinarians, curators, managers, researchers, educators, students, volunteers, facilities staff, and many more may have the time and knowledge to contribute useful information to the evidence-gathering stage of a welfare assessment and so should be consulted where applicable. Managers must consider ways to give their staff regular access to the resources they need in order to gather information and complete assessments. This can be worked into a rota or schedule or happen on a more flexible basis depending on the operational style of the zoo. However, it must be considered that without defined time periods to complete assessments, they can end up repeatedly deferred as more urgent or preferred tasks take precedence.

13.2.3 FINANCIAL SUPPORT

The prospective costs associated with a welfare assessment action plan can be daunting or even prohibitive to some zoos if they set out to achieve perfection. However, as the saying goes, perfection should not stand in the way of good, meaning that if it is not financially viable to achieve all of the actions generated through a welfare assessment, that does not mean that no action should be taken. Even small changes such as adding in an extra feed station or a new training programme can positively impact animal welfare. Perhaps the roof of a building cannot be replaced, but the leak can be patched up.

Managers and budget holders should plan for a dedicated sum of money (however small) that can be allocated to welfare assessment actions each year. This way, the zoo can be sure of achieving regular welfare improvements, year after year. Not all welfare improvements require financial expenditure. Action plans may also include recommendations for further information gathering, reviews of current practice, alterations to routines, or population changes (e.g. export of surplus animals). For those zoos that are able to financially support a welfare assessment programme, funding for regular staff development and equipment to aid data collection will strengthen the outcomes of the programme. Investment in cameras, behaviour monitoring software, or environmental monitoring equipment will improve not only the variety and quality of the data gathered but potentially staff efficiency as well.

13.2.4 ASSESSMENTS

A starting point for a welfare assessment programme is a general assessment format or audit, which aims to provide an overview of the animals' holistic welfare state. Within this overarching assessment, there will likely be more specific and detailed assessments that function as inputs to the audit or are conducted as outputs following recommendations for further information collection. Subsequent sections of this chapter will consider the many and varied assessments that may complement and contribute to a holistic welfare assessment programme, for example, quality of life assessments and event assessments.

All assessments have the following points in common; they ask a relevant question, they generate a robust answer, and then they recommend what to do next. If an assessment follows this format, all other points are open to variation and individual preference. The starting point for a holistic audit is a framework which is used to determine the questions that will be asked within the assessment. There are several framework options (e.g. Five Domains) to refer to, but using a framework-based template ensures that all aspects of an animal's experience are considered consistently across individuals and over time. Generalised holistic audits aim to cover all aspects of the framework to build up an overall picture of the animal's welfare state. By necessity this means that they cover many areas but may lack comprehensive detail; therefore, supplementary specific assessments are of benefit. They can focus on a more limited area of the framework and delve into greater detail to give a more detailed representation of the welfare state. To this end, a zoo's welfare assessment programme will be made up of several complementary assessment templates and processes.

13.2.5 TEMPLATES

It is not necessary for each zoo to create their own bespoke assessment templates, and there are now a variety of freely available examples published in journals and through zoo associations. However, to ensure buy-in from key stakeholders, some zoos may find it useful to adapt an existing template or create their own so that it best meets their objectives and creates a sense of ownership. It will be useful to first gather information about the needs of the stakeholders, which will be varied. Senior management may ask for an assessment system which provides them with clear prioritisation of actions, or one that is simple and easy to review. Operational staff may want a quick process that focuses on the resources available to provide good care to the animals, so that they can put forward suggestions for more efficient and effective working practices. Scientists and academic stakeholders may want to use the opportunity to gather data that allows them to look for trends and patterns that cannot be seen in isolation, thus identifying issues that would otherwise go unnoticed. There is no right or wrong approach to assessment design, though there are stronger and weaker ones. However, the most important criterion is that it gets done, and therefore the staff must be supportive of the process. If a comprehensive and scientifically validated template design never gets completed because it takes too long or is never reviewed by managers because they do not have time to read all of it, then it is not helpful. A compromise, based on availability of staff time and resource, and the value of the data within the assessment must be reached.

13.2.6 INFORMATION COLLECTION

All assessments will require a degree of information collection. This may take the form of summarising records, monitoring enclosure and habitat parameters, or spending time observing the animals. Data collection can be time-consuming and require familiarity with the methodology and equipment required and thus may be off putting to many people who feel they lack the necessary knowledge. It is not necessary or often practical for a single person to collect all data required for an assessment, but the methodology and equipment should be standardised to ensure reliability.

Effective animal welfare assessments require more than just data collection. Auditors must be knowledgeable about the species they are evaluating. Thus, information collection which includes thorough research into each animal's behavioural biology, natural history, and specific needs is also integral to the process, thus allowing for a more objective and informed assessment. This research phase is often the most time-consuming component of the welfare assessment; however, it is integral for ensuring accuracy and promoting meaningful improvements to animal welfare.

Dedicated specialists responsible for oversight and capacity building will be necessary to achieve robust scientific methodologies (Kagan et al., 2015), but this can be developed once and then cascaded and trained through the wider staff body.

Consider that staff time will need to be allocated for data collection and where possible this can be minimised by making use of data already collected routinely (e.g. temperature records, food intake records). Box 13.1 illustrates how keeper data collection practices can be effectively utilised in welfare management audits. This emphasizes their contribution to welfare assessments and the importance of standardized methodologies like body condition scoring in routine animal husbandry. Furthermore, simple protocols such as photographic and video records can also be useful and take minimal time to collect, thus allowing for a large and thorough data set (e.g. photographic record of hoof condition). The period and frequency for data collection or reviews must be agreed, and all persons involved should be trained as to the method used. This will ensure that assessments are consistent between individuals and over time (Jones et al., 2022).

BOX 13.1: Striving for best practice: The keeper's role in welfare assessment

Keepers' in-depth knowledge and experience of the animals in their care is an important component of ensuring optimal animal welfare. Keepers' daily observations and interactions, combined with regular management practices, provide invaluable insights that can form part of the zoo's comprehensive welfare assessment programme. As such, it is essential to incorporate standardised and validated indicators like body condition scoring (BCS), faecal analysis, and behavioural monitoring as part of the keepers' daily records to create an accurate picture of animal welfare. To illustrate this, we use an example outlining best practice for assessing nutritional status with BCS.

Figure 13.2 Neck of a domestic horse (*Equus caballus*) (left) compared to Grevy's zebra (*Equus grevyi*) (right).

BEST PRACTICE EXAMPLE: BCS IN WILD EQUID MANAGEMENT

Equids in managed settings can face nutritional challenges due to changes in forage quality, social dynamics, or health issues. Body Condition Scoring (BCS) is a standardized, semi-quantitative method used to assess an animal's nutritional status by evaluating fat deposition at key anatomical points. This species-specific process is particularly valuable when obtaining live weights is impractical or impossible. BCS is useful for detecting changes in weight and enabling proactive management adjustments. Additionally, BCS provides measurable data for tracking trends over time, allowing for effective evaluation of interventions, such as dietary changes, and promoting informed decision-making in animal care.

KEEPER'S ROLE:

1. Keepers visually assess each equid's body condition using the modified (no palpation) Henneke scale (1–9) (Henneke et al., 1983). This BCS tool is a standardized approach for assessing an equid's nutritional status where one (1) is poor/emaciated, five (5) is moderate (ideal) and nine (9) is extremely fat/obese.
2. Keepers look for changes in fat deposits over the ribs, behind the shoulder, along the neck, withers, and around the tailhead keeping in mind the confirmation of the wild equid (e.g. neck of a Grevy's zebra (*Equus grevyi*) is thicker and may appear fleshy compared to a domestic equid (*Equus caballus*) (Figure 13.2)).
3. BCS scores are recorded in software such as the Zoological Information Management System (ZIMS), a global database used by zoos. For smaller zoos that may not subscribe, this should be included in the equid's individual digital or paper records, along with any relevant observations linked to nutrition, e.g. competition, age-related mastication issues, changes in behaviour, activity levels, or coat condition, which can provide additional insights into an equid's health status.
4. The animal care team must regularly graph and review BCS trends. An example is provided (see Figure 13.3). If BCS falls significantly outside the 'ideal score' or the 'acceptable range' (depending on reproductive status), the team will consider adjustments to the animal's diet, conduct veterinary examinations, or make modifications to its environment.

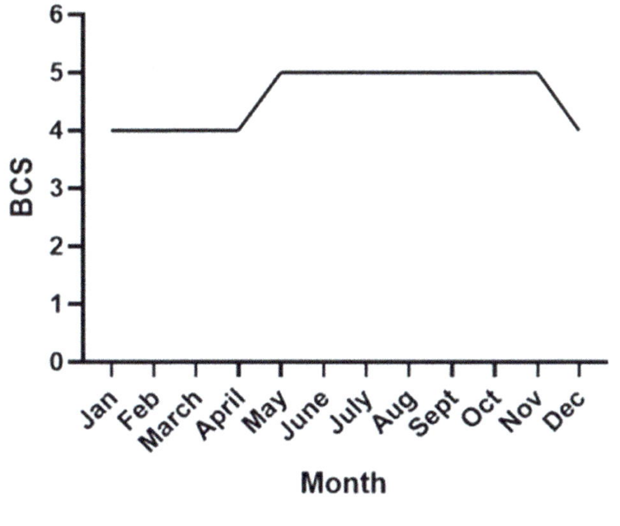

Figure 13.3 Graph illustrating the typical fluctuation in body condition score (BCS), showing a decline over winter into early spring, followed by optimal BCS in late spring and summer. This pattern aligns with the adaptive strategies of temperate grazers, who utilise stored fat reserves during periods of resource scarcity.

13.2.7 ACTION PLANS

An essential component of any animal welfare assessment is the action plan; without this there is little to no purpose in collecting the information. To develop effective action plans, a review and summary of the provided information is required. Drawing in experts from fields across the zoo can be a helpful part of this process. For example, it makes sense for a veterinarian to review the health data that has been collected or for a research specialist to look over the animal behaviour data. If specialists are not able to review the assessments, then staff from different tiers of the organisation can be used. For example, if a junior keeper collects data on the environmental parameters of the habitat, then the team manager will be capable of checking the information and providing a summary of required actions. Action plans should be targeted to improve specific areas of concern and always be considered from the animal needs perspective, not the human (human needs are also important, but an animal welfare assessment is not the primary place to assess them!). Actions require a specific goal, a method, a nominated person/team to complete them, and a suggested time frame. Breaking down large and complicated actions into their individual steps can make them more accessible and demonstrate progress as smaller tasks are completed.

13.2.8 REVIEW PROCESS

A review process has several purposes; firstly, an initial review of the assessment should check for objectivity, inclusion of relevant evidence, and that the agreed methodology has been followed. Secondly, it may be beneficial to gain input from additional stakeholders and specialists such as keepers, section managers, scientists, curators, or vets. These are the people who have specific areas of expertise and may be able to add further information to the assessment.

In some zoos the person(s) responsible for completing an assessment will not be the only stakeholders; therefore, a further review of the assessment by budget holders or decision makers will be necessary for final approval and implementation of the action plan. Decision makers are the people who sign off on improvement actions, and they may vary between different forms of assessment. They can include budget holders, team managers who allocate staff time, veterinarians who sign off treatment plans, managers responsible for oversight of the zoo, or perhaps an ethical review panel. Zoos may wish to consider involving an external reviewer for the purposes of greater objectivity (Kagan et al., 2015). This could be an animal staff member who does not work with the animal being assessed, a member of staff from a non-animal department, or an external ethical review committee.

Follow-up reviews of assessments are then recommended to determine if the agreed actions have been completed and were successful in their objectives. The timeline for a follow-up review will depend on the nature of the assessment and its recommendations. For example, a routine holistic audit with several long-term actions may be repeated annually, but a specific quality of life assessment with short-term actions may be repeated monthly.

13.3 TYPES OF ASSESSMENTS

The following are examples of frequently used and useful assessment types, but an assessment is not limited to these examples. An assessment can be created for any scenario as long as it includes questions, evidence, conclusions, and an action plan. Please note that zoos may use alternative names for each assessment and that the content is more important than the chosen name.

13.3.1 WELFARE AUDITS

A welfare audit provides a holistic overview of all aspects of animal welfare. Typically, an audit is based upon a framework such as the Five Domains, as mentioned earlier in this chapter. Some regulatory frameworks offer additional guidance or stipulate requirements for the content of the audit. For instance, WAZA requires that audits should include both input and output-based measures of animal welfare (WAZA, 2023). Inputs include resources provided by the caregiver, such as a balanced diet, enclosure furnishings, or a training plan. Outputs are indicators measured directly from the animal and include health status, food intake, and activity budget. Without consideration of both inputs and outputs, it is not possible to get a complete picture of the animal's welfare state. The aim is to gain a general overview of the animals' welfare state, identify areas that could be improved, take action, and be used regularly to check for progress. Some regulatory bodies require annual coverage of the entire zoo; however, audits can be done as often as resources allow. Audits once per year allows for routine assessment, and time for actions to be implemented, but repeating an assessment after any significant change is advised. Be sure to consider over what time period data will be collected and assessed to cover seasonal environmental changes, cyclical behaviours, and animal management changes (Jones et al., 2022).

13.3.2 QUALITY OF LIFE

Quality of life refers to the state of an individual, and assessments are applied to those who are suffering from an illness, injury, impairment, or other condition that negatively impacts their physical or mental state, as well as their interactions with their environment and how the individual lives its life (Wemelsfelder, 2007). Quality of Life (QoL) assessments are not compulsory at the time of writing, but are recommended as best practice for applicable animals, and required in more specific conditions, such as for elephants (Proboscidea) (Sach et al., 2019). The purpose of such an assessment is to identify and mitigate any suffering caused by health or behaviour conditions. QoL assessments should not be considered as just a euthanasia decision-making tool but utilised for tracking trends in welfare and assessing if changes to treatment or management are improving welfare. They are frequently utilised for geriatric animals but may also be initiated following a recommended action from a routine welfare audit, or after a diagnosis of a chronic condition.

QoL assessments utilise output-based health and behaviour indicators, such as health status and behaviour repertoire, and should generate animal care inputs as mitigation suggestions, such as changes to enclosure furnishings or diet provision. More so than any other form of assessment, a QoL assessment will require input from veterinarians and animal care staff. To reliably understand the animal's welfare state, it is necessary to collect data over a period of time, using a range of indices (Watters et al., 2015), and therefore these assessments may focus on fewer indicators in greater detail than a more general welfare audit. The frequency of assessment will be variable and can be increased or decreased depending on the severity of the condition and the period needed for mitigation actions to take effect.

13.3.3 ZOO EVENTS AND EXPERIENCES

Increasingly zoos are diversifying their engagement practices with the visiting public and looking to offer events and experiences above and beyond a normal visit to the zoo. To ensure that these experiences are non-stressful and provide positive experiences for the animals it is necessary to evaluate their effects (D'Cruze et al., 2019). Close contact visitor and animal experiences are defined as close proximity between a zoo visitor and an animal in the care of the zoo, either in the absence of, over, or through a safety barrier, and experiences such as these now require welfare assessments for all BIAZA members (BIAZA, 2024).

The purpose of assessment is to identify and mitigate any human-led activities that may cause disturbance, negative welfare impacts, or limit opportunities for positive welfare states and should include an action plan that identifies mechanisms to offer positive welfare experiences to the animals. They can be used for regular events such as keeper experiences, as well as for one off events such as out of hours fundraisers, or even applied to other sources of anthropogenic disturbance such as construction. Indicators for this form of assessment focus on inputs and outputs relating to the environment and animal behaviour, but it may also be useful to include some output indicators of health and nutrition, such as food intake and faecal scores. However, these can be hard to interpret over short time frames when multiple variables may be involved. Collaboration between many staff departments, such as event/project organisers and animal care staff, is fundamental to the accuracy and successful mitigation efforts of these assessments. This form of assessment need only be carried out when events occur, but best practice is to collect data from before, during, and after the event so as to understand the impacts (Clifforde & Harley, 2021). Figure 13.4 serves as a useful illustration of an animal welfare event management

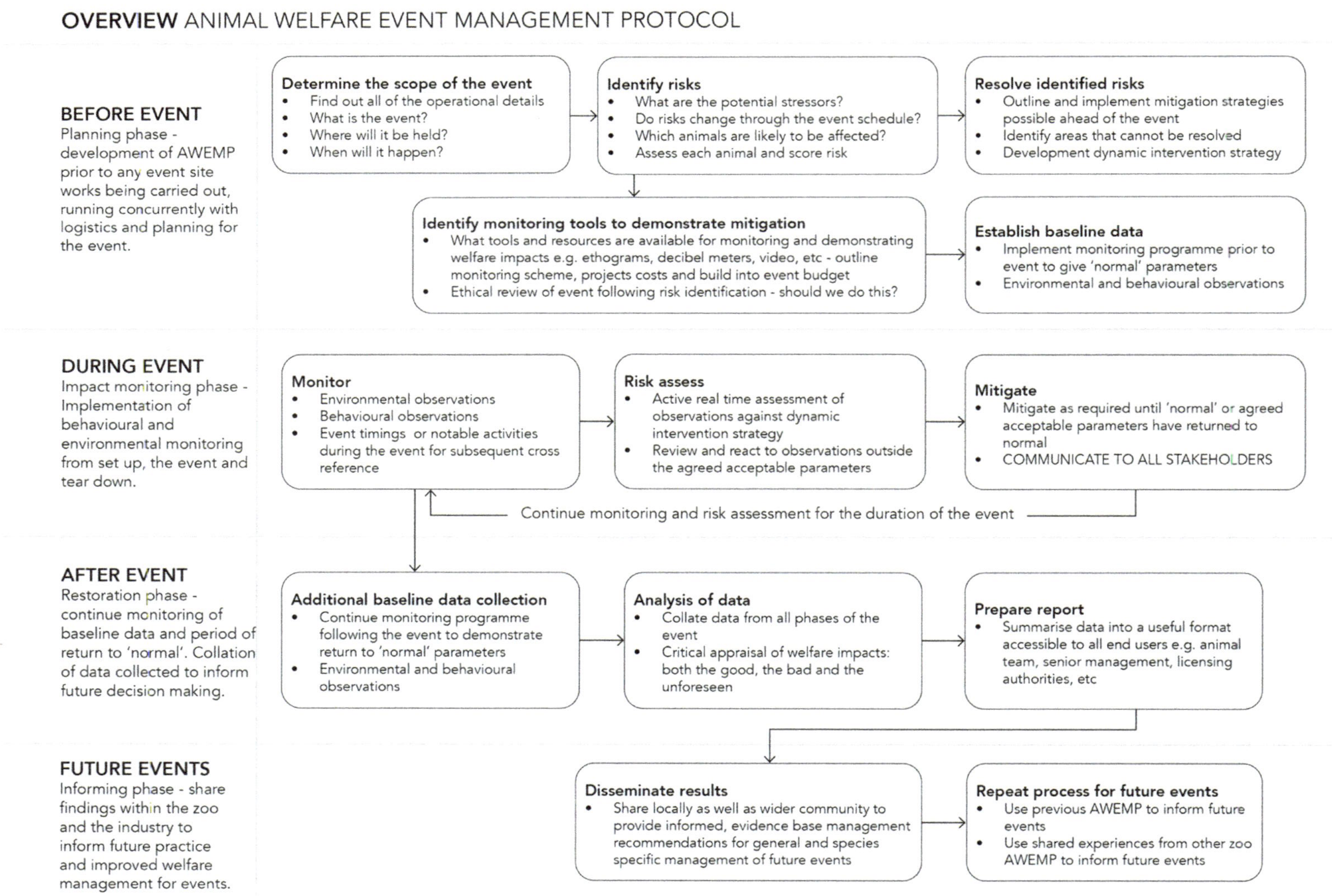

Figure 13.4 This comprehensive protocol outlines the systematic management of animal welfare during events at Knowsley Safari, encompassing three distinct phases: planning, impact monitoring (during), and restoration (post-event). Each phase is designed to prioritise the welfare of the animals while ensuring successful event execution.

protocol outlining the key steps and considerations necessary to protect the welfare of animals during events at the zoo.

13.4 CONCLUSIONS AND FUTURE RECOMMENDATIONS

Animal welfare assessment programmes are not only mandated by specific legislation (e.g. Irish Standards of Modern Zoo Practice (ISMZP)) and accreditation standards (EAZA, BIAZA) but are also an ethical obligation for zoos. The authors have emphasised the importance of adopting a comprehensive approach to animal welfare management, including creating holistic audit templates that can be adapted from existing models or built from scratch to align with the zoo's specific needs. These audits should be complemented by targeted assessments to address specific concerns. Examples of these targeted assessments include (but are not limited to):

- QoL Assessments: These audits can be particularly useful for geriatric animals, those with chronic health conditions, or those exhibiting behavioural challenges.
- Event-based Assessments: These are triggered by specific events, such as the introduction of a new animal, a change in social grouping, construction near an animal habitat, or a Christmas event. They help to assess the impact of these events on animal welfare and implement mitigating measures if necessary.
- Visitor Experience Assessments: These audits focus specifically on animals involved in visitor experiences, such as encounters, feedings, or demonstrations. These assessments evaluate the potential impact of these interactions on the animals, considering factors like the frequency and duration of the experiences, the level of human contact involved, and the animals' behavioural responses. This helps ensure that visitor experiences are managed in a way that prioritises animal welfare and minimises any potential stress or disruption to the animal involved.

By combining regular holistic audits with these targeted assessments, zoos can create a robust and responsive welfare programme that addresses both the general and specific needs of their animals.

Data collection, an underlying aspect of welfare assessment, can be resource-intensive and requires specialised expertise. Standardised methodologies and equipment, as well as trained staff, are integral to the assessment process. Utilising existing data sources, such as temperature records and food intake logs, along with simple collection methods like photos and videos, can simplify and optimise the process. Accurate interpretation and understanding of the collected data rely on thorough research of each animal's behavioural biology, natural history, and current research. This important step, though time-consuming, is essential for a meaningful welfare assessment.

Each audit (irrespective of type) must result in an action plan that outlines specific steps to address any identified concerns. These plans should prioritise the animal's needs and involve input from key staff across the zoo. Approval of the action plan by decision-makers (e.g. budget holders, veterinarians, or ethical review panels) is essential, followed by subsequent reviews to evaluate the effectiveness of implemented mitigations. This iterative process, involving cycles of observation, data collection, analysis, and implementation of adjustments, ensures the continuous improvement of animal welfare. By constantly refining and adapting care practices based on feedback and new information, zoos can create environments that better meet the physical, social, and psychological needs of the animals in their care.

Future recommendations for zoos and the development of their welfare management programme:

1. Staff and Training: Invest in ongoing professional development for animal care staff, ensuring staff have up-to-date knowledge and skills in animal welfare science and assessment techniques. Consider the importance of a dedicated 'Welfare Officer', a position similar to the requirements of conservation, education, and research roles.
2. Embrace Technological Advancements: Continue to explore and integrate emerging technologies, such as wearable sensors and biologging tags into welfare assessment protocols.

BOX 13.2: Bridging research and practice – essential reading in welfare assessment for zoos

- Clifforde, L.M., McDonald, K. & Farmer, H.L. (2022). Animal Welfare Toolkit: Guidelines for Quality of Life Assessment. BIAZA: London, UK.
- Clifforde, L.M., Holmes, L., Richdon, S., Newbolt, J.K., Free, D. & McDonald, K. (2023). Animal Welfare Toolkit: Guidelines for Visitor Animal Experiences Assessment. BIAZA: London, UK.
- O'Brien, S. L., & Cronin, K. A. (2023). Doing better for understudied species: Evaluation and improvement of a species-general animal welfare assessment tool for zoos. *Applied Animal Behaviour Science*, *264*, 105965.
- Miller, Lance J. & Sathya K. Chinnadurai. (2023). "Beyond the Five Freedoms: Animal Welfare at Modern Zoological Facilities" *Animals* 13,11: 1818.
- Powell, D. M., Beetem, D., Breitigan, R., Eyres, A. & Speeg, B. (2024). A perspective on ungulate management and welfare assessment across the traditional zoo to large landscape spectrum. *Zoo Biology*, *43*(1), 5–14.
- Rose, P. (2023). Identifying Essential Elements of Good Giraffe Welfare—Can We Use Knowledge of a Species' Fundamental Needs to Develop Welfare-Focussed Husbandry?. *Journal of Zoological and Botanical Gardens*, *4*(3), 549–566.
- Rose, P. & O'Brien, M. (2020). Welfare assessment for captive Anseriformes: A guide for practitioners and animal keepers. *Animals*, *10*(7), 1132.

These tools can provide a wealth of real-time data, enabling more precise and timely interventions.

3. Expand Research: To develop more refined assessment tools for the animals in our care, we must continue investing in research to better understand the specific needs of the diverse species housed in zoos. This research will inform the development of more comprehensive, evidence-based best practice guidelines, ultimately leading to improved animal welfare.
4. Communicate Transparently: Share the results of welfare assessments with the public and the zoo community. This, in turn, will highlight the zoo's commitment to animal welfare and foster a greater understanding of the complex issues involved in the management of zoo animals.

For further reading on animal welfare assessments, including both general principles and species-specific applications, please refer to Box 13.2.

REFERENCES

BIAZA. (2023). *BIAZA Animal Welfare Policy.* BIAZA, London, UK.

BIAZA. (2024). *BIAZA Close Contact Policy.* BIAZA, London, UK.

Blokhuis, H., Veissier, I., Jones, B. & Miele, M. (2013). The welfare quality® vision. In *Improving Farm Animal Welfare* (pp. 71–89). Wageningen Academic, Wageningen, Netherlands.

Clifforde, L.M. & Harley, J.J. (2021). *Animal Welfare Toolkit: Zoo Events and Animal Management.* BIAZA, London, UK. https://biaza.org.uk/projects/detail/biaza-welfare-toolkit (Access date:10.03.2025).

Diana, A., Salas, M., Pereboom, Z., Mendl, M. & Norton, T. (2021). A systematic review of the use of technology to monitor welfare in zoo animals: Is there space for improvement?. *Animals*, *11*(11), 3048.

D'Cruze, N., Khan, S., Carder, G., Megson, D., Coulthard, E., Norrey, J. & Groves, G. (2019). A global review of animal–visitor interactions in modern zoos and aquariums and their implications for wild animal welfare. *Animals, 9*(6), 332.

Jones, N., Sherwen, S. L., Robbins, R., McLelland, D. J. & Whittaker, A. L. (2022). Welfare assessment tools in zoos: From theory to practice. *Veterinary Sciences, 9*(4), 170.

Henneke, D. R., Potter, G. D., Kreider, J. L. & Yeates, B. F. (1983). Relationship between condition score, physical measurements and body fat percentage in mares. *Equine Veterinary Journal, 15*(4), 371–372.

Kagan, R., Carter, S. & Allard, S. (2015). A universal animal welfare framework for zoos. *Journal of Applied Animal Welfare Science, 18*(sup1), S1–S10.

Mellor, D. J. & Beausoleil, N. J. (2015). Extending the 'Five Domains' model for animal welfare assessment to incorporate positive welfare states. *Animal Welfare, 24*(3), 241–253.

Mellor, D. J., Beausoleil, N. J., Littlewood, K. E., McLean, A. N., McGreevy, P. D., Jones, B. & Wilkins, C. (2020). The 2020 five domains model: Including human–animal interactions in assessments of animal welfare. *Animals, 10*(10), 1870.

Miller, L. J. & Chinnadurai, S. K. (2023). Beyond the five freedoms: Animal welfare at modern zoological facilities. *Animals, 13*(11), 1818.

Sach, F., Tatchley, C., Needham, N. & Pullen, K. (2019). *Guidelines for the Management of Elephants within BIAZA Zoos: Incorporating BIAZA's Policy on the Management of Elephants*. Biaza, London, UK.

Tallo-Parra, O., Salas, M. & Manteca, X. (2023). Zoo animal welfare assessment: where do we stand?. *Animals, 13*(12), 1966.

Ward, S. J., Sherwen, S. & Clark, F. E. (2018). Advances in applied zoo animal welfare science. *Journal of Applied Animal Welfare Science, 21*(sup1), 23–33. https://doi.org/10.1080/10888705.2018.1513842

WAZA. (2023). *The WAZA 2023 Animal Welfare Goal*. http://www.waza.org/priorities/animal-welfare-goal/ (Access date: 10.03.2025)

Watters, J., Sulzner, K., Marrin, D., Huang, S., MacDonald, C., Ostapak, S., Poole, A. & Hayle, H. (2015). *Quality of Life in Geriatric Zoo Animals; WAZA Magazine*. WAZA, Gland, Switzerland.

Wemelsfelder, F. (2007). How animals communicate quality of life: the qualitative assessment of behaviour. *Animal Welfare, 16*(Suppl.), 25–31.

14

The zookeeper perspective on animal welfare

ZOË NEWNHAM AND JACK BOULTWOOD

14.1 INTRODUCTION

Animal welfare is a complex and evolving field, and our awareness of animal welfare states has increased in recent decades. The topic has grown into an important multidisciplinary subject, with significant public concern and social influence regarding the welfare of zoo animals (Budaev et al., 2020). In response to this, zoos have evolved from collections of animals originally kept for food or work, and then to places of entertainment, to modern organisations focused on animal conservation that are required to uphold high levels of welfare (Ward & Sherwen, 2018). To ensure this promotion of good welfare, zookeepers and animal care staff (hereafter "keepers") have played a key role in the mindset shift across the industry over the years. Assessing, managing, and enhancing animal welfare is now a crucial role of keepers, including longer-term considerations such as enclosure design, husbandry and training developments, enrichment, managing population dynamics, and providing correct nutrition.

This chapter examines welfare from the authors' perspective, including the practicalities of conducting welfare assessments, the importance of informed and relevant enclosure design, the applications of traditional and non-traditional enrichment, managing and informing public perception of animal welfare, and the practicalities of and ethical and future considerations of welfare within the zoo industry, including how to overcome, if possible, hurdles to ensuring enhanced welfare states are provided for.

DOI: 10.1201/9781003429425-14

14.2 WHAT DOES WELFARE MEAN TO KEEPERS?

It is generally accepted that welfare comprises multiple behavioural, physical, and psychological elements (Jones et al., 2022). Our expanding understanding and deeper consideration of animal welfare can make it hard for keepers to remain up-to-date due to the presence of multiple conflicting theories and concepts (Liptovszky, 2024). One of the most well-known and accepted definitions, as discussed throughout this book, is Broom's "The welfare of an individual is its state as regards its attempts to cope with its environment" (Broom, 1986). However, how easy is it for this definition to be understood by keepers? Do keepers understand what state constitutes, and what is meant by environment? Do keepers have enough knowledge, experience, and education regarding the species they work with and their physiological, psychological, and behavioural needs on both a species and individual level to be able to identify how well an individual is coping within its environment? Are keepers able to use their knowledge to adapt an individual's environment, both physical and psychological, to improve their welfare? The answers to these questions are complex and often dependent on taxa, zoo, time, and financial constraints.

14.3 PRACTICALITIES OF ASSESSING ANIMAL WELFARE

14.3.1 FORMAL AND INFORMAL ASSESSMENTS

The Department for Environment, Food and Rural Affairs (DEFRA) states that "all zoo stakeholders" should conduct "twice yearly assessments" (DEFRA, 2012). However, as welfare is a continuum and a cumulative experience across an individual's lifetime that changes as they age (Hecht, 2021), assessments conducted twice a year only provide a snapshot of an individual's welfare at the point of assessment and do not give keepers an accurate representation of an individual's continued welfare state. To address this, it is important to note that keepers are able to continuously monitor the welfare of the animals under their care in a variety of ways. All interactions with animals, even those involving routine husbandry in the enclosure, give keepers an opportunity to undertake both formal and informal welfare assessments, involving behavioural, psychological, and physiological indicators of welfare.

14.3.2 EXAMPLES OF INFORMAL ASSESSMENTS

Informal assessments and observations that are conducted multiple times a day during every interaction between keepers and animals allow keepers to build a comprehensive understanding of many aspects of the individual's welfare state.

Informal observations can identify changes in welfare indicators, such as changes in behaviours and reactions to certain stimuli and situations, including how an individual reacts to their keepers, the performance of unexpected behaviours in response to a situation or stimuli that an individual is normally habituated to, changes in social interactions, or how an individual reacts to a novel situation.

Possible indicators may also include changes in preferences, such as those related to nutritional intake. Keepers are not only providing nutrition during feed times but also observing an animal's appetite, eating habits, interactions with food and reactions to conspecifics, if applicable, as well as keepers and the wider environment.

Husbandry and cleaning routines can also offer insights into an individual's welfare state via activity levels and enclosure use. Evidence of stereotypical pacing, for example, can be identified via tracks left in substrates, the causation of which can then be investigated and addressed, as well as possible physical indicators of welfare such as faecal consistency. Enclosure maintenance during husbandry routines not only ensures that the environment is safe and secure but promotes positive welfare states by meeting the animals' ecological and biological requirements and minimising stressors. Recording and adjusting temperatures, humidity and daylight exposures, especially for ectothermic species that are managed in climate-controlled conditions, is especially important. A keen attention to detail throughout all interactions with animals is imperative for all keepers, as many changes may be small or appear insignificant to start and therefore impact accurate reporting.

14.3.3 EXAMPLES OF FORMAL ASSESSMENTS

The informal insights detailed above provide excellent opportunities to build and create an extensive body of welfare-related data which can act as a baseline and be used to identify irregularities in welfare and behaviour during day-to-day interactions. However, this method relies on accurate reporting by keepers. One way to ensure that, regardless of keeper experience and level of knowledge, welfare can be scored objectively and result in robust evaluations is the use of a formal, science-based welfare scoring system (Kagan et al., 2015). Formal welfare assessments are required for licensing and research purposes and are often performed at regular intervals and then assessed under a standardised setting.

Both national and international zoo and aquaria associations such as WAZA (World Association of Zoos and Aquariums), EAZA (European Association of Zoos and Aquaria) and BIAZA (British and Irish Association of Zoos and Aquariums) have published policies and strategic plans regarding formal welfare assessment creation, undertaking and evaluation which zoos can use to inform their practices. Whilst the design and formation of a welfare assessment strategy is beyond the scope of this chapter, this topic is covered in other chapters of this book.

Both ZSL and Wild Planet Trust (WPT) use formal assessments based on frameworks including both input and output measures of welfare states. These assessments score several parameters under key headings such as health, behavioural, environmental and stress and use the combined scores to identify and measure the welfare state of individuals. Marwell Wildlife adapted the Animal Welfare Assessment Grid (AWAG) (Justice et al., 2017), from its original basis in laboratory settings (Honess & Wolfensohn, 2010), that measures multiple factors under four key parameters, – 1) physical indicators, including scores for body condition score and faecal scores, 2) psychological indicators, including presence of abnormal repetitive behaviours (ARBs) and response to human interaction, 3) environmental indicators, including scores for enclosure complexity and ability to express natural behaviours and 4) procedural indicators, relating to the impact of veterinary procedures. Scoring these factors on a daily or weekly basis results in a long term, and where possible, lifetime individual welfare record where factors that impact welfare can continually be identified and addressed.

Although most available literature on welfare assessments concerns individual animals, and this is arguably the most effective method for monitoring welfare, scoring all individuals daily can be time-consuming and logistically challenging. For some species in large social groups, identification of individuals is not always possible, and therefore group welfare assessments may be more beneficial (Harley & Clark, 2019). Group assessments can allow keepers to identify factors that impact the group as a whole or identify what percentage of the group are impacted and what can be done to address this.

14.3.4 EFFECTIVE WELFARE ASSESSMENTSA COMBINATION OF BOTH FORMAL AND INFORMAL

Welfare concerns noticed by keepers during informal observations often lead to more formal assessments to be conducted, as depicted in Figure 14.1. A combination of both informal and formal assessments gives a wide breadth of information of standardised results complemented with intricate details regarding individuals and can also provide insights to more cryptic or less studied species. This is especially important when assessing welfare with standardised formal toolkits, which are usually designed for mammals. Gathering extra informal information on understudied taxa such as birds and lower vertebrates offers flexibility in the assessments to tailor it further for the benefit of the species being assessed. Without this flexibility, some aspects of welfare can be exaggerated, underestimated or misunderstood resulting in incorrect conclusions to be drawn. An example of this relates to the value of space for species such as crocodilians (*Crocodilia* spp.). Naturally they are usually a species that conserves energy and are therefore typically viewed as having limited activity budgets. Welfare scoring systems designed for mammals using activity budgets as part of the

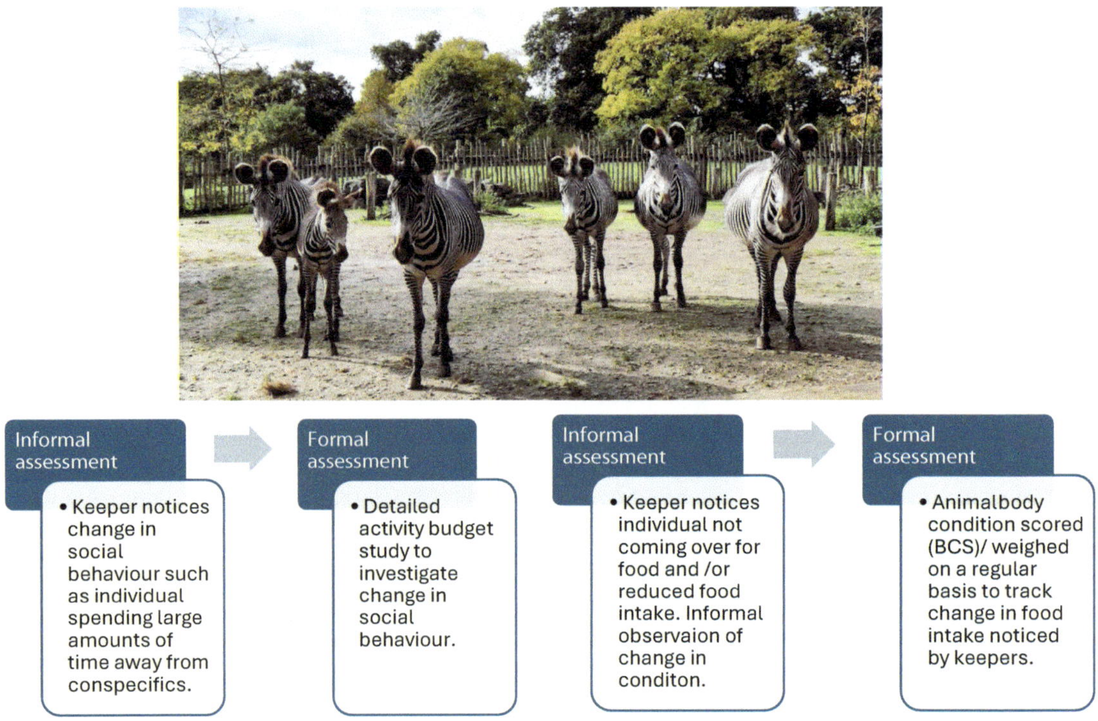

Figure 14.1 Welfare concerns noticed by keepers during informal observations often leads to more formal assessments to be conducted.

assessment would therefore lead to incorrect welfare scoring if used on non-mammalian species such as crocodilians; low activity budgets may be a welfare indicator for a mammalian species, but not for crocodilians. Furthermore, these incorrect conclusions could lead to welfare-related budgets not being allocated for this species, who, despite having low activity budgets still require a complex and spacious environment.

It is important to note that for keepers completing welfare assessments with the overall aim of identifying how to promote positive welfare, the presence of short-term negative states does not necessarily equate to negative welfare overall (Jones et al., 2022). In the wild, individuals are presented with challenges that result in negative welfare states such as hunger, thirst, predation, and competition for resources or mates. Short-term exposure to such negative states in the zoo can benefit an individual's health, fitness, life experience, and overall welfare state (Mason, 2010) and can help to retain and conserve natural behaviours and adaptive traits. For example, captive meerkats (*Suricata suricatta*) show the ability to distinguish between potential predator calls and retain natural cooperative behaviours when played recorded audio of predator calls in the zoo (Schneider et al., 2021), a situation that could be deemed detrimental to welfare by new keepers and naïve public as it has the potential to result in a short-term stress response from the meerkats. If the need for this natural behavioural repertoire is not explained to ensure further understanding, incorrect conclusions regarding welfare states could be drawn.

14.4 CHALLENGES OF WELFARE ASSESSMENTS

14.4.1 UNDERSTANDING WHAT WELFARE IS AND AVOIDING BIAS

BIAZA has instigated a Welfare Policy and Welfare Toolkit to help its members choose and implement an assessment framework (Harley & Clark, 2019) however, even in these documents, the lack of a welfare definition is evident, with no one definition mentioned. This can create confusion for keepers on how to design, conduct, and evaluate welfare assessments, and which indicators to include in these assessments. Zoos with the available resources have begun to employ welfare-specific scientific staff to assist with designing and conducting formalised welfare assessments to address this issue, although it must be noted that this is not a financially available position for all zoos.

Once successfully designed, both formal and informal assessments can be affected by keeper bias. Keepers develop strong bonds with certain animals or species, and this has the potential to influence and bias their perception of the animal's welfare, or lead to downplaying or overselling of welfare concerns. Furthermore, what one keeper denotes as positive or negative welfare may not correlate with the thoughts of other keepers or zoo personnel. Mehrkam & Dorey (2015) found that keepers can be biased towards individual animals and towards a species as a whole, and Gaalema et al. (2011) found that food choice preferences of giant pandas (*Ailuropoda melanoleuca*) and African elephants (*Loxodonta africana*) did not correlate with what keepers thought preferences for these species would be. Therefore, objectivity and honesty are required when scientific welfare assessments are conducted.

However, there is evidence that keepers can be effective welfare scorers (Brouwers & Duchateau, 2021), and that they can be a reliable source of information regarding welfare indicators, especially keepers with in-depth knowledge of the species and individuals under their care. Keepers with extensive knowledge of groups and individuals are able to identify even small changes in behaviour patterns and other potential indicators of welfare, which may be missed by less experienced keepers, thus making assessments more accurate or robust. Individual animals will all perceive their environment differently and therefore cope in different ways, even if all individuals are housed in the same enclosure with consistent husbandry parameters, what is considered "normal" or "typical" may differ greatly between individuals; therefore, expert knowledge is imperative to successful welfare assessments.

Recognising and mitigating possible bias is key to ensuring objective welfare assessments are performed, and areas where improvements to process (i.e. the welfare scoring itself), and the result (i.e. changes to the animal's environment based on the assessment) are conducted based on impartial and objective evidence.

This can be achieved by including multiple assessors, such as species-specific specialists, behaviour specialists, keepers, research students, veterinary or animal health staff, and independent (i.e. non-zoo focused individuals) to combine multiple viewpoints. A variety of experiences and backgrounds from the assessors can create scope for requirements or issues that may have been overlooked or not thought of by someone who is well established within the zoo industry. Impartial assessors with specialised knowledge and experience are also just as valuable as they are able to offer greater depth of knowledge on possible key indicators of welfare state and give species-specific knowledge or advice to those with more limited understanding of a species' needs.

Collaboration between these assessors is key to providing an improved welfare state within a captive setting, creating a comprehensive, broad, and holistic judgement on animal welfare. Keepers should not feel like including multiple assessors is an indicator that their experience or scores are not valid; rather, that this practice can result in more well-rounded and in-depth data that can be missed by fewer assessors.

14.4.2 KEEPER WORKLOAD

Although formal and informal assessments can provide unique and beneficial insights, a key limitation of many formal welfare assessments is the keeper time needed to design, implement, and conduct them. A keeper's biggest pressure is time constraint, as they have a wide range of responsibilities (animal care, guest interactions, research, record keeping, and wider zoo operations) to complete daily. These duties combine to form part of the high standard of welfare modern zoos should aim to achieve, but can make completing lengthy, formal welfare assessments difficult often resulting in poor completion rates or incorrect reporting.

Despite daily welfare assessments appearing daunting and time consuming initially, if designed efficiently, with a consistent method applied, they can provide accurate and detailed data that zoos can use as evidence to help implement welfare advances. Efficient management of zoo personnel and adequate staffing levels, with streamlining of tasks or jobs that need to be completed, allows for more time to be given to welfare scoring and the development of better husbandry practice. Web based applications, such as the AWAG, or well designed and quick to use formats, can help keepers conduct these assessments effectively and increase completion rates. Work overload is highlighted as a key stressor for animal care staff (Burns, 2020), so time efficient assessments and research methods are also imperative to support keeper wellbeing as well as completion rates.

Once designed, a limitation of all forms of welfare assessment is keeper working hours. Keepers are only on site for a limited proportion of an animal's daily time budget, and this can mean that, even for diurnal species who are active when keepers are present, our knowledge of what these species do when keepers are not present is lacking. This can make generating robust and valid conclusions regarding their welfare state challenging. Advances in technology, including the use of camera traps and LED bulbs that emit long-wavelength red and infrared lights (Baskir et al., 2021), have allowed keepers to study the activity patterns of species over a 24-hour time-period with findings used to assess and improve welfare where needed. Research across a wide range of supposedly diurnal species, from Asian elephants (*Elephas maximus*) (Evison et al., 2020), lesser flamingos (*Phoeniconaias minor*) (Rose et al., 2022), giraffe (*Giraffa camelopardalis*) (Burger et al., 2020), and snub-nosed monkeys (*Rhinopithecus brelichi*) (Tan et al., 2013), indicate these species display more nocturnal and crepuscular activity than expected. Therefore 24-hour behavioural and welfare needs should be considered by keepers and husbandry protocols. The advances in technology should also be used to consider the behavioural and welfare needs of nocturnal species whose natural circadian rhythms have historically made it difficult, and not always time efficient, for keepers to complete accurate welfare assessments.

14.4.3 FINANCE AND RESOURCES

Good keepers should be continually trying to provide a high standard of welfare on a daily basis during every interaction they have with the species under their care, however, even small improvements to housing, husbandry and management protocols etc., require time and resources to implement. With many zoos facing financial, time and resource limitations, the feasibility of addressing welfare issues that have been highlighted from welfare assessments can be difficult. This shortfall can impact not only animal welfare, but keeper wellbeing and motivation. Knowing what is needed to improve the lives of animals under your care but being unable to act on these issues can result in a sense of hopelessness or frustration. Zoos should aim to successfully finance operations (including those focussed on animal management, welfare and human resources) in a sustainable way, enabling flexibility, resilience and easy reactivity to future needs or policy changes, especially when attempting to manage welfare-challenging species.

14.4.4 CONTINUOUS IMPROVEMENT OF WELFARE

Undertaking welfare assessments should be an ongoing commitment and continuous cycle of conducting assessments, implementing evidence-based changes to improve or maintain welfare, evaluating outcomes, and starting the cycle again (Figure 14.2).

A welfare assessment or change in management conducted in response to a specific welfare concern, or a one-time welfare assessment can highlight the response to a single event but will not provide enough information for a continuous assessment over the animal's lifetime. Regular reviews and updates to welfare practices are needed based on scientific research, industry standards, animal status, and assessment findings. Animal welfare continuously evolves, and keepers are required to keep up to date with modern standards and research. This requires a proactive approach to welfare management, ongoing professional development, species-specific research, and relationship building with other zoos. This will not only facilitate science-based assessments but also allow keepers to accurately educate visitors and the wider public on the importance of animal welfare. Ultimately, despite challenges and limitations, there is clear evidence that keepers have the desire, ability, knowledge, and skills to improve welfare, and the use of welfare assessments can help keepers to determine a priority action list, with resources allocated where possible.

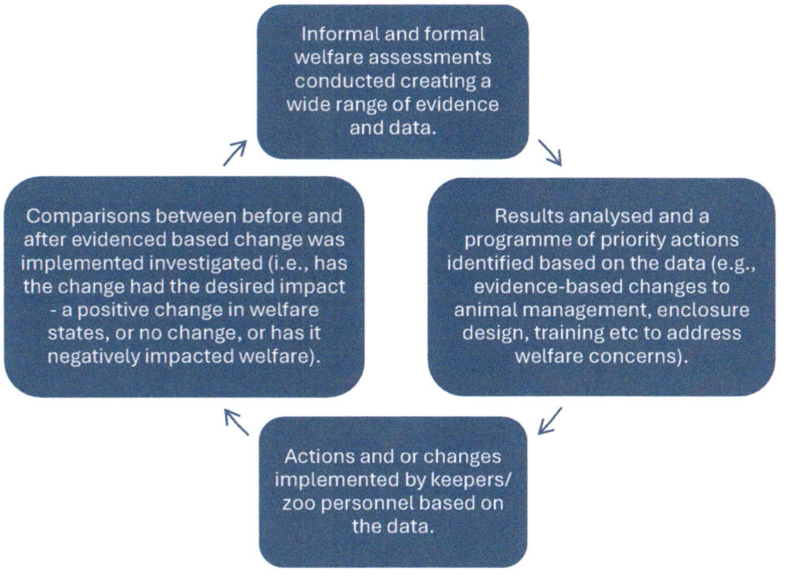

Figure 14.2 Undertaking and combining both formal and informal assessments should be an ongoing commitment and continuous cycle of conducting assessments, implementing evidence-based changes to improve or maintain welfare, evaluating outcomes, and starting the cycle again.

14.5 ANIMAL WELFARE AND KEEPER EDUCATION

14.5.1 KEEPER EDUCATION

Despite the importance of science-based welfare assessments and the need for continued keeper education, formal and ongoing education, especially regarding animal welfare, has not always historically been part of a keeper's training. Furthermore, education programmes are unable to fully cover the behavioural ecology, both wild and captive, of all taxa a keeper may work with in detail, and, similar to ethics and welfare, are topics that are not always part of the primary training that keepers receive (Bacon et al., 2023).

For many species, research is lacking and this further limits available resources for keepers to use for developing husbandry and welfare assessments. A lack of research can be caused by multiple factors, including a taxon bias in research towards mammals and vertebrates; however, even for mammals, a species' natural habitat can limit the feasibility of research. For example, until recently, most research regarding the natural behaviour of okapi (*Okapia johnstoni*) comprised of only three studies undertaken in situ due to the natural habitat challenges, elusive behaviour, and political unrest in range states of the species (Michaels et al., 2022).

Even when research is available, using a comparison between wild and captive behavioural ecology as a welfare indicator can be misleading as species can show changes to their behavioural and biological phenotypes due to differences between the natural habitat a species has evolved to inhabit and their captive environment (Howell & Cheyne, 2019). Zoo-bred regent honeyeaters (*Anthochaera phrygia*) show undeveloped or different variants of wild song in captivity (Tripovich et al., 2021) – a factor keepers should consider as part of their husbandry and welfare scoring, as this could negatively influence the breeding success both in captivity and of released individuals if part of conservation programmes. Mallorcan midwife toads (*Alytes muletensis*) showed reduced predator defence and developed at a slower rate after 12 generations in captivity, and these factors will not only impact animal welfare but potentially decrease the success of future release programmes (Kraaijeveld-Smit et al., 2006).

Welfare-related training offered to keepers also has an important impact on keeper well-being and job satisfaction. Keepers offered animal welfare training are more likely to report higher levels of job satisfaction (DeSmet & Ogle, 2022) which links closely to increased job satisfaction found in keepers who perceive that they provide high levels of welfare to their animals, enabled by the training in the first place. This correlates with increased positive keeper and animal relationships that can significantly and positively influence zoo animal welfare.

14.5.2 PUBLIC ENGAGEMENT AND EDUCATION

Managing both animal welfare and the public expectation of said welfare can be complex. Keepers must consider the public's perception of the animals under their care, as this plays a critical role in shaping the public's attitude when receiving and understanding welfare and conservation messaging. Engaging visitors through education programmes, experiences, and keeper talks can provide an excellent opportunity to reveal the complexities of such operations, and the efforts keepers expend to ensure a high level of welfare and good public education. Keeper talks, animal encounters, and demonstrations in the form of enrichment activities, training sessions, and feeding sessions highlight the daily efforts made to promote welfare and can be used to address possible welfare concerns the public may have. An example of this is in the explanation of ARBs. Behaviours such as head bobbing and perceived stereotypic pacing can be distressing for visitors and keepers alike and flag concerns about the welfare state of the animal. ARBs can indicate a current or historic underlying welfare issue, so understanding and addressing the stressor (or latent effect of a stressor) and any associated welfare implications is part of a keeper's role, as well as explaining such abnormal activities in an objective but accessible way to visitors. Management of ARBs can involve a multifaceted approach, including introducing novel and behaviourally satisfying enrichment activities, encouraging of natural behaviours by modifying enclosures to provide a more stimulating environment, implementing stimulating positive reinforcement training and targeted veterinary care and ongoing welfare assessments to document the timeframe of ARB performance. It can be useful to educate visitors that the ARB being presented has been noted by keepers and is being addressed as this can help to create transparent communication with visitors, thus making the zoo more open and accountable, whilst highlighting the challenges of managing wild animals. Keeper talks have a large, positive effect on public perception and information retention due to a keeper's knowledge about the species and specific individuals they care for (Lyon et al., 2024). This then creates moments for visitors to foster a deeper understanding of, and appreciation for, the work that keepers undertake, as well as the needs of the animals.

14.6 METHODS USED TO PROMOTE POSITIVE WELFARE OUTPUTS

14.6.1 ENRICHMENT PROVISION AND DESIGN

Enrichment is an ever-evolving cornerstone of modern keeping, with the aim to enhance the physical and mental well-being of animals via implementing changes within their environment (Bartolomé et al., 2023). Traditionally, enrichment has focused on providing inputs such as novel objects that (hopefully) encourage natural or biologically relevant behaviours, such as puzzle feeders, foraging boxes, hanging feeders, and scented objects and has often been mammal-centric. However, the importance of enrichment to welfare is now known across all taxa including reptiles and birds (Hoehfurtner et al., 2021; Rachinas-Lopes et al., 2024; Rickman et al., 2025) and now focuses on promoting positive behavioural outcomes (Vicino et al., 2022). Alongside this, modern zoos are now factoring enrichment into species-specific naturalistic enclosure design, reducing the need for reliance solely on traditional input-based enrichment items.

Providing a complex, dynamic and naturalistic environment which includes factors such as biologically relevant plants, multiple substrates, microclimates, seasonal temperature fluctuations, frequent changes to enclosure furnishings, basking opportunities, changes to water quality and (where possible) live prey items, creates a biologically relevant and enriching environment. Including factors which can be changed or are under constant fluctuation further encourages natural behaviours such as foraging, territoriality, social and breeding behaviours and thermoregulation – all important aspects of a species behavioural repertoire.

However, large- scale enclosure modifications and/or creating complex, dynamic enclosures for all species require substantial finances and resources and therefore may not be achievable on a wide scale or on a regular basis for all zoos. In these instances, keepers are able to achieve positive environmental change on a smaller, short-term scale via the use of "traditional" species-specific environmental enrichment items. Additionally, enrichment has been shown to encourage activity and interaction within the environment (Ward et al., 2024), so it should therefore be considered even for those individuals in large, complex naturalistic enclosures.

Whether designing a novel enrichment object, modifying or designing enclosures (or both), keepers should focus on encouraging natural behaviours. For example, to encourage natural hunting behaviours, a scent trail leading to a food source can encourage specific hunting tactics. For mammalian carnivores this could come in the form of a meat source tied to an anchor point at the end of a blood trail, promoting the performance of natural behaviours and filling out their time budget in a biologically relevant way. For insectivorous species in the UK, natural feeding behaviours can be more easily replicated via the feeding of live insects which can either be offered with tongs to facilitate health checks, or scatter fed to allow an insectivore to hunt for its own food replicating a natural hunting experience. Although beyond the scope of this chapter, it is important to note that the welfare of live food items such as invertebrates can be a controversial topic, and one which requires further, in-depth research.

Historically, "traditional" enrichment items have often been artificial looking. Although Jacobson et al. (2017) and Salas et al. (2021) showed visitors viewing enrichment use by chimpanzees (*Pan troglodytes*) and ring-tailed lemurs (*Lemur catta*) respectively were not concerned with the design of the enrichment and were aware that artificial items can be used to elect natural or biologically relevant behaviours thus improving welfare, it can be argued that using naturalistic looking items can present more educational messaging potential and fit the aesthetics of the naturalistic environment modern zoos aim to create. Where possible, behaviours elected from an artificial item can often be encouraged by a more naturalistic design. In Figure 14.3, red river hogs (*Potamochoerus porcus*) are encouraged to move plastic feeders around to release food items encouraging foraging and rooting behaviours; however, these behaviours can also be encouraged via a more naturalistic design of food items placed inside a hollowed-out log.

Evidence-based enclosure design and enrichment has benefited greatly from innovations in technology. For example, UV (ultraviolet radiation) provision has improved from low-output on-off systems to high-output sources with an incremental light level to mimic the rise and fall of the sun, creating more naturalistic enclosures. Automated feed systems on variable timers can help avoid set feeding schedules, replicating more naturalistic and unpredictable feeding cycles. Where possible, linking technological innovation and the zoo can create opportunities to further advance our understanding of, and ability to influence, positive welfare states.

14.6.2 EVALUATING ENRICHMENT EFFECTIVENESS

Keepers should monitor the animals' responses to enrichment, both items and enclosure design, ensuring that the enrichment and environment are fulfilling or promoting biologically relevant behaviours and not causing negative stressors. A key consideration should be given to ensure the health and safety of the individuals that are interacting with the enrichment, as enrichment that only results in frustration or is not safe for the participant can result in negative associations and a worsened welfare state.

Consideration should be given when introducing enrichment items to a neophobic species or to a social group to ensure issues such as fear, aggression, or fighting over an item do not lead to displacement, injury, or death. Although multiple animals may safely interact with the same item (even if multiple items are given), ensuring there are multiple enrichment items, or a large enough item for all individuals to interact with, can help to mitigate potential negative situations (Figure 14.3).

Figure 14.3 Although multiple animals may safely interact with the same item (even if multiple items are given), ensuring there are multiple enrichment items, or a large enough item for all individuals to interact with, can help to mitigate negative situations in social species. Here, multiple feeding devices are used for a group of red river hogs. Although promoting natural feeding behaviours, the feeding balls (left) are artificial in aesthetic; however, the same behaviours can be encouraged via more naturalistic items that keepers can design and make, such as log feeders in which food can be placed and that require movement to release the food (middle, right).

14.6.3 ENCLOSURE DESIGN: BALANCING SPACE AND NATURAL BEHAVIOUR

Enclosure design and adaptations are pivotal in creating optimal welfare states. Other chapters in this book examine the link between welfare, enclosure design, space, and complexity in more detail, but it is important to highlight the impact keepers can have on enclosure design.

For example, for browsing species, adaptations to enclosures to increase browse provision, such as leaves, twigs, forbs as feed, can increase positive welfare. Increasing the number of browse hangers in the okapi enclosures at Marwell Wildlife, as seen in Figure 14.4, has allowed keepers to better replicate the natural feeding ecology of the species, providing browse at multiple times of the day in different areas of the enclosure, encouraging locomotion between foraging areas and reducing the risk of abnormal repetitive behaviours (ARBs) developing. These browse hangers, using rope, metal tubing, and winches, are relatively cheap and easy to install by keepers, showing that keepers can potentially make significant improvements to animal welfare regardless of time or financial constraints.

Alongside this, to increase browse provision further, plantations have been established. This, however, is not easily achieved by all zoos. Zoos that have the land space to create browse plantations; horticulture-based land managed to grow edible browse species (Fidgett, 2023), or the financial resources to buy in browse are likely to be able to address any welfare concerns regarding browse more easily than zoos with fewer resources. Plantations also take time to establish and become viable, but starting projects such as these which can improve the welfare of future generations of captive wild animals is just as important to consider as those that current keepers can see immediate welfare improvements for.

It is imperative that enclosure designs and future plans take both keepers and animal management into account. Whilst it is important that enclosures promote natural behaviours, enclosures also need to be practical for keepers to access, manage, and service. If it is not practical and safe for the keepers to manage the animals in, welfare for both animals and keepers may decline over time as optimal standards will be hard to achieve and sustain. Keepers should have a key input in the design of new enclosures and modifications to existing ones. Keepers at Marwell Wildlife were involved in the design process of the scimitar horned oryx (*Oryx dammah*) house, including the design for the raceway built to improve the welfare and ease of handling of individuals during key veterinary procedures such as injections. This has improved the process from using a heavy and hard to move board with limited accessibility to a purpose built, multiuse

Figure 14.4 Adaptations to enclosures by keepers to improve animal welfare. Browse pulleys designed and installed by keepers in the okapi enclosures at Marwell Wildlife help keepers to increase browse provision in a way that promotes natural feeding behaviours as well as helping keepers to work more efficiently by making it easier and more convenient to hang browse.

area that the oryx have access to and walk through every day to encourage habituation (Figure 14.5).

The size of an enclosure is often commented on as an indicator of enclosure suitability, but bigger is not always better. An enclosure that is complex and meets the needs of a species can be more stimulating and increase opportunities to perform natural behaviours, keep or gain physical condition, create microclimates, and provide security for inhabitants over large, empty spaces.

Keepers are key to providing complex environments and can achieve this through regular enclosure modifications and updates; renewing resources such as branching, substrates, and enclosure furnishings. For example, for an arboreal species such as green anoles (*Anolis carolinensis*), an enclosure with a complex system of branching and climbing opportunities that are regularly changed alongside diverse planting, substrates, and enrichment provides multiple options for exercise, social interactions, security, and mental stimulation compared to a vast empty space (Figure 14.6). Even species such as white rhino (*Ceratotherium simum*), where a large open space for grazing has historically been the basis for enclosure design, can benefit from a complex enclosure that includes differing substrates, landscaping, logs and rocks, wallows, and grazing opportunities.

14.7 CONCLUSIONS

Providing the highest level of welfare for all species should be the aim for all zoos; however, welfare considerations are complex, including balancing animal welfare considerations with available resources, new research and information, transparency with visitors and visitor expectations, and the role of keepers and keeper experience in navigating these challenges.

Keepers and regular visitors will have a perceived understanding of the animals they care for and visit over extended periods of time. Across an individual's lifetime, their welfare status will change, impacted by different life stages, illness, or psychological states. During these periods, ethical situations may arise where the welfare of the animal must take priority over any emotional or personal attachment keepers or visitors may have. Keeper bias related to welfare can be addressed using scientific, evidence-based welfare

Figure 14.5 Implementing keeper knowledge during enclosure design can lead to improvements in welfare and ease of handling of individuals during key veterinary procedures. Keepers at Marwell Wildlife were involved in the design process of scimitar-horned oryx housing. Moving from the use of a heavy and hard to move board with limited accessibility (top) to a purpose-built, multi-use area animals have access to and walk through every day to encourage habituation (middle). Doors at each end can be closed when the desired individual is in the raceway, with one side on runners allowing for space to be reduced safely, securing the individual, and multiple removable panels increasing accessibility on both sides, allowing for safe handling for keepers and veterinarians (bottom – seen from above). Note: risk assessments and safe handling measures are conducted and discussed prior to all animal interactions.

Figure 14.6 A naturalistic enclosure for green anoles that is designed to promote natural behaviours and positive welfare from species-specific considerations. Diverse planting allows for background matching, an important behaviour for a cryptic species

assessments conducted by multiple assessors, and visitor expectations managed through clear communication and education.

Keepers are essential in providing high levels of welfare and conducting and then implementing changes based on welfare assessments. Every interaction a keeper has with an individual is a chance to monitor and assess welfare. Therefore, keepers must possess strong observation skills and in-depth species-specific and individual animal knowledge. Keeping up with and conducting research, especially on underrepresented species and taxa, is a key way for keepers to do this.

Creating and maintaining enclosures designed to replicate a species' natural habitat with diverse planting, landscaping, and furnishings is also a key way to promote high levels of welfare; however, resources and funding are needed to achieve this, a limiting factor in many zoos. Keepers, therefore, must be resourceful, implementing changes to improve welfare with what is available to them. Keepers also have to be resourceful with their time, being responsible for a wide range of tasks on a daily basis, including but not limited to husbandry tasks, training, enrichment, maintenance and upkeep of enclosures, conducting research, implementing welfare assessments, and visitor education.

Despite high workloads, time, financial and resource limitations, keepers are fundamental to moving zoo animal welfare standards in a positive direction. The welfare of zoo animals would be far lower without the dedication and hard work of good zookeepers currently in the industry and those working towards it in the future.

REFERENCES

Bacon, H., Vigors, B., Shaw, D. J., Waran, N., Dwyer, C. M. & Bell, C. (2023). Zookeepers–The most important animal in the zoo?. *Journal of Applied Animal Welfare Science, 26*(4), 634–646.

Bartolomé, A., Carazo, P. & Font, E. (2023). Environmental enrichment for reptiles in European zoos: Current status and perspectives. *Animal Welfare, 32*, e48.

Baskir, E. A., Hazen, R. E. & Powell, D. M. (2021). Documenting nocturnal activity of dragon-headed katydids (Lesina blanchardi) under artificial light. *Zoo Biology, 40*(4), 342–345.

Broom, D. M. (1986). Indicators of poor welfare. *British Veterinary Journal, 142*(6), 524–526.

Brouwers, S. & Duchateau, M. J. (2021). Feasibility and validity of the Animal Welfare Assessment Grid to monitor the welfare of zoo-housed gorillas *Gorilla gorilla gorilla*. *Journal of Zoo and Aquarium Research, 9*(4), 208–217.

Budaev, S., Kristiansen, T. S., Giske, J. & Eliassen, S. (2020). Computational animal welfare: Towards cognitive architecture models of animal sentience, emotion and wellbeing. *Royal Society Open Science*, *7*(12), 201886.

Burger, A. L., Fennessy, J., Fennessy, S. & Dierkes, P. W. (2020). Nightly selection of resting sites and group behavior reveal antipredator strategies in giraffe. *Ecology and Evolution*, *10*(6), 2917–2927.

Burns, D. (2020). *Working with wildlife: The effects of stressors and resources on burnout and engagement for animal caretakers*. MSc thesis: University of Tennessee at Chattanooga, Chattanooga, USA.

DEFRA. (2012). *Zoos Expert Committee Handbook*. https://assets.publishing.service.gov.uk/government/uploads/system/uploads/attach ment_data/file/69611/pb13815-zoos-expert-committee-handbook1.pdf (Access date: 21.12.24).

DeSmet, A. & Ogle, B. (2022). The influence of welfare and bonds with animals on the job satisfaction of felid keepers in North America. *Zoo Biology*, *41*(3), 193–199.

Evison, E., McKenzie, A. & Holmes, L. (2020). Social and environmental impacts on sleep in captive Asian elephants (Elephas maximus). *Zoo Biology*, *39*(6), 397–404.

Fidgett, A. L. (2023). Browse selection and management. *In:* Miller, E. Lamberski, N. & Calle, P. (eds). *Fowler's Zoo and Wild Animal Medicine Current Therapy, Volume 10*. Elsevier: St Louis, MO. pp. 239–242.

Gaalema, D. E., Perdue, B. M. & Kelling, A. S. (2011). Food preference, keeper ratings, and reinforcer effectiveness in exotic animals: The value of systematic testing. *Journal of Applied Animal Welfare Science*, *14*(1), 33–41.

Harley, J. & Clark, F. E. (eds). (2019). *Animal Welfare Toolkit*. BIAZA: London, UK.

Hecht, L. (2021). The importance of considering age when quantifying wild animals' welfare. *Biological Reviews*, *96*(6), 2602–2616.

Hoehfurtner, T., Wilkinson, A., Nagabaskaran, G. & Burman, O. H. (2021). Does the provision of environmental enrichment affect the behaviour and welfare of captive snakes?. *Applied Animal Behaviour Science*, *239*, 105324.

Honess, P. & Wolfensohn, S. (2010). The extended welfare assessment grid: A matrix for the assessment of welfare and cumulative suffering in experimental animals. *Alternatives to Laboratory Animals*, *38*(3), 205–212.

Howell, C. P. & Cheyne, S. M. (2019). Complexities of using wild versus captive activity budget comparisons for assessing captive primate welfare. *Journal of Applied Animal Welfare Science*, *22*(1), 78–96.

Jacobson, S. L., Hopper, L. M., Shender, M. A., Ross, S. R., Leahy, M. & McNernie, J. (2017). Zoo visitors' perceptions of chimpanzee welfare are not affected by the provision of artificial environmental enrichment devices in a naturalistic exhibit. *Journal of Zoo and Aquarium Research*, *5*(1), 56–61.

Jones, N., Sherwen, S. L., Robbins, R., McLelland, D. J. & Whittaker, A. L. (2022). Welfare assessment tools in zoos: from theory to practice. *Veterinary Sciences*, *9*(4), 170.

Justice, W. S. M., O'Brien, M. F., Szyszka, O., Shotton, J., Gilmour, J. E. M., Riordan, P. & Wolfensohn, S. (2017). Adaptation of the Animal Welfare Assessment Grid (AWAG) for monitoring animal welfare in zoological collections. *Veterinary Record*, *181*(6), 143–143.

Kagan, R., Carter, S. & Allard, S. (2015). A universal animal welfare framework for zoos. *Journal of Applied Animal Welfare Science*, *18*(sup1), S1–S10.

Kraaijeveld-Smit, F. J., Griffiths, R. A., Moore, R. D. & Beebee, T. J. (2006). Captive breeding and the fitness of reintroduced species: A test of the responses to predators in a threatened amphibian. *Journal of Applied Ecology*, *43*(2), 360–365.

Liptovszky, M. (2024). Advancing zoo animal welfare through data science: Scaling up continuous improvement efforts. *Frontiers in Veterinary Science*, *11*, 1313182.

Lyon, C. E., Alba, A. C., Burgess, A., Weibel, C. J., Miller, D., Alligood, C. & Ferrie, G. M. (2024). Visitor engagement with zoo staff and interpretive demonstrations increases visitor stay time in two free-flight aviaries. *Zoo Biology*, *43*(2), 136–148.

Mason, G. J. (2010). Species differences in responses to captivity: Stress, welfare and the comparative method. *Trends in Ecology & Evolution*, *25*(12), 713–721.

Mehrkam, L. R. & Dorey, N. R. (2015). Preference assessments in the zoo: Keeper and staff predictions of enrichment preferences across species. *Zoo Biology*, *34*(5), 418–430.

Michaels, C. J., Jakobsen, L., & Newnham, Z. (2022). Behavioural biology for the evidence-based keeper. In P. Rose (Ed.), *The Behavioural Biology of Zoo Animals* (pp. 323–330). CRC Press: Boca Raton, FL.

Rachinas-Lopes, P., Rocha, I. C., Dias, T., Tavares, M., Neto, R., Flanagan, C. & Neves, J. (2024). The effects of enrichment on zoo-housed scarlet ibis behavior. *Animals*, *14*(13), 1903.

Rickman, E. L., Wilkinson, A., Pike, T. W. & Burman, O. H. (2025). The impact of enriched housing on the behaviour and welfare of captive leopard geckos (Eublepharis macularius). *Applied Animal Behaviour Science*, *283*, 106487.

Rose, P. E., Chapman, J., Brereton, J. E. & Riley, L. M. (2022). What's black and white and pink all over? Lesser Flamingo Nocturnal behaviour captured by remote cameras. *Journal of Zoological and Botanical Gardens*, *3*(4), 624–640.

Salas, M., Laméris, D. W., Depoortere, A., Plessers, L. & Verspeek, J. (2021). Zoo visitor attitudes are more influenced by animal behaviour than environmental enrichment appearance. *Animals*, *11*(7), 1971.

Schneider, S., Goettlich, S., Diercks, C. & Dierkes, P. W. (2021). Discrimination of acoustic stimuli and maintenance of graded alarm call structure in captive meerkats. *Animals*, *11*(11), 3064.

Tan, C. L., Yang, Y. & Niu, K. (2013). Into the night: camera traps reveal nocturnal activity in a presumptive diurnal primate, Rhinopithecus brelichi. *Primates*, *54*, 1–6.

Tripovich, J. S., Popovic, G., Elphinstone, A., Ingwersen, D., Johnson, G., Schmelitschek, E. & Pitcher, B. J. (2021). Born to be wild: Evaluating the zoo-based regent honeyeater breed for release program to optimise individual success and conservation outcomes in the wild. *Frontiers in Conservation Science*, *2*, 669563.

Vicino, G. A., Sheftel, J. J. & Radosevich, L. M. (2022). Enrichment is simple, that's the problem: Using outcome-based husbandry to shift from enrichment to experience. *Animals*, *12*(10), 1293.

Ward, S. J. & Sherwen, S. L. (2018). Zoo animals. *In:* Hosey, G. & Melfi, V. (eds). *Anthrozoology: Human-animal interactions in domesticated and wild animals*. Oxford University Press: Oxford, UK, pp. 81–103.

Ward, S. J., Hosey, G., Williams, E. & Bailey, R. (2024). Enrichment and animal age, not biological variables, predict positive welfare indicators in zoo-housed carnivores. *Applied Animal Behaviour Science*, *270*, 106006.

15

One Health and welfare – veterinary perspectives

MATTHEW FIDDES AND NATHALIE WISSINK-ARGILAGA

15.1 INTRODUCTION

Animal welfare encompasses not only the physical health of the animals (including the absence of disease and injuries) but also their behaviour and emotions (Fraser et al., 1997). Physical health is an important part when assessing animal welfare; however, the absence of disease, pain and injuries does not always equate to good welfare. There are numerous clear examples where animals might be doing well from a physical point of view and present no medical issues; however, their psychological needs are not met. It is important to not just consider the physical health but also the mental state of the animals. Eventually, poor mental state can end up affecting the physical health of the animal. Animals under our care should be able to engage in highly motivated behaviours that contribute to these animals experiencing positive emotions (Lawrence et al., 2019) and not just be pain-free and healthy. Any disease processes involve husbandry and dietary components, and this chapter discusses the importance in considering the holistic approach to the treatment and prevention of zoological animal disease, in turn supporting good welfare. Promotion of the One Health approach (examined in detail in the following section) in zoos and aquariums, namely the integration of animal health and welfare, human well-being and environmental health, can bring about more effective welfare assessment, understanding and positive outcomes. The case studies presented in this chapter, across a variety of taxa are used to demonstrate how disease states can impact animal welfare and how consideration of veterinary treatment in conjunction with husbandry and environmental factors is key. Specialised veterinary knowledge is not only important for the effective treatment of zoo animal disease and injury, but veterinary involvement in enclosure design, diet, husbandry and education, as part of a multidisciplinary team, can benefit zoo animal welfare.

DOI: 10.1201/9781003429425-15

15.2 THE ROLE OF THE VETERINARY SURGEON IN ZOO ANIMAL WELFARE

As veterinary surgeons (hereafter "vet"), a large part of our job is centred around disease prevention and treatment of injured or ill animals. Having said this, the role of the zoo vet should also encompass participation in welfare assessments and promoting good welfare through the reduction of stress and the provision of opportunities to engage in highly motivating behaviours. Therefore, the vet's role should include providing input into correct husbandry and diet practices that, in turn, will be of benefit for the health and well-being of the animals under our care; as well as protecting them from injury, pain and disease through the development of a robust biosecurity, disease prevention and surveillance programme and adequate treatment of ill animals.

A significant proportion of health issues seen in captive zoo animals have an underlying husbandry or dietary component to them and therefore, it will be of paramount importance to take a holistic approach to these issues, not just focusing on the treatment of the condition, but also on environmental/dietary adjustments that will improve the current issue and prevent future deterioration. We must, therefore, not just be well versed on animal diseases but also natural history, husbandry, nutrition and behaviour of the species we care for. Education of caregivers and keepers in health and welfare, as well as identification of health and behavioural issues, is another important aspect of the vet's role. Ensuring to meet the welfare needs of our zoo animals will be a multidisciplinary effort involving the zoo managers, keeper teams, veterinary teams and management of interactions with the public.

Although our main aim is to improve the health and welfare of our patients, we also need to consider the effect that veterinary interventions might have on the welfare of the animals under our care. Some interventions such as physical examination, sample taking and treatments can be invasive and, although in the long term, will hopefully be of benefit to the patient, and in the short term could cause stress. For this reason, we must endeavour to make veterinary interactions as stress-free as possible, and medical behaviour training and animal habituation to the veterinary teams should be an integral part of zoo husbandry practices. Time and resources will need to be dedicated to this and discussions with zoo managers/owners will be necessary. This will allow for better veterinary care, limiting the stress and negative effects our interventions might have, as well as potentially saving monetary resources due to being able to perform some tasks without the need for sedation or anaesthesia. In modern zoos, medical training takes shape in the form of conscious blood draws, hand injections, conscious ultrasonography amongst others allowing for smooth and stress-free veterinary interventions.

15.3 A ONE HEALTH APPROACH

An often-overlooked concept in the veterinary care of zoos is the One Health approach which is defined as a transdisciplinary approach to sustainable management of complex health problems arising from the interactions of animals, humans and their environment (Cork et al., 2016), see Figure 15.1. This has mainly been focused on research and messaging about zoonotic diseases and the human-wildlife interface (Deem, 2015). Another important aspect of One Health approach is the health benefits for zoo visitors in a largely urbanised and sedentary world (Padda et al., 2018). Consideration must be given to the One Health aspect in a zoo's veterinary programme as there is a human-wildlife interface between animals and keepers and sometimes between animals and members of the public. This needs to be reflected in the biosecurity protocols as well as the disease risk and surveillance programme. Moreover, these public and keeper interactions with zoo animals should also be part of the welfare assessments to ensure they are not only safe from a disease transmission point of view but also from an animal welfare and stress aspect. Examples of these interactions might include handling of certain species in zoo and petting zoos as well as "animal experiences" where members of the public get to be up close and personal with some zoo residents. Regular reviews of these practices from a biosecurity and disease point of view need to be undertaken as well as keeping thorough documentation of these interactions and possible effects on the animals involved. The environment

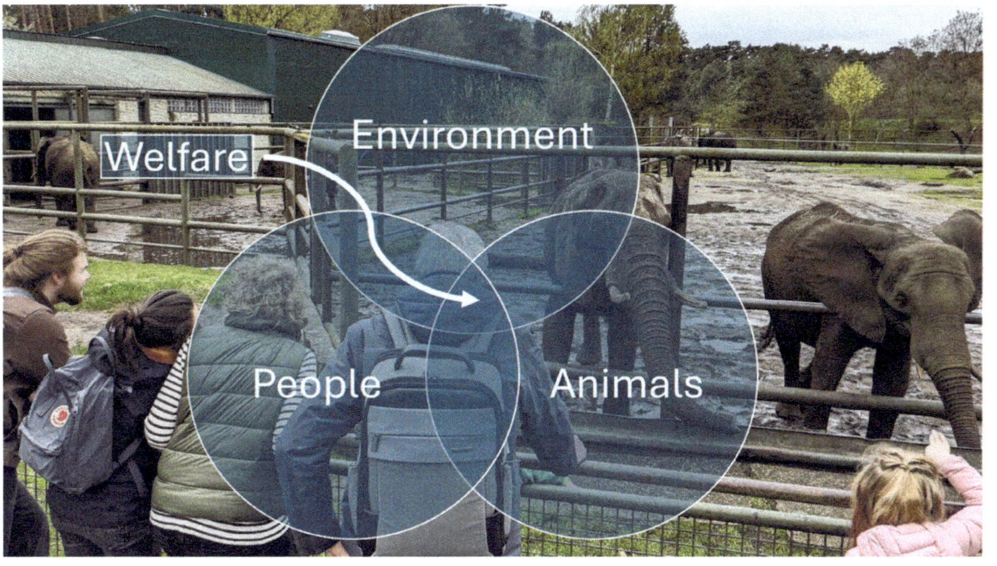

Figure 15.1 The concept of One Health involves the interaction between humans, animals, and the environment. The centre of the diagram, which corresponds to One Health implications for all, will also impact animal welfare, human wellbeing, and the quality of the environment overall. Zoological institutions should consider and evaluate the impact of all interactions between these three factors to fully comprehend how and why animal welfare states change or fluctuate.

is the third element in this approach and is an important factor in the overall well-being of the animals. Providing the correct environment for the species will help to reduce health issues. For example, providing different and adequate substrates to elephants and giraffes will reduce the prevalence of foot problems. However, a correct environment will also promote good welfare opportunities such as providing adequate hiding spots so that animals can retreat from their group and not feel threatened but also so that they can escape from public view.

The presence of the public can have various effects on zoo animals. The presence of people, the noise, and unfamiliar smells might all elicit stress in captive animals and negatively affect their welfare. Issues such as increased aggression, an increase in abnormal behaviours, visitor avoidance and increased vigilance have been documented. However, in other studies, it has been suggested that the effect is not negative (Sherwen et al., 2015) and can even be positive. Sherwen and Hemsworth (2019) explored the implications and opportunities that the presence of visitors offers zoo animal welfare. Steps need to be taken to mitigate the negative effects of the presence of visitors, which can include enriching the environment, for example adding barriers and hiding spots to allow the animals to avoid visual contact with visitors. Education of visitors to minimise noise and movement can also improve captive animal welfare.

15.4 CASE STUDIES

The following case studies present several infectious and non-infectious conditions in different zoological taxa to demonstrate potential welfare considerations surrounding the relationship between animal health, welfare, and veterinary involvement. This includes not only the effect of the disease on individual animals but also issues with concurrent health implications for the wider zoo population, environmental factors, and hence One Health importance.

15.4.1. ULCERATIVE DERMATITIS IN A KOI CARP

Ulcerative dermatitis is a common condition affecting ornamental fish, notably koi (*Cyprinus rubrofuscus*) and goldfish (*Carassius auratus*). It is an important consideration in zoological settings, as well as aquaculture and domestic holdings. The disease is evident by skin lesions, progressing to open ulcers, septicaemia, and death. Infection with bacterial *Aeromonas* species, including atypical *Aeromonas salmonicida*, is regarded as a common cause of the condition (Goodwin & Merry, 2009; Soni et al., 2021), but other bacterial infections are also implicated (Rich & Naguib, 2024). Extensive and chronic skin ulceration should be considered a painful condition and therefore a welfare concern for the individuals affected. The ability of fish to feel pain has been debated, especially the consideration of nociception versus true pain perception and the concept of suffering related to this (Hardy-Smith & Roadknight, 2021). It has however been considered that fish have awareness of negative welfare experiences (Smith, 2023), so whilst specific pain sensation related to ulceration may not have been established, extrapolation from other species should raise the high likelihood of ulceration being a negative welfare factor. In the zoological collection setting, public perception of fish disease, poor health status, and negative behaviours may be lacking compared to, for example, terrestrial mammals where more anthropomorphic, empathic comparisons may be drawn. Skin ulceration in fish, however, is obviously visually evident and so it is likely to be perceived by the zoo visitor as a clear marker of poor health and therefore poor welfare.

Aeromonas salmonicida may be considered a primary pathogen (Park et al., 2020) but other bacteria and *Aeromonas* species may be commensal – in either case, exposure to the bacterium does not necessarily result in clinical signs in all individuals. Husbandry and environmental factors play a key role, many of which can be controlled by those managing the zoological exhibit. As with most fish health conditions, correct water conditions and quality are the most important factors in promoting good fish health (Smith, 2023). Fish nitrogenous waste and its breakdown products require adequate and correctly maintained biological filtration to keep levels down. Ammonia in particular, the main protein excretory product from fish, has many toxic effects on tissues and results in immunocompromise (Xu et al., 2021).

Water temperature is an interesting consideration in this condition. Ulcerative skin disease is usually observed at water temperature of 13–24°C (Watson et al., 2004), so it may be considered a seasonal disease in temperate climates with no artificial temperature control. Koi demonstrate a wide range of thermal tolerance (Yanar et al., 2019) but a preferred temperatures of 18–24°C are often cited (Watson et al., 2004). Colder temperatures may inhibit the fishes' immune system, and similarly, excessively high temperatures may also cause heat stress in koi, making them more vulnerable to infections (Yu et al., 2020). By keeping the fish at its preferred temperature zone, ulcerative skin disease will occur. However, keeping the fish outside this preferred zone may well have a greater negative welfare effect, for example, immunosuppression increasing the prevalence of other infectious diseases or inability to perform normal behaviours (for example, spawning when water temperature approaches 20°C) (Watson et al., 2004). This raises questions as to whether to provide supplementary overwinter heating in colder climates and whether it is acceptable, in welfare terms, to house koi outside in tropical locations or in artificially controlled areas such as tropical houses.

Stocking density is also an important factor in this disease process. High stocking densities (common in koi exhibits) may overload biological filtration and result in poor water quality. Physical damage to the skin from overcrowding may also facilitate bacterial invasion. There are further welfare considerations in the veterinary treatment of ulcerative dermatitis. Debridement of ulcers and topical treatments may be employed, but this requires netting or other catch up of the fish which may result in stress or further damage to skin or protective slime coat. In the UK, for example, fish are not included in the Veterinary Surgeons Act 1966 and so (not withstanding other legislation including the Animal Welfare Act 2006) treatment, including anaesthesia, may be carried out by a lay person not a veterinary surgeon. This raises potential welfare issues, however increased veterinary involvement in zoological collections makes this less of an issue compared to aquaculture and hobbyists. Use of prescription- only medicines, including antibiotics in the UK, would require veterinary input. As a bacterial aetiology is commonly implicated in ulcerative dermatitis, antibiotics will often be used. Antimicrobial resistance is a particular concern in the ornamental fish trade and aquaculture. Widespread and indiscriminate use of antibiotics, including those deemed of importance to human health especially quinolones, has resulted in multi-resistant strains of bacteria (including *Aeromonas*) becoming widespread (Alderman and Hastings, 1998). For this reason, correcting the water quality and husbandry issues should be paramount over antimicrobial therapy. Bacterial culture and antibiotic sensitivity testing should be employed when treating ulcerative dermatitis, but where there is an effective choice available, responsible antibiotic use can bring good welfare by treating the disease process. Veterinary treatment of affected koi may require transfer to quarantine facilities, which raises further welfare implications. Quarantine tanks are unfamiliar surroundings, often devoid of furnishings or substrate, absent of conspecifics and may not have established effective biological filtration which can result in a suboptimal environment and stress to this fish. This may exacerbate the ulcerative dermatitis condition with further welfare compromise to the koi.

Aeromonas can be considered a potential zoonosis with the potential for bacterial infection through open wounds on hands (Igbinosa et al., 2012). Although not thought to be a major concern, the potential for infection of immunocompromised zoo visitors through touch pools should be considered (Figure 15.2). Irrespective of zoonotic risk, the potential ethical and negative welfare implications of touch pools are well known (Smith, 2023). There is however some evidence that koi may seek human interaction (Smith, 2023) and that touch pool interactions may be a positive welfare factor in koi (Farmerie, 2018).

15.4.2 FEATHER DESTRUCTIVE BEHAVIOUR IN PSITTACINE BIRDS

Feather destructive behaviour (FDB), also referred to as 'feather picking', can be considered an abnormal repetitive behaviour or stereotypy. The condition has been likened to the impulse control disorder trichotillomania in humans (Van Zeeland et al., 2009). It is a common condition in captive parrots, with estimates of 10–15% of birds affected (Ebisawa et al., 2021; Kubiak, 2015). Whilst especially seen in pet parrots, the condition is also observed in zoological collections (Woods et al., 2022). FDB is not seen in wild parrots and can be considered a condition related to captivity and warrants a particular welfare focus. Feather destructive behaviour may be regarded as a medical or behavioural condition, but in many cases, both will be implicated. Birds are notorious for hiding signs of clinical disease, and from a welfare perspective, it is important for the zoo vet to rule out medical causes so as not to dismiss the behaviour as a behavioural cause. Possible medical causes are wide and varied (Kubiak, 2015), but a particular One Health consideration is psittacine beak and feather disease virus (BFDV)/circovirus. This virus causes chronic feather abnormalities and feather loss in adult birds which may be implicated in (or indeed confused with) FDB (Figure 15.3). In the UK, it has been detected in free-living ring-neck parakeets (*Psittacula krameri*) that have become established as an invasive species (Sa et al., 2014) with potential disease spread risk back and forth to zoo-housed psittacine birds. BFDV only affects psittacine birds, and so there is no known risk to native UK species (Sa et al., 2014) but this would be a consideration in regions with native parrots. Indeed, in Mauritius, BFDV has negatively affected the previously critically endangered (but now recovering) echo parakeet (*Alexandrinus eques*) population, with the invasive ring-neck parakeet suggested as a viral reservoir (Fogell et al., 2021). The general economic and ecosystem damage caused by invasive ring-neck parakeet populations should also be considered, not just issues related to circovirus (Shiels & Kalodimos, 2019).

Negative welfare implications of FDB to the individual include pain on pulling feathers (or if the condition progresses to self-mutilation, skin and muscle damage) and an inability to effectively thermoregulate. Lack of feathering may also compromise flight or display behaviour. Abnormal repetitive behaviours, including FDB, are generally considered a negative welfare indicator which raises an interesting welfare consideration when treating behavioural causes of FDB. Collars may be fitted to prevent the feather damaging behaviour and it has been suggested that this may disrupt the cycle of feather pulling and related endorphin release (Kubiak, 2015). Collars however significantly impair the bird's ability to move, fly, eat, preen and interact with conspecifics, so the collar may have a greater negative welfare effect than the FDB. Pharmacological treatments involving behaviour-modifying drugs (or 'anti-depressants') may be used (van Zeeland, 2018), but again consideration needs to be given to potential serious

Figure 15.2 Koi carp in a zoological collection's open-topped exhibit. Despite signage prohibiting touching the fish, there is an opportunity for contact between visitors' hands, the fish, and water. This allows for the possibility of transfer of zoonotic infections, and the welfare implications of physical contact with the fish need to be considered.

side effects both medically and behaviourally, where especially marked sedation may be worse behaviourally for the bird than the original FDB. Given limited efficacy in collar use and behaviour modifying drugs and where thorough attempts have been made to address inciting behavioural or husbandry factors, no treatment may be the preferred option. This is not to say the potential negative behavioural consequences should be ignored. Indeed, in humans, trichotillomania is characterised by distress and impaired functioning (Odlaug et al., 2010), although the societal implications of alopecia from pathological hair pulling cannot be directly comparable to psittacine birds.

Higher incidence of FDB in certain species and genera are noted, with higher incidences of up to 30% seen in captive grey parrots (*Psittacus erithacus*) and cockatoos (*Cacatua* sp.) (Ebisawa et al., 2021). Consideration should therefore be given in captive collection planning as to whether these species are more prone to this condition and more likely to experience negative welfare in captivity; housing other species may be a better choice and further work is needed to meet problem species' welfare needs in relation to FDB.

Captive psittacine bird husbandry procedures can influence FDB. Nutrition is a common problem area in parrot care, with selective feeding often associated with inappropriate seed-based diets (including hypovitaminosis A), resulting in immunocompromise and poor skin and feather quality. Feeding techniques in captivity can also be implicated in FDB. A large proportion of wild parrots' activity budget is spent on foraging for food. In

Figure 15.3 A ring-neck parakeet with psittacine beak and feather disease (BFDV) infection. Note the generalised feather abnormalities and feather loss which will affect flight and thermoregulation. The missing feathers on the head are more indicative of a medical cause of feather loss as the bird cannot itself pluck the head feathers with the beak (as seen in behavioural causes of feather destructive behaviour). There is a risk of virus spread to other psittacine birds within a captive collection or to wild psittacine birds.

captivity, there is reduced opportunity for foraging behaviours as food is often provided ad lib in bowls and this can result in boredom and subsequent FDB.

Deflighting, especially wing clipping, is implicated in the increased incidence of FDB (Kubiak, 2015). Hand rearing of parrots is also implicated (Schmid et al., 2006), presumably through lack of social interaction and abnormal relationships with humans. Whilst this practice is commonly employed in the pet trade to achieve 'tame' pets, it is also used in zoological collections, especially in species of conservation value to maximise rearing numbers, and so efforts should be made to minimise negative behavioural effects of hand rearing, for example, 'creche' rearing rather than in isolation. Keeping single parrots is common in the pet trade which again is implicated in deleterious behaviours including FDB. This practice would not be expected in zoological collections but there may be circumstances where parrots may be kept in isolation and so the negative effects of this should be taken into account.

15.4.3 LAMENESS IN GIRAFFE

Chronic lameness and hoof overgrowth is a common condition in adult zoo-housed giraffe (*Giraffa camelopardalis*) (Gage, 2019) and is associated with a variety of pathological processes (Dadone,

2018). Lameness is often related to pain (although mechanical and neurological causes are noted) (Dadone, 2018) and so is an obvious welfare concern for the affected individuals. The pathologies resulting in lameness commonly reported in captive giraffe are almost never seen in free-ranging populations (Dadone et al., 2021) raising welfare and ethical concerns with keeping giraffe in a zoological setting. With giraffe remaining popular and charismatic zoo species and lameness often clearly evident to the zoo visitor (Figure 15.4), there is an obvious need for collections and zoo vets to address this issue to maximise the individual animal welfare and also the positive perception of keeping giraffe in captivity.

Treatment of hoof overgrowth carries a number of challenges. Unlike with domestic ruminants and horses, pathologies may only be picked up at an advanced stage, which has obvious concerns that pain, ill health, and hence welfare compromise has gone on for some time without detection. The mainstay of treatment for hoof overgrowth remains corrective hoof trimming. Hoof trimming is cited as the reason for over 80% of chemical immobilisation procedures performed in captive giraffe (Dadone, 2018). These procedures may require regular repetition, and this presents a welfare conundrum in the stress aspects of immobilisation versus treating the hoof overgrowth. Hoof overgrowth left unchecked can cause (further painful) secondary conditions including ligament injuries and osteoarthritis (Dadone & Foxworth, 2023) and so does warrant treatment. Giraffe anaesthesia is challenging, with significant morbidity and risk of anaesthetic death (Citino & Bush, 2014) and this has been reported for hoof trimming procedures. Death occurs not necessarily during the surgical anaesthesia phase where the

Figure 15.4 A zoo-housed giraffe with chronic hoof overgrowth. Hoof overgrowth and associated lameness can become a painful condition and hence is a chronic welfare concern. Change to diet, increased opportunities for exercise, and improved enclosure substrate (type and variety) can prevent abnormal hoof growth; however, direct trimming of the hooves to reshape them into a natural position would be needed in this case. As giraffe anaesthesia poses a risk to animal survival, is stressful for human personnel involved, and requires substantial environmental alteration to ensure the treatment can be performed as safely as possible, this becomes a One Health and One Welfare challenge. In this case, prevention is definitely better than cure.

animal may be considered unresponsive to pain, but during induction and recovery when the animal is conscious and so presents a welfare concern. Multimodal analgesia, low-level laser therapy, and glucosamine-chondroitin have been used in giraffes (Boothe et al., 2016) which brings welfare benefits in pain management. Analgesia over a number of years has been successful in managing arthritic pain in captive giraffes (Pizzi et al., 2010). In clinical practice, dosing varies widely, and so the need for further preclinical and clinical data on the efficacy and safety of various analgesics is required (Boothe et al., 2016). Development of standardised locomotion scoring allows for improved assessment and response to treatment, similar to that performed in elephants (Proboscidea) (Turner et al., 2023).

Training a giraffe for conscious hoof assessment and hoof trimming has been described (Dadone et al., 2016) and it provides obvious benefits for earlier pathology detection, as well as collection of baseline data including diagnostic radiography. Specialised handling and restraint devices can also facilitate this (Bush, 2003) and hopefully reduce stress plus the need for more risky (full recumbency) immobilisation. Techniques for standing sedation, similar to domestic horses, have been described with less anaesthetic risk compared to "knock down" (Bouts et al., 2017). It should be noted that training takes considerable time and so may well not be appropriate for animals already requiring treatment and will vary by individual animal and institution.

Advances in veterinary diagnostic imaging have allowed a greater understanding of the pathologies involved. Digital radiography, with associated improved speed of image acquisition and image quality, has facilitated data acquisition in captive and free-ranging giraffes (Dadone et al., 2016, 2021) with value in comparing the two populations to benefit one another. Advanced imaging (MRI) has provided new insights in post-mortem examination (Wakeman et al., 2014).

Husbandry practices are suggested to contribute to hoof overgrowth and include insufficient exercise, nutritional imbalances, inappropriate substrate, and trauma (Dadone, 2018). Husbandry changes can therefore be useful alongside veterinary treatment to improve outcomes and overall welfare. Hard substrates, including concrete, are shown to be a significant risk factor for poor foot health in megaherbivores including elephants (Miller et al., 2016) and sand flooring and/or mats, plus removal of larger pebbles, are suggested to manage lameness in giraffe (Dadone, 2018).

Laminitis is recognised in captive giraffes as a cause of lameness and hoof changes (Wiedner et al., 2014) and is likely underdiagnosed in zoo populations (Dadone, 2018). As in domestic ruminants, rumen acidosis is a suggested possible cause of laminitis in giraffe (Dadone, 2018). Wild giraffes mainly consume low-starch browse, whereas captive giraffes' consumption of supplemental feeds over forage can increase the risk of ruminal acidosis (Kearney et al., 2024). Dietary manipulation away from high-starch/low-fibre diets has been suggested to reduce the risk of laminitis (Wiedner et al., 2014) but the logistical challenges of year-round browse provision are considerable, and the differing characteristics and suitability of different forage types are recognised (Clauss et al., 2002). Dietary modification may also bring other positive welfare results including increased rumination and reduction in stereotypies (Depauw et al., 2023; Rose, 2023).

Increasing exercise to improve hoof health is an interesting consideration. Correlation between larger enclosure size and increased locomotor activity has been shown (Veasey et al., 1996). Typically, outdoor enclosures provide more space, but provision of space does not necessarily equate to increased enclosure usage (Saito et al., 2024). Lameness may also result in less willingness to exercise, compounding the problem, as noted in other megavertebrates (Leighty et al., 2009)

Chronic intractable lameness may warrant euthanasia on welfare grounds (Pizzi et al., 2010).

Zoo animal euthanasia is an emotive subject and warrants a close collaborative effort between zoo vets and stakeholders. Quality of Life (QoL) can be difficult to define and measure, and development of QoL scoring systems is useful as a quantitative method of determining effective end of life care, including timing of euthanasia (Campbell-Ward, 2023; Föllmi et al., 2007). This facilitates education and the positive welfare aspect of timely euthanasia to zoo visitors and the wider public.

15.5 CONCLUSIONS

This chapter illustrates the clear role that vets play in improving zoo animal welfare, not only from a disease, pain prevention and treatment perspective but also via involvement in enclosure design, husbandry, dietary practices and keeper education on aspects of welfare and health. Therefore, specialist veterinary knowledge will be an important asset to the zoo manager and private owners to improve their animal's welfare and good relationships between all stakeholders will be paramount. Good health will support good welfare however, this does not equate to good welfare. Therefore, a holistic approach to welfare including not only maintenance of health but other aspects such as the environment, the effect of public presence and good husbandry provision is needed.

The evidence-based approach is essential to the development of effective treatment of diseases that afflict zoo animals but is also the best way of ensuring husbandry and housing is species appropriate. To further a One Health and welfare approach, Box 15.1 outlines key questions that need to be researched to further integrate this concept in zoo animal husbandry, management and medicine.

BOX 15.1: Key research questions to help develop a One Health and Welfare approach in the zoo

- Do zoo vets understand the concept of One Health and if not, why not?
- Do zoo vets understand that good health doesn't always equate to good welfare?
- Should there be more knowledge on preventative measures embedded into day-to-day husbandry?
- Should there be more emphasis placed in integrating welfare within the veterinary interventions and utilise training more to improve/facilitate these interventions?
- Should there be a more collaborative effort between all the teams (vets, keepers, management) to improve all aspects of welfare (nutrition, husbandry, health, wellbeing) in captive zoo species?
- Do zoo vets have enough involvement in animal-human interactions and their consequences?

REFERENCES

Alderman, D. & Hastings, T. (1998). Antibiotic use in aquaculture: development of antibiotic resistance–potential for consumer health risks. *International Journal of Food Science & Technology*, *33*(2), 139–155.

Boothe, M., Kottwitz, J., Harmon, R., Citino, S. B., Zuba, J. R. & Boothe, D. M. (2016). Results of the megaverebrate analgesia survey: giraffe and hippopotamus. *Journal of Zoo and Wildlife Medicine*, *47*(4), 1049–1056.

Bouts, T., Dodds, J., Berry, K., Arif, A., Taylor, P., Routh, A. & Gasthuys, F. (2017). Detomidine and butorphanol for standing sedation in a range of zoo-kept ungulate species. *Journal of Zoo and Wildlife Medicine*, *48*(3), 616–626.

Bush, R. M. (2003). Giraffidae. In: Fowler, M. E. & Miller, R. E. (eds). *Zoo and wild animal medicine: Current therapy 5.* Elsevier: St Louis, MO, pp. 625–633.

Campbell-Ward, M. (2023). Quality-of-Life assessments in zoo animals: not just for the aged and charismatic. *Animals*, *13*(21), 3394.

Citino, S. B. & Bush, M (2014). Giraffidae. In: West, G., Heard, G., & Caulkett, N. (eds). *Zoo animal and wildlife immobilization and anesthesia.* John Wiley & Sons Inc: Ames, IA, pp. 809–821.

Cork, S., Hall, D. & Liljebjelke, K. (2016). *One health case studies: addressing complex problems in a changing world.* 5M Publishing Ltd: Sheffield, UK.

Clauss, M., Lechner-Doll, M., Flach, E., Wisser, J. & Hatt, J.-M. (2002). Digestive tract pathology of captive giraffe: a unifying hypothesis. *Proceedings of the European Association of Zoo and Wildlife Veterinarians, 4*, 99–107.

Dadone, L. (2018). Lameness diagnosis and management in zoo giraffe. In: Miller, R.E., Lamberski, N., & Calle, P. (eds). *Fowler's zoo and wild animal medicine: current therapy 9*, Elsevier: St Louis, MO, pp. 623–629.

Dadone, L. & Foxworth, S. (2023). Footcare for zoo ungulates (Zoo Hoofstock Trim Program). In: Miller, R. E., Calle, P., & Lamberski, N. (eds). *Fowler's zoo and wild animal medicine: current therapy 10*. Elsevier: St Louis, MO, pp. 303–312.

Dadone, L., Foxworth, S., Aruho, R., Schilz, A., Joyet, A., Barrett, M., Morkel, P., Crooks, G., Fennessy, J. & Johnston, M. S. (2021). Foot shape and radiographs of free-ranging Nubian giraffe in Uganda. *PLoS One, 16*(12), e0252929.

Dadone, L. I., Schilz, A., Friedman, S. G., Bredahl, J., Foxworth, S. & Chastain, B. (2016). Training giraffe (Giraffa camelopardalis reticulata) for front foot radiographs and hoof care. *Zoo Biology, 35*(3), 228–236.

Deem, S. L. (2015). Conservation medicine to one health: the role of zoologic veterinarians. In: Miller, R. E. & Fowler, M. E. (eds). *Fowler's zoo and wildlife animal medicine*. Elsevier: St Louis, MO, pp. 698–703.

Depauw, S., Verbist, L., Stevens, J. & Salas, M. (2023). Feeding management of giraffe towards positive welfare. *Journal of Zoo and Aquarium Research, 11*(4), 400–413.

Ebisawa, K., Nakayama, S., Pai, C., Kinoshita, R. & Koie, H. (2021). Prevalence and risk factors for feather-damaging behavior in psittacine birds: analysis of a Japanese nationwide survey. *PLoS One, 16*(7), e0254610.

Farmerie, M. R. (2018). *Impacts of an educational human-animal interaction program on conservation education outcomes and program animal welfare for Koi (Cyprinus carpio) in a modern zoo setting*. PhD thesis. Robert Morris University, Pittsburgh, PA.

Fogell, D. J., Tollington, S., Tatayah, V., Henshaw, S., Naujeer, H., Jones, C., Raisin, C., Greenwood, A. & Groombridge, J. J. (2021). Evolution of beak and feather disease virus across three decades of conservation intervention for population recovery of the Mauritius parakeet. *Diversity, 13*(11), 584.

Föllmi, J., Steiger, A., Walzer, C., Robert, N., Geissbühler, U., Doherr, M. & Wenker, C. (2007). A scoring system to evaluate physical condition and quality of life in geriatric zoo mammals. *Animal Welfare, 16*(3), 309–318.

Fraser, D., Weary, D., Pajor, E. A., Milligan, B. N., Fraser1t, D., Weary, D., Pajor, E. A. & Milligan, B. N. (1997) A scientific conception of animal welfare that reflects ethical concerns. *Animal Welfare, 6*, 187–205.

Gage, L. J. (2019). Giraffe husbandry and welfare. In: Miller, R. E., Lamberski, N., & Calle, P. (eds). *Fowler's zoo and wild animal medicine: current therapy 9*. Elsevier: St Louis, MO, pp. 619–622.

Goodwin, A. E. & Merry, G. E. (2009). Are all koi ulcer cases associated with infection by atypical Aeromonas salmonicida? Polymerase chain reaction assays of koi carp skin swabs submitted by hobbyists. *Journal of Aquatic Animal Health, 21*(2), 98–103.

Hardy-Smith, P. & Roadknight, N. (2021). Fish welfare and One Welfare–A veterinarian's perspective. In: Stephens, T. (ed). *One welfare in practice*. CRC Press: Boca Raton, FL, pp. 255–277.

Igbinosa, I. H., Igumbor, E. U., Aghdasi, F., Tom, M. & Okoh, A. I. (2012). Emerging Aeromonas species infections and their significance in public health. *The Scientific World Journal, 2012*(1), 625023.

Kearney, C. C., Ball, R. L. & Hall, M. B. (2024). Effects of altering diet carbohydrate profile and physical form on zoo-housed giraffe, Giraffa camelopardalis reticulata. *Journal of Animal Physiology and Animal Nutrition, 108*(4), 1119–1133.

Kubiak, M. (2015). Feather plucking in parrots. *In Practice, 37*(2), 87–95.

Lawrence, A. B., Vigors, B. & Sandøe, P. (2019). What is so positive about positive animal welfare: a critical review of the literature. *Animals, 9*, 783.

Leighty, K. A., Soltis, J., Wesolek, C. M., Savage, A., Mellen, J. & Lehnhardt, J. (2009). GPS determination of walking rates in captive African elephants (Loxodonta africana). *Zoo Biology, 28*(1), 16–28.

Miller, M. A., Hogan, J. N. & Meehan, C. L. (2016). Housing and demographic risk factors impacting foot and musculoskeletal health in African elephants (Loxodonta africana) and Asian elephants (Elephas maximus) in North American zoos. *PLoS One, 11*(7), e0155223.

Odlaug, B. L., Kim, S. W. & Grant, J. E. (2010). Quality of life and clinical severity in pathological skin picking and trichotillomania. *Journal of Anxiety Disorders, 24*(8), 823–829.

Padda, H., Niedbalski, A., Tate, E. & Deem, S.L. (2018). Members perceptions of the one health initiative at a zoological institution. *Frontiers in Veterinary Science*, 5(22), 1–8. https://doi.org/10.3389/fvets.2018.00022

Park, S. Y., Han, J. E., Kwon, H., Park, S. C. & Kim, J. H. (2020). Recent insights into Aeromonas salmonicida and its bacteriophages in aquaculture: a comprehensive review. *Journal of Microbiology and Biotechnology, 30*(10), 1443.ĀAAA

Pizzi, R., Cracknell, J. & Dalrymple, L. (2010). Postmortem evaluation of left flank laparoscopic access in an adult female giraffe (Giraffa camelopardalis). *Veterinary medicine international, 2010*(1), 789465.

Rich, A. F. & Naguib, M. (2024). Shewanella xiamenensis-associated ulcerative dermatitis in koi carp (Cyprinus rubrofuscus). *Journal of Fish Diseases, 47*(7), e13942.

Rose, P. (2023). Identifying essential elements of good giraffe welfare - Can we use knowledge of a species' fundamental needs to develop welfare-focussed husbandry? *Journal of Zoological and Botanical Gardens, 4*(3), 549–566.

Sa, R. C., Cunningham, A. A., Dagleish, M. P., Wheelhouse, N., Pocknell, A., Borel, N., Peck, H. L. & Lawson, B. (2014). Psittacine beak and feather disease in a free-living ring-necked parakeet (Psittacula krameri) in Great Britain. *European Journal of Wildlife Research, 60*, 395–398.

Saito, M., Matsunaga, M., Fukuizumi, H. & Nakamichi, M. (2024). Effects of nocturnal outdoor enclosure access on space use and sleep-related behaviour in captive giraffes. *Journal of Zoo and Aquarium Research, 12*(3), 154–162.

Schmid, R., Doherr, M. G. & Steiger, A. (2006). The influence of the breeding method on the behaviour of adult African grey parrots (Psittacus erithacus). *Applied Animal Behaviour Science, 98*(3–4), 293–307.

Sherwen, S. L., Hemsworth, P. H., Butler, K. L., Fanson, K. V. & Magrath, M. J. (2015) Impacts of visitor number on kangaroos housed in free-range exhibits. *Zoo Biology, 34*(4), 287–295.

Sherwen, S. L. & Hemsworth, P. H. (2019) The visitor effect on zoo animals: implications and opportunities for zoo animal welfare. *Animals, 9*(6), 366.

Shiels, A. B. & Kalodimos, N. P. (2019). Biology and impacts of Pacific Island invasive species. 15. Psittacula krameri, the rose-ringed parakeet (Psittaciformes: Psittacidae). *Pacific Science, 73*(4), 421–449.

Smith, S. A. (2023). Fish welfare in public aquariums and zoological collections. *Animals, 13*(16), 2548.

Soni, M., Qureshi, Q. A., Mishra, M., Nishad, C. S., Chhaba, B. & Das, S. A. (2021). Common Aeromonas infections in ornamental fishes: a review. *Bioogical Forum, 13*(2), 433–439.

Turner, A., Masters, N., Pfau, T., Hutchinson, J. R. & Weller, R. (2023). Development and evaluation of a standardized system for the assessment of locomotor health in elephants under human care. *Journal of Zoo and Wildlife Medicine, 54*(3), 529–537.

van Zeeland, Y. (2018). Medication for behavior modification in birds. *Veterinary Clinics: Exotic Animal Practice, 21*(1), 115–149.

Van Zeeland, Y. R., Spruit, B. M., Rodenburg, T. B., Riedstra, B., Van Hierden, Y. M., Buitenhuis, B., Korte, S. M. & Lumeij, J. T. (2009). Feather damaging behaviour in parrots: a review with consideration of comparative aspects. *Applied Animal Behaviour Science, 121*(2), 75–95.

Veasey, J. S., Waran, N. K. & Young, R. J. (1996). On comparing the behaviour of zoo housed animals with wild conspecifics as a welfare indicator, using the giraffe (Giraffa camelopardalis) as a model. *Animal Welfare, 5*(2), 139–153.

Wakeman, K. A., Sanchez, C. R., Lung, N. P., Hersman, J. & Barrett, M. F. (2014). The use of magnetic resonance imaging to better define hoof pathology in the reticulated giraffe (Giraffa camelopardalis reticulata). *Journal of Zoo and Wildlife Medicine*, *45*(3), 668–671.

Watson, C. A., Hill, J. E. & Pouder, D. B. (2004). *Species profile: Koi and goldfish*. Southern Regional Aquaculture Center, Stoneville, MS.

Wiedner, E., Holland, J., Trupkiewicz, J. & Uzal, F. (2014). Severe laminitis in multiple zoo species. *Veterinary Quarterly*, *34*(1), 22–28.

Woods, J. M., Eyer, A. & Miller, L. J. (2022). Bird welfare in zoos and aquariums: general insights across industries. *Journal of Zoological and Botanical Gardens*, *3*(2), 198–222.

Xu, Z., Cao, J., Qin, X., Qiu, W., Mei, J. & Xie, J. (2021). Toxic effects on bioaccumulation, hematological parameters, oxidative stress, immune responses and tissue structure in fish exposed to ammonia nitrogen: a review. *Animals*, *11*(11), 3304.

Yanar, M., Erdoğan, E. & Kumlu, M. (2019). Thermal tolerance of thirteen popular ornamental fish species. *Aquaculture*, *501*, 382–386.

Yu, H., Zhang, C., Xing, W., Li, T., Xu, G., Ma, Z., Jiang, N. & Luo, L. (2020). Comparative study on the non-specific immune response and hsp70 gene expression among three strains of koi (Cyprinus carpio) under acute heat stress. *Aquaculture Reports*, *18*, 100461.

16

Enriching the awkward – promoting welfare in challenging species (reptiles and amphibians)

STEVE NASH

16.1 INTRODUCTION

Despite the chapter heading, the implied awkwardness of enriching reptiles and amphibians comes not from the animals themselves, but from a historically entrenched view of these two vertebrate classes which continues to influence their captive welfare. Perhaps more so than for other groups of terrestrial vertebrates, their welfare is often compromised by unconscious bias, taxonomic distance, and outdated approaches to husbandry, alongside a lack of clarity on how to accurately assess the welfare state of individuals. These factors are further exacerbated by the speciose nature of these two taxonomic classes and the consequent diversity of form, function, and need that such variety creates. Challenges such as these provide multiple opportunities for caregivers to devise and implement new approaches to captive husbandry that enable these behaviourally complex animals to prosper. These changes in approach will not only help to banish the aforementioned awkward truth, but will move welfare forwards for captive individuals, as well as realise the potential for increasing the success of conservation initiatives for the survival of their wild relatives.

Reptiles and amphibians are commonly grouped together under the collective heading of herptiles, a term derived from the Greek 'herpeton' meaning creeping thing, and the basis of herpetology (the scientific study of non-avian reptiles and amphibians). This historic association, which dates back to the work of Linnaeus, infers a similarity and relatedness between these two classes that is not entirely helpful when viewed through the lens of modern taxonomy and the implications that this infers. Along with fish, reptiles and amphibians are commonly referred to as 'lower vertebrates', a

DOI: 10.1201/9781003429425-16

phraseology that often reinforces a level of unconscious bias towards how these animals live and behave, with subsequent consequences regarding their husbandry and perceived welfare in captivity. The phylogenetic distance of these taxa from humans can also result in a lowered intuition on the part of the keeper regarding the recognition of behavioural abnormalities, the assessment of welfare, and the appropriateness of the captive environment as a whole.

While research on reptile and amphibian welfare and behaviour is growing, this still remains sparse in comparison to the body of work available for mammals and birds. As a consequence, much of the available information on husbandry for many species is based on anecdote, hearsay, and dogma, rather than factual evidence. In some cases, this approach may well lead to success but, in others, its acceptance may promote sub-optimal methodologies. Given the huge diversity of species within these two classes, the definition of normal behaviour is challenging to explain. Basic ecological knowledge for many commonly kept species is limited, and this has clear implications for our understanding of lower vertebrate ethology – a hallmark of which is clear information on what an animal does in both field and captivity (Burghardt, 2013). Closely related species may occupy significantly different environments, resulting in distinct differences in their evolved behaviours and, hence, their husbandry needs. For example, the waxy monkey frog (*Phyllomedusa sauvagii*) (Figure 16.1) displays physiological and behavioural adaptations that allow it to survive in arid environments that appear far removed from the habitats usually frequented by many other tree frogs. Furthermore, even within a species, behaviour may vary substantially, for example with age. In amphibians, where a distinct larval stage forms part of their development, this shift is self-evident, but it also occurs with some frequency in reptiles, with neonate and juvenile specimens occupying markedly different niches with subsequent differences in behaviour. Carter et al. (2021) demonstrate such a shift with *Shinisaurus crocodilurus* (a semi-aquatic Chinese lizard), with individuals climbing more and showing a greater tendency to perch in more open conditions as they mature, but examples can readily be found elsewhere; juvenile Komodo dragons (*Varanus komodoensis*) are largely arboreal, for example, unlike the predominantly terrestrial adults.

A growing body of research provides evidence to show that reptiles are sentient beings, capable of experiencing a wide range of emotional states, with Lambert et al. (2019) identifying anxiety, distress, fear, and excitement among these states. One of the challenges in understanding the behavioural complexity of reptiles and amphibians is ensuring that the issue being explored takes account of the sensory abilities and physiological processes that underpin how these creatures interact with their surroundings. Snakes, for example, experience the world largely through chemosensory and tactile means and therefore require a very different approach to enrichment compared with a more visually oriented mammal or bird.

Perhaps the most persistent (and pernicious) issue is the widely held belief that reptiles and amphibians 'don't do much', a view that can often be reinforced by a visit to a reptile house where many of the residents appear as stationary automatons, sitting motionless in their enclosures. It is this false belief that leads many captive specimens to live lives of compromised welfare and leads many keepers to mistakenly assume that this behavioural inertia means that their animals do not require the same consideration for enrichment that would be taken for granted in other taxa. In addition, this misunderstanding may result in keepers failing to recognise behavioural indicators of poor (or indeed positive) welfare. A compromised captive life can create a state of chronic stress, which reveals itself through a wide range of behaviours, including anorexia, hypoactivity, hyperactivity, 'aggression', and interactions with transparent boundaries, all of which have the potential for being misinterpreted, attributed to other causes, or even, in some instances, accepted as normal for the species.

16.2 CREATING AND MANAGING DYNAMIC HABITATS

Accurate information on the natural history and lifestyle of a species is essential if one is to provide a captive environment that allows an animal to demonstrate a full repertoire of species-appropriate behaviours. For example, the Moroccan spiny-tailed lizard (*Uromastyx nigriventris*) inhabits a deceptively simple landscape (as shown in Figure

Figure 16.1 The waxy monkey frog's (*Phyllomedusa sauvagii*) unique physiology allows it to survive in more arid environments where other tree frogs could not. For example, cutaneous lipid excretions result in their waxy coating, preventing evaporative water loss through the skin. Such species highlight the need for species-specific knowledge to inform correct husbandry. (Photo credit: A. Trimmings).

16.2) yet will require a detailed understanding of its ecology and behaviour to maintain high welfare standards. Unfortunately, for many species of reptiles and amphibians, this information is sorely lacking; a situation that requires keepers to rely heavily on assumption, inference, and guesswork, as well as the sharing of successful experience from others.

Reptile and amphibian keepers are encouraged to spend some time in nature, observing the wild counterparts of their captive charges and witnessing first-hand the extensive behavioural repertoires that are evident for all to see. Behavioural divergence of captive animals from their wild progenitors is well-documented (for example see McPhee, 2003), although this typically manifests as an absence of normal behaviour rather than the expression of abnormal or deviant ones (Michaels et al., 2014) as may often be encountered in other taxa. The extent of these repertoires in wild specimens tells us immediately that the apparent lack of behavioural diversity in captivity is not because these animals do not have the capacity to engage in such displays, but that the conditions provided in captivity do not allow them to realise their full self.

16.2.1 ENVIRONMENTAL REQUIREMENTS

Being ectotherms, it has long been known that the physiology, ecology, and behaviour of reptiles and amphibians are underpinned by the thermal

Figure 16.2 The environmental parameters to which wild animals are exposed are largely unknown, even for commonly maintained species. This wild Moroccan spiny-tailed lizard (*Uromastyx nigriventris*) inhabits a deceptively simple landscape but still requires a detailed understanding of its ecology and behaviour to maintain to high welfare standards. (Photo credit: D. Kane).

environment; however, recent research has highlighted the complexities of this interplay. The extent to which the wider electromagnetic spectrum fundamentally influences the behavioural response of an animal to its surroundings provides multiple opportunities for further investigation, and it is clear that there is far more to a reptile's activities than simply shuttling between hot and cold. A captive environment that fails to provide the thermal and structural complexity required by a given species will constrain behavioural expression as the animal struggles to meet its core metabolic needs. The importance of heterogeneity is demonstrated by Rose et al. (2014) who show that the provision of a choice enables animals to utilise their environment in a way that befits their needs and, by extension, facilitates the behavioural optimisation of physiological processes.

If a reptile's wild behaviour is a consequence of a complex relationship between 'light' and 'heat', the requirement to provide comparable complexity in captivity is obvious. The provision of a thermal gradient or thermal matrix is an essential component of the successful maintenance of those species that thermoregulate; however, the importance of comparable gradients/matrices for other parameters is rarely considered. We must ensure that, thermal variation, we provide variations in light intensity, UV radiation, and humidity as well. Such parameters will fluctuate throughout the day as well as throughout the year, and all of them will influence the behavioural regime of a wild animal in nature. Incorporating these gradients into an enclosure is not difficult. Heat and light gradients are commonly provided in a linear manner, by having the source of heat and light at one end of the enclosure. In such circumstances, the closer the animal gets to the source, the hotter/brighter its immediate environment becomes, thus providing a basic gradient. Nature, however, is more complex than this, and keepers should explore opportunities for increasing options beyond the obvious.

Enclosures housing basking species typically contain a single basking lamp or cluster of lamps. Keepers should explore the use of multiple basking locations, particularly if the enclosure holds multiple individuals. Such a situation can give rise to covert overcrowding, whereby animals in a shared space do not have equal access to all features of that space. Even if only one specimen is present, multiple opportunities to acquire heat in different ways should be viewed as positive additions to an enclosure, provided, of course, that the addition of multiple heat sources does not create a homogenous temperature throughout. Notably, the larger the enclosure, the less the likelihood there is of this occurring, yet another reason for viewing the provision of space as a fundamental element of the welfare of these animals.

16.2.2 ENCLOSURE SIZE AND COMPLEXITY

One of the most enduring husbandry myths encountered with captive reptiles, particularly snakes, surrounds enclosure size, with a proportion of keepers still adhering to the idea that these animals do not benefit from access to a spacious enclosure. Although a growing body of research clearly refutes this assertion (for example, Hoehfurtner et al. (2021) found that larger enclosures directly correlated with higher welfare for captive snakes), its perseverance is indicative of the entrenched beliefs that persist within herpetoculture. The sedentary nature of many captive reptiles is often used to justify the use of smaller enclosures, yet this view fails to accommodate the reality that, in nature, sedentary behaviour is typically transient and should certainly not be used as the sole measure for determining spatial needs.

Quite apart from the obvious benefits in terms of freedom of movement and the reduction in health issues caused by enforced immobility that larger enclosures provide, such a space also enables a wider degree of complexity and choice to be offered. Precise environmental conditions may be harder to achieve in larger enclosures, and it is likely that historical difficulties in doing so were a key contributing factor to the erroneous idea that small is somehow better. It should be apparent that the caveat to large spaces is that it must represent functional space; arranged and furnished in a way that meets the needs of the inhabitant. Large, open spaces can create adverse environments for secretive species, for example, many of which would be actively predated upon in nature. Active movement in such species involves an ongoing trade-off between staying safe and achieving thermoregulatory needs alongside other key behaviours such as foraging and finding a mate. Such a lack of available, accurate data highlights the importance of providing captive animals with complexity and choice in their environment. In the absence of detail, keepers are urged to apply caution and to aim to provide a range of conditions that the animal can then choose to inhabit accordingly. A large enclosure, providing a range of environmental opportunities, allows an animal to select according to its preference depending upon its physiological and behavioural needs at that time. Conversely, poorly structured, smaller enclosures, or those with an otherwise restricted range of opportunities, can force an animal to endure a homogenous environment with no possibility of agency or control over the space, or their response to it.

Providing different substrates and surfaces is one way to create more diverse conditions; different materials have different thermal and hydric properties. Orchid bark or potting compost have very different moisture-holding and heat-retentive properties to rocks, gravel or paper towels, for example. The provision of elevated hides and furnishings can also increase the breadth of available parameters and will often be utilised by species of squamate reptile, even if they are typically considered terrestrial. Such additions also increase the functional space of an enclosure by utilising otherwise 'dead' areas.

16.2.3 THE ROYAL PYTHON PARADOX: A CASE STUDY

Field observations for many species may be lacking, so keepers are often required to interpret what is available rather than relating it directly. Reviewing field data for royal python (*Python regius*) provides interesting insights to inform captive welfare and promote not just species-specific approaches to enrichment but approaches that may vary with sex and ontogeny. Luiselli and Angelici (1998) demonstrated that, on average, wild-caught female specimens were larger than males. This differentiation extended beyond size however, as faecal analysis from these specimens also indicated a dietary diversification between the sexes. The smaller males feed predominantly on birds and arboreal mammals, whereas the larger (and more terrestrial) females have a diet comprised largely of ground-dwelling species. Juveniles of both sexes tended to feed primarily on birds. Later work by Luiselli

(2006) described differentiated ectoparasite loads between the sexes, further evidencing that the species exhibits niche differentiation as well as dietary differentiation. We therefore have field evidence to suggest that juvenile and adult male royal pythons have a stronger tendency towards arboreal behaviour than their larger female counterparts, and a subsequent variance in prey species. These facts alone should inspire keepers of this widely kept species to adopt new approaches to enclosure design and feeding strategies. An example of such a design can be seen in Figure 16.3.

However, a quick search on the internet will reveal an abundance of websites that promote a very different approach to husbandry; one that bases itself on the assertion that these are animals that spend a large proportion of their time in underground burrows or termite mounds. This assertion is used to justify the keeping of this species in size-restricted, overly simplistic, intensive racking systems. Whilst this approach is undoubtedly convenient for the keeper, it is vastly removed from addressing the realities of what this species actually requires to carry out even a basic range of behaviours.

16.2.4 FORM OVER FUNCTION

There is a growing tendency to display reptiles and amphibians in intricately designed naturalistic exhibits, featuring detailed theming, natural substrates, and live planting. When viewed from an educational perspective, these 'living dioramas' can be hugely effective in engaging zoo visitors with conservation topics, illustrating the diversity and importance of specific species and habitats. When properly appointed, such environments can provide exemplary homes for their inhabitants, as is shown in Figure 16.4. We must, however, ensure that this 'naturalness' is functional as well as aesthetic, with furnishings playing a functional role within the animal's environment. It is worth

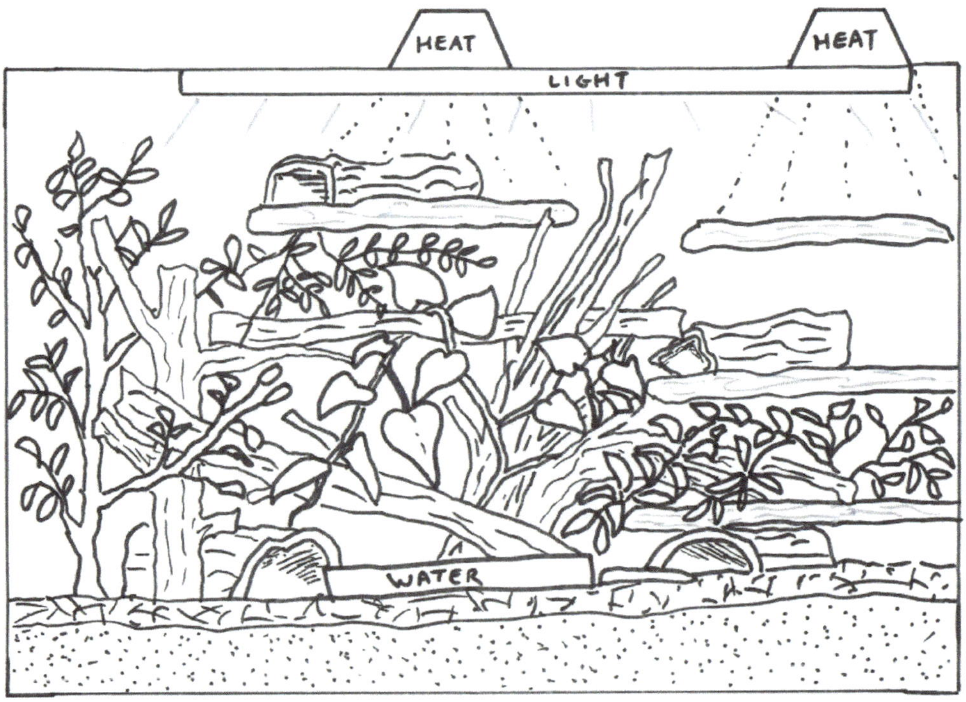

Figure 16.3 An example enclosure for royal python (*Python regius*) that takes natural history evidence into account. A network of interconnected branches, ledges, and hides enables the animals to utilise all aspects of their available space. Additional burrow hides (not pictured) provide subterranean options, whilst full spectrum lighting and a choice of basking areas encourage thermoregulatory behaviour.

Figure 16.4 This Amazon Basin tree boa (*Corallus batesii*), lives in a beautifully appointed vivarium that enhances visitor appreciation of snakes in general, as well as the welfare of the individual animal. (Photo credit: A. Trimmings)

noting that these complex on-show enclosures are often in contrast to the overly simplistic, size-restricted enclosures found in the off-show areas of many zoos. The inference of this discrepancy is that the on-show enclosures are primarily furnished for display and engagement, rather than because of the potential welfare benefits which they may provide. If welfare were the primary driver, one would expect that off-show enclosures would be similarly furnished.

Reptiles and amphibians are generally maintained under conditions of close captivity, by which we mean self-contained enclosures where all aspects of the animal's life are controlled by a carer. It is inevitable that such conditions will lack the dynamism of open enclosures exposed to, for example, climatic variation and sensory stimuli from external factors, and this lack of variation provides a challenge for keepers to overcome. The idea of "controlled deprivation" (Burghardt, 1996) is central to most captive scenarios for these and other taxa, and recognises the inescapable limitations of a captive life. Dynamic captive environments are invariably easier to achieve in larger enclosures, further reason, if reason be needed, to provide spacious environments for captive specimens. Keepers should consider how features of the captive environment can be tweaked and changed to provide appropriate stimuli to the inhabitants, either through changes to enclosure layout and furnishings, or enhanced provision of light, heat, and humidity gradients. Providing 'dawn' and 'dusk' light settings, alongside basking spots that move throughout the day, tracking the natural cycle of the sun, could be implemented. Although the majority of captive reptiles and amphibians are housed indoors, the benefits of outdoor housing (where local conditions allow) should certainly be explored, with such provision often allowing greater space as well as a greatly enhanced sensory and biotic environment. This is frequently seen with many chelonian species and, although rarely encountered with other reptile taxa, is eminently achievable with dedication and a clear understanding of an animal's environmental needs, as seen in in Figure 16.5.

Figure 16.5 Housing animals outdoors can provide multiple benefits in terms of sensory stimulation and access to natural environmental conditions. Here, a Mexican alligator lizard (*Abronia taeniata*), explores its outdoor enclosure during a UK summer. (Photo credit: A. Trimmings)

16.3 ETHOLOGICALLY INFORMED HUSBANDRY AND ENVIRONMENTAL ENRICHMENT

Looking beyond providing the right habitat, species-focused husbandry is key to delivering the welfare needs of these animals and, while environmental enrichment is often more attributed to the management of mammals and birds, this can be equally important for the welfare of captive herptiles. As well as assisting with the creation of more dynamic spaces, the use of novel objects, for example, provides ample opportunity for enrichment. Novel items provide clear avenues for olfactory and other sensory enrichment as well as adding interest to otherwise static environments. Leaf litter, for example, can introduce new scents and provide foraging options through the concealment of food items. Adding a handful of soiled rodent bedding to a snake enclosure can also promote exploratory foraging. Play behaviours have been documented in numerous taxa (see Burghardt (2013) for a comprehensive review) and the addition of certain objects will lead to periods of interaction between the animals and the new additions; for example, balls for aquatic chelonians (Burghardt, 1996) and water currents in crocodilians (Dinets, 2015).

It's worth considering too that, although often viewed primarily as a means of enhancing welfare, enrichment can fulfil a vital role in ex-situ husbandry programmes (see Burghardt, 2013;

Michaels et al., 2014). Given the current conservation status of many reptiles and amphibians, and the recognised need for ex-situ management for many taxa, there is an urgent need to develop husbandry protocols that enable the expression of behaviours that may aid reintroduction success. Behavioural diversity should therefore be considered alongside genetic and species diversity when setting conservation priorities (Angeloni et al., 2010; Browne et al., 2011).

16.3.1 ENCOURAGING FORAGING

Promoting natural feeding and foraging behaviours likely reflects one of the most frequently seen approaches for enriching captive reptiles and amphibians (Bartolomé et al., 2023). The use of live invertebrate food for many species provides an avenue for enrichment that can keep many species gainfully occupied throughout their active hours (see Hurme et al. (2003) for an example with dendrobatid frogs (Dendrobatidae)). Note, though that the welfare of these live prey animals should not be overlooked, and this is explored further in Chapter 19. Foraging times can be extended by using 'slow release' feeders, or similar, where food items are released at intervals throughout the day rather than all at once. Many species can be encouraged to display a range of hunting and food extraction behaviours, with varanid lizards being popular subjects for research on this matter. Although the use of live vertebrate prey is not usually suitable for these animals due to welfare impacts on the prey species, keepers can still replicate elements of hunting or prey location behaviours through the use of scent trails and hiding food items around an enclosure. In comparing food-based versus scent-based enrichment, Waterman et al. (2021) found that scent-based enrichment was just as effective as food-based methods in increasing natural behaviours across three species (Komodo dragon, emerald tree monitors, *Varanus prasinus*, and crocodile monitors, *V. salvadorii*).

Further dietary novelty can be introduced by varying the range of food items offered. It is fair to assume that the majority of rodent-feeding snakes live their entire lives in captivity, feeding on one or two prey species: a far cry from the range of options available in the wild. Quite apart from the likely nutritional benefits of a varied diet, different food items suggest a range of hunting and prey capture strategies in use by wild specimens and represent a wider scope of options for how food is offered in captivity. Strategies for herbivorous and omnivorous species should also reflect how such animals secure their food in the wild.

16.3.2 THE IMPORTANCE OF SOCIAL INTERACTIONS

The social lives of reptiles and amphibians are poorly documented, although it is becoming increasingly apparent that they are far more complex than many people have previously assumed. That said, social networks and hierarchies are well documented in many lizards and crocodilians, and it seems reasonable to assume that, in other taxa, where individuals often aggregate, some form of similar interaction takes place. For example, Skinner et al. (2024) illustrated that social settings in garter snakes (*Thamnophis* spp.) can influence the behavioural plasticity of individuals, whilst the spawning aggregations of many amphibians provide multiple examples of auditory, tactile, visual, and chemosensory interactions between both potential mates and competing individuals.

Many taxa are maintained in solitary situations on the basis of 'aggression' seen between individuals kept in close proximity to each other. Chameleons (*Chamaeleonidae* spp.) are a commonly kept example of this, being rare to see these animals kept with other conspecifics, despite evidence suggesting that social interaction in young chameleons plays a key role in behavioural development (Ballen et al., 2014). Wild chameleons do not live in isolation, and the interactions between neighbours form a key part of their natural behavioural repertoires. Although the permanent presence of others may be detrimental to welfare, transient interactions in controlled environments may be beneficial. Therefore, the aggressive behaviours often seen in captivity should be viewed as a challenge to overcome, rather than something to avoid for reasons of convenience and space.

16.3.3 BEHAVIOUR TRAINING AS A HUSBANDRY TOOL

While training has long been common practice for mammals and birds housed in zoos, an increasing number of studies are highlighting the multitude of welfare benefits for herptiles, although data for amphibians is still largely undocumented. The focus of most efforts has been geared towards certain reptile taxa, with crocodilians, tortoises, and varanid lizards being the most frequently reported subjects, demonstrating capacity for these animals to learn and respond in previously unexpected ways, and showing benefits such as the reduction of stress during health procedures or enclosure cleaning. Target training to improve welfare during animal moves and veterinary procedures is becoming increasingly common and has the added advantage of creating safer processes for staff as well.

The ectothermic metabolism of reptiles creates particular challenges that require creative thinking on the part of the practitioner. One of the evolutionary benefits of ectothermy is that feeding frequency is less than for comparably sized endotherms; a contributing factor to the success of many reptiles in harsh environments where endotherms often fail to prosper. This decreased reliance on regular food intake does, however, mean that the use of food rewards in training schedules can be less effective, with alternative forms of reinforcement being required in many instances. The vocal repertoire of crocodilians provides one means of resolving such a challenge, as does the use of tactile rewards for chelonians. In addition, just as a reptile's general behavioural repertoire is inextricably linked to the environmental parameters of its surroundings, the response of said reptile to training is similarly impacted. As has been mentioned already, these factors extend beyond the traditional elements of temperature and should include light intensity, wavelength, humidity, and seasonality as a minimum. UVA light for example influences colour perception and appetence, both vital components in many traditional training regimes and discrimination trials. Box 16.1 poses some important thoughts on survival compared to thriving for captive reptiles and amphibians, highlighting the need for a shift change in how we consider "what care is" for these species.

16.4 CONCLUSIONS AND FUTURE CONSIDERATIONS

Keepers of reptiles and amphibians are faced with a significant challenge. For the species in their care, they must identify their pertinent needs (environmental, social, ecological etc.), disentangle fact and folklore from the available information on husbandry, and create an environment that enables the fullest expression of a behavioural repertoire that may be largely unknown. Alongside this, we must learn to identify reliable indicators of welfare that enable the ongoing evaluation of provision and implement appropriate means of either rectifying or replicating what they see. That this is achievable is without doubt; one can find fantastic examples of intricate, ethologically informed captive environments in many zoos around the world. Continuing the steady stream of evidence remains a vital role for current and future caregivers, ensuring that these incredible animals receive the same levels of consideration afforded to other, more well-studied taxa. Finally, it is likely that an increasing number of reptile and amphibian species will require some element of captive husbandry to underpin future conservation efforts, the success of which will undoubtedly be enhanced by an evidence-led approach to husbandry that recognises their rich diversity and complex lives. Box 16.2 provides more information on enhancing welfare in captive reptiles and amphibians in the form of useful suggested reading.

BOX 16.1: A thought to consider – striving for better practice

Many reptiles and amphibians will survive under relatively simple captive conditions that provide for their basic environmental needs, with some even breeding under such conditions. We must remember, however, that survival and reproduction are not accurate indicators of good welfare; these merely indicate that certain core parameters are present. This is particularly pertinent for herptiles, with many species being bred and sold commercially for the exotic pet trade where cost and convenience often underpin the choice of husbandry method. It should be the aim of every conscientious caregiver to 'move the dial forward' and to actively engage in a process that enhances the welfare of the species they work with. Striving for best practice requires the constant questioning of accepted practice, even if the methods currently in use appear to produce acceptable results.

BOX 16.2: Suggested further reading

Taken collectively, the following suggestions for directed reading provide further guidance on how captive caregivers might inform their approach to improving the welfare of their charges. They consider the concept of 'folklore husbandry', a methodology for exploring pertinent elements of natural history, an overview of ethologically informed husbandry, and a discussion on the use of behavioural criteria to evaluate welfare state.

- Arbuckle, K. (2013). Folklore husbandry and a philosophical model for the design of captive management regimes. *Herpetological Review*, *44*, 448–452.
- Loughman, Z. J. (2020). Utilization of natural history information in evidence-based herpetoculture: A proposed protocol and case study with Hydrodynastes gigas (false water cobra). *Animals*, *10*(11), 2021.
- Rose, P. (ed.). (2022). *The behavioural biology of zoo animals*. Boca Raton, FL: CRC Press.
- Warwick, C., Arena, P., Lindley, S., Jessop, M., & Steedman, C. (2013). Assessing reptile welfare using behavioural criteria. *In Practice*, *35*(3), 123–131.

REFERENCES

Angeloni L., Crooks K.R., & Blumstein D.T. (2010). Conservation and Behavior: Introduction. In: Breed M.D., & Moore J. (eds.) *Encyclopedia of Animal Behavior*, Academic Press, Oxford, UK, Vol. 1, pp. 377–381.

Ballen, C., Shine, R., & Olsson, M. (2014). Effects of early social isolation on the behaviour and performance of juvenile lizards, Chamaeleo calyptratus. *Animal Behaviour*, *88*, 1–6.

Bartolomé, A., Carazo, P., & Font, E. (2023). Environmental enrichment for reptiles in European zoos: Current status and perspectives. *Animal Welfare*, *32*, e48.

Browne, R. K., Wolfram, K., García, G., Bagaturov, M. F., & Pereboom, Z. J. J. M. (2011). Zoo-based amphibian research and conservation breeding programs. *Amphibian and Reptile Conservation*, *5*(3), 1–14.

Burghardt, G. M., Ward, B., & Rosscoe, R. (1996). Problem of reptile play: Environmental enrichment and play behavior in a captive Nile soft-shelled turtle, Trionyx triunguis. *Zoo Biology*, *15*(3), 223–238.

Burghardt, G. M. (2013). Environmental enrichment and cognitive complexity in reptiles and amphibians: Concepts, review, and implications for captive populations. *Applied Animal Behaviour Science*, *147*(3–4), 286–298.

Carter, K. C., Hicks, J. J., Kane, D., Tapley, B., & Michaels, C. J. (2021). Age-dependent enclosure use in juvenile Chinese crocodile lizards, Shinisaurus crocodilurus crocodilurus. *Journal of Zoological and Botanical Gardens*, *2*(3), 406–415.

Dinets, V. (2015). Play behavior in crocodilians. *Animal Behavior and Cognition*, *2*(1), 49–55.

Hoehfurtner, T., Wilkinson, A., Walker, M., & Burman, O. H. (2021). Does enclosure size influence the behaviour & welfare of captive snakes *(Pantherophis guttatus)*?. *Applied Animal Behaviour Science*, *243*, 105435.

Hurme, K., Gonzalez, K., Halvorsen, M., Foster, B., Moore, D., & Chepko-Sade, B. D. (2003). Environmental enrichment for dendrobatid frogs. *Journal of Applied Animal Welfare Science*, *6*(4), 285–299.

Lambert, H., Carder, G., & D'Cruze, N. (2019). Given the cold shoulder: A review of the scientific literature for evidence of reptile sentience. *Animals*, *9*(10), 821.

Luiselli, L., & Angelici, F. M. (1998). Sexual size dimorphism and natural history traits are correlated with intersexual dietary divergence in royal pythons (Python regius) from the rainforests of southeastern Nigeria. *Italian Journal of Zoology*, *65*(2), 183–185.

Luiselli, L. (2006). Why do males and females of Python regius differ in ectoparasite load?. *Amphibia-Reptilia*, *27*(3), 469–471.

McPhee, M. E. (2003). Generations in captivity increases behavioural variance: Considerations for captive breeding and reintroduction programs. *Biological Conservation*, *115*, 71–77.

Michaels, C. J., Gini, B. F., & Preziosi, R. F. (2014). The importance of natural history and species-specific approaches in amphibian ex-situ conservation. *The Herpetological Journal*, *24*(3), 135–145.

Rose, P., Evans, C., Coffin, R., Miller, R., & Nash, S. (2014). Using student-centred research to evidence-base exhibition of reptiles and amphibians: Three species-specific case studies. *Journal of Zoo and Aquarium Research*, *2*(1), 25–32.

Skinner, M., Nagabaskaran, G., Gantert, T., & Miller, N. (2024). Bolder together: Conformity drives behavioral plasticity in eastern gartersnakes. *Animal Cognition*, *27*(2). https://doi.org/10.1007/s10071-024-01859-5

Waterman, J. O., McNally, R., Harrold, D., Cook, M., Garcia, G., Fidgett, A. L., & Holmes, L. (2021). Evaluating environmental enrichment methods in three zoo-housed Varanidae lizard species. *Journal of Zoological and Botanical Gardens*, *2*(4), 716–727.

17

Enriching the awkward – promoting welfare in challenging species (aquarium fish)

CHLOE H. STEVENS

17.1 INTRODUCTION

Fishes have been popular additions to zoological collections since the opening of the first public aquarium in London Zoo in 1853 (Brunner, 1964). Now, public aquaria (hereafter simply 'aquaria') are widespread attractions, existing as part of larger zoos, wildlife parks, museums, and as standalone facilities. The aquarium is often one of the most species-rich parts of a zoological collection – a single large aquarium may house hundreds of different fish species, and it is likely that the total number of different species represented in aquaria across the world is in the thousands. However, this only represents a fraction of fish species; fishes represent an immensely large and diverse taxonomic group, with over 35,000 extant species recorded (FishBase, 2024). The range of habitats in which they are found is almost as diverse, ranging from small, ephemeral freshwater pools on the African savannah to Arctic seas and ocean depths of over 8,000 m. This diversity creates a significant challenge for safeguarding the welfare of the animals housed. Captive environments (Figure 17.1) should reflect the specifics of these diverse habitats (including factors like water temperature, salinity, pH, flow rate, light levels, plant life, substratum, and more) and support the varied physiologies and behaviours of their inhabitants, often with somewhat limited information available about these factors, and similarly limited information on species-specific indicators of good or poor welfare.

There are further challenges associated with providing fishes with good welfare in the aquarium beyond understanding the details of their wild habitats. The study of fish welfare has tended to receive less attention than welfare in other vertebrates, in part due to debates over whether fishes

DOI: 10.1201/9781003429425-17

are sentient and therefore experience welfare at all. Although interest in fish welfare is now increasing, there are still a relatively limited number of fish welfare studies. Those that exist largely concern aquaculture species, and to a lesser extent, species used in scientific research and testing, where the purpose of the study may be more related to maximising production, achieving higher-quality flesh, or producing higher numbers of embryos rather than achieving better welfare for welfare's sake. Even within the range of studies on ornamental fish welfare, bony fishes are severely underrepresented compared with cartilaginous fishes like sharks and rays (Elasmobranchii), despite bony fishes (Osteichthyes) making up the vast majority of fish species (Oldfield & Bonano, 2023).

Beyond these more practical matters, fish welfare is also regarded as a challenging topic for various social and cultural reasons. Fishes inhabit an underwater world that we rarely visit, lack facial expressions, and make vocalisations that we cannot hear without specialised equipment. Many species live in shoals, and as such are often viewed as a group, rather than as individuals. Myths about their capabilities, such as the goldfish (*Carassius auratus*) 'three-second memory' persist. It is likely that these reasons make it harder for humans to relate to fishes in the way we do other animals, and thus have contributed to attitudes that often regard fishes as less deserving of moral status.

Considering all of these factors, perhaps it is not surprising that fishes are regarded as challenging animals when it comes to promoting and maintaining good welfare. However, as attitudes towards fishes and their capabilities change, and research into fish welfare gathers pace, we can perhaps be hopeful that there will be broader acceptance of the fact that fishes are deserving of the same moral consideration afforded to other vertebrate species. In this chapter, I will consider what 'good welfare'

Figure 17.1 Examples of the diversity of underwater habitat replication seen in aquariums within public zoological facilities that all evidence knowledge of wild environmental biotic and abiotic factors with which the inhabitants of the exhibit would naturally interact. Top left; a tropical marine coral reef with a habitat structure dominated by hard (Scleractinia) and soft (Alcyonacea) corals. Top right; replication of an expansive, sandy seabed that is the specific habitat in which garden eels (Heterocongrinae) construct their burrows. Bottom left; dense planting in a freshwater tropical exhibit housing a large school of cardinal tetras (*Paracheirodon axelrodi*). Bottom right; replication of a network of caves within large boulders on a lakebed that lemon cichlids (*Neolamprologus leleupi*), a resident of Lake Tanganyika, would live in and around.

may look like for fishes, and outline some of the key welfare challenges that may impact fish welfare in the aquarium. I will then discuss how environmental enrichment can help promote good welfare, both by creating opportunities for positive experiences and by helping to mitigate some of the welfare challenges that aquarium species may face. Given the vast diversity of fishes, this discussion will be very broad, and it should be recognised that in practice, all of these considerations for welfare must be informed by species-specific knowledge as much as possible.

17.2 WHAT IS FISH WELFARE?

There are various definitions, concepts, and frameworks of animal welfare, and no clear scientific consensus over which is most appropriate. However, although early concepts of animal welfare focused largely on the biological functioning of an animal, many now agree that there should be at least some consideration of an animal's subjective affective states (its 'feelings'). This has made the topic of fish welfare a contentious issue, as the application of welfare concepts to fishes has been complicated by debates over whether fishes have the capacity to experience subjective states – in other words, whether fishes are sentient.

The debate over fish sentience has largely focused on discussions of whether fish are capable of feeling pain. Pain perception is recognised as being a subjective experience related to, but distinct from, the detection of noxious stimuli by sensory neurons (nociception) (IASP, 2021). Thus, sentience is required for pain perception. In humans, pain is processed in the neocortex (Tan & Kuner, 2021) – a brain structure that fishes (and all non-mammals) lack. Some argue that this means that fish cannot feel pain, and also point to a lack of evidence for the existence of nociceptors in cartilaginous fishes (Key, 2015; Rose et al., 2014). Their opponents maintain that fishes possess sufficient neural structures for them to experience pain and argue that behavioural changes, associative learning, and the modifying effect of analgesics after exposure to potentially painful stimuli provide further evidence for fish pain perception (Brown, 2015; Sneddon, 2003, 2015). Fish pain and sentience advocates also highlight complex behaviours and cognitive abilities demonstrated by some fish species. For example, some Lake Tanganyika cichlids (*Astatotilapia burtoni*) have been shown to be capable of using transitive inference (a form of logical reasoning) to deduce their place in a dominance hierarchy by observing fights between others (Grosenick et al., 2007). Some species show numerical abilities (Petrazzini et al., 2015), perform interspecific cooperative hunting (Bshary et al., 2006), or recognise individual humans (Miller et al., 2023). The cleaner wrasse (*Labroides dimidiatus*) has even been shown to be capable of passing the mirror test, a test that is often presented as providing evidence of not only sentience, but self-awareness (Kohda et al., 2019). However, some parties remain unconvinced and have argued that many of these behaviours can be explained without the need for sentience (Key, 2015). The fish pain and sentience debate is not settled, although it seems there are a growing number of voices that maintain that fishes are sentient, and it is my own position that many species of fish are likely to be sentient. Even where there is uncertainty, many point to the application of the precautionary principle to give fishes the 'benefit of the doubt' and treat them as though they are sentient.

So, what does good welfare look like for fishes? Accepting the position that fishes are sentient means accepting the possibility that fishes can experience positive affective states as well as negative ones, and thus places the responsibility on those of us housing fish in captivity to try to create and promote these positive experiences. Unfortunately, current knowledge about what might constitute positive welfare for fishes is sparse, as are validated indicators of positive welfare. Providing fishes with species-appropriate conditions, such as appropriate physical environments (particularly appropriate water quality parameters), appropriate social settings, and appropriate diets are all likely to be essential, but even with all of these in place, we are unlikely to be able to create more than a neutral state of welfare. Fife-Cook and Franks (2019) have suggested that positive experiences for fishes might be created through the provision of opportunities for free choice, control and exploration, and that one way to create these opportunities is through the provision of environmental enrichment. Environmental enrichment may therefore be the best tool currently available to us to provide fishes with good welfare (Figure 17.2). To progress beyond this, we will need to further develop our understanding of what good welfare means for fishes, how best we can measure it, and what kinds of welfare indicators are most useful tankside.

17.3 WELFARE (AND ITS CHALLENGES) IN THE AQUARIUM

Most of the research into fish welfare has focused on fishes in aquaculture settings, and to a lesser extent, fishes used in scientific research. Unlike these two settings, aquaria might be considered to be at somewhat of an advantage, given that one of the key focuses of the modern zoo should be to promote animal welfare for its own sake (Rose & Riley, 2022), rather than because good welfare helps produce better quality meat or data. However, this does not mean that housing fish in the aquarium is without its own challenges. Many of these challenges are common across different captive fish settings and will only be briefly discussed, but some are specific to aquaria. Most importantly, all are impacted by a lack of available information about species-specific environmental requirements, behaviours and welfare indicators.

17.3.1 GENERAL HOUSING AND HUSBANDRY CONSIDERATIONS

Fishes are in constant direct contact with water, and as such are vulnerable to poor water quality. Maintaining water quality parameters, such as temperature, pH, salinity, conductivity, hardness, and others that are appropriate to the species is key for good care and husbandry of captive fishes. Different species vary in their range of tolerance, as well as their robustness in response to changes to these conditions. However, not all changes to these parameters are negative for welfare – changes in water quality parameters after rainfall can stimulate various behaviours in the wild, including spawning, in some species. Therefore, a detailed understanding of not only the optimal conditions for a species but how that species responds to change is important to maintain welfare.

Figure 17.2 An example of species-specific environmental enrichment provided to fish in a public aquarium. In this case, archerfish (*Toxotes jaculatrix*) are provided with a suspended live-food feeder that can be filled with crickets or locusts. As the insects leave the container, the archerfish can utilise their unique "spitting" hunting strategy – aiming jets at water at potential prey to knock it into the water for consumption. Welfare is promoted by enabling choice (of behaviours to perform), control (over when to forage), and exploration (of different dimensions of the aquatic and aerial environment). To view the enrichment in situ, visit: https://youtu.be/c5ojebaHzfE

Other physical features of a fish tank, including size and shape, lighting and photoperiod, and the presence of enrichment will all contribute to the appropriateness of the environment, and each needs to be guided by the conditions each species experiences in the wild, where possible. Ideally, only species for which we can provide all of their needs would be housed in captive environments, but where these species are housed, other opportunities to have positive experiences to mitigate the effects of these potential welfare challenges are particularly important.

Fish welfare in the aquarium is also affected by the diets provided. The optimal nutritional requirements of many species are not known, so fish health must be carefully monitored to try and identify deficiencies where better information is unavailable. Some captive species require live prey for optimal welfare, and many may require some variation in diet to meet their needs. Feeding frequency can also impact welfare, either by influencing feeding behaviour, or because providing food has the potential to create positive experiences for fish, and so should be considered with this possibility in mind (see below).

17.3.2 SOCIAL ENVIRONMENT

Fishes live in a wide range of different social structures in the wild – from species that are generally solitary to those that live in large, organised schools (Helfman et al., 2009). The most basic requirement for maintaining welfare through social environments is therefore to ensure that fishes have access to an appropriate number of conspecifics and have space to engage in meaningful social associations and interactions (Figure 17.3). A more complex issue for aquaria is that fishes may be housed in large, multi-species assemblages, in contrast to other captive settings like aquaculture and scientific research. This presents a clear challenge for maintaining welfare as species must be carefully selected to ensure they are compatible tank-mates and to minimise behaviours like aggression and harassment. Even where species co-exist in the wild, there are potential welfare issues that may arise if predators and prey are housed together – these must be carefully managed using appropriate feeding regimes and tank design to minimise predation on tank-mates (Smith, 2023).

17.3.3 HUMAN-ANIMAL INTERACTIONS

Fishes in the aquarium experience a variety of interactions with their caretakers, such as feeding, cleaning, or capture for veterinary purposes. These interactions have the potential to cause disturbances that negatively affect welfare. For example, it has been shown that cleaning increases stress and decreases positive social behaviours in zebrafish (*Danio rerio*) (Powell et al., 2021), and that zebrafish can also distinguish between different human caretakers and respond differently to them based

Figure 17.3 Space to shoal is important. For pelagic species that travel large distances in enormous shoals and coordinate their movements closely, open space and the recreation of water currents are essential to enable such group-level movement to occur in the aquarium. Fish within the shoal gain welfare benefits from being close to conspecifics and engaging in collective movement; therefore, they should only be maintained by aquariums that can afford them the space and volume of water.

on their roles (Miller et al., 2023). However, such interactions may also create the potential for positive experiences; some species have been shown to seek out human interactions, which may be used to promote better welfare (Fife-Cook & Franks, 2021).

A greater challenge for aquarium fishes is the presence of visitors. Relatively few studies have been conducted on the impact of visitors on fish welfare, and those that have been conducted often do not reveal clear responses, but the increased risk of noise and other disturbances means that these interactions are likely to negatively affect welfare. For example, noise has been shown to be highly stressful for a number of fish species, with effects on both behavioural and physiological responses (Cox et al., 2018). Fishes in aquaria may also be housed in interactive exhibits, such as touch pools. The increased degree of interaction within these exhibits presents an even higher risk of disturbance or other negative welfare impacts due to issues like mishandling, exposure to substances on visitors' hands, or transfer of pathogens (Smith, 2023). Although it is possible that some species might benefit from interactions with humans, trained caretakers are much better placed to provide these interactions in a way that does not compromise welfare; consequently, touch pools and other interactive exhibits should be discouraged.

17.4 PROMOTING WELFARE THROUGH ENVIRONMENTAL ENRICHMENT

Environmental enrichment (hereafter 'enrichment') is a term that, like welfare, has been described in a variety of ways in the literature; it can refer to any modification made to the environment of a captive animal with the aim of improving their welfare, or only to those modifications which have been demonstrated to have a clear welfare benefit (see Chapter 5). Understanding the definition in use in welfare studies is crucial; it is not uncommon to see or hear claims that fish do not need or benefit from enrichment (e.g. Woodward et al., 2019), rather than that the modification tested was simply not enriching for that particular species. Even if we accept that modifications must have been demonstrated to benefit welfare to be considered enrichment, there are difficulties; if a modification conveys a welfare benefit to an animal, at what point can we say that modification constitutes enrichment, rather than simply a requirement for that species? As understanding of good fish welfare develops, it is likely that many of the modifications that might currently be considered enriching will become part of the basic husbandry and care standards for that species, and caretakers will have to think about how else they might be able to enrich animal environments and experiences. This process of continual improvement is essential if we are to maximise the welfare of animals in captivity

17.4.1 PHYSICAL ENRICHMENT

For fishes, physical enrichment could include substrate on the tank floor, plants or corals, rocks, logs, pipes, or other ornaments. There is good evidence from a variety of species that these objects benefit welfare; they give animals 'what they want' (Dawkins, 2003), and can reduce stress, increase the display of natural behaviours, reduce aggression, increase positive social behaviours, and promote better learning and cognition (Näslund & Johnsson, 2016; Stevens et al., 2021). However, objects should not simply be provided at random; modifications need to be tested and validated to show they are truly enriching before long-term or widespread use.

A more interesting question is: why do fishes want access to these items? Some may be valued as they would represent high-quality habitats for breeding in the wild; others may increase feelings of safety or security in tank inhabitants, either by providing refugia from disturbances or aggressive tank mates or camouflage from potential predators. This may, in turn, increase the ability of fishes to cope with stressful events. Beyond this, the additional complexity created by the addition of these objects can help make environments more heterogeneous. This creates more opportunities for exploration, and by helping to create zones or microhabitats within a tank, allows animals much more free choice and control over their lives than a barren tank – factors likely to be essential for good welfare in any captive animal.

17.4.2 SOCIAL ENRICHMENT

As a minimum, social fishes should have access to enough other individuals to support the formation of social structures like those that would be seen in the wild, and ideally with a relevant distribution of ages and sexes in the group. In the wild, some species may naturally join different groups at different life stages, but it is not clear whether this would promote welfare – it is possible that maintaining stable social groups is better for welfare as disruption to social hierarchies can be stressful and promote fighting, as occurs in a number of cichlid species. Stable social groups may also promote further welfare benefits – some species, like zebrafish, have been shown to recognise and respond differently to familiar individuals (Madeira & Oliveira, 2017), while others, like guppies (*Poecilia reticulata*) have been shown to develop preferential associations with others (suggested by some to represent 'friendships') (Heathcote et al., 2017). A better understanding of these kinds of social dynamics in different fish species is likely to be needed before we can say that fishes are housed in socially enriched environments.

The potential for interspecies interactions to promote welfare merits further exploration. The presence of multiple species in an exhibit is certainly likely to add an extra layer of complexity, and different species are likely to interact, even outside of predator-prey interactions – cooperative behaviours, like hunting, have been observed between different species - groupers (*Plectropomus pessuliferus*) and moray eels (*Gymnothorax javanicus*) (Bshary et al., 2006). Some evidence from home aquaria suggests that mixing species can have positive effects on welfare. For example, the presence of angelfish (*Pterophyllum scalare*) reduced aggression among smaller shoaling species (Sloman et al., 2011) – but much more research is needed to understand how to create socially enriched environments in multi-species tanks.

17.4.3 NUTRITIONAL ENRICHMENT

Any diet provided to a captive animal should ideally be nutritionally complete and support optimal biological functioning. To be enriching, diets therefore need to provide something more. For example, providing a more varied diet is likely to create complexity that benefits welfare, and providing different foods may also allow caretakers to stimulate different kinds of hunting or foraging behaviour. It is possible that providing particularly palatable foods as treats could be enriching. Unfortunately, little research has been conducted on these potential effects on fishes.

Diets may also provide opportunities for improved welfare through the frequency and method of delivery. Most species have particular times of day or night when they prefer to hunt or forage in the wild, and generally, feeding schedules in captivity should reflect this (López-Olmeda et al., 2012). Evidence from aquaculture species also suggests that welfare can be improved through the use of self-feeders (e.g. Attia et al., 2012) – these may provide fishes with some control over their environment, which may benefit their psychological wellbeing. Extra complexity may also be created by feeding fishes from different locations in the tank.

Another approach that may benefit welfare relates to the predictability of feeding. Kleiber et al. (2022) found that pairing food provision with a visual cue (bubbles) reduced stress and aggression in rainbow trout (*Oncorhynchus mykiss*). If adding some predictability to feeding increases anticipation, this may be a positive experience (a feeling of excitement), although delays to food provision may then also cause psychological stress or frustration.

17.4.4 SENSORY ENRICHMENT

Fishes in the wild are exposed to huge amounts of sensory input, so biologically relevant sensory stimuli are likely to increase environmental complexity, behavioural diversity, and welfare. The impacts of lighting on fish health and behaviour has been studied in various species (Arechavala-Lopez et al., 2021), and lighting that mimics wild conditions may be enriching, as may other visual stimuli such as colourful structures or objects. It may be possible to also provide stimulation within other sensory modalities. For example, auditory enrichment has been found to reduce stress in zebrafish (Barcellos et al., 2018), and some recent results suggest that tactile stimulation might be perceived as positive in Nile tilapia (*Oreochromis niloticus*) (Gauy et al., 2021). More research is needed to fully understand the potential of sensory stimulation to convey welfare benefits.

17.4.5 OCCUPATIONAL ENRICHMENT

Occupational enrichments for fishes might include the use of feeding devices that require interaction or manipulation to release food, water currents for fishes to swim against (Arechavala-Lopez et al., 2021), or training. These types of enrichment have all been used in aquarium settings, although not necessarily for all fish species - for example, training in the may be more common among large fish species like zebra sharks (*Stegostoma fasciatum*) (Marranzino, 2013). There may therefore be scope to increase the amount of enrichment provided for fishes by trialling these approaches among a wider range of fish species. Although it may be difficult to train species that are housed in large groups to perform some behaviours (e.g. target training), it may be possible to introduce other forms of training, such as training fishes to occupy a particular area of a tank. However, there is still much scope for this area to be explored in fishes.

17.5 CONCLUSIONS AND FUTURE CONSIDERATIONS

Interest in fish welfare, and particularly positive welfare appears to be growing, and aquaria are the ideal settings to develop this field further. More information is particularly needed on the needs and psychological welfare of bony fish (Oldfield & Bonano, 2023), but exploration of how modern concepts of animal welfare can be applied to all fishes, and what general good fish welfare looks like is also essential to move the field forward. Considering how fish engage with the physical aspects of their environment (Figure 17.4) and what this means for this health, behavioural repertoires and psychological outputs should be considered across species. Utilisation of the observational welfare assessment technique, Qualitative Behavioural Assessment (QBA) has been used to identify welfare outputs presented by farmed salmon, *Salmo salar* (Wiese et al., 2023) and such approaches should be adapted for use in zoological collections. Further development of how we provide enrichment for fish (Box 17.1) and what the evidence basis is for this, would be extremely helpful to future enhancements of welfare. Combining developments to husbandry (i.e. use of environmental enrichment) with assessment of welfare outputs (i.e. QBA) could give a more complete illustration of what welfare-focussed husbandry manifests as for captive fishes.

BOX 17.1: Environmental enrichment and striving for best practice

It is likely that all fish species will benefit from some form of environmental enrichment. The types of enrichment that are most appropriate will depend on the wild behaviours and habitats of the species – a species that spends most of its time in the open ocean is unlikely to benefit from a heavily planted tank – but a sufficiently good enrichment programme should ideally include multiple types of enrichment (physical, social, dietary, sensory, or occupational enrichment). Most importantly, rather than simply helping to reduce the impact of negative experiences, such as providing fishes with refuges from disturbance, enrichment should aim to create positive experiences and positive affective states for fishes. These could be excitement in anticipation of feeding, interest through providing tasks to occupy fishes, choice through exploration of different microhabitats or a range of social companions, and overall, a degree of control over their own lives.

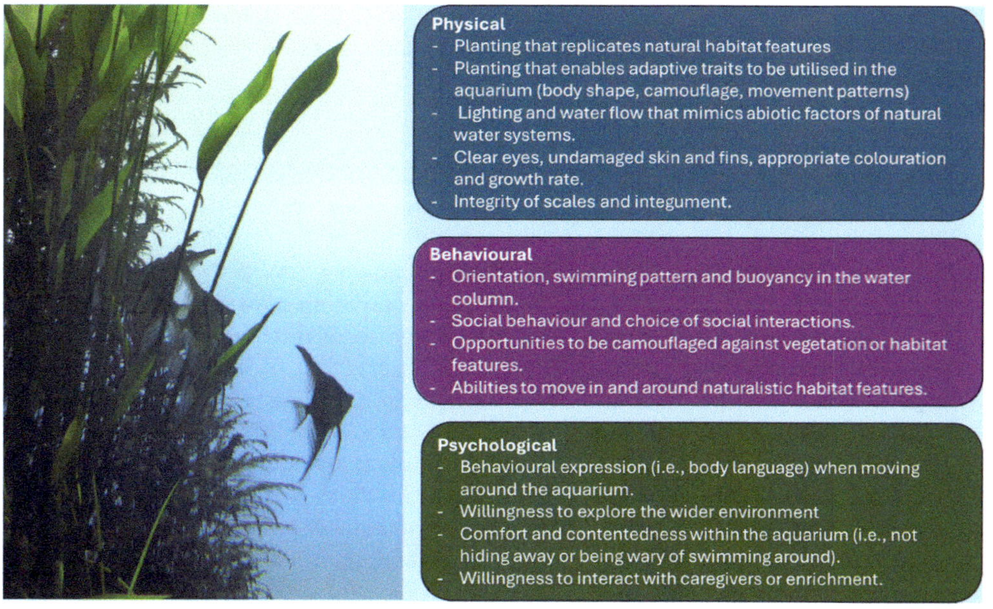

Figure 17.4 Freshwater angelfish utilise aquarium planting that is aligned to their compressed body shape. The angelfish have evolved to live amongst dense but tall vegetation, and this is illustrated by the flattened morphology and vertical side stripes. As welfare comprises physical, behavioural, and psychological components, we can look at how fish interact with their environment, their tank mates, their caregivers, and with their husbandry to understand how these will all impact prevailing welfare states.

There has been some consideration of this in the tentative suggestions of certain behaviours that may be positive welfare indicators, such as 'heightened shoaling' in zebrafish (Franks et al., 2018), but further indicators will be necessary to assess whether attempts to provide good welfare for fishes are effective. Consideration of the available literature (Box 17.2) needs to continue in earnest, alongside empirical research that has tested welfare measures and hence added to such literature, to ensure that what we provide for fishes under human care meets their behavioural, physical, and psychological needs.

BOX 17.2: Useful references to enhance understanding of these fish welfare topics and issues

fair-fish database: https://fair-fish-database.net/

Corcoran, M. (2015). Environmental enrichment for aquatic animals. *The veterinary clinics of North America. Exotic Animal Practice, 18*(2), 305–321.

Kristiansen, T. S., Fernö, A., Pavlidis, M. A. & van de Vis, H. (Eds.). (2020). *The welfare of fish.* Cham, Switzerland: Springer.

Monreal-Pawlowsky, T., Vaicekauskaitė, R., Membrive, G. P., Delfour, F. & Manteca, X. (2021). Goal-oriented behavioural and environmental enrichment in aquarium species. *Journal of Zoo and Aquarium Research, 9*(4), 273–280.

Smith, S. A. (2023). Fish welfare in public aquariums and zoological collections. *Animals, 13*(16), 2548.

Young, R. J. (2003). *Environmental enrichment for captive animals.* Oxford, UK: Blackwell Science.

REFERENCES

Arechavala-Lopez, P., Cabrera-Álvarez, M. J., Maia, C. M. & Saraiva, J. L. (2021). Environmental enrichment in fish aquaculture: A review of fundamental and practical aspects. *Reviews in Aquaculture, 14*(2), 704–728.

Attia, J., Millot, S., Di-Poï, C., Bégout, M.-L., Noble, C., Sanchez-Vazquez, F. J., Terova, G., Saroglia, M. & Damsgård, B. (2012). Demand feeding and welfare in farmed fish. *Fish Physiology and Biochemistry, 38*(1), 107–118.

Barcellos, H. H. A., Koakoski, G., Chaulet, F., Kirsten, K. S., Kreutz, L. C., Kalueff, A. V. & Barcellos, L. J. G. (2018). The effects of auditory enrichment on zebrafish behavior and physiology. *PeerJ, 6*, e5162.

Brown, C. (2015). Fish intelligence, sentience and ethics. *Animal Cognition, 18*(1), 1–17.

Brunner, B. (1964). *The ocean at home: An illustrated history of the aquarium*. New York, NY: Princeton Architectural Press.

Bshary, R., Hohner, A., Ait-el-Djoudi, K. & Fricke, H. (2006). Interspecific communicative and coordinated hunting between groupers and giant moray eels in the Red Sea. *PLoS Biology, 4*(12), e431.

Cox, K., Brennan, L. P., Gerwing, T. G., Dudas, S. E. & Juanes, F. (2018). Sound the alarm: A meta-analysis on the effect of aquatic noise on fish behavior and physiology. *Global Change Biology, 24*(7), 3105–3116.

Dawkins, M. S. (2003). Behaviour as a tool in the assessment of animal welfare. *Zoology, 106*(4), 383–387.

Fife-Cook, I., & Franks, B. (2019). Positive welfare for fishes: Rationale and areas for future study. *Fishes, 4*(2), 31.

Fife-Cook, I., & Franks, B. (2021). Koi (*Cyprinus rubrofuscus*) seek out tactile interaction with humans: General patterns and individual differences. *Animals, 11*(3), 706.

FishBase. (2024). *FishBase*. https://www.fishbase.org

Franks, B., Graham, C. & von Keyserlingk, M. A. G. (2018). Is heightened-shoaling a good candidate for positive emotional behavior in zebrafish? *Animals, 8*(9), 152.

Gauy, A. C. D. S., Bolognesi, M. C., Martins, G. D. & Gonçalves-de-Freitas, E. (2021). Preference and Motivation Tests for Body Tactile Stimulation in Fish. *Animals, 11*(7), 2042.

Grosenick, L., Clement, T. S. & Fernald, R. D. (2007). Fish can infer social rank by observation alone. *Nature, 445*(7126), 429–432.

Heathcote, R. J. P., Darden, S. K., Franks, D. W., Ramnarine, I. W. & Croft, D. P. (2017). Fear of predation drives stable and differentiated social relationships in guppies. *Scientific Reports, 7*(1), 41679.

Helfman, G., Collette, B. B., Facey, D. E. & Bowen, B. W. (2009). *The diversity of fishes: Biology, evolution, and ecology*. Oxford, UK: John Wiley & Sons.

International Association for the Study of Pain (IASP) (2021). *Terminology*. https://www.iasp-pain.org/resources/terminology/

Key, B. (2015). Fish do not feel pain and its implications for understanding phenomenal consciousness. *Biology & Philosophy, 30*(2), 149–165.

Kleiber, A., Le-Calvez, J.-M., Kerneis, T., Batard, A., Goardon, L., Labbé, L., Brunet, V., Ferreira, V. H. B., Guesdon, V., Calandreau, L. & Colson, V. (2022). Positive effects of bubbles as a feeding predictor on behaviour of farmed rainbow trout. *Scientific Reports, 12*(1), 11368.

Kohda, M., Hotta, T., Takeyama, T., Awata, S., Tanaka, H., Asai, J.-Y. & Jordan, A. L. (2019). If a fish can pass the mark test, what are the implications for consciousness and self-awareness testing in animals? *PLoS Biology, 17*(2), e3000021.

López-Olmeda, J. F., Noble, C. & Sánchez-Vázquez, F. J. (2012). Does feeding time affect fish welfare? *Fish Physiology and Biochemistry, 38*(1), 143–152.

Madeira, N. & Oliveira, R. F. (2017). Long-term social recognition memory in zebrafish. *Zebrafish, 14*(4), 305–310.

Marranzino, A. (2013). The use of positive reinforcement in training zebra sharks (*Stegostoma fasciatum*). *Journal of Applied Animal Welfare Science, 16*(3), 239–253.

Miller, S. L., Leri, F., Pushinsky, A., Franks, B., DePasquale, C. & Braithwaite, V. A. (2023). Zebrafish (*Danio rerio*) distinguish between two human caretakers and their associated roles within a captive environment. *Applied Animal Behaviour Science, 267,* 106053.

Näslund, J. & Johnsson, J. I. (2016). Environmental enrichment for fish in captive environments: Effects of physical structures and substrates. *Fish and Fisheries, 17*(1), 1–30.

Oldfield, R. G. & Bonano, P. E. (2023). Psychological and social well-being of bony fishes in zoos and aquariums. *Zoo Biology, 42*(2), 185–193.

Petrazzini, M. E. M., Lucon-Xiccato, T., Agrillo, C. & Bisazza, A. (2015). Use of ordinal information by fish. *Scientific Reports, 5,* 15497.

Powell, C., von Keyserlingk, M. A. G. & Franks, B. (2021). Tank cleaning temporarily increases stress and decreases affiliative behavior in zebrafish. *Applied Animal Behaviour Science, 242,* 105414.

Rose, J. D., Arlinghaus, R., Cooke, S. J., Diggles, B. K., Sawynok, W., Stevens, E. D. & Wynne, C. D. L. (2014). Can fish really feel pain? *Fish and Fisheries, 15*(1), 97–133.

Rose, P. E. & Riley, L. M. (2022). Expanding the role of the future zoo: Wellbeing should become the fifth aim for modern zoos. *Frontiers in Psychology, 13,* 1018722.

Sloman, K. A., Baldwin, L., McMahon, S. & Snellgrove, D. (2011). The effects of mixed-species assemblage on the behaviour and welfare of fish held in home aquaria. *Applied Animal Behaviour Science, 135*(1–2), 160–168.

Smith, S. A. (2023). Fish welfare in public aquariums and zoological collections. *Animals, 13*(16), 2548.

Sneddon, L. U. (2003). The evidence for pain in fish: the use of morphine as an analgesic. *Applied Animal Behaviour Science, 83*(2), 153–162.

Sneddon, L. U. (2015). Pain in aquatic animals. *Journal of Experimental Biology, 218*(7), 967–976.

Stevens, C. H., Reed, B. T. & Hawkins, P. (2021). Enrichment for laboratory zebrafish - A review of the evidence and the challenges. *Animals, 11*(3), 698.

Tan, L. L. & Kuner, R. (2021). Neocortical circuits in pain and pain relief. *Nature Reviews: Neuroscience, 22*(8), 458–471.

Wiese, T. R., Planellas, S. R., Betancor, M., Haskell, M., Jarvis, S. & Davie, A. (2023). Qualitative behavioural assessment as a welfare indicator for farmed Atlantic salmon (*Salmo salar*) in response to a stressful challenge. *Frontiers in Veterinary Science, 10,* 1260090.

Woodward, M. A., Winder, L. A. & Watt, P. J. (2019). Enrichment increases aggression in zebrafish. *Fishes, 4*(1), 22.

18

Animal welfare and policy

MICHELLE O'BRIEN AND CAMPBELL MURN

18.1 INTRODUCTION

A discussion of policy and how it relates to animal welfare in zoos can, at first glance, appear a dry and unexciting topic. However, the policies produced by governments, legislative bodies and other regulatory organisations underpin the frameworks on which animal welfare standards in zoological organisations are based. Policies state what can and, perhaps more importantly, cannot be done with respect to the welfare and management of animals living in zoos. They also allow action to be taken, sometimes punitively, when the welfare standards in zoos do not reach required or expected levels. Without these policies, and their enforcement by regulatory bodies, combined with compliance of licensed zoos, there would be a significant reduction in the welfare of animals in zoos. In short, policy provides the basis for upholding animal welfare standards in zoological facilities.

In this chapter, we focus on policies related to zoo animal welfare in Great Britain (England, Scotland and Wales) and the United Kingdom (UK), which includes both Great Britain and Northern Ireland, drawing comparisons with other regions and their zoo associations in Europe, North America and Australasia/Oceania. Welfare policy is governed both by the UK (Westminster) government and the devolved administrations in Scotland, Wales and Northern Ireland. The word 'policy' itself can have some ambiguity, and because in this chapter we are not only concerned with government policy, we have adopted the broad definition of policy used by the Centers for Disease Control and Prevention (CDC) – a law, regulation, procedure, administrative action, incentive, or voluntary practice of governments and other institutions (CDC, 2015). Importantly, we consider the focus of a policy to be a drive towards a desired state or set of conditions, and this is paramount in terms of animal welfare.

Each country will have both laws and policies covering the requirements of zoos that house captive animals. In England, Scotland and Wales, zoos are regulated under the Zoo Licensing Act (1981) if they are open to the public (paying or non-paying) on seven or more days in a 12-month period (although some dispensations can apply) and are displaying non-domesticated animals. Therefore, a zoo is classed as "any establishment where wild animals are exhibited to the public, apart from circuses and pet shops," and wild animals are "any animals that aren't normally domesticated in Great Britain" (see https://www.gov.uk/guidance/keeping-zoo-animals for more information). The Secretary of State's Standards of Modern Zoo Practice for Great Britain (hereafter SSSMZP

DOI: 10.1201/9781003429425-18

or 'the Standards') specify the minimum standards that zoos in England, Wales and Scotland must meet in order to be licensed zoo premises. The existing Standards (as of 2024) include both requirements that *should* be fulfilled (i.e. fulfilment of these requirements is ideal) and those that *must* be fulfilled (i.e. licensing cannot be achieved without compliance).

The authors and editors of this chapter are aware of the 2025 published amendments to the UK Standards of Modern Zoo Practice (to be enforced from 2027); however, due to delays in the eventual publication of these new Standards and the timeframe of writing this chapter for this book, it was not possible to discuss the new Standards in the final book. Although this chapter is not region-specific, we recommend the reader to visit the UK Government website (www.gov.uk) and search for "Standards of Modern Zoo Practice" to familiarise themselves with the changes to zoo licence practice and zoo welfare for Great Britain (England, Wales, Scotland) that will come into force from 2027.

Zoo Licence Holders are ultimately *responsible* for a licensed zoo meeting the requirements of the SSSMZP (termed Standards of Modern Zoo Practice or SMZP from 2027), including all aspects of animal care and welfare. In reality, the expectation is that zoo operators (who are often the license holders themselves), keepers, staff and any persons for the time being in charge of the animals will meet these requirements too, in addition to the license holder. However, because there are so many different types and sizes of zoos, not all components of the SSSMZPs are relevant to all zoos equally.

Worldwide, there are various member organisations for zoological institutions, and these organisations each have animal welfare policies to which their members must adhere (Box 18.1). Membership of some of these organisations can be gained by paying an annual fee, whereas accreditation (in most cases) is gained by an institution satisfactorily fulfilling the husbandry, welfare, and other criteria set out by the membership organisation. In virtually all cases, a zoo will gain accredited member status only after the completion of a successful inspection by inspectors chosen and trained by the membership organisation. Accreditation provides a greater level of confidence that the institution is fulfilling the required criteria than membership, which can be gained without inspection in many cases.

BOX 18.1: Animal welfare policy requirements for members of various zoo associations

Specific animal welfare policies of regional organisations include:

British and Irish Association of Zoos and Aquariums - BIAZA (biaza.org.uk)
Members are required to have an institution-specific welfare policy (referring to the BIAZA Welfare Toolkit, 2019) as well as adhere to the standards required in the BIAZA Welfare Policy 2023 (BIAZA, 2023).

European Association of Zoos and Aquaria - EAZA (eaza.net)
Members are also required to ensure that the welfare of animals within their organisations adheres to their welfare policies, including the EAZA Standards for the Accommodation and Care of Animals in Zoos and Aquariums (AZA, 2024a).

Pan-African Association of Zoos and Aquaria - PAAZA (zoosafrica.com)
Aims for "Sustainable conservation through best welfare practices."

The Association of Zoos and Aquaria - AZA (aza.org)
North America has a Strategic Framework for the Well-being of Animals (AZA, 2024a) that incorporates a position statement consisting of five strategic priorities related to supporting their commitment to the well-being of animals. Further, AZA also requires its member zoos to adhere to the Association's Twelve Guiding Principles of Animal Wellbeing (AZA, 2024b).

The World Association of Zoos and Aquaria - WAZA (waza.org)
WAZA provides a zoo and aquarium animal welfare strategy (Mellor et al., 2015) and also states a Code of Ethics that it expects member zoos to adhere to. These members also need to adhere to their welfare evaluation process, which includes specific elements approved by WAZA (WAZA, 2023).

The Zoo and Aquarium Association Australasia (ZAA) (zooaquarium.org.au)
ZAA policy for its members states that the organisation "believes that all zoos and aquariums must ensure the animals under their care live a good life by promoting positive welfare to the highest possible standard for the duration of their lifetime" (ZAA, 2012).

The zoo associations (both accreditation and membership bodies) shown in Box 18.1 provide extensive policy frameworks relating to animal welfare, and adherence to them is not only an expectation but a requirement for membership or renewed accreditation. As zoo associations have existing (or are developing) accreditation processes that are designed to ensure that both existing and aspiring member zoos meet the association's standards, particularly in relation to zoo animal welfare, these association standards are increasingly aligned with methods for animal welfare assessment. The review of animal welfare policy can form a key part of an accreditation visit by the zoo association. The standards of these associations provide more detail on zoo animal welfare expectations, with a requirement for members to meet these higher standards, over and above any minimum requirements prescribed by government legislation. For example, in North America, AZA standards and policies (see https://www.aza.org/board-approved-policies-and-position-statements for more information) are much more stringent than the federal Animal Welfare Act, and in Britain and Ireland, the requirements for BIAZA membership aim to have the same stance.

Despite the number of zoo membership associations globally (as of 2024, EAZA is the largest association with more than 400 members in nearly 50 countries), there are many zoos that are not association members. In Britain and Ireland, for example, out of the more than 350 licensed zoos, just over 100 are members of BIAZA and 49 are members of EAZA (some zoos are members of both associations). For zoos that are not members of national or regional zoo organisations, underlying government policies and regulations aim to protect the welfare of zoo animals. These legally enforceable statutes are the principal protection afforded to the welfare of zoo animals, which highlights the importance of the key animal welfare role of policy such as the SSSMZP.

18.2 ANIMAL TRANSPORT

Beyond the welfare requirements for animals living within a zoo, the transport of animals between zoos, both within and between countries, carries potentially significant risks to animal welfare. Several different policies govern the transport of animals within and between countries, and international transportation can become extremely complex. For example, policies must be adhered to from the originating country, the destination country, and also for those through which an animal may be transported en route to their final destination.

BOX 18.2: Policies (Orders and Regulations) covering the movement of zoo animals in Great Britain:

- England Welfare of Animals (Transport) (England) Order 2006 (WATEO)
- Scotland Welfare of Animals (Transport) (Scotland) Regulations 2006 (as amended)

Wales Welfare of Animals (Transport) (Wales) Order 2007 (as amended)

The Orders and Regulations listed in Box 18.2 provide examples of how legislation aims to prevent the transport of animals in ways that may cause injury or unnecessary suffering and thus uphold minimum welfare standards. These documents also stipulate that animals are transported in the correct receptacles and under conditions (e.g. space, temperature, ventilation) appropriate for the species concerned. While these requirements aim to meet basic welfare requirements for zoo animals during transport, there is minimal provision and no specific policies for the psychological well-being of these animals. In addition to those policies listed in Box 18.2, the SSSMZP outlines a far more detailed set of requirements for the transport of zoo animals, dealing with the competence of relevant personnel, fitness and health of the animals, catching and methods of restraint, and the overall transport setting, which must comply with International Air Transport Association (IATA) regulations (see more information later in this chapter). While the policies described in Box 18.2 do not apply specifically to the transport of invertebrate animals (such as insects, crustaceans, cephalopods or molluscs), in England, WATEO provides a general duty of care towards invertebrate animals aimed at protecting them from injury or unnecessary suffering.

All these orders and regulations apply more broadly to anyone transporting live vertebrate animals for an economic activity. These policies consider both the transport of zoo animals and the movement of animals for conservation activities (for example, moving animals intended for release to the wild to their release site) to be economic activities. Further afield, within the European Union (EU), the Balai Directive (Article 4 of Council Directive 92/65/EEC) governs animal import and export between zoos in European Member States. This legislation covers various species of animals not covered by other legislation and does not apply to domestic animals (see https://www.gov.uk/government/publications/balai-directive-registration-of-premises-to-export-animals for more information). Premises need to be either registered or approved by the authorities to ensure that facilities are free from disease and that isolation and quarantine facilities are adequate for the movement of animals between institutions, thereby strengthening welfare requirements for the species held on the premises.

In order for zoos in Great Britain, Jersey and the Isle of Man to be able to export captive birds or ungulates to the EU or Northern Ireland, they must undergo inspection by the Animal and Plant Health Agency (APHA) and gain approval as Confined Establishments (defined as permanent premises such as zoos or laboratories approved to export captive-bred ungulates or birds to the EU or Northern Ireland). This approval includes ensuring the presence of adequate quarantine facilities, as well as appropriate levels of biosecurity, disease screening, animal observations and record keeping, all of which all link to ensuring good welfare states in those animals held in such areas prior to export.

The reality of Brexit in 2020, by which the UK left the EU single market and Customs Union (BBC, 2020), is the creation of numerous delays and difficulties associated with moving zoo animals between the UK and European collections. Data from Species360's Zoological Information Management System (ZIMS) (see https://species360.org for more information) show that before Brexit, approximately 1,400 animals registered on ZIMS were transported each year across the border (in both directions). In the year post-Brexit, the number of transfers dropped to just 57. This was in part due to a lack of existing paperwork for zoo species for both export and import into European countries and has led to a number of animal welfare issues for a variety of species, including:

- It has become difficult to move animals to form breeding pairs for the European Ex situ Programme (EEP) studbooks, and therefore the ability to breed endangered species and maintain genetic management is compromised.
- Lone animals have not been able to be transported, and their need to be housed with other members of their species is not realised.
- Animals that can no longer be provided with satisfactory housing are unable to be exported to collections better able to provide the requirements for their welfare.
- Some animals have been secured in a transport crate and transported to airports only to be turned around at the aircraft due to confusing paperwork. This has resulted in increased transit time, stress, and potentially unnecessary sedation for that individual animal.
- If paperwork has been deemed insufficient at an airport or border post and an animal needs to be returned to its point of origin, it may need to be transported more than once within a relatively short period, leading to unnecessary stress to that individual (and cost to the transporting collection).

Pre-Brexit, CITES-listed animals for whom road transport provided an overall improved welfare experience compared to air transport were often transported from the Republic of Ireland (ROI) and Northern Ireland (NI) through the UK to the EU. However, since Brexit, a NI or ROI zoo would need both a CITES import permit for the UK and then an export certificate to move from the UK to the EU, which they would not be able to obtain as they are not based in the UK (BIAZA, 2024). These animals from NI or ROI must therefore currently be transported using a method, such as air travel, which may be less conducive to upholding positive welfare states during movements and transportation.

Without harmonisation of UK and EU veterinary and animal health legislation, alongside increased compatibility of health certificates and approval of institutions, it is difficult to see how this situation will improve to provide pre-Brexit levels of animal transfers. For example, disease testing requirements before export for certain diseases and certain countries (e.g. rabies and tuberculosis testing for export to France) can lead to welfare concerns when animals are required to undergo repeated sedation for testing requirements (APHA, 2022) and/or extended periods of indoor quarantine.

A further issue compounding the difficulties in moving animals to Europe post-Brexit is the lack of Border Control Posts (BCP) open to Live Animals Other (LAO), which includes zoo animals, in both the UK and EU. These BCPs are designated and approved in line with GB legislation to carry out inspections of animals and animal products arriving at the GB border to protect animal and public health and animal welfare. This has welfare implications for those animals being transported that require inspection, as these animals may need to travel further distances, with an associated increase in transit time, to reach an appropriate BCP for inspection. Currently, some countries in Europe do not have an appropriately licensed centre within their borders, so any associated increases in transit times (required in order to travel to a BCP) may risk exceeding those stipulated within animal transport policies or the SSSMZP.

There are several regulations related to the transport of animals by air, set out by IATA. When moving animals, zoo operators in England, Scotland and Wales must comply with IATA's Live Animal Regulations (IATA, 2024). These Live Animal Regulations aim to ensure that all live animals transported by air are handled and transported in a way that prioritises their welfare and that they travel in a safe and healthy condition.

The IATA regulations include the type and size of crate that different species require. Although these regulations benefit the welfare of transported animals, occasionally problems can arise. For example, if there is an unexpected change to the type of aircraft an animal is to be transported on, this may mean that a certain size of crate can no longer fit inside the cargo hold. In these cases, the animal may reach the airport only for keepers to find that due to the cargo space issue, the animal is no longer able to be transported and must return to its point of origin, along with the associated welfare concerns that this may cause (e.g. isolation requirements, repeated travel and potentially repeated sedation).

18.3 ACQUISITION AND RELINQUISHMENT

In general, before any animal is acquired, there is a requirement for zoos to carry out due diligence regarding the source of the animal(s) and also that sufficient expertise and appropriate accommodation are in place at the receiving zoo. The due diligence process must ensure that the acquisition does not contribute to a poor welfare experience for the animals and that the associated animal movements are not contributing to illegal acquisitions.

18.3.1 CITES REQUIREMENTS

For all animal acquisitions, licensed zoos must comply with the requirements of CITES (the Convention on International Trade in Endangered Species of Wild Fauna and Flora), which is an international agreement between governments aimed at ensuring that the international trade in specimens of wild species (including zoo animals) does not threaten the survival of the species. CITES regulations govern the import, export, sale, and other commercial use, including display, of species listed in its Appendices and/or Annexes as applicable (CITES, 2023).

18.3.2 ACQUISITIONS

Before being subject to CITES regulations, zoos in the UK that are acquiring animals must comply with the SSSMZP relating to acquisitions, in particular animals from wild populations. Wild-caught acquisitions are not permitted under the UK's Zoo Licensing Act and the SSSMZP unless there is a clear justification (such as *ex situ* conservation management) for doing so. All acquisitions from the wild must follow requirements in the Standards that relate specifically to minimising stress and negative welfare outcomes for wild-caught animals, in addition to ensuring the acquisitions do not negatively affect wild populations. Furthermore, all zoo animal transfers into the UK must undergo due diligence to ensure that acquisitions from another country do not contribute to any practices that may contravene the SSSMZP.

Individual zoological collections may also have their own internal policies and standards related to acquisitions, specifically for taxa where it may be difficult to determine the exact source of stock in all instances. For example, where invertebrates such as butterflies (Lepidoptera) are sourced from dealers, it can be challenging to determine the exact primary source of all individuals. Where only a small number of zoos hold certain species, prevention of inbreeding may necessitate the acquisition of animals from private breeders or other non-zoo sources and can, in some cases, exacerbate the issue if records are not available of the original sourcing (or at a minimum, data confirming that animals are not wild caught). Wild-caught individuals may also carry diseases that are not commonly reported in captive individuals, and any lack of information regarding sourcing may pose a welfare risk to other individuals in the collection if quarantine procedures are insufficient for screening for these diseases.

18.3.3 DISPOSITIONS

Similar to that required for acquisitions, due diligence is needed to ensure that the party receiving animals being relinquished from a zoo has the necessary expertise and accommodation to maintain animal welfare. This is a requirement not only within the SSSMZP but also within the policies (see Box 18.1) by which members of zoo associations must abide. The sale or relinquishment of zoo animals into the private sector can be fraught with animal welfare concerns, given the relative lack of policy regulating private ownership of exotic animals. In the UK, to sell animals as pets (i.e. into the private sector), licensed zoos must hold pet-selling licences. These policies, which are all centred on animal welfare regulations, include the England Animal Welfare (Licensing of Activities Involving Animals) (England) Regulations 2018, Scotland Animal Welfare (Licensing of Activities Involving Animals) (Scotland) Regulations 2021, and the Wales Animal Welfare (Licensing of Activities Involving Animals) (Wales) Regulations 2021.

18.4 NOTIFIABLE DISEASES

Stringent biosecurity protocols and other key aspects of animal health (such as appropriate quarantine facilities, holding areas for new individuals, and isolation areas for sick or injured individuals) all provide a strong foundation for achieving overall good welfare inputs in the zoo. Aligning animal health protocols with welfare considerations can include having awareness of notifiable diseases and ensuring there is surveillance in place to protect against them. Notifiable diseases are those that represent a high risk of significant damage to human and/or livestock health and/or economic processes (Sanderson & Hartley, 2011). Such diseases are subject to national or international control policies. In the UK, the Animal Health Act (1981), or an Order made under this Act, lists all notifiable diseases. It is a legal requirement to report notifiable diseases to the appropriate government agency, even when there is only a suspicion of their presence and before the confirmation of any test results. Notifiable diseases fall under one of two categories: endemic (those already existing in a country) or exotic (those not normally found in a country). Policies related to the management of notifiable diseases vary between countries. In Great Britain, the Department for Environment, Food and Rural Affairs (Defra) and APHA detail the policies and procedures that must be followed if a notifiable disease is found in zoo animals and how it must be reported (APHA, 2019).

The management of both suspected and confirmed notifiable disease outbreaks under the policies described above can have welfare implications for both the animals infected with a disease and those found to have been in contact with infected animals. APHA works to contain the disease on the premises, to limit any further spread from the premises, and to investigate where the disease originated. The agency also conducts monitoring to ascertain any further spread of the disease and eventually determine that it has been eradicated. Biosecurity measures will be increased in both scenarios, and in certain cases, this may lead to welfare concerns if animals are confined in smaller areas and therefore at greater stocking density than would usually be the case. In the case of species such as wildfowl (Anseriformes), animals may not have access to swimming water, which can lead to health issues over time – for example, lameness in diving ducks that should not spend long periods out of the water.

Depending on the disease and species concerned, testing will usually take place if the relevant veterinary personnel suspect disease presence. This testing can involve taking a variety of samples depending on the disease suspected, and in certain cases (such as diseases in invertebrates or fish) may require some animals to be euthanised to enable appropriate sampling. Catching and handling animals for sample collection is stressful, and the risk of compromising welfare will almost certainly increase where there is a need for large numbers of samples, or if people who lack sufficient familiarity with the species concerned carry out the handling. For species in the zoo, disease testing can be particularly stressful if animals are not used to being handled or if sedation is required for samples to be taken.

Upon confirmation of a notifiable disease, control strategies may include isolation and/or depopulation. Although euthanasia is not in itself a welfare concern when carried out by qualified and experienced personnel, the stress involved with multiple handling events for individual animals can adversely affect animal welfare. The method of euthanasia employed when a number of animals need to be euthanised at the same time can also have welfare implications, depending on species and circumstances. Some euthanasia methods employed for some farmed species, for example gassing, would not be acceptable in the zoo.

Movement restrictions around infected premises are an additional control measure in the event of a confirmed notifiable disease. These restrictions can have welfare implications if the inability to move animals off the zoo premises leads to overcrowding or the formation of undesirable group structures. A variety of contraceptive methods may also need to be employed depending on the species, and these can have a welfare effect both by the need for anaesthetics in some species to deploy implants or by increasing the risk of certain reproductive diseases such as pyometra (an infection of the uterus) in some individuals. This may also affect the ability of certain individuals of conservation importance to pass on their genetics to future members of their species (Sayers, 2018).

In the case of certain notifiable diseases, vaccines are available for the commercial animal sectors. In some cases, these vaccines can also be used in zoo species; however, the need for further catching and handling events (as most vaccines require repeat administration) must be balanced against the level of protection they may provide for species in which their effectiveness has not been tested. Housing in biosecure units during periods of heightened disease risk may be considered preferable for certain species when compared to the need for repeated catching for vaccination (e.g. for large flocks of flamingos, Phoenicopteriformes).

In the EU, the Transmissible Diseases Handbook produced by the European Association of Zoo and Wildlife Vets and EAZA (EAZA/EAZW, 2024) provides further information regarding policies related to the management of diseases. In North America, similar guidance exists in the Infectious Disease Manual produced by the American Association of Zoo Veterinarians Animal Health and Welfare Committee (AAZV, 2023). If culling of animals in the zoo is necessary, then this can also have a welfare effect on the species overall. Zoos often house individuals of genetic importance for conservation breeding programmes, and the loss of such individuals that may be genetically important or underrepresented can have negative effects on the genetic management of a species and potentially compromise long-term population management goals.

18.5 CONCLUSIONS AND FINAL THOUGHTS

The standards and policies of member organisations in various regions of the globe aim to increase the welfare standards in zoological collections. However, not every zoological collection is a member of these organisations, and therefore it is the legal policies and frameworks of individual countries that must include sufficient minimum welfare standards for all zoological taxa and the provision for the inspection and legal enforcement of these standards.

In Great Britain, the zoo licensing process is realised and implemented through the SSSMZP. After several previous iterations and a revised version in review at the time of writing (November 2024), these standards represent the practical application of policy that underpins animal welfare standards in zoos across Great Britain. The current standards base their welfare provision on the "Five Freedoms" that were originally constructed for farm animals, which are now represented as the "Five Welfare Needs" of zoo animals. Although the standards provide the baseline of zoo animal welfare requirements, the nuances required for individuals and monitoring of welfare within the collection are addressed in detail only for certain species (e.g. elephants, Proboscidae). In the renewed standards, welfare must remain at the forefront of the requirements for licensed zoos, and evidence-based standards of zoo animal care must be the norm moving forward.

Although an evidence base should always inform policy, such evidence is not always available in cases where there is no research into the welfare of certain taxa or existing research does not provide sufficient detail. Moving forward, the production of more species-specific methods of welfare assessment will help to inform policies aimed at improving welfare across all taxa.

Fundamentally, despite policy providing the basis for zoo animal welfare, the translation of robust policies into optimum animal welfare relies heavily on the effectiveness of the inspection process and subsequent regulatory enforcement. Even after a zoo license inspection is complete, in Great Britain, the onus for subsequent adherence to the Standards and/or implementing directions identified during the inspection falls to the relevant local authority. There can be complexities involved in the process of inspection and subsequent enforcement, which require training and experience for both zoo inspectors and especially local authority personnel with responsibility for licensed zoos.

Where only one or two licensed zoo premises exist within the jurisdiction of a local authority, the available knowledge, training and experience for effective post-inspection and interim inspection monitoring of the zoo premises may be limited. Furthermore, if the small number of licensed zoos within a local authority's jurisdiction contains specialist taxa, there may be a need for additional training of the relevant local authority personnel responsible for zoo inspections and zoo animal welfare assessment. Such inspections and assessments are particularly relevant for "informal" yearly inspections where only local authority personnel are present, interim inspections (midway through a zoo licence period) and for smaller zoos with 14(2) dispensations under the UK's Zoo Licensing Act (a zoo with a 14(2) dispensation is one that has fewer than 50 hazardous and/or conservation-sensitive species in the collection). "Conservation-sensitive" relates to species categorised as VU or worse on the IUCN Red List and "hazardous" are Category 2 species or above according to the SSSMZP. The main distinction for inspection purposes is that only one Secretary of State Inspector is required here, rather than the two needed for non-dispensation zoos.

Despite the wide range of orders, regulations and policies, the alignment of policy to optimum animal welfare outcomes remains a moving target. Obstacles to implementation result from a complex and diverse operational landscape, while new developments in methods of animal husbandry and management may mean that a new policy may start becoming out of date from the moment of its production, and before even its implementation. As we progress further into the 21st century, the urgency and need for integrating the most up-to-date research and innovation into zoo animal welfare policy remains as challenging as ever.

REFERENCES

AAZV – American Association of Zoo Vets. (2023). Infectious disease manual. https://www.aazv.org/page/754 [Accessed September 2024]

APHA – Animal and Plant Health Agency. (2019). Notifiable diseases in animals. https://www.gov.uk/government/collections/notifiable-diseases-in-animals [Accessed August 2024]

APHA. (2022). Health certificate for the import into, and transit within France and French overseas departments for non-domestic carnivores: certificate 8651. https://www.gov.uk/export-https://www.gov.uk/export-health-certificates/health-certificate-for-the-import-into-and-transit-within-france-and-french-overseas-departments-for-non-domestic-carnivores-certificate-8651 [Accessed August 2024]

Animal Health Act. (1981). c.22. https://www.legislation.gov.uk/ukpga/1981/22 [Accessed August 2024]

AZA. (2024a). Strategic framework for the wellbeing of animals. The Association of Zoos and Aquariums. https://assets.speakcdn.com/assets/2332/aza_animal-wellbeing-strategy.pdf [Accessed August 2024]

AZA. (2024b). The guiding principles of AZA animal wellbeing culture. The Association of Zoos and Aquariums. https://www.aza.org/assets/2332/aza_animal_wellbeing_guiding_principles.pdf [Accessed August 2024]

BBC. (2020). Brexit: What you need to know about the UK leaving the EU. https://www.bbc.co.uk/news/uk-politics-32810887 [Accessed November 20204]

BIAZA. (2023). BIAZA Welfare Policy. https://biaza.org.uk/policies-guidelines [Accessed September 2024]

BIAZA. (2024). BIAZA cross-border animal transfers update January 2024. https://biaza.org.uk/members/ [Accessed August 2024]

CDC. (2015). Definition of policy. https://www.cdc.gov/policy/paeo/process/definition.html [Accessed August 2024)

CITES. (2023). CITES appendices I, II & III. https://cites.org/eng/app/appendices.php [Accessed April 2024].

EAZA/EAZWV – European Association of Zoos and Aquaria/European Association of Zoo and Wildlife Veterinarians. (2024). EU animal health law and confined establishments. https://www.eazwv.org/page/inf_handbook [Accessed August 2024].

IATA - International Air Transport Association. (2024). Live Animals Regulations (LAR). https://www.iata.org/en/publications/store/live-animals-regulations [Accessed April 2024].

Mellor, D. J., Hunt, S. & Gusset, M. (Eds). (2015). *Caring for Wildlife: The World Zoo and Aquarium Animal Welfare Strategy.* Gland, Switzerland: WAZA.

Sanderson, S. & Hartley, M. (2011). Disease contingency planning for zoos. https://www.eazwv.org/page/resources [Accessed October 2024]

Sayers, G. (2018). Help! We're under TB restrictions; what now? *Proceedings of the British Veterinary Zoological Society,* p. 52. https://www.cabidigitallibrary.org/doi/pdf/10.5555/20193073408

WAZA. (2023). WAZA's approach to animal welfare. https://www.waza.org/priorities/waza-code-of-ethics/our-approach-to-animal-welfare/ [Accessed September 2024]

ZAA. (2012). Zoo and Aquarium Association Australasia Animal Welfare Policy. https://zooaquarium.org.au/common/Uploaded%20files/Policies/ZAA_Policy_Animal_Welfare.pdf [Accessed September 2024]

19

From mantids to mantis shrimp – what about invertebrate welfare?

SARAH PELLET AND STEVEN A. TRIM

19.1 INTRODUCTION

Invertebrates have a fundamental role in the world's ecosystems, as they comprise over 90% of extant species. However, this is inversely proportional to the importance given to them in most zoological collections, and our understanding of their welfare needs is less than that of other taxa. This is exemplified by the absence of the majority of invertebrates from legislation and even several legal definitions of animals. As with all animals in captivity, invertebrates need opportunities to thrive and the ability to self-maintain, being provided with the right environment to allow choice and control and the expression of species-specific behaviours (Greggor et al., 2018). What differs for invertebrates, however, is our understanding of their biology, especially their effective state. This chapter will focus on the health and welfare of captive invertebrates while highlighting important husbandry practices for example species and discussing areas of potential research that are essential to further our understanding of their requirements.

19.2 BIOLOGICAL FUNCTIONING

19.2.1 HEALTH STATUS: STRUGGLING, SURVIVING, SUBSISTING, THRIVING

Invertebrates in collections are used as display animals, often in interactive exhibits, and they can perform a critical role in species conservation, such as the *Partula* snail (Zachariah, 2022). In a collection there may be a wide range of invertebrate species from Cnidaria and Crustacea to

DOI: 10.1201/9781003429425-19

Insecta, Lepidoptera, and Chelicerata. For assessment of invertebrate welfare within a collection, animal-based indicators, such as food consumption, growth, and mobility, can be assessed. In addition to this, resource-based indicators can include food offered, provision of water quality, enclosure size, and space that can be utilised. There is no single pointer to assess welfare, and multiple factors should be used, continuously monitored and reviewed. Historically, for most species, there have been minimal preventative health measures performed other than meticulous care to the environment, husbandry, optimum food quality, preparation, supplementation, and removal of all contaminants. Unlike many vertebrate species, there are few published husbandry guidelines, but some do exist, such as the *Guidelines for the Care and Welfare of Cephalopods in Research* (Fiorito et al., 2015). For cephalopods, indicators such as skin colour, skin texture, and integrity can be used to help determine welfare state. Other species' reactions can be clearer; most New World tarantulas (Theraphosidae) can flick urticating hairs from their opisthosoma in response to a perceived threat. This will result in observed alopecia on the opisthosoma with pink cuticle, indicating that a stressful event has occurred which may need addressing. Conversely, opisthosomal alopecia with a darkened colour is likely to be pre-ecdysis, or shedding, as newly formed hairs are underlying the cuticle (Pizzi & Kennedy, 2022), both illustrated in Figure 19.1. This example highlights the importance of being able to identify subtle differences to understand the stress responses and, subsequently, the welfare state of many invertebrate species.

19.2.2 ENVIRONMENTAL FACTORS

Before housing, the thermal range for the individual species should be identified and provided for within the enclosure, using a maximum-minimum thermometer for close monitoring of environmental temperatures. Temperature extremes should be avoided, and a temperature gradient across the enclosure should be provided to allow thermoregulation as per species-specific guidelines. Appropriate use of full-spectrum lighting (including ultraviolet) should be provided depending on the species, but not all invertebrates will need this. For example, tarantulas are photophobic and are currently understood not to require additional lighting, which has been shown to induce stress-like behaviour and physiology. King baboon spiders (*Pelinobius muticus*) exposed to full-spectrum lighting showed no difference in burrowing, but significantly more webbing was observed (Somerville et al., 2017). In the same study, cortisol levels were significantly raised after exposure to full-spectrum lighting, providing evidence of detectable cortisol in arachnid haemolymph and changes in concentration in response to environmental factors. In contrast, some true spiders such as *Nephila edulis*, are heliophilous (Liebsch et al., 2020).

19.2.3 GROWTH AND REPRODUCTION

Traditionally, reproductive success measured at the population level was considered the key indicator of animal welfare for captive animals, assuming the welfare to be good if there was longevity

Figure 19.1 Alopecia in a tarantula (Theraphosidae) in response to repeated adverse events (left) versus alopecia in a tarantula prior to ecdysis (pre-moult) (right).

and successful production of offspring per the animals' breeding lifetime. There has been a change in focus from population-level welfare parameters such as reproductive success and longevity to the subjective experience of an individual (DiVincenti et al., 2023). There may at times be discrepancies between individual animal welfare and population welfare in zoos and aquariums, and this is not unusual in invertebrate keeping, where often there are many individuals kept within the enclosure. Invertebrates should be provided with the means necessary to thrive, being able to grow and successfully reproduce. Institutions holding invertebrates can encounter unique challenges in balancing individual welfare and population welfare when faced with breeding, animal acquisition, and conservation aims. Using isopods as an example, some species can be bred in captivity effortlessly, such as *Porcellio scaber*, provided they have the appropriate set up. Other species are more challenging, such as *Merulanella* spp. or *Cubaris* spp. There are many welfare challenges with collecting wild animals for display, but breeding in captivity for many species has been a priority for captive institutions, and recent successes have led to the potential for acquisition of animals through captive breeding programs.

19.3 AFFECTIVE STATE

19.3.1 EMOTION AND MOOD-BASED EXPERIENCES

It is a recent concept to consider that invertebrates may display some form of emotion, as historically the general consensus has been that invertebrates do not have the neural requirements to do this. Defining emotions as temporary states, to include subjective, cognitive, behavioural, and physiological phenomena that are triggered by environmental stimuli, is often associated with human experience. Research on invertebrates has found that there is accumulating evidence in favour of cognitive abilities, showing that sea slugs (Aplysiidae), bees (Apidae), crayfish (Astacidea), pond snails (Lymnaeidae), crabs (Brachyura), ants (Formicidae), and fruit flies (Drosophilidae) display many cognitive, behavioural, and/or physiological signs that indicate internal states suggestive of emotions (Perry & Baciadonna, 2017). Cassill et al. (2016) discussed the possibility of positive and pleasurable emotions in fire ants (*Solenopsis invicta*), where they observed them moving their abdomens up and down at around 45° in response to attending to brood and consuming sugar water. It has been suggested that this behaviour may be similar to facial expressions and body postures indicating happiness in humans and other mammals, though more research is required to validate these claims. There is divided opinion as to whether to describe what is seen in invertebrates as 'emotions', particularly as we are unable to directly measure their subjective component. As we are unaware of its neurological basis, we cannot dismiss the idea that invertebrates have some basic form of subjective experience, and we must take this into account when considering invertebrate welfare.

19.3.2 PAIN AND FEAR

Considering the feeling of pain/nociception, a lack of evidence to support this does not mean that invertebrates are incapable of experiencing it, particularly when it is difficult to distinguish between responses of nociception, pain, and stress. It is therefore prudent to assume that invertebrates could possibly feel pain, and we need to make provisions to alleviate this. Studies have been performed on decapod crustaceans and cephalopods, but more research is essential across the invertebrate taxa. The question still asked is whether invertebrates experience pain or are simply just demonstrating a reflexive response to a noxious stimulus. Defining the ability to perceive pain across many species, especially in invertebrates, is challenging. Some taxa lack a defined nervous system, but others, such as cephalopods and arthropods, have well-developed nervous systems and brains with complex behaviours, including the ability to respond to environmental cues (Tarsitano & Jackson, 1997). Some invertebrate species also respond to exogenous opioids with opioid systems having a functional role in nociception (Nathaniel et al., 2010; Lewbart, 2012). Furthermore, Sladky (2014) reported that tarantulas respond to noxious thermal stimuli similar to vertebrates, by withdrawing and rubbing their limb if a hypodermic needle is inserted into the exoskeleton. Opioid administration (morphine sulphate 50–100 -mg/

kg intracoelomically; butorphanol tartrate administered at 20 mg/kg intracoelomically) minimised responses to noxious stimuli. In another study, shore crabs (*Carcinus maenas*) were shown to actively avoid electrical shocks even if this means breaking cover, suggesting that they may be perceiving pain (Magee & Elwood, 2013). Appropriate anaesthesia and analgesia should therefore be considered when treating invertebrates within collections to prevent any response to noxious stimuli, though hypothermia is not considered appropriate for analgesia as it merely slows responses (Pizzi & Kennedy, 2022).

Studies in the sea slug (*Aplysia californica*), the fruit fly (*Drosophila melanogaster*) and the nematode (*Caenorhabditis elegans*) have shown that the basic mechanisms of learning and memory are conserved. Fear-based learning, using electric shock in both *Drosophila* and *Aplysia*, demonstrated that short term changes in behaviour are elicited because of serotonin- or dopamine-relaying aversive stimuli which leads to changes in intracellular calcium levels and neurotransmitter release (Pribadi et al., 2022). It is commonly accepted that fear-related behaviours can be observed in predator-prey relationships, such as using the scent of cats in rodent studies. Certain invertebrate species have also demonstrated fear learning; for example, *Aplysia californica* when attacked and bitten by the spiny lobster (*Panulirus interruptus*) multiple times, will have long-term sensitisation of the withdrawal reflex of both head and tail-mantle which lasts for at least 24 hours. It is therefore essential to reflect that invertebrates may be capable of both pain and fear, but more research is needed to further investigate the extent of this.

19.3.3 DEMEANOUR AND CONTENTMENT

Invertebrates are often overlooked when considering demeanour and contentment. This welfare indicator should not be discounted and should be considered in many invertebrate settings. Studies have shown that some species of invertebrates may display cognitive capacities as well as having individual profiles, which is similar to the personalities seen in vertebrates (Horvath et al., 2013). Theraphosids, for example, have been shown to benefit from behavioural enrichment and that web construction is a sign of the spiders' well-being in captivity (Bennie et al., 2011). A pioneering MSc thesis by Blenderman (2020) has already determined that Mantis shrimps (Stomatopods) exhibit individual variation in 16 traits, especially boldness and exploration, demonstrating personality.

19.4 HOUSING AND HUSBANDRY

19.4.1 REPLICATING THE NATURAL ENVIRONMENT

Although often neglected in conversations surrounding the importance of the captive environment for zoo animals, progress has thankfully been made in invertebrate housing provisions. As with other taxa such as mammals and birds, invertebrate enclosures should allow opportunities to exhibit natural behaviours, thus providing space to move, natural materials to interact with, and opportunities for socialisation depending on the species. Positive examples include millipedes (Diplopoda) and beetles (Coleoptera) now widely being given climbing and basking opportunities. Arboreal Theraphosid spiders that are maintained in enclosures with enrichment, including artificial plants and vertical hides, display positive behaviours compared with their counterparts housed in un-enriched environments, who were more aggressive and showed a stronger reaction to a perceived threat (Bennie et al., 2011). For nocturnal species such as scorpions (Scorpionoidea), we can also allow caregivers better husbandry opportunities by housing these in reverse lighting environments, which will consequently also allow for better educational opportunities for visitors.

Enclosure size and design usually vary, and individual species' requirements must be taken into consideration. Enclosures should be escape-proof and made from materials such as acrylic, clear plastic or glass to allow sufficient cleaning but also adequate ventilation. Ventilation holes must be smaller than the least compressible part of the animal, such as the prosoma in spiders, to prevent escape, and ideally smaller than the leg diameter to prevent entrapment of invertebrate limbs. Some species, such as terrestrial tarantulas, have scopulae on the plantar aspects of the tarsus and metatarsus,

Figure 19.2 Leaf-cutter ants (*Atta cephalotes*) using a rope walkway to move between food resources and their colony.

plus two to three tarsal claws, common in many arthropods. If these claws get caught in mesh-top tanks, then autotomy of the limb may occur. That said, mesh in these enclosures can also be beneficial for some species, as mantids (Mantidae) and phasmids (Phasmatidae) will use this to assist with ecdysis.

As zoo enclosures continue to evolve, some progress is being made towards larger, open enclosures with minimal barriers between them and the public. Some invertebrate species can be housed successfully in such spaces, with leaf-cutter ants (*Atta cephalotes*) being a perfect example. These animals can be provided with suspended walkways over relatively large areas, using rope or branches, allowing them to move between different food sources and their colony, as they would do in their natural environment (Figure 19.2).

19.4.2 INVERTEBRATE HANDLING: AN UNNATURAL INTERACTION

Wild-living invertebrates will likely perceive being picked up by another species as the prelude to predation, so we need to consider how these animals may experience handling by humans, bearing in mind the potentials for learning and acclimation as discussed above. Many collections offer interactive guest experiences to engage in public interest for the species and conservation efforts, by using ambassador animals or providing facilities where people can interact with animals, often in educational handling settings or allowing visitors access to touch-pools. While this provides greater public engagement and educational experiences which could ultimately benefit the species, it is essential to provide the choice for the animals to participate in such interactions. Providing space for animals to move out of reach of handling and designing in features to the enclosure such as barriers to protect against visual and auditory stimuli is essential.

Handling guidelines recommending invertebrate species for use in education have been produced by the British and Irish Association of Zoos and Aquariums (BIAZA) Terrestrial Invertebrate Working Group (BIAZA, 2014). With an emphasis on animal welfare, these guidelines suggest more suitable species to use in handling interactions (see Table 19.1). Millipedes may be used in certain situations, though over-handling can be stressful or even fatal, so it is essential to ensure rotation of animals. These animals can also excrete a defensive

Table 19.1 Species-specific welfare considerations

Species	Species specific information	Welfare considerations
Malagasy hissing cockroach (*Gromphadorhina* spp.).	Very good for handling. Rarely bites.	Avoid using cockroaches that excessively hiss, as this may be a sign of stress.
Thorny stick insect (*Aretaon asperrimus*). Macleay's spectre (*Extatasoma tiaratum*). Giant spiny stick insect (*Eurycantha calcarata*) (female).	Good for handling. Good for showing individuals attached to a branch. Do not use male giant spiny stick insects as animals have large spines on legs.	Allow them to remain attached to branches if possible, during a demonstration. If they do need to be handled, be patient when moving them and allow them to walk onto a hand. Avoid causing trauma to limbs by picking them up from a branch.
Giant African land snails (*Achatina/Archacatina* spp.).	Good for supervised handling by people with wet hands or gloves.	Avoid touching tentacles as this may cause stress. Do not forcibly pick up snails if they are attached onto a smooth surface as this can hurt them.
Giant African millipede (*Archispirostreptus gigas*). Tanzanian red-legged millipede (*Epibolus pulchripes*). Chocolate millipede (*Ophistreptus guineensis*).	They rarely bite and are often docile.	Over handling can be fatal. They can excrete a defensive fluid. Avoid digging up millipedes as they may be in the process of ecdysis. If a millipede is curled into a ball, do not use as this is a defensive posture.

fluid, presumably as a reaction to perceived predation, so care must be taken. This has been reported in Madagascar, where wild millipedes produce a toxic secretion as a response to being picked up and bitten by red-fronted lemurs (*Eulemur rufifrons*), while the lemurs anoint themselves with the secretion to use it as a natural pesticide (Peckre et al., 2018). This toxin secretion is likely an anti-predatory response, suggesting that handling is stressful for this species.

Handling of arachnids is not advised due to the risk of injury to the animal, including autotomy through rough handling, getting caught in clothing or being dropped. Approximately 90% of new world tarantula species can kick off urticating hairs, which are mostly distributed over the dorsal opisthosoma, in response to a perceived threat or stress. These can cause skin and corneal irritation. Despite this, many institutions still offer sessions to overcome arachnophobia, and it is worth considering the potential benefits of a possible increased empathy for these species as a result. Spiders must however never be handled during a moult including the pre- to post-moult periods, due to their delicate cuticle. It takes 16 to 20 days before the new cuticle reaches maximum strength (Pizzi & Kennedy, 2022).

Any individual invertebrate should not be handled more than five times over a seven-day period, not more than once a day and less than 20 minutes under direct supervision with trained staff. Handlers must wash and dry their hands before handling; nicotine was once used as an insecticide and is harmful to invertebrates, therefore smokers may pose a risk to the animals. Ideally gloves should be worn to handle invertebrates to protect animals from nicotine, cosmetics and medication exposure (see Figure 19.3). Care must be taken if using powdered gloves, as the powder can affect respiration if the spiracles are blocked. It should also be noted that powdered gloves can also be an allergen risk to the wearer. All species must be kept in suitable climatic conditions and any species taken for outreach must be transported in

Figure 19.3 Correct handling of a giant African land snail (*Achatina fulica*).

appropriate temporary enclosures,thereby ensuring all welfare requirements are met.

Aquatic invertebrate inhabitants in touch pools often include whelks (*Busycotypus canaliculatus, Busycon carica, Buccinum undatum*), hermit crabs (*Pagurus* spp.), Atlantic horseshoe crabs (*Limulus polyphemus*), limpets (*Lottia* spp.) and periwinkle snails (*Littorina littorea*) (Hadfield & Tuxbury, 2019). Animals must be provided with suitable areas to retreat and be out of reach from guests and ideally out of sight. Provision of suitable enclosures off show has been demonstrated to improve animal welfare and health; therefore, animals should be rotated to allow time away from public interactions. Water systems must be a consideration for this, as having off show holding enclosures on the same water system will allow transit without acclimatisation concerns. The disadvantage of this, however, is that should there be a contamination in the public area, it will also affect the off-show enclosures. Animal health and welfare must be assessed continuously, comparing with data and observations from conspecifics in display-only enclosures.

19.4.3 REPLICATING A NATURAL DIET

Correct nutrition and water must be provided depending on the species. Historically, the use of cotton wool or sponges was used in water dishes of terrestrial invertebrates to prevent prey from drowning and to allow access to water. However, the use of water-soaked sponges should be avoided as these may act as an unhygienic substrate for bacterial and fungal growth. Herbivorous invertebrates must be transitioned onto available food plants, and their appetite monitored carefully. Thus, it must be remembered that many herbivores in collections are not eating a comparable diet to their wild counterparts. All vegetation must be thoroughly washed to ensure absence of insecticides or pesticides. Predators, including mantids and tarantulas, require live food as their natural diet, and it is an important enrichment for many species. Whether herbivorous or carnivorous, the body condition of all animals should be monitored carefully, and feeding frequency adjusted to this; for example, overweight tarantulas have a distended opisthosoma which they drag on the substrate and thus can suffer abrasions.

While considering the diets for captive invertebrates, we must consider that some species exist in captivity primarily to be used as live food, such as locusts (*Schistocerca gregaria*), mealworms, the larval form of *Tenebrio molitor*, Dubia roaches (*Blaptica dubia*) and crickets (*Acheta domesticus* or *Gryllus bimaculatus*). As with all other species, the same possibility for suffering still exists and we have an ethical responsibility to take their welfare into consideration (Keller, 2017). Although these live feed animals provide food for other species and often provide enrichment benefits, it is also recommended to take the welfare of the live invertebrate prey into consideration. Provision of optimum husbandry is as important here as it is for any other captive animal in the zoo. This includes maintaining a suitable environment for the species' needs and ensuring the environment is kept clean and free from faecal accumulation and mould. The enclosure should be regularly cleaned and contain provision of hides or areas to climb (depending on the species) to allow demonstration of their natural behaviour. For terrestrial species, cardboard egg trays/boxes can be used to increase the surface area of an enclosure and can easily be disposed of and replaced; however, more naturalistic enclosures are encouraged (see Figure 19.4). While the majority of live-fed invertebrates are used for species in terrestrial enclosures, the same principles should be applied for aquatic live food. Should caregivers

Figure 19.4 A naturalistic enclosure for feeder locusts (*Schistocerca gregaria*).

expect that live-fed invertebrates may remain in an enclosure for any length of time, provision of food and water is essential for these animals (not just the 'primary' species), ensuring multiple feeding stations are available (Cooper & Williams, 2014).

19.4.4 ABILITY TO EXHIBIT WILD BEHAVIOURS, WITH OR WITHOUT OTHER ANIMALS

All species should be housed as they would in their natural habitat. For example, free-living tarantulas are not generally social and, in captivity, should be housed on their own (Marnell, 2016). Yet, Foelix (1996) reported that there were approximately 20 species of spiders within the *Poecilotheria* genus living together socially in the wild. In theory, these species could be maintained together in captivity if raised together as spiderlings. Caregivers should be mindful, however that cannibalism often occurs, as it probably does in the wild, so individual housing is still advised in captivity. To meet the potential need to be housed with their own species, some invertebrates may need symbiotic species, and all need protection from chronic predator stress. We are a considerable way off understanding the invertebrates' perception of their environment, but we these factors should be considered as they usually are for vertebrates.

In addition to housing with or without conspecifics or other species, consideration must also be given to the ability for roaming and/or flight for animals kept in collections compared with free-living counterparts, and their resulting ability to exhibit normal behaviour. One example of this is Lepidoptera in butterfly houses. As with most species, butterflies need space, appropriate temperature, humidity and ventilation. Further research is required but, as a general rule, one adult butterfly per square metre is recommended and the ceiling height for netting should ideally be 3.0 metres (Cooper & Gordon, 2022). This allows space for flight and allow butterflies to exhibit natural behaviours as far as possible. It should be noted that some butterfly species migrate and, therefore, the captive environment would restrict this wild behaviour.

19.4.5 PROTECTION FROM PAIN, SUFFERING, INJURY AND DISEASE

As with other taxa, it is important to be mindful of pain and fear-based learning when considering the welfare needs of invertebrates, particularly when it comes to euthanasia. Invertebrates are becoming increasingly popular in zoological collections and, as such, there is a growing need to understand and document ethically acceptable guidelines for humane euthanasia for these taxa. The methods used to perform euthanasia must be irreversible, and should minimise pain, distress and suffering. Euthanasia is suggested if the animal would be suffering if kept alive. Reasons for welfare-derived euthanasia for captive invertebrates could be severe dysecdysis, for example, commonly seen in phasmids and mantids, or a disease that cannot be resolved. In the UK, most invertebrates are not governed by the same legislation as vertebrate species, but humane treatment is important on welfare and ethical grounds. The Animal Welfare (Sentience) Act 2022 in the UK does recognise some species (lobsters, octopus, crabs, and other decapod crustaceans) as sentient beings, meaning these animals are protected by law.

Recommended methods of euthanasia are based on a two-step method in which, firstly, anaesthesia or chemical means are administered, followed by a secondary physical (such as freezing or pithing) or chemical (hyperkalosis, or immersion in 70% alcohol or 10% formalin) method. The first step must be insensible, and the second step must be terminal, with destruction of the brain or major ganglia. Freezing or submersion into a preservative is not an acceptable method as a first step for euthanasia. Hypoxic death of aquatic invertebrates by removing them from water is also not acceptable, and anaesthesia within the water (correctly aerated and buffered appropriately) must be considered for aquatic species.

If the reason for euthanasia is for disposal of surplus stock, then it may be appropriate to produce a protocol for the disposal of excess ova, or prevention of breeding for population management as a preventative measure, rather than relying on euthanasia. With this, euthanasia can be an important management tool if used to prevent potential suffering through inadequate care due to population levels exceeding maximum capacity. Bear in mind that determining death in some invertebrate species may be challenging, however, and reviews on the methods of how to perform euthanasia is beyond the scope of this chapter. For more information on this topic, see further recommended reading listed in Box 19.1.

BOX 19.1: Useful sources on invertebrate species' husbandry needs and welfare requirements

- AVMA *Guidelines for the Euthanasia of Animals* (2020). https://www.avma.org/sites/default/files/2020-02/Guidelines-on-Euthanasia-2020.pdf Accessed 03/08/2024
- Bennie, N. A. C, Loaring, C. D., Bennie, M. M. G. & Trim, S. A. (2012). An effective method for terrestrial arthropod euthanasia. *Journal of Experimental Biology, 215*, 4237–4241.
- Braun, M. E., Heatley, J. J. & Chitty, J. (2006). Clinical techniques of invertebrates. *Veterinary Clinics Exotic Animal Practice, 9*(2), 205–221.
- Brereton, J. E. (2022). The behavioural biology of invertebrates. In Rose, P. E. (Ed.), *The behavioural biology of zoo animals*. Boca Raton, FL: CRC Press.
- Cooper, J. E. (2004). Invertebrate care. *Veterinary Clinics of North America – Exotic Animal Practice, 7*(2), 473–486.
- Cooper, J. E. (2011). Anaesthesia, analgesia, and euthanasia of invertebrates. *Institute for Laboratory Animal Research, 52*, 196–204.
- De Voe, R. S. (2009). Captive invertebrate nutrition. *Veterinary Clinics of North America: Exotic Animal Practice, 12*, 349–360.
- Lewbart, G.A. (2021).*Invertebrate Medicine*(3rd ed.). Hoboken, NJ: Wiley Blackwell.
- Mather, J. A. (2001). Animal suffering: an invertebrate perspective. *Journal of Applied Animal Welfare Science, 4*(2), 151–156.

19.5 CONCLUSIONS AND FUTURE CONSIDERATIONS

Large numbers of invertebrate species are kept in both private and zoological collections and, as caretakers of these animals, we have an ethical obligation to ensure good welfare. Institutions have a responsibility for providing the best possible care to any animal, which should include the necessary space, suitable nutrition, and opportunities for both mental and physical stimulation. Further research is needed in these areas and should be broadened to include taxa largely representative of those in zoological collections. There are valuable opportunities for research questions in areas that we are even yet to ask, providing answers which would undoubtedly not only contribute to better lives for these animals in captivity, but at the same time increase public awareness about the importance of protecting invertebrates and their habitats (Box 19.2).

BOX 19.2: Invertebrate welfare: 'Striving for Best Practice'

- Multiple factors of invertebrate welfare should be used, continuously monitored and reviewed. We are still learning the true welfare needs of many species.
- Optimum husbandry, to include species-specific temperature ranges, photoperiod, lighting and high-quality nutrition is essential.
- Provision of veterinary care both for diagnosis, treatment and humane euthanasia is growing.
- As information on pain and nociception is expanding, it has to be assumed that invertebrates may be capable of both pain and fear. We have a duty of care to address this and minimise suffering.
- More research is essential across the invertebrate taxa to expand our knowledge. We need to continue to engage public awareness about the importance of protecting invertebrates and their habitats. Institutions have come a long way in doing this and collectively can make an impact to improve the welfare for the more 'forgotten', yet vital species.

REFERENCES

Bennie, M., Loaring, C. & Trim, S. (2011). Laboratory husbandry of arboreal tarantulas (Theraphosidae) and evaluation of environmental enrichment. *Animal technology: Journal of the Institute of Animal Technicians*, *10*(3), 163–169.

BIAZA Terrestrial Invertebrate Working Group (TIWG). (2014). Recommended invertebrates for education use species list & handling guidelines. BIAZA. https://biaza.org.uk/resources/animal-husbandry/invertebrates (Access Date 09/07/24).

Blenderman, J. P. (2020). Personality in Stomatopod Crustaceans. Stephen F. Austin State University, https://scholarworks.sfasu.edu/etds/302 (Access Date 04/08/24).

Cassill, D., Ford, K., Huynh, L., Shiffman, D. & Vinson, S. B. (2016). A study on abdominal wagging in the fire ant, *Solenopsis invicta*, with speculation on its meaning. *Journal of Bioeconomics*, *18*, 159–167.

Cooper, J. E. & Williams, D. L. (2014). The feeding of live food to exotic pets: Issues of welfare and ethics. *Journal of Exotic Pet Medicine*, *23*, 244–249.

Cooper, J. E. & Gordon, I. J. (2022) Butterfly Houses. In Lewbart, G. A. (Ed.), *Invertebrate Medicine* (3rd ed., pp. 521–535). Hoboken, NJ: Wiley Blackwell.

DiVincenti, L. Jr., McDowell, A. & Herrelko, E. S. (2023). Integrating individual animal and population welfare in zoos and aquariums. *Animals*, *13*(10), 1577.

Fiorito, G., Affuso, A., Basil, J., Cole, A., de Girolamo, P., D'Angelo, L., Dickel, L., Gestal, C., Grasso, F., Kuba, M., Mark, F., Melillo, D., Osoria, D., Perkins, K., Ponte, G., Shashar, N., Smith, D., Smith, J. & Andrews, P. L. R. (2015). Guidelines for the care and welfare of cephalopods in research –A consensus based on an initiative by CephRes, FELASA and the boyd group. *Laboratory Animals*, *49*(2), 1–90.

Foelix, R. F. (1996). Functional anatomy. In Foelix, R. F. (Ed.), *Biology of spiders* (pp. 12–37). New York, NY: Oxford University Press.

Greggor, A. L., Vicino, G. A., Swaisgood, R. R., Fidgett, A., Brenner, D. J., Kinney, M. E., Farabaugh, S. M., Masuda, B. M. & Lamberski, N. (2018). Animal welfare in conservation breeding: Applications and challenges. *Frontiers in Veterinary Science*, *5*, 323.

Hadfield, C. & Tuxbury, K. A. (2019). Touch-pools: The other side of the hand. In Miller, R. E., Lamberski, N., & Calle, P. (Eds.), *Fowler's zoo and wild animal medicine current therapy* (Vol. 9, pp. 334–337) St Louis, MO: Elsevier.

Horvath, K., Angeletti, D., Nascetti, G. & Carere, C. (2013). Invertebrate welfare: an overlooked issue. *Annali dell'Istituto Superiore di Sanità*, *49*(1), 9–17.

Keller, M. (2017). Feeding live invertebrate prey in zoos and aquaria: Are there welfare concerns? *Zoo Biology, 36*(5), 316–322.

Lewbart, G. A. (2012). Clinical anaesthesia and analgesia in invertebrates. *Journal of Exotic Pet Medicine, 21*(1), 59–70.

Liebsch, C., Fliess, M., Kuhbier, J. W., Vogt, P. M. & Straus, S. (2020). Nephila edulis—breeding and care under laboratory conditions. *Development Genes and Evolution, 230*, 203–211.

Magee, B. and Elwood, R. W. (2013). Shock avoidance by discrimination learning in the shore crab (*Carcinus maenas*) is consistent with a key criterion for pain. *Journal of Experimental Biology, 216*(3), 353–358.

Marnell, C. (2016). Tarantula and hermit crab emergency care. *Veterinary Clinics: Exotic Animal Practice, 19*(2), 627–646.

Nathaniel, T. I., Panksepp, J. & Huber, R. (2010). Effects of a single and repeated morphine treatment on conditioned and unconditioned behavioural sensitisation in Crayfish. *Behavioural Brain Research, 207*(2), 310–320.

Peckre, L. R., Defolie, C., Kappeler, P. M. & Fichtel, C. (2018). Potential self-medication using millipede secretions in red-fronted lemurs: Combining anointment and ingestion for a joint action against gastrointestinal parasites? *Primates, 59*, 483–494.

Perry, C.J. & Baciadonna, L. (2017). Studying emotion in invertebrates: What has been done, what can be measured and what they can provide. *Journal of Experimental Biology, 220*(21), 3856–3868.

Pizzi, R. & Kennedy, B. (2022). Spiders. In Lewbart, G. A. (Ed.), *Invertebrate Medicine* (3rd ed., pp. 301–347). Hoboken, NJ: Wiley Blackwell.

Pribadi, A. K. & Chalasani, S. H. (2022). Fear conditioning in invertebrates. *Frontiers in Behavioral Neuroscience, 16*, 1008818.

Sladky, K. K. (2014). Current understanding of fish and invertebrate anaesthesia and analgesia. In *Proceedings of the Association of Reptilian and Amphibian Veterinarians* (pp. 122–124).

Somerville, S., Baker, S., Baines, F., Trim, S. A. & Trim, C. M. (2017). Measuring cortisol levels in theraphosids and scorpions. *Proceedings of the Veterinary Invertebrate Society Summer Scientific Meeting*, 4.

Tarsitano, M. S. & Jackson, R. R. (1997). Araneophagic jumping spiders discriminate between detour routes that do and do not lead to prey. *Animal Behaviour, 53*(2), 257–266.

Zachariah, T. T. (2022). Invertebrate Animal Welfare. In Lewbart, G. A. (Ed.), *Invertebrate medicine* (3rd ed., pp. 669–675). Hoboken, NJ: Wiley Blackwell.

20

Considering the wild – opportunities and challenges of natural behaviour and zoo welfare

ROBERT KELLY

20.1 INTRODUCTION

For zoos and aquariums (hereafter "zoos"), it is imperative to ensure that the welfare of all animals being accommodated is maintained to the highest achievable standard, both from a moral perspective and to uphold the zoo's key aims (Rose & Riley, 2022), as well as to ensure positive public perception. It is also vital to gain an understanding of how factors relating to the captive environment, such as human disturbance or spatial restrictions, influence animal welfare states and how animals cope with such potential disturbances. To ensure that the welfare of animals in captive environments is being promoted or maintained at an acceptable standard, zoos rely on evidence-based assessments of welfare and are obligated by policy and legislation to measure welfare objectively. The Standards of Modern Zoo Practice (SMZP) assert benchmarks that UK zoos are expected to adhere to, for example. Of these, zoos must safeguard the welfare of the animals that they keep – animals must have "the ability to exhibit normal behaviour patterns" and be "protected from pain, suffering, injury, and disease". Such requirements are taken from the Five Welfare Needs set out in the Animal Welfare Act (2006) for England and Wales, and from the Animal Health and Welfare (Scotland) Act 2006 for Scotland. Similar requirements can be found in other parts of the world, for example, in continental Europe and in the USA. Under the internationally recognised Five Domains Model framework, developed by Mellor (2016), positive affective states (such as pleasure and comfort) can be promoted by enabling opportunities for behaviours that align with such emotional outputs, for example, foraging, exploration, or rest (Fureix & Meagher, 2015; Neville et al., 2024).

In spite of the increase in zoo research outputs, noted in other chapters of this book, making reliable and repeatable assessments of zoo animal welfare is still challenging, partially due to a lack

DOI: 10.1201/9781003429425-20

of knowledge around the specific needs of many zoo-housed species (Wolfensohn et al., 2018). Zoo professionals and management staff strive to use techniques that can be used to reliably and accurately assess the welfare of their animals, including behavioural observations and physiological or psychological measures, amongst others. However, incorporating multiple welfare indices, such as measuring blood glucocorticoids and heart and/or respiratory rates, can be expensive and complex to assess (Tallo-Parra et al., 2023), particularly given the range of species held in zoos. Further, acquiring samples to analyse – particularly direct sampling of blood – may cause stress hormone levels to elevate due to the invasive restraint methods required for sample collection. Similarly, other physiological welfare measures that rely on measuring autonomic responses (e.g. heart rate or body temperature) may require restraining an animal and can contribute to increased stress levels (Wolfensohn et al., 2018). Due to these challenges, there can be dissatisfaction around the use of such welfare measures, which in themselves may implicate the welfare of the animals being assessed. This highlights a need to validate welfare indices that mitigate such negative impacts of sample collection and to provide results that more accurately inform on an animal's current welfare. Furthermore, stringent policies and legislation governing zoos in the UK and Europe require that welfare be assessed through minimally disruptive methods, such as behavioural indicators, rather than invasive techniques that could potentially be detrimental to welfare.

It is clear that assessing how animals are coping or feeling, or just generally living within the zoo environment, necessitates valid, objective, and repeatable non-invasive methods of data collection and interpretation. One such approach involves observing natural or species-typical behaviours, as these are indicative of an animal's well-being and can assess whether normal biological functioning is being maintained in captivity. Natural, or species-typical behaviours performed by captive animals may be defined as those exhibited under natural conditions which promote normal biological functioning (Bracke & Hopster, 2006), or use measurable parameters such as reproductive success.

Natural behaviour provides a powerful, non-invasive welfare indicator, offering insights into whether the captive environment meets a species' immediate needs. By evaluating how well individuals express behaviours typical of their wild counterparts, zoos can assess not only the well-being of animals in captivity but also ensure that these populations remain viable for conservation efforts (McPhee & Carlstead, 2010). For instance, implementing multiple off-exhibit areas for the elusive and timid fishing cat (*Prionailurus viverrinus*) provides areas for refuge that enable them to escape from perceived threats, thus reducing stress and increasing reproductive success (Fazio et al., 2020). To understand the appropriateness of natural behaviours, or the ability of captive animals to express behaviours that are typical of wild conspecifics, comparisons may be drawn with the activity of wild conspecifics by constructing time-activity budgets, which measure the performance of pre-determined behaviours over a period of time. However, natural behaviour as a welfare indicator is under scrutiny (Browning, 2022; Veit & Browning, 2021), and questions surrounding whether such measures should be applied in the zoo to make informed judgements on animal welfare have been raised. Therefore, this chapter will address some of the benefits of using different measures of natural behaviour and the relevance of comparing zoo with wild behaviour patterns (as a benchmark to inform the welfare states of animals), whilst also acknowledging potential challenges and limitations.

20.2 EXPLORING NATURAL BEHAVIOUR AS A WELFARE INDICATOR

This book defines welfare as the state of the individual when coping with its environment, remembering that "environment" is both internal and external and coping links to changes in homeostatic mechanisms. When measuring the welfare state of an animal, it is vital that the methods used address this objectively, as assumptions around the perception of behaviour can lead to an incorrect diagnosis of the animals' welfare (Wolfensohn et al., 2018). Thus, scientific grounding is needed for evidence-based assessments. Although welfare indicators are broadly categorised as environment-based or animal-based, those which are animal-based include behavioural measures, and are often favoured due to providing direct information on how the animal is coping within its immediate environment (Manteca et al., 2016).

Dawkins (2003) argues that two key questions can be posed to counter issues associated with the subjectivity around the use of natural behaviour welfare measures which integrate physical health and motivational states. The first question: "Is the animal physically healthy?" An effective approach to measure physical health with natural behaviour is the use of qualitative behavioural assessments (QBA). In applying QBA, behaviour is observed to capture behavioural expression and assess an animal's perceived experiences, such as 'anxious' and 'contented' (Wemelsfelder, 2007). Thus, if animals can perform their repertoire of natural behaviours, they may be more likely to display feelings and emotions that indicate positive welfare (e.g. comfort or relaxation). QBA has served as a meaningful welfare indicator in farm animals (Phythian et al., 2016), but is yet to be employed across a range of taxa in the zoo environment (Patel et al., 2019).

It should not be assumed that a behaviour observed in the wild is inherently linked to positive welfare and therefore should be encouraged in a zoo setting. In fact, some natural behaviours are problematic in captivity and should be discouraged due to the potential negative welfare impacts. For instance, disturbances associated with captive living (e.g. continued exposure to visitors) may induce increased levels of vigilance, aggression and avoidance behaviours (Sherwen & Hemsworth, 2019) in species including Galapagos tortoises (*Chelonoidis nigra*), koalas (*Phascolarctos cinereus*) and little penguins (*Eudyptula minor*) (Freeland et al., 2020; Larsen et al., 2014; Sherwen et al., 2015). Whilst these behaviours are typically linked to survivability in the wild, deviation in their expression - such as altered frequency or duration - from patterns observed in wild counterparts may signal an inability to cope, potentially leading to chronic stress, impaired reproductive success, and compromised welfare (Fischer & Romero, 2019; Morgan & Tromborg, 2007; Salas et al., 2024). However, carefully planned husbandry or enrichment provision can utilise the functional traits associated with these behaviours to gain advantages from performing them, such as providing hiding areas in enclosures which enable refuge from perceived threats. This promotes eustress, an adaptive response to stress that benefits the animal and indicates an ability to cope when faced with environmental challenge (Rose & Lewton, 2025). The implications of this approach are significant for conservation, particularly when keeping populations intended for release into the wild, as it ensures they possess the behavioural traits necessary for survival that are characteristic of their wild conspecifics.

The second question posed by Dawkins (2003), "Does the animal have what it wants?" can be used to assess whether highly motivated behaviours, such as feeding, reproduction and rest, which are associated with positive welfare states, can be satiated. To facilitate the performance of species-specific and motivated behaviours, environmental enrichment can be integrated into husbandry practices. However, for more elusive species or those where natural history is poorly understood, applying effective enrichment programmes can prove challenging due to a paucity of literature that indicates which natural behaviours are beneficial in promoting welfare. Research has demonstrated such complexities in armadillos (Cingulata). Kelly and Rose (2024) found that implementation of physical enrichment (plastic ball, cardboard tube and cardboard box) promoted exploratory behaviour but not time spent foraging, with temperature having an overall larger effect on behaviour performance. For species where empirical evidence is lacking, research confirming which aspects of husbandry and housing that best support species-typical behavioural repertoires is needed. Some natural behaviours that species would normally perform in the wild are considered maladaptive in captivity. For example, despite having a motivation to perform migratory behaviour, it cannot be fulfilled in the captive environment (Learmonth, 2019). This may induce abnormal repetitive behaviour (ARB), leading to increased stress. For example, an increase in pacing behaviour has been observed in giant pandas (*Ailuropoda melanoleuca*) as a result of a motivational urge to travel that is not possible when housed in the zoo (Gandia et al., 2023) which can lead to frustration and a decrease in overall welfare state. Therefore, natural behaviours associated with biologically functional benefits, defined as the positive outcomes that enhance an animal's ability to survive, reproduce, or maintain its overall well-being (Peck et al., 2004) may be considered a better representative of welfare states. For example, motivated states such as foraging or maintenance that occur in pursuit of a goal, and this goal can be reached in captivity (Learmonth, 2019; Wolfensohn et al., 2018).

20.3 COMPARISONS OF WILD AND CAPTIVE BEHAVIOURS TO ASSESS ANIMAL WELFARE

The concept of natural living can be defined as "providing opportunities for animals to engage in natural, species-specific behaviours" (Learmonth, 2019) and highlights the importance of providing environments that enable the performance of natural behaviours to promote the well-being of animals. Using a zoo's rockhopper penguin (*Eudyptes moseleyi*) exhibit as an example, Figure 20.1 illustrates how the concept of natural living can be incorporated into zoo enclosure design. However, this approach requires having a clear understanding on which behaviours contribute to the behavioural repertoires of free-living wild animals, and currently there is a lack of sufficient data across many taxa. When comparisons can accurately be made, behaviour performance in captive animals that reflects that observed in wild conspecifics may be used as a potential welfare indicator.

Comparing time-activity budgets between wild and captive contexts can help identify which constraints in the captive environment cause differences in behaviours. For example, in birds, flight-restraint may increase abnormal behaviour performance in captive waterbirds (Rose et al., 2022), and morphological changes may implicate natural behaviours including courtship displays and social interactions (Rose et al., 2024). Time-activity budgets can also prove valuable in determining to what extent the frequency and diversity of behaviours observed in wild animals have diverged from those in captivity and help to understand the appropriateness of an animal's environment to meet its needs. For example, research on zoo-housed primates compared time-activity budgets with wild populations to determine the effects of captivity on natural behaviour patterns. Research has found that non-enriched captive chimpanzees (*Pan troglodytes*) spend less time foraging and more time resting compared to wild conspecifics (Yamanashi & Hayashi, 2011). Such differences may indicate boredom due to reduced environmental complexity, which is a known risk factor for impoverished animal welfare states (Meagher, 2019).

However, some natural behaviours, although observed in wild counterparts, are induced by stress, including avoidance behaviours to escape from threats (Browning, 2022). In the zoo, avoidance behaviours may result from increased visitor

Figure 20.1 Applying the concept of natural living in zoo enclosure design using a rockhopper penguin exhibit at a large UK zoo as an example.

presence that can increase glucocorticoid concentrations in arboreal primates (Hosey et al., 2023) with subsequent decreases in welfare state. Additionally, just as natural behaviours like predator avoidance may negatively affect welfare states, certain unnatural behaviours that provide environmental challenges – such as primates participating in computer game tasks (Clark et al., 2019) – can have a positive impact on welfare by promoting play behaviour and increasing foraging time. Therefore, behaviours should be considered carefully prior to encouraging their performance in the zoo.

In some species, natural behaviours that serve to aid survival or are valuable in the wild (e.g. vigilance to scan the surroundings for predators) may either reduce in frequency or even be thwarted due to a decrease in the stimuli that encourages the behaviour (e.g. lack of predator presence in the zoo). Despite the potential of such behaviours being underrepresented in zoo-housed animals, it does not necessarily mean that animal welfare is being compromised. We should therefore be posing questions on how the individual animal is spending its time instead. For example, is the natural behaviour being replaced with other actions that are indicative of positive welfare states (such as foraging or play) or is it replaced with behaviours that may indicate boredom or stress (e.g. ARBs or stereotypies) that are often associated with poorer welfare? Natural behaviours that enhance welfare, and thus should be promoted in captivity, vary significantly between species. Table 20.1 illustrates examples of some of these behaviours for a range of species, supported by recommendations for encouraging them in zoos.

20.4 MEASURING BEHAVIOURAL FLEXIBILITY AND BEHAVIOURAL DIVERSITY TO PROMOTE NATURAL LIVING IN THE ZOO

Two concepts that rely on measures of natural behaviour performance can be combined and applied in the zoo environment to provide a holistic assessment of animal welfare states: behavioural flexibility and behavioural diversity. Behavioural flexibility is defined as the ability to shift response patterns (or strategies) after changes in environmental conditions (Ragozzino & Rozman, 2007). Behavioural diversity is defined as a measure of the richness of behaviour (number of behaviours) as well as the evenness (frequency of each behaviour) (Miller et al., 2020). The application of behavioural diversity is becoming more prevalent as a positive indicator of welfare in captive animals, including zoos (Delfour et al., 2021; Hall et al., 2021); however, research into the behavioural flexibility of zoo animals is scarce.

Certain aspects of an animal's behavioural ecology, including migration, sociality, dietary selection, and diel activity patterns, may be challenging to provide for and cannot be easily replicated in the zoo. For example, migratory species are exposed to rapidly changing habitat complexity due to their traversal through different landscapes, which is difficult to mimic in captivity. Although it may therefore be suggested that migratory species could be predisposed to negative welfare in captivity, dynamic behavioural adaptation, or the ability of animals to modify their behaviour in response to environmental heterogeneity (e.g. geological and ecological variation), may facilitate the development of alternative behaviours (Kelly et al., 2025). However, when animals, irrespective of their behavioural ecology, are accommodated in barren environments (that do not meet their behavioural needs or motivations) or they become exposed to unnatural stimuli such as visitor presence, they may develop inflexible behavioural repertoires (Burn, 2017; McPhee & Carlstead, 2010), and perform ARBs as an attempt to cope. Comparing baseline wild time-activity with captive conditions provides insight into the degree of behavioural flexibility that zoo animals can achieve.

Like behavioural flexibility, the application of behavioural diversity scores can be used to assess the effectiveness and appropriateness of enclosure complexity. For example, when implementing changes to an enclosure or providing a form of environmental enrichment (EE), if an increase in behavioural diversity is observed following the changes enabling wider behavioural choice in the animals' environment, then it could be concluded that they are beneficial to the welfare of the animal. Further, through offering environmental complexity or EE that facilitates the expression of natural, species-typical behaviours, unnatural behaviours

Table 20.1 Examples of natural behaviours that may promote welfare states in species and how to promote them in the zoo.

Species	Behaviour	How to promote in the zoo
African lion (*Panthera leo*)	Foraging	Low frequency (once per week) carcass feeding (Höttges et al., 2019).
	Resting	Sufficient refuge areas in enclosure, or outdoor shaded areas that provide visual barriers (Goswami et al., 2023).
Galapagos giant tortoise	Basking/ thermoregulatory	Provide adequate lighting within the enclosure and a thermogradient. Access to multiple heat stations to enable isolation and prevent aggression with conspecifics (Fieschi-Méric et al., 2022).
Giraffe (*Giraffa camelopardalis*)	Rumination	Provide sufficient areas for resting, as this behaviour occurs with rumination and ensure provision of palatable forage and browse (Rose, 2023).
Polar bear (*Ursus maritimus*)	Foraging	Irregular feeding patterns and scent-based stimuli to promote searching behaviours (Fernandez, 2021).
	Swimming	Access to large pool within the enclosure that, in itself provides environmental variety, and opportunities for aquatic-based enrichment to be utilised.
Red-tailed black cockatoo (*Calyptorhynchus banksii*)	Flight	Covered aviaries to enable flight and perching/ branches throughout enclosure.
	Foraging	Food-based enrichment devices, e.g. millet discs (Fangmeier et al., 2020).
Ring-tailed lemur (*Lemur catta*)	Foraging	Scatter feeding, browse provision, food-based enrichment devices (e.g. food boxes/baskets).
	Affiliative	House with an appropriate number of conspecifics.

including ARBs which are associated with negative welfare may be reduced or eliminated. Conversely, if an animal is restricted from performing natural behaviours, or is unable to fulfil its motivations, ARB may predominate time-activity budgets and behavioural diversity will reduce. Previous research in chimpanzees has demonstrated that behavioural diversity can be used as valid measure of animal welfare, with higher levels of behavioural diversity negatively correlated with concentrations of faecal glucocorticoid metabolites (Hall et al., 2021).

Despite the benefits in these approaches for measuring welfare, they are not without challenges. Behavioural flexibility in captivity has scarcely been investigated (particularly in birds, reptiles, amphibians and fish) and there is paucity of research into the time-activity budgets of species in the wild. It is evident that larger charismatic mammals still currently dominate the literature, and applied welfare research that provides insight into reliable welfare indicators is still inadequate (Tallo-Parra et al., 2023). To highlight the current bias in behavioural research, Figure 20.2 illustrates published time-activity budget studies across a range of taxonomic orders in the wild and the zoo. Wild data displayed are for animals of wild origin, due to behavioural differences observed in

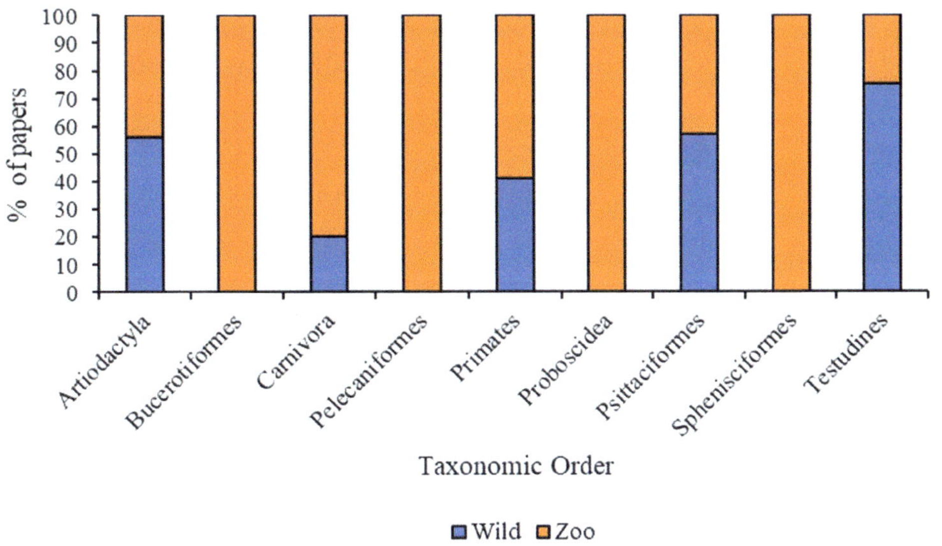

Figure 20.2 A sample of 69 papers available on the Web of Science© and/or Google Scholar© database, published from 1972 to 2023 on time-activity budgets in the wild and the zoo across a range of taxonomic Orders, demonstrates the discrepancy between behavioural research output in the wild and zoo contexts.

captive-origin wild species (Ain, 2023; Dunston et al., 2016). Understanding the key behaviours essential to species in the wild is crucial for accurately assessing behavioural flexibility, as it helps determine to what extent they are exhibited in captive animals. Additionally, comparing time-activity budgets in the wild with those in captivity can reveal how behaviours are partitioned throughout the day, helping to assess whether biologically functional behaviours in the wild are replaced by those indicative of poor welfare in captivity (e.g. ARB). The application of behavioural flexibility can guide zoo staff and animal caretakers in optimising enclosure design and husbandry practices to meet the behavioural needs of species. Furthermore, it ensures the preservation of natural behaviours crucial for fitness and reproductive success (e.g. foraging and nest building), which are essential for supporting conservation efforts in zoo populations.

It must be noted that behavioural diversity as a welfare measure should be applied with caution. When quantifying behavioural diversity, the Shannon Diversity Index is commonly used. The calculation assumes that all behaviours are of equal valence and do not account for their functional role for the animal (Watters et al., 2021). Behaviours associated with negative welfare states, such as ARB, should not be included in behavioural diversity calculations as this could indicate a higher level of behavioural diversity, indicating that the animal is coping, when indeed the performance of ARB may suggest otherwise. Therefore, similar to behavioural flexibility, key ecologically relevant, species-specific behaviours that are conducive to positive welfare must first be determined before using behavioural diversity as a welfare measure (Miller et al., 2020). With knowledge of wild behavioural ecology, this can be achieved and ensures that the management practice of endangered ex-situ populations (e.g. red pandas, *Ailurus fulgens*) (Spiezio et al., 2022) can be maintained to support zoo conservation efforts.

Improving our understanding of natural behaviours and combining measures of behavioural flexibility and diversity can provide more reliable assessments of captive animal welfare. Such information can offer zoo professionals and captive animal caretakers insight into the suitability of exhibits to meet species' needs and which behaviours animals may suffer in the absence of appropriate care and resource provision. Further, conserving species-specific natural behaviours with biological significance (e.g. foraging and reproduction) that aid species survival will serve to ensure the viability of zoo populations with conservation potential (McPhee & Carlstead, 2010).

20.5 CONSIDERATIONS FOR ANIMAL BEHAVIOUR AS A WELFARE INDICATOR

This chapter has demonstrated how different behavioural approaches, informed by natural history and evidence-based knowledge, can be acquired to develop an understanding of species-specific natural behaviours and serve as reliable and valid indicators of welfare in zoo-housed animals. Using foraging behaviour in a zoo-housed colobus monkey (*Colobus guereza*) as an example, Figure 20.3 illustrates how information from a species' natural behaviour can be utilised to inform welfare states in captive animals.

The aforementioned behavioural concepts: behavioural flexibility and diversity should be combined to assess whether behavioural performance in captivity reflects the species in the wild. Additionally, they will inform which behaviours the individual is motivated to perform, and that they are associated with positive valence, validated through QBA. This approach exemplifies an integrative and holistic framework that employs natural behaviour as a key metric for assessing animal welfare.

As discussed throughout this chapter, there are still challenges associated with assessing welfare based on natural behaviours. However, many of these challenges can be addressed with a combined effort from both researchers and zoo professionals. The current arguments against natural behaviour as a welfare measure and what can be done to mitigate these are shown in Box 20.1. Captive animal care staff should create and share standardised QBAs, particularly for understudied species, to enable holistic welfare assessments that evaluate an animal's emotional state in relation to the performance of natural behaviours. Further research to determine wild (baseline) time-activity budgets for species with limited data will help to categorise their levels of behavioural flexibility and explain how well they adapt to the constraints of captive living. Additionally, examining species-specific behaviours before including them in calculations for behavioural diversity will result in more accurate assessments of welfare states.

Figure 20.3 Taking information from species' natural history and what is observed in the zoo environment to apply behavioural measures and make an assessment of positive welfare indicators, using foraging behaviour in a zoo-housed colobus monkey as an example.

To summarise the points discussed within this chapter, Table 20.2 outlines the key opportunities and current challenges associated with using behavioural measures to assess animal welfare in the zoo. Whilst these methods offer valuable opportunities for monitoring and improving captive animal welfare, challenges still exist. To enhance the effectiveness of these methods, efforts should focus on filling data gaps, particularly for species in the wild where data are scarce and ensuring that the behavioural indicators chosen accurately reflect positive welfare. Furthermore, addressing issues of subjectivity in qualitative assessments and expanding the use of these methods across a broader range of species could improve their utility in zoo settings.

Table 20.2 The opportunities and current challenges associated with using behavioural measures to assess animal welfare, summarized from the key discussion points in this review.

	How to measure	Opportunities	Current challenges
Behavioural diversity	Collect baseline activity budgets in captivity and calculate using Shannon diversity index.	Non-invasive. Shows what animals are motivated to do in the zoo. Approach has been validated in some species.	Appropriate inclusion of behaviours in calculations to quantify positive welfare.
Behavioural flexibility	Compare wild and captive baseline activity budgets using effect size calculations (e.g. Hedges' g).	Non-invasive. Inform on adaptation to restrictions in the zoo. Better understanding of behavioural coping mechanisms employed in the zoo.	Lack of wild data in under-represented species, e.g. lack of wild origin wild time-activity budgets.
Observations of wild behaviours to inform captive animal welfare	Collect baseline activity budgets of wild populations (in-situ or remotely) over a 24hour period.	Non-invasive. Determine the appropriateness of captive environments. Understanding of animals' motivations. General understanding of how animals spend their time in the zoo compared to the wild, i.e. determine whether normal wild behaviours replaced with those associated with positive or negative welfare.	Difficulty in collecting data in wild, free-living animals. Questions around validity, i.e. using all wild natural behaviours (even if associated with stress) to inform good welfare.
Qualitative behavioural assessment	Develop species-specific emotional state descriptors (e.g. comfort and agitation) and conduct observations in the zoo.	Non-invasive. Assesses the emotional state, or valence of an animal when natural behaviours are performed.	Subjectivity in assessments. Lack of standardised criteria for assessments.

BOX 20.1: The current arguments against utilising natural behaviour as a measure of animal welfare and the strategies to mitigate these issues.

Argument 1: Natural behaviours such as predator avoidance or fighting that represent a struggle to survive do not indicate positive welfare (Veit and Browning, 2021).

Solution 1: Low levels of some of these natural behaviours may promote eustress, which is beneficial for populations (or progeny of populations) that are destined for wild release. However, the level of the behaviour should not far exceed that seen in wild counterparts, which can be determined by comparing wild and captive time-activity budgets for the species.

Argument 2: Some natural behaviours are maladaptive in captivity and may cause stress, for example, migratory behaviour which cannot be fulfilled in the captive environment (Learmonth, 2019).

Solution 2: To prevent the occurrence natural behaviours that are maladaptive, promote other natural behaviours (e.g. foraging or engagement with enrichment) that occur in pursuit of a goal that can be achieved in captivity and are associated with functional benefits for the species.

Argument 3: If only natural behaviours are indicative of positive welfare, then what about some unnatural behaviours, such as participation in computer game tasks in primates that may promote welfare ?(Clark et al., 2019).

Solution 3: Determine whether these behaviours are associated with biologically functional benefits (e.g. resource utilisation) and use QBA to understand whether they are associated with positive valence.

Argument 4: By including all natural behaviours within behavioural diversity calculations, they are considered of equal valence and do not account for the functional role of the behaviour to the animal (Watters et al., 2021).

Solution 4: Only include behaviours that animals are motivated to perform and that are indicative of positive welfare (e.g. foraging and play) within calculations of behavioural diversity. Behaviours that suggest negative welfare or a struggle to cope (e.g. ARB and aggression) should be excluded from calculations.

Argument 5: Behavioural research in wild animals is still predominantly focused on charismatic mammal species (Tallo-Parra et al., 2023), thus we have a limited understanding of behaviours that should be encouraged in the zoo.

Solution 5: Expand wild behavioural research to include understudied and elusive species (e.g. birds, reptiles, amphibians) to enhance knowledge of activity patterns and species-specific behaviours that indicate positive welfare states. Understanding wild natural behaviour patterns ensures that zoo professionals can strive to maintain wild behavioural repertoires in captive populations to improve their viability for conservation programmes.

20.6 CONCLUSIONS

Despite current contention around the use of natural behaviour to inform on animal welfare states, with a carefully formulated approach and scrutiny of observed behaviours, reliable assessments of welfare and suitability of species-specific enclosure design can be achieved. Utilising a range of behavioural indices can be fruitful for gaining insights into the welfare of an animal without necessitating the use of more invasive methods, including various physiological measures (e.g. blood cortisol analyses and autonomic responses). Further, accurately measuring and providing the opportunity for captive animals to perform species-specific natural behaviours is vital for maintaining the evolutionary integrity of populations, thus supporting the wider conservation efforts of zoos. Further, as zoos are governed by policy that necessitates non-invasive welfare measures, we must strive to continue the development of behavioural measures through combined efforts from researchers and zoo professionals to validate behavioural measures. By studying how animals behave in the wild, especially those species that are less studied, zoo professionals will be better able to assess their well-being and understand how to meet their needs in captivity.

REFERENCES

Ain, N.-U. (2023). Effect of seasonal variations on the behavior of captive Lions (Panthera leo) and Tigers (Panthera tigris). *Pakistan Journal of Zoology, 56*(4), 1961. https://doi.org/10.17582/journal.pjz/20221127151159

Bracke, M. B. M., & Hopster, H. (2006). Assessing the importance of natural behavior for animal welfare. *Journal of Agricultural and Environmental Ethics, 19*(1), 77–89. https://doi.org/10.1007/s10806-005-4493-7

Browning, H. (2022). The measurability of subjective animal welfare. *Journal of Consciousness Studies, 29*(3), 150–179. https://doi.org/10.53765/20512201.29.3.150

Burn, C. C. (2017). Bestial boredom: A biological perspective on animal boredom and suggestions for its scientific investigation. *Animal Behaviour, 130*, 141–151. https://doi.org/10.1016/j.anbehav.2017.06.006

Clark, F. E., Gray, S. I., Bennett, P., Mason, L. J., & Burgess, K. V. (2019). High-tech and tactile: Cognitive enrichment for zoo-housed gorillas. *Frontiers in Psychology, 10*, 1574.

Dawkins, M. S. (2003). Behaviour as a tool in the assessment of animal welfare. *Zoology, 106*(4), 383-387.

Delfour, F., Vaicekauskaite, R., García-Párraga, D., Pilenga, C., Serres, A., Brasseur, I., Pascaud, A., Perlado-Campos, E., Sánchez-Contreras, G. J., & Baumgartner, K. (2021). Behavioural diversity study in bottlenose dolphin (Tursiops truncatus) groups and its implications for welfare assessments. *Animals, 11*(6), 1715.

Dunston, E. J., Abell, J., Doyle, R. E., Evershed, M., & Freire, R. (2016). Exploring African lion (Panthera leo) behavioural phenotypes: Individual differences and correlations between sociality, boldness and behaviour. *Journal of Ethology, 34*(3), 277–290. https://doi.org/10.1007/s10164-016-0473-9

Fangmeier, M. L., Burns, A. L., Melfi, V. A., & Meade, J. (2020). Foraging enrichment alleviates oral repetitive behaviors in captive red-tailed black cockatoos (Calyptorhynchus banksii). *Zoo Biology, 39*(1), 3–12.

Fazio, J. M., Freeman, E. W., Bauer, E., Rockwood, L., Brown, J. L., Hope, K., Siegal-Willott, J., & Parsons, E. C. M. (2020). Longitudinal fecal hormone monitoring of adrenocortical function in zoo housed fishing cats (Prionailurus viverrinus) during institutional transfers and breeding introductions. *PLoS One, 15*(3), e0230239. https://doi.org/10.1371/journal.pone.0230239

Fernandez, E. J. (2021). Appetitive search behaviors and stereotypies in polar bears (Ursus maritimus). *Behavioural Processes, 182*, 104299.

Fieschi-Méric, L., Ellis, C., Servini, F., Tapley, B., & Michaels, C. J. (2022). An improvement in enclosure design can positively impact welfare, reduce aggressiveness and stabilise hierarchy in captive galapagos giant tortoises. *Journal of Zoological and Botanical Gardens, 3*(4), 499–512.

Fischer, C. P., & Romero, L. M. (2019). Chronic captivity stress in wild animals is highly species-specific. *Conservation physiology, 7*(1), coz093.

Freeland, L., Ellis, C., & Michaels, C. J. (2020). Documenting aggression, dominance and the impacts of visitor interaction on Galápagos tortoises (Chelonoidis nigra) in a zoo setting. *Animals, 10*(4), 699.

Fureix, C., & Meagher, R. K. (2015). What can inactivity (in its various forms) reveal about affective states in non-human animals? A review. *Applied Animal Behaviour Science, 171*, 8–24. https://doi.org/10.1016/j.applanim.2015.08.036

Gandia, K. M., Herrelko, E. S., Kessler, S. E., & Buchanan-Smith, H. M. (2023). Understanding circadian and circannual behavioral cycles of captive giant pandas (Ailuropoda melanoleuca) can help to promote good welfare. *Animals, 13*(15), 2401. https://doi.org/10.3390/ani13152401

Goswami, S., Tyagi, P. C., Malik, P. K., & Gupta, B. K. (2023). Effects of enclosure complexity and visitor presence on the welfare of Asiatic lions. *Applied Animal Behaviour Science, 260*, 105853.

Hall, K., Bryant, J., Staley, M., Whitham, J., & Miller, L. (2021). Behavioural diversity as a potential welfare indicator for professionally managed chimpanzees (Pan troglodytes): Exploring variations in calculating diversity using species-specific behaviours. *Animal Welfare, 30*(4), 381–392.

Hosey, G., Ward, S., & Melfi, V. (2023). The effect of visitors on the behaviour of zoo-housed primates: A test of four hypotheses. *Applied Animal Behaviour Science, 263*, 105938.

Höttges, N., Hjelm, M., Hård, T., & Laska, M. (2019). How does feeding regime affect behaviour and activity in captive African lions (Panthera leo)? *Journal of Zoo and Aquarium Research, 7*(3), 117–125.

Kelly, R., Freeman, M., & Rose, P. (2025). What behavior is important behavior? A systematic review of how wild and zoo-housed animals differ in their time-activity budgets. *Frontiers in Ethology, 4*, 1517294.

Kelly, R., & Rose, P. E. (2024). Assessing the impact of environmental enrichment on behavior in understudied armadillo species: A case study. *Zoo Biology, 43*(1), 100–109.

Larsen, M. J., Sherwen, S. L., & Rault, J.-L. (2014). Number of nearby visitors and noise level affect vigilance in captive koalas. *Applied Animal Behaviour Science, 154*, 76–82.

Learmonth, M. J. (2019). Dilemmas for natural living concepts of zoo animal welfare. *Animals, 9*(6), 318. https://doi.org/10.3390/ani9060318

Manteca, X., Amat, M., Salas, M., & Temple, D. (2016). Animal-based indicators to assess welfare in zoo animals. *CABI Reviews, 11*, 1–10. https://doi.org/10.1079/PAVSNNR201611010

McPhee, M. E., & Carlstead, K. (2010). The importance of maintaining natural behaviors in captive mammals. *Wild Mammals in Captivity: Principles and Techniques for Zoo Management, 2*, 303–313.

Meagher, R. (2019). Is boredom an animal welfare concern? *Animal Welfare, 28*(1), 21–32.

Mellor, D. J. (2016). Updating animal welfare thinking: Moving beyond the "Five Freedoms" towards "a Life Worth Living". *Animals, 6*(3), 21.

Miller, L., Vicino, G., Sheftel, J., & Lauderdale, L. (2020). Behavioral diversity as a potential indicator of positive animal welfare. *Animals, 10*(7), 1211. https://doi.org/10.3390/ani10071211

Morgan, K. N., & Tromborg, C. T. (2007). Sources of stress in captivity. *Applied Animal Behaviour Science, 102*(3–4), 262–302.

Neville, V., Finnegan, E., Paul, E. S., Davidson, M., Dayan, P., & Mendl, M. (2024). You are how you eat: Foraging behavior as a potential novel marker of rat affective state. *Affective Science, 5*(3), 232–245.

Patel, F., Wemelsfelder, F., & Ward, S. J. (2019). Using qualitative behaviour assessment to investigate human-animal relationships in zoo-housed giraffes (Giraffa camelopardalis). *Animals, 9*(6), 381.

Peck, L. S., Webb, K. E., & Bailey, D. M. (2004). Extreme sensitivity of biological function to temperature in Antarctic marine species. *Functional Ecology, 18*(5), 625–630.

Phythian, C. J., Michalopoulou, E., Cripps, P. J., Duncan, J. S., & Wemelsfelder, F. (2016). On-farm qualitative behaviour assessment in sheep: Repeated measurements across time, and association with physical indicators of flock health and welfare. *Applied Animal Behaviour Science, 175*, 23–31.

Ragozzino, M. E., & Rozman, S. (2007). The effect of rat anterior cingulate inactivation on cognitive flexibility. *Behavioral Neuroscience, 121*(4), 698.

Rose, P. (2023). Identifying essential elements of good giraffe welfare—can we use knowledge of a species' fundamental needs to develop welfare-focussed husbandry? *Journal of Zoological and Botanical Gardens, 4*(3), 549–566.

Rose, P., Freeman, M., Hickey, I., Kelly, R., & Greenwell, P. (2024). Considering what animals "need to do" in enclosure design: Questions on bird flight and aviaries. *Birds, 5*(3), 586–603.

Rose, P., Roper, A., Banks, S., Giorgio, C., Timms, M., Vaughan, P., Hatch, S., Halpin, S., Thomas, J., & O'Brien, M. (2022). Evaluation of the time-activity budgets of captive ducks (Anatidae) compared to wild counterparts. *Applied Animal Behaviour Science, 251*, 105626.

Rose, P. E., & Lewton, J. (2025). Key concepts for enhancing zoo animal welfare: Coping, comfort, choice, control, challenge, and compassion. *Journal of Applied Animal Welfare Science, 28*(3), 497–514.

Rose, P. E., & Riley, L. M. (2022). Expanding the role of the future zoo: Wellbeing should become the fifth aim for modern zoos. *Frontiers in Psychology, 13*, 1018722.

Salas, M., Tallo-Parra, O., & Manteca, X. (2024). Evidence-based zoo animal welfare assessment: Putting science into practice. *Journal of Zoo and Aquarium Research, 12*(4), 205–211.

Sherwen, S. L., & Hemsworth, P. H. (2019). The visitor effect on zoo animals: Implications and opportunities for zoo animal welfare. *Animals, 9*(6), 366.

Sherwen, S. L., Magrath, M. J., Butler, K. L., & Hemsworth, P. H. (2015). Little penguins, Eudyptula minor, show increased avoidance, aggression and vigilance in response to zoo visitors. *Applied Animal Behaviour Science, 168*, 71–76.

Spiezio, C., Altamura, M., Weerman, J., & Regaiolli, B. (2022). Behaviour of zoo-housed red pandas (Ailurus fulgens): A case-study testing the behavioural variety index. *Journal of Zoological and Botanical Gardens, 3*(2), 223–237.

Tallo-Parra, O., Salas, M., & Manteca, X. (2023). Zoo animal welfare assessment: Where do we stand? *Animals, 13*(12), 1966. https://doi.org/10.3390/ani13121966

Veit, W., & Browning, H. (2021). Perspectival pluralism for animal welfare. *European Journal for Philosophy of Science, 11*(1), 9. https://doi.org/10.1007/s13194-020-00322-9

Watters, J. V., Krebs, B. L., & Eschmann, C. L. (2021). Assessing animal welfare with behavior: Onward with caution. *Journal of Zoological and Botanical Gardens, 2*(1), 75–87.

Wemelsfelder, F. (2007). How animals communicate quality of life: The qualitative assessment of behaviour. *Animal Welfare, 16*(S1), 25–31.

Wolfensohn, S., Shotton, J., Bowley, H., Davies, S., Thompson, S., & Justice, W. (2018). Assessment of welfare in zoo animals: Towards optimum quality of life. *Animals, 8*(7), 110. https://doi.org/10.3390/ani8070110

Yamanashi, Y., & Hayashi, M. (2011). Assessing the effects of cognitive experiments on the welfare of captive chimpanzees (*Pan troglodytes*) by direct comparison of activity budget between wild and captive chimpanzees. *American Journal of Primatology, 73*(12), 1231–1238. https://doi.org/10.1002/ajp.20995

21

Promoting zoo animal welfare to the future professional

MARIANNE FREEMAN AND KERRY HUNT

21.1 INTRODUCTION

As has been shown throughout this book, zoo animal welfare is an evolving scientific discipline. The European Professional Zookeeper Qualification Framework (EPZQF) places animal welfare as a key job-specific competency, highlighting its importance to the industry. This chapter considers how zoo animal welfare has been, and could further be, embedded into a lifelong learning approach. Broom (2005) identified that prior to 1986, there were only two animal welfare-focused courses. This number increased to over 100 in 2005, and a UCAS (Universities and Colleges Admission Service) search in September 2023 returned 196 courses in the UK alone. Such a change highlights the increasing interest in the study of animal welfare, which mirrors changes in public opinion. As these courses have progressed, teaching has evolved from the physiological and behavioural starting points of animal welfare to the incorporation of wider disciplines including ethics, philosophy, psychology, and sociology, as promoting an effective understanding of animal welfare requires not only an appreciation of animal welfare science but also evaluation of the social bias that informs behaviour and policy (i.e. cultural and ethical philosophy) (MacKay et al., 2016). With increasing public pressure and critical opinion, the future zoo professional needs an understanding of animal welfare to be able to develop an evidence-based approach to all aspects of zoo operations that they are involved in.

Educating future professionals requires a multifaceted approach, with interest often first being piqued in informal environments like at home or during family visits to zoos and aquariums, before being more formally approached in college and university settings. This chapter discusses how the different stakeholders in animal welfare education contribute towards developing a well-rounded welfare-savvy professional, and it concludes with a focus on the future considerations for teaching welfare.

21.2 AT HOME AND IN THE COMMUNITY

Home, community and cultural environments shape empathetic values and emotional connections; a prerequisite to understanding animal

DOI: 10.1201/9781003429425-21

needs. Many believe that targeting primary school-aged children through in-school sessions is the best way to achieve empathy and connection with animals (and the wider natural world) at an early age. However, children's experiences are often already shaped by this age. Urban children as young as 4 years old tend to use an anthropocentric model, where animal behaviour is compared or explained by its relation to human behaviour, whilst rural children have a better understanding of the animal's biological needs (Ross et al., 2003). Children who have had frequent opportunities to engage with wildlife and spend more time in nature are more likely to form connections that develop interest and empathy towards animals. Equally, children who form connections with a pet show increased empathy and awareness of sentience in that animal (Muldoon et al., 2016). Early years cognitive development is shaped by observing others through social learning and setting good animal welfare examples. Working within communities can be a good place to start to raise animal welfare awareness, advocacy, and changing social norms.

Zoo animal welfare campaigns such as *#sharesavvy* from Battersea Park Children's Zoo (https://www.batterseaparkzoo.co.uk/conservation/sharesavvy/) have the capacity to gain a wide reach, with many different institutions alongside other zoos and aquariums (charities, colleges, etc.) signing up and sharing the information to encourage those on social media to keep animal welfare in mind when sharing footage of animals online. Wild animal welfare is growing in interest, and zoos' roles extend to the community and the nature within it, as well as the animals that they care for. Zoos Victoria's "When Balloons Fly" campaign reached large audiences and resulted in societal change to balloon release events (Mellish et al., 2019).

Zoo education can reach a greater range of people when online sessions are offered to the community, inspiring the public to make a change. An online session from Auckland Zoo reached over 22,000 people in just a few months, and included conservation actions related to responsible pet ownership (as evaluated through resource downloads) and also improved the connectivity of the zoo to the community (Thomas, 2020a). It is, however, important to consider social and cultural differences in attitudes – targeting specific areas and understanding the community's existing beliefs towards animals can make for more effective and sustainable community-based interventions.

21.3 IN ZOOS

Industry bodies like EAZA, BIAZA and WAZA have had education integrated into their member guidelines for many years; however, the focus of these was, until recently, predominantly on biological and conservation topics, with little-to-no specific mention of the teaching of animal welfare. In 2020 WAZA and International Zoo Educators published a new strategy, including a greater focus on teaching animal welfare as part of conservation education (Thomas, 2020b). This is being integrated into other regional association strategies as they get updated. Most zoos have educational sessions that include consideration of animal welfare as part of their education provision. Figure 21.1 highlights some of the welfare focused sessions that zoos are currently doing.

Zoo education covers a range of subjects and ages, including formal education programmes for schools, colleges, etc., and informal education aimed at day-to-day visitors. A range of experiences reinforcing connections to animals helps with developing an interest in welfare, and zoos are well placed to offer experiences to all ages and levels of knowledge. "Ocean Todds" are sessions aimed at 2- to 4-year-olds run by the National Marine Aquarium (National Marine Aquarium, N.D.), with a focus on giving children experiences to understand the difference between themselves and the world around them and to ultimately develop an interest in biology. Young et al. (2018) stated that accurate empathy is based on knowledge of an animal's natural history, not projected assumptions. Thus, early understanding of biological requirements allows for more suitable responses to welfare issues in later life. An example of interpretation linking biological requirements to animal welfare is the Exploratorium at Colchester Zoo (Figure 21.2).

Having a parent/guardian present helps to reinforce what was learnt in sessions, as when parents are involved with their children's education, there is an increase in positive associations when learning, with children demonstrating better use of vocabulary and increased task-persistence (El

Figure 21.1 Formal and informal zoo and aquarium educational provision that includes mention of animal welfare. These were correct as of June 2024.

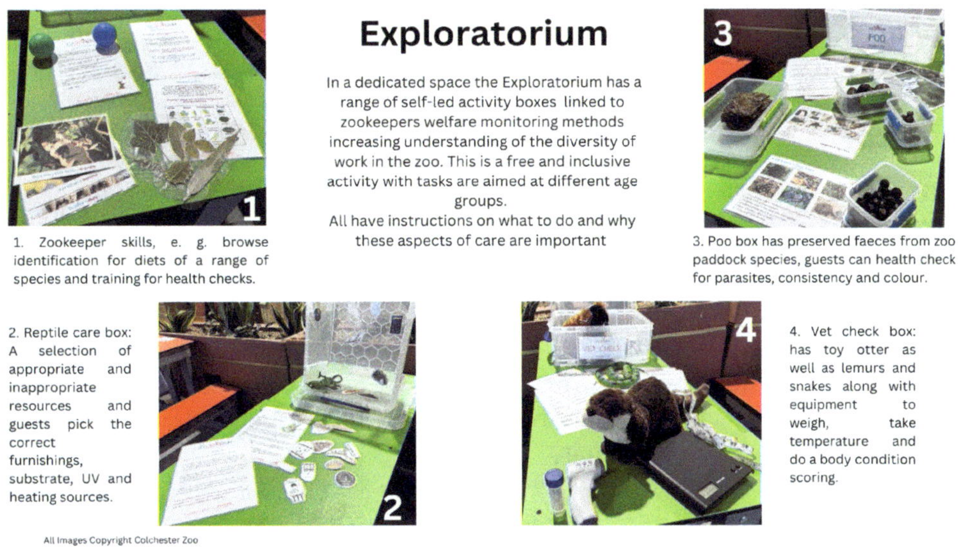

Figure 21.2 Examples of education provisions that Colchester Zoo have available to inform visitors of the ways welfare is monitored for different species.

Nokali et al., 2010). When children are too old for the National Marine Aquariums "Ocean Todds," they can join "Mini Marine Biologist" sessions. Emulating such an approach could allow a zoo to encourage engagement early and then, by having a progressive education programme that evolves as learners do, maintain such interest in nature and engagement with the zoo throughout childhood and beyond.

Research has shown that educational interventions with zoo guests that focus on how visitor behaviours can influence the welfare of zoo species have led to a subsequent change in their behaviour. Collins et al. (2019) showed that educational interventions reduced a range of undesirable behaviours, such as shouting, banging on glass, and trying to feed animals, in school-aged children (6–12 years old). The intervention was a presentation that described the biology of the focal species, which included discussion on the difference between life in the zoo and in the wild, how environmental enrichment can be used to elicit certain behaviours, and how visitor behaviour can impact these behaviours. Collins et al. (2019) also included a hands-on activity where children helped to make enrichment. The study states this activity was multifunctional by:

1. Improving animal welfare by encouraging specific behaviours.
2. Increasing (positive) animal activity levels to stimulate children's learning.

This intervention resulted in a significant reduction in the mean rate of negative behaviours performed by guests, suggesting that zoo education can facilitate behavioural change, at least in the short-term.

Engaging with teenagers can spark and maintain an interest in animal welfare that can progress towards a future career in this field. There are several zoo initiatives catering for those wishing to pursue their knowledge outside of formal school settings, including:

1. In October 2022, the Zoological Society London (ZSL) worked with The Prince's Trust to run a "get started with zoos and conservation" course for 16–25-year-olds. Sessions included how to design a suitable habitat for one of the zoo's species, with consideration of encouraging natural behaviours. The Prince's Trust covers course costs and reasonable travel fees to make attending these courses more accessible in person because, commonly, cost and lack of local access are cited as barriers to engagement with zoos.
2. The Royal Zoological Society of Scotland (RZSS) ZEBRA course is designed for 16 years+ and includes four modules, one solely focused on animal welfare (Cockburn et al., 2022). When evaluated, Nature Connectedness Index (NCI, a tool that evaluates how connected people are to nature) increased after the course, especially for species that the learners directly studied during the course.
3. To explore connections with less charismatic species, Farmerie et al. (2023) created a week-long human-animal interaction course working with koi carp (*Cyprinus rubrofuscus*), and this documented an increase in the human-animal bond, improved perceptions of fish, and a better understanding of fish welfare after the course. Outcomes applied not only to the participants but also to their family members, demonstrating how perceptions of fish can be improved through modes of social learning between family group members.

Zoo visitors might not understand the welfare needs of captive wild animals. Melfi et al. (2004) found no relationship between what visitors thought wild tigers (*Panthera tigris*) spent their time doing and what behaviours were considered indicative of good welfare. Even though respondents were aware that tigers would spend large portions of the day resting, they felt that more active behaviours were indicative of good welfare. This ties in with other studies that suggest visitors might have a misconception of what constitutes good welfare in zoo-housed animals. Considering the future zoo professional, misinterpretation of the welfare of animals in zoos by others in their social group, as well as in the media, could deter those with an interest in animal welfare from engaging with the zoo industry.

Other studies have investigated behaviour and attitude changes to livestock welfare through comparisons and engagement with captive wild animal welfare (e.g. Okabe et al., 2024) Evaluating these impacts can often be difficult as those who are more likely to engage with such sessions often already have an interest in, and thus potentially a greater understanding of, animal welfare needs.

21.4 SCHOOLS

Muldoon et al. (2016) argue that early-stage formal animal welfare education is necessary to ensure an ability to identify and differentiate the needs of species and the communication cues animals give to indicate their desires and welfare states. There have been several initiatives to try to build animal welfare into the curriculum in the UK, either focusing on "One Welfare," which is the interconnected wellbeing of humans, animals, and the environment to promote practices that benefit all three, or through the UNESCO pillar of education; Learning to Live Together (Muldoon & Williams, 2021) or within topics such as Healthy Relationships within Personal, Social, Health and Economic (PSHE) on the English National Curriculum.

At the time of writing the focus of animal welfare educational programmes in schools has been through charities, such as the Pet Education Partnership and initiatives such as the British Veterinary Nursing Association's welfare of pets in schools toolkit, the Royal Society for the Prevention of Cruelty to Animals (RSPCA) Compassion Class and the People's Dispensary for Sick Animals (PDSA) PetWise School Award. Theory of Change models (Figure 21.3) can be used to analyse impact and have been employed to demonstrate the abilities of animal welfare educational interventions to convey key messages and improve attitudes, beliefs and behaviours related to animal welfare in children after an educational intervention that focussed on common farm and companion species (Baatz et al., 2020; Jamieson et al., 2012).

Further research into long-term behavioural changes and methods that foster empathy and compassion with a wider range of species is needed. Such measures could focus on key children-animal interactions (Table 21.1).

However, regular interventions are needed to retain pro-welfare behaviours, so reinforcing these initial connections in different situations is important. Muldoon & Williams (2021) postulate that teachers may feel that they lack expertise to teach about animal welfare or that they have too much to focus on in other areas of society, culture, and the environment. Younger school-age children naturally engage with the topic, but maintaining interest in adolescence and beyond is equally important and could be achieved by training teachers in sustainability and One Health/Welfare curricula.

There are many examples of zoos partnering with teachers. For example, the RZSS and "Living Links to Human Evolution" partnership offers a variety of continuing professional development (CPD) training and additional resources for secondary school teachers. BIAZA and the Department of Education Ireland have partnered in junior level teaching of Science, Technology, Engineering, Arts, and Maths (STEAM) with sessions on biodiversity and sustainability. The precedence from these partnerships paves the way for further development of schemes such as Zoo Animal Welfare Education and Pet Education Partnerships through One Health/Welfare, and its links to sustainability.

21.5 LIFELONG LEARNING

Lack of knowledge has been highlighted as one of the main animal welfare issues (Rioja-Lang et al., 2020), and for those entering animal professions in

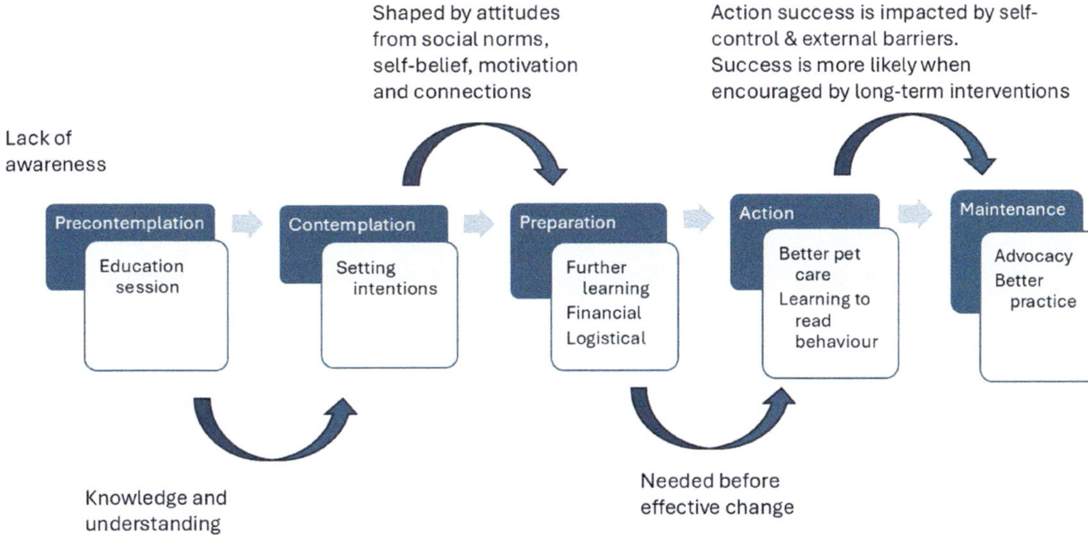

Figure 21.3 Transtheoretical model of change for animal welfare (Adapted from Prochaska and DiClemente, 1983).

Table 21.1 Interactions that children may have with animals and suggested scales to measure impact

Children/animal interaction dimensions	Measurable outcomes
(1) Pet ownership and attachment	The Pet Attachment and Life Impact Scale (PALS) or the Lexington Attachment of Pets Scale (LAPS). Nature Connected Index (NCI), Inclusion of Nature in Self (INS).
(2) Knowledge, attitudes and beliefs about animals	Children's beliefs about animals' minds (Child BAM), Coleman Dog Attitude Scale (CDAS) and testing knowledge and attitude aims of the intervention.
(3) Empathy and socio-emotional measures	Index of Empathy in children and adolescents (IECA). Sussex-Oxford Compassion Scale (SOCS-O), Pro-social tenancies revised (PTM-R).
(4) Human-animal interaction behaviours	Human Animal Interaction Scale (HAIS), Children's Attitudes and Behaviours towards Animals (CABTA).

Adapted from Muldoon and Williams (2021).

the future, colleges and universities offer a range of animal science, veterinary science, and zoo profession courses with animal welfare embedded within them. However, this is not a compulsory element for all courses. When surveyed, most animal science students agreed that welfare was an important element of the course, which was best taught through ethical discussions and practical application (Mijares et al., 2021). This active learning pedagogical strategy is known to improve learning. However, when active learning involves hands-on contact with animals, animal welfare assessment frameworks should be considered to ensure positive welfare states in ambassador animals involved in teaching (Thomas, 2020b). There are examples of institutional change within zoos; for example, Zoos Victoria redesigned their educational programmes to give choice and control (fundamental pillars of support for attaining good welfare states) to the animals – students are encouraged to observe rather than touch (Priestley et al., 2023). However, opponents to this change in teaching strategy argue that best practice handling and restraint in further and higher education courses is important for long-term skill development in future professionals. With well-developed student observational skills and appropriate staff-to-student ratios, animal welfare and assent to participate can still be practised while affording opportunities for the learner to gain important skills. Currently, the practicality and feasibility of land-based colleges moving to a similar model to Zoos Victoria, without the loss of safe and humane handling skills in the next generation of professionals, needs further consideration.

Research into alternatives, i.e. virtual reality training as a starting point before students work with live animals, alongside training educational ambassador animals for voluntary handling, is needed to uphold good welfare (and be effective in education) in the longer term. The welfare of animals in higher education is coming under more scrutiny, with several countries in Europe as well as New Zealand and Australia asking institutions to sign up to transparency agreements, showing their commitment to openness and transparency in the use of animals in research (EARA, 2023).

Within the veterinary field, it has been noted that, globally, animal welfare education for veterinary surgeons and paraprofessionals is often not taught sufficiently or applied in practice. However, since the implementation of graduate competencies by the Federation of Veterinarians Europe and European Association of Establishments for Veterinary Education, the level of animal welfare science education has increased in European veterinary graduates (De Briyne et al., 2020). EPZQF also has Animal Welfare as a core topic in zookeeper competencies (Table 21.2).

The differences in the professional standards competencies (in Table 21.2) between zoo and veterinary professions may, in part, be due to their specialised roles; even so there is scope to further develop the competencies for a zoo graduate. Like veterinarians, zoo professionals need to be able to consider the societal, ethical, and legal contexts

Table 21.2 Animal welfare competencies from EPZQF (zookeepers.eu/resources/) and animal welfare curriculum for European veterinarians (De Briyne et al., 2020)

Animal welfare competencies for zookeepers (from EPZQF)	**Day one competencies for veterinary graduates**
2.10.1 Theoretical Animal Welfare	Appraise concepts and frameworks of animal welfare.
2.10.2 Applied Animal Welfare	Evaluate the biological basis of welfare requirements.
2.10.3 Animal Welfare Assessment	Apply an animal welfare assessment to various categories of animals.
	Formulate and communicate an informed view on animal welfare matters.
	Place animal welfare in societal and ethical contexts.
	Place animal welfare in legal and professional context.

of animal welfare and be able to communicate informed views on animal welfare issues. Both professional competences include animal welfare theory, in particular, animal welfare frameworks, such as those adopted by WAZA; however, education on these is not always globally accessible, and even when the research is available, it may be misinterpreted (Bacon et al., 2021). Equally, there is not always consensus as to the best framework to follow (Hampton et al., 2023). Thus, teaching the skills to continually develop and apply new animal welfare knowledge is as important as current subject knowledge to progress into the profession (Dwyer et al., 2021).

Knowing what to do and how to do it are two different things and it is important that the skills and behaviours learnt are maintained into the workplace and throughout a career (Dwyer et al., 2021). On-the-job training allows practical application of welfare in daily animal management situations. However, if there is no structure nor opportunities for reflection, such training may not result in changes to behaviours or dispel any existing myths. Research into best mechanisms for transferring knowledge into practice, suggests that good coaching helps improve uptake and application of new practice (Joyce & Showers, 2002). Mentees adapted strategies more readily when coached compared with those who just observed, they were also more able to retain information and skills over time. Organisations, such as the Wild Animal Welfare/Southeast Asian Zoos and Aquarium Association partnership, work with zoos that do not have the in-house expertise or local availability of animal welfare education. Outside of formal education, zoos work with animal welfare organisations. BIAZA works with the Animal Behaviour and Training Council on an Animal Trainer Accreditation Course for zoo professionals, and the EAZA academy run a range of courses covering quality of life and animal welfare assessments, along with several openly available welfare webinars. Zoo staff also regularly present at animal welfare conferences, helping educate those working in other areas of industry about zoo animal welfare.

Zoos can engage the cognitive purposes of education through working in in-situ conservation projects to help with understanding of behaviour or veterinary treatment helping to further develop professionals. Lisbon Zoo provides veterinary expertise as part of the reintroduction of Iberian lynx (*Lynx pardinus*) to Portugal and zoo staff worked with in-situ staff on development of breeding practices providing veterinary expertise and strategies of conservation education for island foxes (*Urocyon littoralis*) to encourage support for the species from local residents (Coonan et al., 2010).

Continual development through online resources can be another alternative. Recently there has been an increase in online resources by regional associations (e.g. EAZA) as well as higher education driven initiatives such as Wild Animal Welfare via University of Edinburgh (Bacon et al., 2021) and Animal Welfare Massive Open Online Courses (MOOCs) (MacKay et al., 2016). Though these online resources have increased, there have been debates as to whether controversial topics can be delivered well via an online medium and concerns around the quality control for online education, due to high dropouts and lack of evaluation (Li, 2019). Free online resources can be a supplementary way to pass on animal welfare information to satisfy a need and desire for more information, but they are not always accessible to all as there may be other economic barriers, as whilst the courses are free, they require regular access to a suitable electronic device and the internet (MacKay et al., 2016). When promoting animal welfare to the professional it is important to consider any cultural differences that exist. For example, in some cultures there is an expectation for leadership role models and "top-down" approach to animal welfare whereas in others there is an ethical value of personal responsibility (Bacon et al., 2021).

Identifying the barriers that limit animal welfare provision by staff and evaluating the solutions to aid implementation of best practice in animal welfare is key to continuing improvement in zoo animal welfare. This should include an evaluation of the culture of the organisation to identify if behaviours change due to cultural norms and loss of idealism (Dwyer et al., 2021). Building resilience and preventing burnout is essential to prevent the loss of qualified and experienced staff or the loss of any creativity and sense of control within the existing staff and opportunities for continual education and development along with the encouragement to utilise this knowledge can incentivise staff. More effort should be given to empowering zoo professionals, ensuring that a zoo's leadership team recognises the barriers that exist in making change and implementing regulation that provides a focus on the changes that are needed (Bacon et al., 2023). Over the long-term a more robust education in animal welfare, for those entering the industry and those already within it, will result in a stronger appreciation of the multifaceted approach needed for managing and improving captive wild animal welfare.

21.6 CONCLUSIONS AND FUTURE CONSIDERATIONS

This chapter has highlighted that concern for, and promotion of, animal welfare education is a societal issue and that all people should be educated to appreciate the importance of animal welfare starting from a young age and inspiring them through their surrounding environment using developmentally appropriate messaging (Box 21.1). These young people may be the next generation of decision-makers with the power to promote good animal welfare standards, and thus they will ensure changes are made with the best interests of animals in mind. It is important that it is not just knowledge that is focused on in the range of outlets for animal welfare education but also skills to implement management practices to improve welfare and the ability to critique and reflect on new information as it arises. The future professional may be caretakers, managers, health professionals, researchers, policy advocates or educators themselves and all will have their part to play in passing on their knowledge and skills, to inspire the next generation.

Zoos, themselves, can be sources for animal welfare education across many stages of a future professional's lifetime. However, they are not the only source of animal welfare education which can also be formed from personal and community experiences. Formal education from early years through to university level education offers the opportunity to develop a deeper understanding of the theory and application of animal welfare. Collaboration between organisations can help to enhance formal animal welfare education and zoos can contribute their resources for capacity building for in-situ conservation projects and teacher training as well as development of online resources. Universities and independent organisations can also work with zoos, in areas where formal animal welfare education is less available, to help develop the skills of staff as well as online resource development. Researchers can work with zoos to investigate the impacts of any animal welfare interventions.

It is essential that we continue to measure, monitor and amend any animal welfare interventions to ensure that the most effective pedagogical approaches are being used to engage with all people on the topic of animal welfare to facilitate long-term changes that improve welfare, however, these educational interventions should include consideration of the welfare of the ambassador animals. Given the link with animal welfare and human wellbeing, zoos must ensure that staff are also thriving and cultures in all relevant institutions should be continually examined to determine and address barriers to best practice care that may exist. Animal welfare education should include human wellbeing education through resilience training to help combat issues of compassion fatigue and eco-anxiety to reduce the impact of emotional coping affecting behavioural change. For the future professional Box 21.2 highlights the online resources that can help in career development. Inspiring and motivating staff to good animal welfare is essential to keep driving this field forward.

BOX 21.1: Developmentally Appropriate Animal Welfare. Consideration of how concepts of animal welfare should be presented based on cognitive, emotional and social maturity, to promote better understanding and engagement

- 2–4-year-olds: Develop empathy with a focus on understanding companion animal body language and patience with interacting with animals, as well as respect for an animal's feelings.
- 5–7-year-olds: Introduce the Five Welfare Needs. Adapt examples from pet/domestic animals that children may be familiar with to zoo animals.
- 8–12-year-olds: Outline the key points of the Animal Welfare Act 2006, with focus on how this links to the practical aspects of caring for animals.
- 13–16-year-olds: One Welfare to consider links between human and animal welfare and the wider environment. Encourage work experience placements and identify career options. Develop practical skills in enrichment design/ building.
- 16–18-year-olds: Build on One Welfare and sustainability, discuss how to advance from current practices to improve welfare.
- 18+: Focus on developing an ability to evaluate the scientific literature on animal welfare. Understanding of methodological design and how findings can be applied in practice.
- Adults in Industry: Keeping up to date with emerging evidence-based practice, regular reflection on current practice.

BOX 21.2: A list of useful resources for those involved in learning and teaching that focusses on animal welfare education and staff wellbeing

- Wild Welfare: online modules cover a range of topics related to captive wild animal welfare. Modules provided in multiple languages and endorsed by WAZA https://wildwelfare.org/resources-elearn-programme
- EAZA Academy will often have courses or workshops related to animal welfare as well as other topics. These are either online or hosted by EAZA members. www.eaza.net/academy-courses/
- EAZA Animal Welfare Webinars, sponsored by the International Fund for Animal Welfare, you can access previous sessions from a range of zoo professionals on topics including nocturnal behaviour, nutrition and use of technology to enhance welfare on EAZA's YouTube channel. https://www.youtube.com/EAZAvideo
- WTS Vets United Animal Welfare Teaching modules to support academic institutions and continuous professional development. https://teaching-animal-welfare.org
- GRAZE a proactive programming and support service for zoo professionals https://zoomentalhealthsupport.com/

REFERENCES

Baatz, A., Anderson, K. L., Casey, R., Kyle, M., McMillan, K. M., Upjohn, M., & Sevenoaks, H. (2020). Education as a tool for improving canine welfare: Evaluating the effect of an education workshop on attitudes to responsible dog ownership and canine welfare in a sample of Key Stage 2 children in the United Kingdom. *PLoS One, 15*(4), e0230832.

Bacon, H., Vigors, B., Shaw, D. J., Waran, N., Dwyer, C. M., & Bell, C. (2021). Is animal welfare an internationally understood concept in the zoo world? Thematic analysis of two regional groups of zoo staff. *Animals, 11*(7), 2059.

Bacon, H., Vigors, B., Shaw, D. J., Waran, N., Dwyer, C. M., & Bell, C. (2023). Zookeepers – The most important animal in the zoo? *Journal of Applied Animal Welfare Science, 26*(4), 634–646.

Broom, D. M. (2005). Animal welfare education: Development and prospects. *Journal of Veterinary Medical Education, 32*(4), 438–441.

Cockburn, B., Hughes, J., & Seymour, L. (2022). RZSS Zoo studies: Converting Onsite Learning to online learning. *International Zoo Educators Association Journal, 58,* 21–26.

Collins, C., Quirke, T., McKeown, S., Flannery, K., Kennedy, D., & O'Riordan, R. (2019). Zoological education: Can it change behaviour? *Applied Animal Behaviour Science, 220,* 104857.

Coonan, T. J., Varsik, A., Lynch, C., & Schwemm, C. A. (2010). Cooperative conservation: Zoos and in situ captive breeding for endangered Island fox Urocyon littoralis ssp. *International Zoo Yearbook, 44*(1), 165–172.

De Briyne, N., Vidović, J., Morton, D. B., & Magalhães-Sant'Ana, M. (2020). Evolution of the teaching of animal welfare science, ethics and law in European veterinary schools (2012–2019). *Animals, 10*(7), 1238.

Dwyer, C., Bacon, H., Coombs, T., & Langford, F. (2021). Educating the animal welfare practitioners of the future. In R. Sommerville (Ed.), *Changing human behaviour to enhance animal welfare* (pp. 65–81). CABI International.

El Nokali, N. E., Bachman, H. J., & Votruba-Drzal, E. (2010). Parent involvement and children's academic and social development in elementary school. *Child Development, 81*(3), 988–1005.

European Animal Research Association. (2023). https://www.eara.eu/transparency-agreements. (Access Date 01/03/2025).

European Professional Zookeeper Qualification Framework (EPZQF). (2025). Framework. https://www.zookeepers.eu/framework/area-2-animal-management/2-10-animal-welfare/#2.10.3 (Access Date 01/03/2025).

Farmerie, M., Zyhier, S., Mehall, S., Duffy-Zimmer, S., Locke, J., McNamara, J., Kodama, T., & Hopkins, M. (2023). Making waves and changing youth perception about fish through an educational Human-Animal Interaction (HAI) program with Koi: The HAI CEWHALL program. *International Zoo Educators Association Journal, 59,* 61–68.

Hampton, J. O., Hemsworth, L. M., Hemsworth, P. H., Hyndman, T. H., & Sandøe, P. (2023). Rethinking the utility of the five domains model. *Animal Welfare, 32,* e62.

Jamieson, J., Reiss, M. J., Allen, D., Asher, L., Wathes, C. M., & Abeyesinghe, S. M. (2012). Measuring the success of a farm animal welfare education event. *Animal Welfare, 21*(1), 65–75.

Joyce, B. R., & Showers, B. (2002). Student achievement through staff development. In *Designing training and peer coaching: Our needs for learning.* Alexandria, VA: Association for Supervision & Curriculum Deve (ASCD).

Li, Y. (2019). *MOOCs in higher education: Opportunities and challenges.* 2019 5th International Conference on Humanities and Social Science Research (ICHSSR 2019), 48–55. https://www.atlantis-press.com/proceedings/ichssr-19/125909086 (Access Date 01/03/2025).

MacKay, J. R. D., Langford, F., & Waran, N. (2016). Massive open online courses as a tool for global animal welfare education. *Journal of Veterinary Medical Education, 43*(3), 287–301.

Melfi, V. A., McCormick, W., & Gibbs, A. (2004). A preliminary assessment of how zoo visitors evaluate animal welfare according to enclosure style and the expression of behavior. *Anthrozoös, 17*(2), 98–108.

Mellish, S., Pearson, E. L., McLeod, E. M., Tuckey, M. R., & Ryan, J. C. (2019). What goes up must come down: An evaluation of a zoo conservation-education program for balloon litter on visitor understanding, attitudes, and behaviour. *Journal of Sustainable Tourism, 27*(9), 1393–1415.

Mijares, S., Sullivan, P., Cramer, C., Román-Muñiz, N., & Edwards-Callaway, L. (2021). Perceptions of animal welfare and animal welfare curricula offered for undergraduate and graduate students in animal science departments in the United States. *Translational Animal Science, 5*(4), txab222.

Muldoon, J. C., Williams, J. M., & Lawrence, A. (2016). Exploring children's perspectives on the welfare needs of pet animals. *Anthrozoös, 29*(3), 357–375.

Muldoon, J., & Williams, J. (2021). The challenges and future development of animal welfare education in the UK. *Animal Welfare, 30*(2), 197–209.

Okabe, K., Kawamura, A., Uetake, K., & Kase, C. (2024). Enhancing Japanese visitors' interest in livestock welfare through zoo carcass feeding activities. *Journal of Zoo and Aquarium Research, 12*(2), 112–120.

Priestley, C., Jaensch, S., & McLeod, E. (2023). Evaluating the effectiveness of welfare-focused animal encounters to deliver education programs. *International Zoo Educators Association Journal, 59*, 69–76.

Prochaska, J., & DiClemente, C. (1983). Stages and processes of selfchange of smoking: toward an integrative model of change. *Journal of Consulting & Clinical Psychology, 51*(3), 390–395.

Rioja-Lang, F., Bacon, H., Connor, M., & Dwyer, C. M. (2020). Prioritisation of animal welfare issues in the UK using expert consensus. *Veterinary Record, 187*(12), 490–490.

Ross, N., Medin, D., Coley, J. D., & Atran, S. (2003). Cultural and experiential differences in the development of folkbiological induction. *Cognitive Development, 18*(1), 25–47.

Thomas, S. (2020a). *Community conservation engagement through a pandemic.* Driving Conservation Action, International Zoo Educators Association 25th Biannual Conference, Online.

Thomas, S. (2020b). *Social change for conservation: The world zoo and aquarium conservation education strategy.* WAZA Executive Office. https://www.waza.org/wp-content/uploads/2024/02/10.06_WZACES_spreads_20mbFINAL-compressed.pdf (Access Date 01/03/2025).

Young, A., Khalil, K. A., & Wharton, J. (2018). Empathy for animals: A review of the existing literature. *Curator: The Museum Journal, 61*(2), 327–343.

22

Welfare in an online world – communicating zoo and animal welfare science through social media

RICARDO LEMOS DE FIGUEIREDO AND KYLE BANTON-JONES

22.1 INTRODUCTION: THE NEED FOR ZOO ANIMAL WELFARE SCIENCE COMMUNICATION

Zoo animal welfare is a fast-evolving field of science, and best practices for animal care are in constant need of updation. New research regarding the welfare of zoo animals is regularly published and it is challenging to stay up to date with all the information that frequently comes out. For example, a quick search for "zoo animal welfare" across all Web of Science (webofscience.com) databases returns 1161 results for the period 2020 to 2024. Those research outcomes are the key for an evidence-based approach to animal care, as they contain the information needed to improve the husbandry and management, and consequently welfare, of the animals in our care. However, those zoo practitioners with the ability to apply such outcomes to animal care practices may struggle to access or keep up with this information as it changes and evolves. Searching for, filtering, reading and processing scientific papers can be time-consuming and demanding tasks, and time availability has previously been reported as

DOI: 10.1201/9781003429425-22

an obstacle for engagement with research by zoo staff (Anderson et al., 2010). The advancement of evidence-based animal care practices is therefore dependent upon effective animal welfare science communication targeting zoo professionals.

Communicating animal welfare to a wider audience can also be beneficial to zoos and aquariums by improving public awareness of the investment and advances they have made, and are making, in animal welfare science. Public attitudes towards zoos are influenced by perceptions of animal welfare (Phillips et al., 2024), but those perceptions are often outdated and fail to reflect the efforts and progress made to improve the well-being of wild animals kept in human care. Furthermore, improving public knowledge and awareness of animal welfare may influence people's attitudes towards potentially welfare-compromising practices such as exotic pet ownership and some forms of wildlife tourism (Naito et al., 2024).

There are many platforms and methods that can be used for animal welfare science communication. These include more traditional approaches like oral and/or poster presentations at scientific conferences, to more informal types of communication such as magazine articles, online blogs, social media posts, and podcasting. In this chapter, social media and podcasting are introduced as avenues for science communication. Their benefits and limitations are discussed, with real-life examples of their applications to zoo animal welfare science communication.

22.2 SOCIAL MEDIA AS A PLATFORM FOR SCIENCE COMMUNICATION

22.2.1 OPPORTUNITIES AND CHALLENGES

It is estimated that over 60% of the world's population uses social media, which corresponds to over 5 billion people, with projections suggesting that number will increase in the following years (Kemp, 2024). On average, people aged 16–64 spend over two hours a day on social media, which is more than the time they spend reading press media, listening to music streaming services or radio (Kemp, 2024). Zoo-related social media pages and groups, such as those of zoo and aquarium associations, are popular and heavily followed, with some having many thousands of followers across social media platforms, as shown in Table 22.1. Thus, social media has much potential as a platform for zoo animal welfare science communication, and research has shown that social media communications can improve knowledge and awareness of animal welfare issues (Locke et al., 2023). Although there are limited data on social media use by zoo professionals specifically, previous research suggests they are active users (Daniels et al., 2021), making social media a suitable option for communications targeting zoo professionals as well as wider audiences.

Although zoos regularly use social media as part of their marketing strategy, they may be overlooking the potential of social media for education and science communication, including animal welfare-focused content. For example, an investigation into the use of social media by British and Irish zoos revealed that although zoos frequently used social media to engage with the public, the number of education- or conservation-focused posts was limited (Rose et al., 2018).

On the downside, social media's popularity and wide use as a platform for communication and marketing can lead to an online saturation of content from various sources and covering a diversity of topics. The quality and credibility of that content are variable, and misinformation is a widespread issue (Chowdhury et al., 2023). This makes it challenging to get scientific content through to the audience, as it must compete with all the other information and misinformation on there. Furthermore, some social media posts can lead to heated and sometimes uncivil discussions (Goyanes et al., 2021), making them difficult to manage, and zoos and animal welfare are both topics that can generate controversy and lead to such discussions.

For zoos in particular, another consideration and challenge in the use of social media is the potential negative impact of some of the imagery they share, or enable the sharing of, such as close encounters and interactions between humans and zoo animals. Although such experiences sometimes have the goal of enabling emotional bonds between humans and nature, and can be associated with educational content, the sharing of those images may inadvertently promote perceptions

Table 22.1 Approximate number of followers/members of zoo-related social media pages, including regional and international zoo and aquarium associations and individually managed groups (Accessed in January 2025).

Organisation	Facebook	LinkedIn	Instagram
World Association of Zoos and Aquariums (WAZA)	37,400	25,000	9,500
European Association of Zoos and Aquaria (EAZA)	21,900	20,200	11,500
Association of Zoos and Aquariums (AZA)	169,500	53,800	54,900
Zoo and Aquarium Association (ZAA) Australasia	5,700	5,400	NA
Southeast Asia Association of Zoos and Aquariums (SEAZA)	2,300	NA	NA
Asociación Latinoamericana de Parques Zoológicos y Acuarios (ALPZA)	17,600	800	8,800
British and Irish Association of Zoos and Aquariums (BIAZA)	31,600	20,800	7,500
The Zoo Scientist	14,400	20,200	4,500
Wild Welfare	7,600	1,500	2,400
Zoo Keepers Europe	15,800	NA	NA
World Zoo Keepers	11,000	NA	NA

and practices that are harmful to both animal welfare and conservation (e.g. exotic pet ownership and the illegal wildlife trade; Shaw et al., 2022). As a result, many zoos have developed social media policies and image guidelines to prevent the sharing of imagery that may send the wrong message and have negative impacts on animal welfare.

22.2.2 KEY PRINCIPLES OF SCIENCE COMMUNICATION

These challenges can be addressed with a strategic and evidence-based approach to science communication. Rajput (2007) proposed a list of principles to improve the impact and effectiveness of science communication, applicable to social media and other platforms, which state that scientific content produced for communication should be:

- Authentic and factual: evidence-based information from reliable sources, which should be clearly stated.
- Concise and simple: avoid word-heavy content and technical terminology when possible.
- Creative, engaging and interactive: content should be attention-grabbing and motivate the audience to engage and interact with it and enable their interest and ability to further explore the topic.
- Audience- and goal-oriented: content should be produced and shared with specific goals and in line with the target audience's motivations, interests, needs and cultural context.
- Reinforced through repetition, enabling the audience to assimilate and remember information.

Additionally, when communicating through social media, inclusive and respectful language should be used, avoiding condescension and inflammatory language. While social media offers opportunities to share creative and fun scientific content, its accessibility must also be considered. The colours used should be accessible, images and videos should have captions/descriptions, and text broken down into bullet points or smaller paragraphs.

Another important consideration is what social media platform(s) to use for communication. There are many platforms to choose from and the most suitable option depends on the content, goals and target audience of the communication. Facebook is still one of the most popular social media platforms, with over 3 billion monthly active users (Kemp, 2024). It is more personal than other social media platforms, and a good option to build groups and communities. Instagram and TikTok are also widely popular, with 1.65 and 1.56 billion monthly active users, respectively. While Instagram is a good option for visual storytelling by focusing on images, photos and short videos, TikTok is mostly video-based and relies on emotional engagement and humour. YouTube is also one of the most popular online platforms, used for video sharing and viewing with users spending over 28 hours on it per month on average (Kemp, 2024). Many zoos use YouTube to engage with large numbers of viewers, although there is a focus on entertainment over educational content (Llewellyn and Rose, 2021). LinkedIn, on the other hand, is ideal for professional networking, with over a billion monthly active users, most of them between the ages of 25 and 34 (Kemp, 2024). It is important to note that most social media users engage with more than one platform and therefore an overlap in audience is possible when using multiple platforms for communication. For example, most Facebook users are also on Instagram and TikTok (Kemp, 2024). Therefore, if using multiple platforms, content should be tailored and adapted for each platform in turn.

These principles apply to science communication in general and can therefore be used when developing social media content about zoo and animal welfare science. But what does zoo/animal welfare science communication look like in practice?

22.3 THE ZOO SCIENTIST: AN ONLINE PLATFORM FOR ZOO SCIENCE COMMUNICATION

The Zoo Scientist is an online platform for sharing evidence-based information and resources about zoo conservation, research, education and animal welfare. Animal welfare science is one of the main themes of The Zoo Scientist, with much of its content focusing on welfare-related topics such as animal care and husbandry, environmental enrichment and human-animal interactions. With a main target audience that consists of zoo and wildlife professionals, as well as students who aspire to pursue zoo-related careers, and thousands of followers across platforms (Table 22.1), The Zoo Scientist plays an important role in disseminating animal welfare science to those who can apply it.

22.3.1 THE ZOO SCIENTIST ON SOCIAL MEDIA

The Zoo Scientist is present on Facebook, LinkedIn and Instagram. Despite an increase in popularity of other social media platforms, Facebook is still one of the most popular and widely used (Kemp, 2024), and The Zoo Scientist's Facebook page, which was the first of its social media channels to be created, reached over 90,000 people in 2024. However, it is on LinkedIn that The Zoo Scientist is the most popular, as despite being the most recent of its social media accounts, it has the most followers (Table 22.1) and has reached almost 200,000 people in 2024. Although contemporary trends in social media use suggest that LinkedIn is generally not as popular as other social media platforms (Kemp, 2024) the success of The Zoo Scientist on LinkedIn is potentially due to the professional focus of this platform (Kemp, 2024). It is likely the more serious scientific content and career resources shared by The Zoo Scientist appeal to an audience of professionals and aspiring professionals. On the other hand, despite Instagram's popularity among users all over the world (Kemp, 2024), it is the least followed of The Zoo Scientist's social media platforms (Table 22.1), with under half the reach of the Facebook account in 2024. This trend is observed in several of the accounts that share zoo-related content, although some seem to be popular on Instagram as well, such as the Association of Zoos and Aquariums and the Latin American Association of Zoos and Aquariums (Table 22.1). It is possible that the visually rich content regularly shared by these accounts contributes to the popularity of their Instagram pages. Therefore, adjusting content for Instagram with increased use of engaging imagery and visual stories may be an important step to further increase the efficiency of this platform in animal welfare science communication.

When generating content for The Zoo Scientist, a format is used that aims to quickly grab users' attention as they scroll down their social media feed. A cover image is featured which explains the type of publication/content (e.g. scientific paper or news) and, when sharing a scientific publication, the title and citation of the study are also included, alongside the journal it was published in, and a reference to The Zoo Scientist as the content creator. Most importantly, the cover image also features an eye-catching image that may include a graphical abstract of the study, when available, photo(s) of the target species (images of the actual animal participants in the research are prioritised, if possible, with the appropriate credits included), and sometimes graphs or other elements of the study (e.g. methods: photos/schematic representations of enrichment used) if they are accessible and clear enough on their own without the need for much explanatory text (Figure 22.1). Accompanying this cover image is a brief overview of the paper, including a quick introduction to what the study is about, key methods such as sample size and animals included, and the main results and conclusions. This format allows readers to quickly process the highlights of the study while further encouraging them to read the abstract and/or the full paper, by providing a link and stating whether the paper is open access. Finally, at the end, hashtags that are relevant to the content are also included (Figure 22.1), to further improve the visibility of the post when the topic the post relates to is searched for (Purba et al., 2020).

Although The Zoo Scientist shares a variety of content, including zoo conservation news and zoo-related career resources, animal welfare is one of the most frequently featured topics. In 2023 and 2024, over a third of the posts shared (excluding job vacancies) were animal welfare-related, with almost 50 scientific papers summarised and shared on topics such as the evaluation of husbandry and enrichment methods and human-animal interactions. This content is popular among The Zoo Scientist's followers, with some of the posts reaching thousands of people (e.g. the post featured in Figure 22.1 was among the most popular in 2024 and reached almost 10,000 people across platforms). This is a potential indicator that The Zoo Scientist is effectively communicating zoo animal welfare science to its audience.

Most of the content shared across The Zoo Scientist's social media platforms is also featured and stored on its webpage (www.thezooscientist.com). The webpage works as a digital library for zoo science and conservation content, compiling, storing, and organising a collection of zoo-relevant resources that would otherwise be scattered across multiple sources. The homepage directs the user to all sections of the website while highlighting the main topics of information covered on the webpage, including a section on animal welfare (Figure 22.2). The latest updates are also featured on the homepage, highlighting and facilitating access to the latest content that was shared.

The 'Animal Welfare' section features resources such as best practice guidelines and welfare assessment tools, as well as subsections on environmental enrichment and animal training, each featuring information regarding the theory underpinning these topics, expert testimonials, and practical examples. The content featuring scientific papers and studies that is shared on social media is also added to a "Scientific publications" section, where they are organised according to topic, helping users search through and locate relevant research. The webpage is a beneficial platform for The Zoo Scientist as it helps keep the social media posts more succinct and user-friendly, by providing an alternative location for further information. Furthermore, having a webpage likely improves the credibility of The Zoo Scientist, making it easier for users to find it online, and easily learn more about who is behind the page and the history of it.

22.3.2 THE ZOO SCIENTIST'S LIMITATIONS AND ROOM FOR IMPROVEMENT

The Zoo Scientist is limited by low human and financial resources, as it is funded only by its founder and director, who is also the only regular and long-term editor and content creator of the page. While keeping The Zoo Scientist active across platforms, with regular posts and content generation, is vital for its continuous growth and popularity (Purba et al., 2020), it can be challenging with limited resources. Many social media platforms fail to grow in popularity due to low activity, irregular posting, and a saturation of pages trying to reach the same audience with similar content. For those wishing to get involved in zoo and animal welfare science communication, before creating their own page, they should consider whether there are already other pages filling that niche. If so, an alternative option could be to get involved with those existing pages and contribute to an already established and/or growing social media platform. For those interested in starting their own project, and who believe there is a gap in the social media market for the page they want to create, then they should create a clear plan including what they want

Figure 22.1 Examples of high-reaching welfare-related post shared on The Zoo Scientist's social media platforms, with a description of its main features. The post is based on research by Ramont et al. (2024).

to focus on, who they want to target, and ensure they can keep the page active and up to date.

Despite The Zoo Scientist's popularity, its effectiveness as a platform for zoo and animal welfare science communication can be optimised with further implementation of the principles introduced in Section 22.2.2. The following priorities for improvement were identified:

- Segment target audience further and tailor content for different social media platforms to more efficiently target different segments of the audience.
- Improve graphic design by ensuring an evidence-based and consistent use of font, colour scheme and imagery, not only to increase the attractiveness of content but also its accessibility.
- Develop strategy for content sharing (what and when to share) instead of relying on opportunistic posting of content.
- Improve interactivity of content and encourage the audience to engage with the page's content as well as each other.
- Tailor content to reflect and target a wider diversity of cultural contexts.

Another challenge to address is the heterogeneous engagement with the content. While posts featuring mammalian species seem to receive more attention, a preference that has previously been detected in both scientists and the public (Prokop et al., 2022), some topics are also more popular (e.g.

Figure 22.2 Screenshot of The Zoo Scientist's homepage (www.thezooscientist.com), accessed on 29th January 2025.

environmental enrichment) than others. Although challenging to increase the attractiveness of content focusing on non-charismatic species and topics, a potential strategy could be to actively and directly share such content with specific relevant audiences (e.g. content featuring reptiles and amphibians could be emailed and/or posted on herpetology pages and online groups). Developing direct communication channels with groups such as zoo keeping associations (e.g. ABWAK – British and Irish, or AICAS – Iberian associations of zoo and wild animal keepers) could also facilitate the direct dissemination of relevant content with one of the Zoo Scientist's main target audiences. For example, ABWAK's publication "Ratel" has a popular feature in collaboration with the BIAZA (British and Irish Association of Zoos and Aquariums) Research Committee that includes sharing a summary of scientific papers that are relevant to animal husbandry, management, and welfare. The Zoo Scientist could explore similar collaborations while utilising its online resources and social media platforms.

Despite these limitations and areas for future improvement, The Zoo Scientist has grown to be a popular online platform for disseminating zoo and animal welfare science. However, while The Zoo Scientist focuses mostly on visual and written communication, verbal communication can be as useful and effective, and this chapter will now focus on podcasting as another platform for science communication.

22.4 PODCASTS: A UNIQUE PLATFORM FOR SCIENCE COMMUNICATION

Podcasting is a unique and increasingly popular method of communication and entertainment. In a recent 2023 publication, Edison Research found that over half of the population of the United Stateshave listened to a podcast and over 40% have listened to a podcast in the last week (Edison Research, 2023). As a result of this surge of listeners globally, there has also been a surge in individual podcasts, with over three million shows for people to choose from. Genres range from talk shows, comedy, political commentary, true crime and of course, science communication, which is the primary focus of this chapter.

22.4.1 ADVANTAGES OF PODCASTING FOR SCIENCE COMMUNICATION

Podcasting has several characteristics that make it a unique and effective platform for science communication, with advantages over more traditional media such as journals, blogs and radio.

One of the main characteristics of podcasting is the use of a personal and conversational tone. Similar to blogs and social media pages, podcasts allow the content creator to adopt a more informal tone when expressing complex scientific topics. This makes the show more approachable to the listener who does not necessarily have the background to read through a scientific article. Many scientific podcasts have an interview format which allows a more conversational tone to the whole episode and topics that are discussed. This also allows the interviewer to ask questions directly to an expert on the topic which can facilitate an exploration of complex topics, making the content easier to process for the listener (Kinkaid et al., 2023).

Podcasts also offer flexibility in format and length. For each of the three million podcasts, there is a unique format, length of episode and style of the host/hosts. Some podcasts are only a few minutes long while others are over three hours long. Some podcasts have one host while others many hosts. Podcasts can be done in the style of an interview or instead focussing on one host speaking to their audience. Some podcasts focus on storytelling and incorporate characters and sound effects. This flexibility allows the host to select a format and length that is better suited to their goals and target audience, while also giving the audience increased opportunities to choose a format that appeal to their preferences and interests.

Podcasting provides unique opportunities for engagement with the audience, as hosts can utilise a variety of tools, such as personal stories and thought-provoking questions. These can encourage the audience to develop a more personal and relatable view of the host and/or interviewee, which can lead to increased engagement (McLung and Johnston, 2010). Accessibility and ease of listening are also advantages of podcasting, with hundreds of thousands of podcasts freely accessible across multiple mobile apps. Furthermore, unlike written communication, podcasts can be listened to while engaging in other tasks, such as driving, walking or doing house chores.

22.4.2 CHALLENGES AND LIMITATIONS OF PODCASTING FOR SCIENCE COMMUNICATION

On the other hand, podcasting also involves unique challenges and limitations. With so many options of podcasts for listeners to choose from, covering a variety of topics, formats and styles, it can be challenging to attract listeners and maintain their interest throughout. Communication of scientific topics can also be difficult during podcasts, as unlike with other media, they rely on verbal communication without the use of visual aids.

Technical barriers such as equipment and sound quality can also be an issue in podcasting, despite the variety of accessible technology that is now available. Audio equipment such as microphones, audio drivers, premium editing software and video conferencing tools are often necessary to achieve a professional podcasting quality. Furthermore, podcasts may also shift towards pseudo-science and anecdotes if not appropriately prepared and developed. Producers and hosts must therefore take extra time and care to ensure the accuracy and credibility of the information shared throughout the entire podcast (section 22.2.2).

22.4.3 EFFECTIVE STRATEGIES FOR SCIENCE PODCASTS

Creating a science podcast that is informative, scientifically accurate and entertaining enough to attract and retain listeners is a challenging task. Therefore, science podcasts rely on careful planning and effective strategies to succeed in the long-term.

Crafting compelling content is probably the most important factor to consider when attempting to produce a successful science podcast. It will often be the deciding factor for listeners: do they listen to the full episode or turn the podcast off prematurely? In order to stand out in the crowd of podcasts, it is important to consider the niche occupied by the podcast and what makes it more or less unique. If the podcast's niche is too small, there is a risk of not reaching a large enough audience, but if it is too broad it will likely be outcompeted by larger shows with established audiences and celebrity hosts (McGregor, 2022). For example, if the show only talks about hyena welfare in human care, that might be too specific, whereas talking about the welfare of all animals in all captive settings may be too broad, depending on competing shows. A podcast's niche must also align with the knowledge of the host(s), allowing them to speak confidently and in a way that is engaging while efficiently retaining listeners.

Choosing the right format for the audience is also important. Guest interviews are one of the most popular podcast formats. But which guests should be interviewed on a podcast? Guests should be chosen according to their credibility, expertise and ability to speak about a specific topic that is interesting for both the host and the audience. Guests should therefore be knowledgeable and entertaining and be able to have a natural and engaging conversation with the host. Another important consideration in this format is about asking the right questions. Questions must facilitate healthy discussion and learning, and should be planned before the podcast, by researching the guest that is to be interviewed. Follow-up questions must also be formulated during the interview. An alternative format is the single host-style podcast, when there is only one host speaking for the duration of the podcast. This format can be challenging as it typically requires extensive research and places the entire responsibility of engaging and entertaining an audience on a single person (Hew, 2009). Another popular format of podcast is to have a second host, who shares the responsibilities of engaging and entertaining the audience. This format can lead to more dynamic conversations between the co-hosts, which is often preferred by the listeners.

Also crucial for a successful science podcast is to enhance listener engagement and building a community around the podcast. Listener engagement is critical when developing and maintaining an interested and committed audience that will listen to the podcast episodes over the long term. One of the best ways to enhance listener engagement is to solicit feedback and participation from listeners, which allows the listener to become part of the

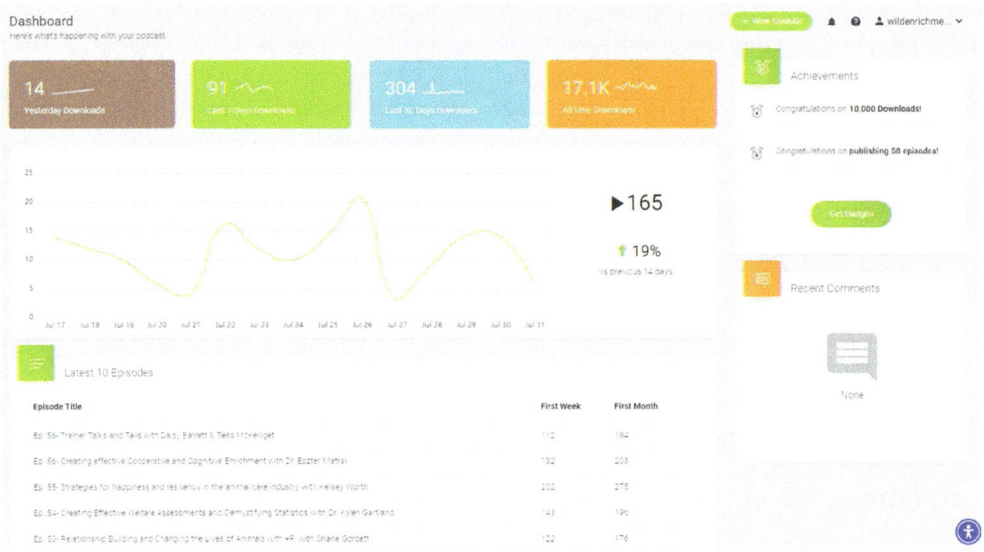

Figure 22.3 Dashboard of the Wild Enrichment podcast, accessed on 31st July 2024. The total number of podcast downloads is featured, as well as number of downloads the day before, as well as 7 and 30 days before the date of access. The Wild Enrichment podcast is hosted by Podbean.

show. One of the most effective ways to do this is to leverage social media and engage with fans and listeners on various social media platforms. Podcasts are often pre-recorded, limiting opportunities for live engagement with the audience. By linking a podcast to a social media account, listeners can more easily send messages, interact with the podcast's content and provide feedback. Feedback can also be actively solicited during the podcast, which will aid content development and improve the audience's engagement with the show.

Analysing feedback from the audience is an important step to understand what the listeners are responding to and what can be improved. Metrics provided by podcasting hosts, such as Podbean (Figure 22.3) including number of downloads, listeners, returning listeners, the amount of time they spent listening, etc. This allows the host to accurately receive feedback from listeners and refine the format, topics and niche accordingly.

22.5 WILD ENRICHMENT: PODCASTING ANIMAL WELFARE SCIENCE

The Wild Enrichment podcast focuses on animal welfare, training, and enrichment. Each episode explores concepts related to behavioural husbandry and the dynamic field of animal welfare. It features interviews with animal care professionals as well as educational episodes about animal care concepts.

Wild Enrichment was created with the goal of exploring multiple communication tools to convey animal welfare science to a wider audience. It focused on blog articles for several years, and the first episodes of the podcast were based on reading those blogs in a single host format. This allowed audiences to assimilate blog articles alongside their daily routines, instead of having to sit and read through text. The podcast also allowed for further exploration of the content within the blog posts. However, as previously discussed, the lack of visual aids and the frequent inability to access information within scientific papers were challenges.

The podcast was refined over time and became a more effective communication tool. Currently, the podcast has 58 episodes, which have been downloaded over 17,000 times, as seen in Figure 3. The format changed to a more traditional interview style, which allowed for greater networking within the field of animal welfare. The guests invited onto the podcast are all experts in the topics the podcast focuses on, which overlaps with the host's interests, experiences, and projects. For example, episode 54 features guest Dr Kylen Gartland, Manager of

Animal Welfare Science at Detroit Zoo, who joined the podcast to discuss the use of statistics as a tool for animal welfare assessments.

Producing and hosting a podcast provides networking opportunities and promotes skill development in public speaking, listening, and communicating. The main audience of the Wild Enrichment podcast consists of animal care professionals, and this podcast provides them with opportunities to learn about animal welfare science from experts in the field.

22.6 THE ZOO SCIENTIST AND WILD ENRICHMENT AS EXAMPLES FOR ZOO AND AQUARIUM SOCIAL MEDIA PRESENCE AND IMPACT

The Zoo Scientist and Wild Enrichment are examples of how social media and podcasting can be effective tools in animal welfare science communication. The Zoo Scientist relies on visual and written communication, with attention-grabbing and quick-to-read summaries of scientific papers and other animal welfare resources. On the other hand, Wild Enrichment focuses on verbal communication by featuring interviews with animal welfare experts. While these pages mainly target zoo professionals or aspiring zoo professionals, there is potential for zoos and zoo-related organisations, such as zoo associations, to similarly utilise their social media channels to educate their staff, visitors, and the wider public on animal welfare topics. Regular animal welfare features, such as image- or video-based posts summarising relevant welfare studies and podcasts with interviews of their own animal husbandry and welfare experts, would allow zoos to effectively communicate animal welfare science while improving their educational impact. This could result in more informed and up-to-date staff on animal welfare matters, and visitors and followers who are more aware of animal welfare considerations and potentially less likely to engage in practices with a negative impact on the welfare of both domestic and wild animals.

22.7 CONCLUSIONS

With the constant evolution of animal welfare science, it is difficult for zoo professionals to stay informed and up-to-date with best practices. Communicating animal welfare science to those professionals is therefore crucial, as they are the ones able to apply that updated evidence to animal care practices and consequently improve the welfare of animals in human care. Social media, which is now used by most of the world's population, and podcasts, which are increasingly popular, can both be effective platforms for zoo and animal welfare science communication. However, that effectiveness depends upon carefully developed scientific content that follows the key principles of science communication, so the information can stand out in the overcrowded digital world where misinformation is widely spread. The Zoo Scientist, an online platform for zoo science, and the Wild Enrichment, an animal welfare-focused podcast, were presented as real-life examples of zoo and animal welfare science communication in practice. Although both have limitations and room for improvement, they have reached thousands of people with their welfare-related posts, including zoo professionals, and have therefore been successful in disseminating animal welfare science to a relevant target audience. Zoos and zoo-related organisations should utilise their own social media channels in similar ways to further engage their own staff, visitors, and the wider public with animal welfare science.

REFERENCES

Anderson, U.S., Maple, T.L. & Bloomsmith, M.A. (2010). Factors facilitating research: A survey of zoo and aquarium professionals. *Zoo Biology*, 29, 663–675.

Chowdhury, A., Kabir, K.H., Abdulai, A.-R. & Alam, M.F. (2023). Systematic review of misinformation in social and online media for the development of an analytical framework for Agri-Food sector. *Sustainability*, 15, 4753.

Daniels, C., Cheyne, S., Waters, S. & Svensson, M. (2021). Professional primate keepers and online primate imagery: An assessment of knowledge and attitudes. *Journal of Zoo and Aquarium Research*, 9(4), 259–265.

Edison Research. (2023). *The Podcast Consumer 2023*. Retrieved from: https://www.edisonresearch.com/wp-content/uploads/2023/03/The-Podcast-Consumer-2023-1.pdf. (Access Date 01/03/2025).

Goyanes, M., Borah, P. & Gil de Zúñiga, G. (2021). Social media filtering and democracy: Effects of social media news use and uncivil political discussions on social media unfriending. *Computers in Human Behavior*, 120, 106759.

Hew, K. F. (2009). Use of audio podcast in K-12 and higher education: A review of research topics and methodologies. *Educational Technology Research and Development*, 57(3), 333–357.

Kemp, S. (2024). Digital 2024: 5 billion social media users. Retrieved from: https://www.meltwater.com/en/blog/digital-2024. (Access Date 01/03/2025).

Kinkaid, E. D., Bulger, M. E. & Lee, A. H. (2023). How academic podcasting can change academia and its relationship with society: A conversation and guide. *Frontiers in Communication*, 8, 1090112.

Llewellyn, T. & Rose, P.E. (2021). Education is entertainment? Zoo science communication on YouTube. *Journal of Zoological & Botanical Gardens, 2*, 250–264.

Locke, S., Hiltbrand, K., Corbitt, K., Richburg, D., Shannon, D., Rodning, S.P., Sawyer, J.T. & Mulvaney, D. (2023). Instagram as a tool of diffusion for the livestock industry. *Journal of Applied Communications*, 107(3), 1–14.

McClung, S. & Johnson, K. (2010). Examining the motives of podcast users. *Journal of Radio and Audio Media*, 17(1), 82–95.

McGregor, H. (2022). Podcast studies. Oxford: Oxford University Press. https://oxfordre.com/literature/view/10.1093/acrefore/9780190201098.001.0001/acrefore-9780190201098-e-1338. (Access Date 01/03/2025).

Naito, R., Chan, K. M. A. & Zhao, J. (2024). Combating the exotic pet trade: Effects of conservation messaging on attitudes, demands, and civic intentions. *Conservation Science and Practice*, 6(2), e13078.

Phillips, N., Maréchal, L., Ventura, B. & Cooper, J. (2024). How does perception of zoo animal welfare influence public attitudes, experiences, and behavioural intentions? A mixed-methods systematic review. *bioRxiv*. https://doi.org/10.1101/2024.03.20.585889.

Prokop, P., Masarovič, R., Hajdúchová, S., Ježová, Z., Zvaríková, M. & Fedor, P. (2022). Prioritisation of charismatic animals in major conservation journals measured by the altmetric attention score. *Sustainability, 14*(24), 17029.

Purba, K. R., Asirvatham, D. & Murugesan, R. K. (2020). Analysis and prediction of instagram users popularity using reression techniques based on metadata, media and hashtags analysis. *Engineering Letters*, 28(3), 1–8.

Ramont, M., Principe, N., Prostko, R., Watts, J., Chinnadurai, S. K., & Miller, L. J. (2025). The provision of browse and its impacts on the health and welfare of animals at the zoo: A review. *Zoo Biology, 44*(2), 105–125.

Rajput, A. S. D. (2007). Effective science communication: Evaluation and appropriation of basic principles. *Indian Journal of Science Communication*, 6(1), 17–23.

Rose, P. E., Hunt, K. A. & Riley, L. M. (2018). Animals in an online world; An evaluation of how zoological collections use social media. *Journal of Zoo and Aquarium Research, 6*(2), 57–62.

Shaw, M. N., McLeod, E. M., Borrie, W. T., & Miller, K. K. (2022). Human positioning in close-encounter photographs and the effect on public perceptions of zoo animals. *Animals, 12*(1), 11.

23

Population management, records and animal welfare

JOHN E. ANDREWS AND ANDREW MOONEY

23.1 INTRODUCTION

Cooperative population management of zoo and aquarium species arose some 40 years ago as a response to conservation crises in the 1970s. The initial goals for managing zoo and aquarium (hereafter "zoos") populations were to minimise collecting animals from the wild and to ensure animals in human care were responsibly and scientifically managed into the future. Modern population management today includes several key pillars of other scientific disciplines (Figure 23.1). Managers need information on how populations change over time, what drives changes, how animals are related to each other, and husbandry data to create tailored plans for a given species to be successful. Detailed animal records are vital to collect and analyse to support population management goals. Population management relies on data from a variety of sources as a basis for analysis and making recommendations for breeding or the transfer of animals between zoos. Animal welfare has grown in prominence in recent decades for managing zoo animals, and individual animal welfare is contingent on sound population planning among zoos (DiVincenti et al., 2023).

Before commencing population management, many types of data from various sources are needed. The modern zoo collects a massive amount of information on individuals or groups of animals in its care. Common data collected on individual animals include individual identifiers (e.g. tags, transponders), body measurements (e.g. weights, lengths), parent and rearing information, as well as travel history (e.g. which other zoos an animal lived in) (Table 23.1). Medical data are also meticulously recorded on disease and injury presence, including treatments received, as well as causes of mortality. Newer applications like ZooMonitor are making the collection, visualisation, and sharing of behavioural data more accessible and available for use in making management decisions (Wark et al., 2019). Studbook databases are the most common data tool used for population management in zoos, with a deep history and widely accepted standards for keeping them.

A significant amount of time and money has been invested to standardise data collection and entry across zoos for broader applications to

DOI: 10.1201/9781003429425-23

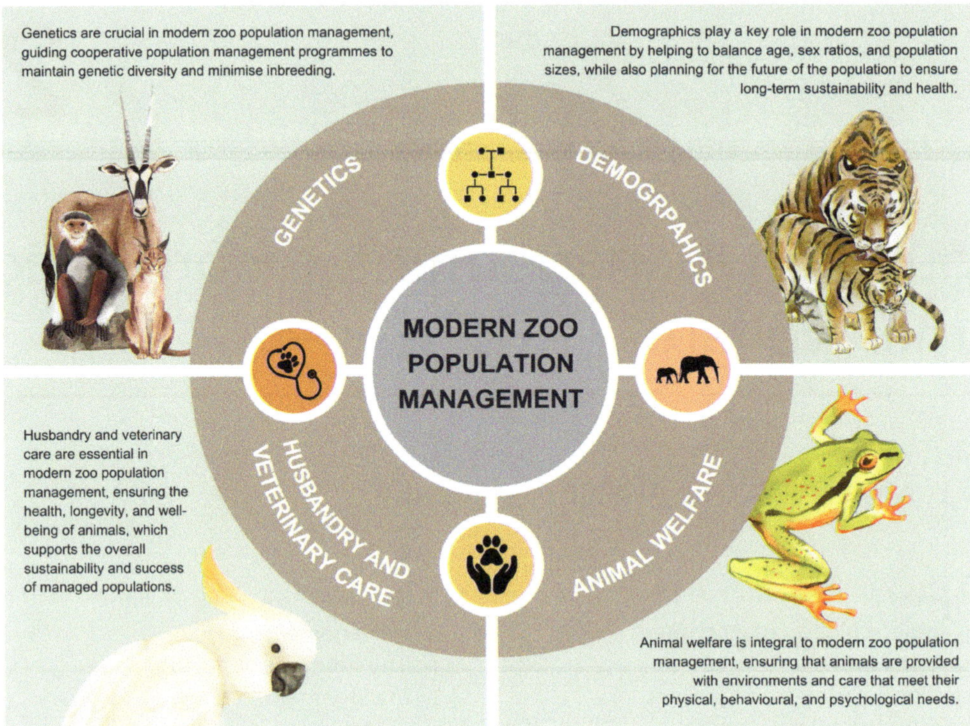

Figure 23.1 Four key components of modern zoo population management are genetics, demographics, husbandry and veterinary care, and animal welfare. Genetics supports inbreeding avoidance and retains diversity; demographics balance age, sex ratios, and population sizes; husbandry and veterinary care ensure animal health and longevity; and animal welfare addresses the physical, behavioural, and psychological needs of animals. Collectively, these elements help ensure healthy and sustainable zoo and aquarium populations.

improve population management and animal welfare (Barber, 2009). Companies like Species360 (www.species360.org) have the Zoological Information Management System (ZIMS) to provide a standard structure for data entry and storage for individual facilities that can be globally shared. ZIMS has many modules, including ZIMS for Husbandry, Medical, and most recently, Studbooks. A studbook database is a population-level database that collects information on a single species from various zoos across a particular scope. Studbooks have been used in various formats or software historically, but the most advanced programmes now use ZIMS for Studbooks. The major advantage of globally available zoo data is the ease of sharing, contributing, and the opportunity for meta-analysis (Barber, 2009). Combining various data types and sources allows for more powerful analyses and a greater ability to detect trends in behaviour or population changes over time, which can positively feed back into improving population management and individual welfare.

With population analysis on studbooks, scientists can often detect potential issues that may be affecting a population unknowingly. Historically, when data were harder to share, zoos may have experienced phenomena with a particular species and been unaware of how common or rare they may be among other zoos. Demographic trends may highlight populations that are crashing and in need of stabilising, and genetic analysis can reveal inbreeding or over-represented genes that can be helped with management. Population data can highlight important questions that can lead to recommendations for advancing husbandry practices, ultimately feeding positively into individual welfare.

Table 23.1 Overview of the types of data (and sources of such data) which are important to consider when managing animals in zoos and aquariums

Data categories	Data example(s)	Potential source(s)
Husbandry	-Taxonomy -Birth/hatch date -Animal weights -Rearing history -Individual identifiers -Parentage -Physical holder history -Reproductive data -Longevity	Zoo Information Management System (ZIMS); Tracks Software
Veterinary care	-Medical history and symptoms -Medical results (e.g. blood tests) -Prescribed medication and dosage -Medical outcomes -Contraception -Biological samples -Necropsy reports -Morbidity and mortality analyses	ZIMS; AZA Reproductive Management Center/EAZA Reproductive Management Group Contraception Database
Animal welfare	-Body condition score -Health status -All occurrence and interval behavioural data -Activity budgets -Habitat usage	ZIMS; ZooMonitor; Animal Behaviour Pro
Population	-Birth and death records -Parentage -Relatedness -Gene diversity -Age distribution -Mean kinship -Breeding and transfer Recommendations -Outcomes (e.g. population growth rate)	ZIMS; PMCTrack; PMx

Note: Also included are specific examples of data which can be collected, and potential sources to record and manage these data.

23.2 LINKING POPULATION MANAGEMENT TO INDIVIDUAL WELFARE

Studbook databases are the primary tool used in population management to analyse data and evaluate population status. Records held in studbooks come from various sources, such as basic husbandry information, parentage and pedigrees, transfer records, and space for customised information depending on the species. Many examples in cooperative population management illustrate how various population challenges in zoos can be identified and mitigated using animal data, and often future welfare challenges can be predicted by population analysis.

Demographic planning is a major part of population management, analysing trends in the past to better predict what is possible in the future. Population structures can provide insight into current or future welfare challenges. For example, a growing population is somewhat pyramidal in shape (Figure 23.2, left) and has a small proportion of older animals. A declining population, with a top-heavy shape, may suggest future strains on individual animal welfare as older animals continue to age and require geriatric care (Figure 23.2, right). Population analysis then signals a future welfare concern. Demographic data can also reveal trends only noticeable when looking at a population as a whole. For example, larger lizard species, like some iguanas (Iguanidae), can often become egg-bound if isolated from male conspecifics, leading to elevated mortality rates for females in reproductive age classes. Managers quickly developed recommendations to house all females with or near a male conspecific to stimulate egg laying, regardless of being recommended to produce offspring (Pearson et al., 2023). Here, elevated mortality rates in a population identified the need to change husbandry recommendations and improved female welfare by allowing more expression of natural reproductive behaviours, regardless and reducing mortality.

Genetic goals of population management generally involve retaining or increasing gene diversity and minimising inbreeding as much as possible. Inbreeding depressions are possible in zoo populations as they are often constrained by limited space and small population sizes. An inbreeding depression often results in increased infant mortality, smaller clutch sizes, physical deformities, decreased longevity,

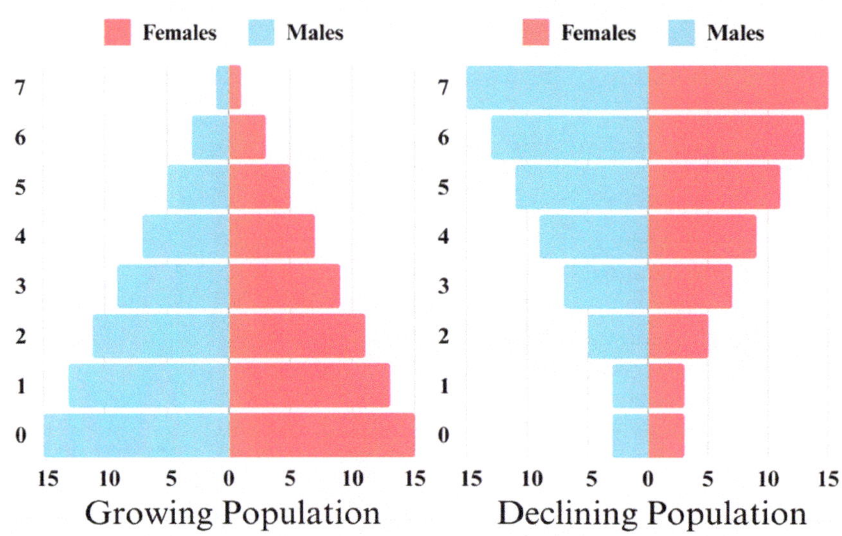

Figure 23.2 Age structures with hypothetical data depicting a growing population (left) and a declining population (right). Males are shown on the left in blue and females on the right in pink, with age structures measured in years shown on the y-axis.

and lower reproductive success, among others (Ralls and Ballou, 1986). Using studbooks and molecular data can help to identify inbreeding depressions and lead to changes in management strategies to mitigate inbreeding and enhance welfare. Hammerly et al. (2016) showed that repairing birds that were closely related using molecular data led to a reduction in mean parental relatedness of offspring and increased clutch survival after only a few breeding seasons. Breeding animals with low inbreeding helps produce more robust animals better able to survive, adapt, and thrive in zoo environments.

Population analysis can also reveal or help document husbandry challenges. Vet advisors are common roles filled for cooperative population management programmes in zoos and often conduct mortality and morbidity studies summarising common diagnoses and causes of mortality. Compiling data from many zoos on a species to summarise these has led to many examples of husbandry improvements made via recommendations in cooperative breeding programmes. For instance, recommending hand-rearing protocols to mitigate infant mortality or establishing medical surveillance programmes and husbandry recommendations for common ailments.

Indeed, data are paramount to informing population management in zoos, and combining data from multiple sources can produce powerful results. Benefits to a population, whether from demographic, genetic, or individual data, can all have direct or indirect impacts on individual animal welfare. Population management has always considered the welfare of individuals to some extent when making breeding and transfer recommendations; however, as welfare grows in importance for zoo communities and advances in data availability continue including animal welfare, concerns have become more expansive and pervasive in population management. For more detail on balancing welfare and population management see Box 23.1.

23.3 GOOD POPULATION MANAGEMENT INCORPORATES ANIMAL WELFARE

Individual animal welfare is a critical aspect of modern cooperative population management programmes in zoos. Although the primary goal of such programmes is species sustainability, the welfare of the individuals concerned is often key to ensuring the overall success of a programme (Greggor et al., 2018). Detailed individual records on health, behaviour, breeding, and life history can allow for more informed decision-making, both enhancing the quality of life for individual animals and contributing to conservation and engagement goals of programmes.

By their nature, cooperative population management programmes require inter- and intra-zoo transfer of individuals to maintain a genetically diverse and demographically stable population. However, the availability of globally shared records in ZIMS has shown that increasing the frequency of such transfers can impact the health and welfare of individuals. For example, such transfers have been identified as a potential risk factor for the development of Macropod Progressive Periodontal Disease, or "lumpy jaw", in zoo-housed kangaroos and wallabies (Macropodidae) (Rendle et al., 2020). Often-fatal, this disease is difficult to treat and has been linked to the age and sex of individuals. Conversely, transferring flamingos (Phoenicopteriformes) between zoos has been shown to result in increased reproductive success across several species, suggesting a positive impact of inter-zoo transfers when managed correctly (Mooney et al., 2023). Identifying and understanding individual and species-specific responses to the necessary population management recommendations are key to ensuring both individual health and welfare and improving management practices to support the overall sustainability of zoo populations.

Natural behaviours associated with the reproductive cycle, such as courtship displays, mating, and rearing offspring, are often considered critical behavioural needs, and allowing and planning for their expression is often reflective of positive animal welfare (Greggor et al., 2018). Breeding and transfer recommendations aimed at maintaining or growing populations began with focus on genetics and demography, making it difficult to account for natural behaviours adequately. This is especially true when we know little about the breeding reproductive biology of the species concerned. However, the availability of globally shared zoo records helps to fill in these gaps and presents solutions to these issues at several levels. For example, recent studies investigating more than 100 species of zoo-housed felids revealed clear seasonality in many species

BOX 23.1: Balancing welfare and population management: Addressing misconceptions in zoo animal care

Common misconceptions of a lack of connection. A common misconception held about zoo population management is that individual animal welfare is not well considered, if at all, when making breeding and transfer recommendations. Animals have to move and breed to sustain populations and management facilitates this. Modern population management considers individual welfare of animals on a regular basis discussing every individual animal or group within a programme in great detail. However, the goals of population management and individual welfare may often be at odds (DiVincenti et al., 2023) and trying to manage both involves including multiple stakeholder perspectives.

- Breeding recommendations require a balance of population management priorities and individual welfare.
  - Older animals may be genetically important, but at risk of reproductive pathologies due to senescence.
  - Species with naturally aggressive breeding behaviour may involve high risk to one mate or the other requiring discussions with all stakeholders.
- Transfer recommendations - Before recommending transfers, many factors are considered depending on the species. Examples include:
  - Timing of transfers to match weather with species biology.
  - Capacity of receiving zoos to provide similar enrichment programmes, exhibit size or type, and group structure to support incoming animals.
  - Sex, age or social needs of animals being transferred.
- Social recommendations - Making recommendations for social animals can also involve trade-offs. Population management recommendations may aim to resolve group dynamic challenges; however, short term stress of transfers and introductions are considered against potential long term benefits. Is short term stress worth long term welfare improvement?
- Reproductive and welfare recommendations - Population analyses often raise questions about behavioural syndromes or reproductive status. Welfare assessments, or reproductive exams may be recommended to investigate common or specific questions to improve individual welfare and help animals accomplish population goals.

Welfare has always been a part of population management and often the biggest challenge to incorporating this in management is effective communication. Making sure all pertinent stakeholders are involved in recommendations (e.g. animal care staff, population coordinators, veterinarians, welfare scientists, researchers, curators) is vital to ensure welfare is properly considered for population management.

(Heldstab et al., 2018), with clear individual welfare and population management implications, such as when it is most appropriate to introduce or transfer animals. Further research on tigers (*Panthera tigris*) has also revealed subspecies-specific variation in mortality and reproductive patterns, pointing out the need to consider subspecies independently when planning and managing populations (Tidière et al., 2021). Furthermore, the incorporation of individual mate choice has been shown to enhance reproductive success in several species, such as the giant panda, *Ailuropoda melanoleuca* (Martin-Wintle et al., 2015). Holistic approaches to population management may be more logistically difficult, but they offer unique opportunities to enhance individual welfare, while also increasing the efficacy of cooperative population management programmes, many of which suffer from low reproductive success, threatening long-term sustainability (Bauman et al., 2019). Animal records, therefore, have immense potential and are indispensable in informing animal welfare and effective population management, and increasing our understanding of wildlife as a whole.

Caring for zoo animals is a lifelong process, and understanding the life history of zoo animals is an important part of the management of individuals, and the success of cooperative population management programmes. However, basic life history characteristics, such as longevity and senescence, are often poorly understood for most species, which not only impacts management in zoos, but also may impact how we conserve species in the wild. Comprehensive demographic information is only available for 1.3% of all known mammals, birds, reptiles, and amphibians (Conde et al., 2019). The lack of basic information on things like birth rates and survival can make the management of species extremely difficult. Availability of long-term data from zoos can help fill in some of these gaps, allowing for more appropriate and informed management recommendations. Using data from Species360 member institutions globally, species-specific demographic knowledge has been shown to increase from 1.3% of species to almost 10% (Conde et al., 2019). For example, detailed zoo records from 133,818 individual parrots (Psittaciformes), representing 244 species, produced estimations of species-specific life expectancies (Smeele et al., 2022), which would be almost impossible to calculate based purely on observations in the wild. Understanding species-specific life expectancies in parrots plays an important role in how we manage both individuals and species, as many show short active reproductive periods, with an often longer post-reproductive period (Young et al., 2012).

23.4 REPRODUCTIVE AND POPULATION MANAGEMENT

The goal of cooperative population management programmes is to maintain genetically and demographically healthy populations of species. Under the most careful reproductive management strategies, however, animals in excess of holding capacity can arise, and their management and welfare are growing areas of concern (Browning, 2018). In this instance, surplus animals are defined as those deemed of less genetic and/or demographic priority compared to other individuals in a population. Limited space available in zoos to house individual animals means that population managers must carefully choose which species to maintain, based on demographic and genetic goals of each programme. Consequently, the use of space to house these animals can be somewhat limiting (Penfold et al., 2014).

Several factors can lead to production of additional animals, ranging from uncontrolled breeding, to the requirement for ongoing reproduction to maintain fertility in some species (Powell and Ardaiolo, 2016). Effective population management should consider space available in zoos and avoid overbreeding; however, excess animals are an inevitable outcome for many populations. For example, equal sex ratios at birth can lead to an abundance of males in many polygynous species, such as primates and hoofstock. Similarly, the prioritisation of genetically under-represented animals for reproduction can result in genetically over-represented individuals that are then not prioritised for breeding (Browning, 2018). Other arguments suggest that allowing individuals to express natural reproductive and parental behaviours should be seen as an enriching and welfare-enhancing management strategy in itself, even if this results in more animals than the population needs to meet its sustainability goals.

Although the generation of excess animals is an inevitable outcome of population management, how we manage them is an ongoing debate involving both individual welfare, and broader population management goals. When considering surplus animal management, there are three general approaches: keep them in the managed population, remove them from the managed population and send them to another facility, or management euthanasia (often referred to as "culling") (Browning, 2018). Each option comes with trade-offs. For example, keeping extra animals in a population removes space for more genetically or demographically prioritized animals, and could threaten the sustainability of a population (Penfold et al., 2014). Similarly, removing animals from a zoo-managed population and sending them to other facilities, could potentially compromise the welfare of the individual if the receiver does not have sufficiently high standards of care. In such instances, management euthanasia could be an accepted alternative rather than compromising individual welfare (AZA, 2016; EAZA, 2023; ZAA, 2015). However, in all cases, decisions around management euthanasia are not made lightly, and are often a last resort once all other management options have been exhausted, and no viable alternatives are available which can ensure the welfare of the individual (Browning, 2018).

Management euthanasia, while often discussed within the context of animal welfare, is not an animal welfare issue. Under management euthanasia, the animal does not suffer and can enjoy excellent welfare throughout its life, up to and including the moment it is euthanized (Browning, 2018). However, this is a constantly evolving science, and there have been suggestions that any definition of welfare should be expanded to encompass individual longevity and opportunities for future positive welfare states (Browning, 2018). In such instances, management euthanasia would become less desirable and population management would have to adapt accordingly, favouring alternative methods to manage population size. However, for many species managed in zoo populations, life expectancy is often much higher than wild counterparts (Tidière et al., 2016). Therefore, the argument could be made that unnecessarily prolonging longevity under human care could negatively impact the welfare of the individual.

Although a variety of methods exist which can help to manage reproduction and population size, these come with their own risks and side effects (Asa and Porton, 2010). For example, physical separation of animals is the simplest way we can prevent reproduction and limit surplus animals. Separation can result in social animals being housed in isolation or in inappropriate social groupings, potentially causing behavioural side effects, and removing opportunities to engage in natural reproductive behaviours, which could be seen as an indicator of negative welfare (Penfold et al., 2014). Similarly, the use of various permanent and reversible contraceptive methods, ranging from steroid hormones to sterilisation, has been highly effective, but has also been shown to have both behavioural and physiological impacts on individual animals (Penfold et al., 2014). Contraception impacts can range from altering secondary sexual characteristics, to altering reproductive behaviour, and can result in potential long-term reproductive health issues and infertility (Asa and Porton, 2010; Penfold et al., 2014).

In reality, the efficacy and side effects of most contraceptive methods are poorly understood for the majority of wildlife species (Asa and Porton, 2010), which is not surprising given the diversity and scope of species and reproductive strategies being managed in zoos. However, progress is being made in this area, with groups such as the Association of Zoos and Aquariums (AZA) Reproductive Management Center (RMC) and European Association of Zoos and Aquaria (EAZA) Reproductive Management Group (RMG) dedicated to gathering knowledge on the use, efficacy and impacts of contraceptives in zoo-housed animals (Saint Louis Zoo, 2024; EAZA RMG, 2024). The RMC and RMG jointly maintain the Contraception Database, a repository containing more than 50,000 records on efficacy, dosing, and reversibility for a wide range of contraceptives and species. Using standardised data available in the Contraception Database and ZIMS, the AZA RMC and EAZA RMG can analyse population trends over time, allowing for the production of species-specific guidelines to support individual zoos and population managers.

The importance of contraception as a reproductive management tool is only going to grow in the future, as ethical and moral questions surrounding the use of management euthanasia as a population management tool intensify (Browning, 2018). Therefore, further research is needed to identify the most appropriate methods of species-specific contraception for individuals, and their potential behavioural and physiological impacts.

23.5 THE FUTURE OF POPULATION MANAGEMENT, RECORDS AND WELFARE

Presently, intensive population management most commonly practised by zoos focuses on the individual level and relies on individual data and pedigrees (Ballou et al., 2010); however, population management is expanding in new ways to serve more types of populations. Group-managed species are often left out of population management or welfare assessment due to complicated data and lack of assessment tools, however, a global cooperation called the Group Management Initiative is working to build accessible data tools and standards to make management more accessible (Rodriguez-Clark et al., 2022). Incorporating molecular data into pedigrees and genetic analyses have been widely recognized as a needed component to improve population management by resolving unknown pedigree or taxonomic uncertainty (Norman et al., 2018). People are an often overlooked variable in population management within zoos. PMCTrack is an online database created to survey animal managers and store data on recommendation fulfilment in AZA and facilitate communication of animal needs. Much has been learned from studying patterns in fulfilment within AZA, including factors influencing successful transfer recommendations, and learning common reasons for unfulfilled breeding recommendations (Faust et al., 2019; Gray et al., 2023).

Although population management often happens at a regional level, each zoo is responsible for deciding which species are part of their institutional species plans, and for the welfare of all individuals under their care. Species plans for zoos should be created by following structured and strategic processes that balance conservation, education, research, and public engagement goals. Often, these are designed to ensure that species housed in zoos are representative of biodiversity, contribute to conservation, and are well-suited to the individual zoo's environment.

Zoos that are part of regional associations, like AZA and EAZA, often consult taxon-specific Regional Collection Plans (RCPs) to determine which species to include in their institutional species plans. RCPs will outline discrete roles for a species within a zoo, such as conservation or education roles (IUCN, 2014). RCPs are then balanced against a zoo's mission and objectives, which often aligns with larger conservation goals, such as those set by regional (AZA, EAZA) or global (World Association of Zoos and Aquariums, WAZA) associations.

Species selection for a zoo involves competing demands. While conservation priorities are paramount, zoos must consider the educational value of a species, visitor appeal, and the practicality of its care. For instance, iconic species like elephants (Proboscidae) and big cats (Felidae) attract visitors and often disproportionately contribute to a zoo's financial sustainability, yet these animals require extensive investment in resources and space, and raise the most welfare concerns (Mooney et al., 2020). Conversely, smaller or less well-known species may be easier to care for and still fill critical conservation roles but might not draw as much public interest (Carr, 2016).

Considering the welfare of individuals and species is a relatively new addition to institutional species planning, which is only going to grow as environments change and our understanding of animal needs and welfare grows (Harley, 2022). For example, Minnesota Zoo decided to no longer keep musk ox (*Ovibos moschatus*) as a direct result of welfare concerns caused by increasing summer temperatures over the past decade that were associated with heat-related issues seen in the musk ox (Minnesota Zoo, 2021).

As modern zoos emphasise the importance of providing environments that meet the physical, psychological, and social needs of animals, welfare considerations may increasingly limit the species and populations that zoos can responsibly keep. In balancing these competing demands, zoos aim to create species plans which are dynamic and responsive to changing conservation needs, and advancements in animal welfare science. The importance of welfare cannot be overstated in this respect, as it directly impacts the success of conservation efforts and the educational value of the zoo experience.

23.6 CONCLUSIONS

Animal welfare is central to effective population management and record-keeping in zoos, shaping how animals are cared for and how populations are maintained. Prioritizing welfare requires detailed records on individual health, behaviour, genetics, and life history to inform decisions that balance the needs of individual animals with broader population goals. Welfare considerations influence key management strategies, such as breeding and transfers, ensuring natural behaviours and species-specific needs are met while minimizing stress and health risks. Tools like ZIMS and studbooks enable the collection and analysis of welfare-related data, helping identify trends, mitigate challenges like inbreeding or reproductive challenges, and refine husbandry practices. By harmonizing welfare considerations, meticulous record-keeping, and population management, zoos can simultaneously improve the quality of life for individual animals and ensure the long-term sustainability of managed populations.

REFERENCES

Asa, C. S. & Porton, I. J. (2010). Contraception as a management tool for controlling surplus animals. In: Kleiman, D. G., Thompson, K. V., & Baer, C. K. (eds). *Wild mammals in captivity: principles and techniques for zoo management*. University of Chicago Press, Chicago, IL, pp. 469–482.

Association of Zoos and Aquariums (AZA). (2016). *AZA policy on responsible population management*. https://assets.speakcdn.com/assets/2332/aza_policy_on_responsible_population_management_final.pdf [Retrieved August 31, 2024]

Ballou, J. D., Lees, C., Faust, L. J., Long, S., Lynch, C., Bingaman Lackey, L. & Foose, T. J. (2010). Demographic and genetic management of captive populations. In: Kleiman, D. G., Thompson, K. V., & Baer, C. K. (eds). *Wild mammals in captivity: principles and techniques for zoo management*. University of Chicago Press, Chicago, IL, pp. 219–252.

Barber, J. C. (2009). Programmatic approaches to assessing and improving animal welfare in zoos and aquariums. *Zoo Biology, 28*, 519–530.

Bauman, K., Sahrmann, J., Franklin, A., Asa, C., Agnew, M., Traylor-Holzer, K. & Powell, D. (2019). Reproductive Viability Analysis (RVA) as a new tool for ex situ population management. *Zoo Biology, 38*(1), 55–66.

Browning, H. (2018). No room at the zoo: Management euthanasia and animal welfare. *Journal of Agricultural and Environmental Ethics, 31*(4), 483–498.

Carr, N. (2016). An analysis of zoo visitors' favourite and least favourite animals. *Tourism Management Perspectives, 20*, 70–76.

Conde, D. A., Staerk, J., Colchero, F., da Silva, R., Schöley, J., Baden, H. M., & Vaupel, J. W. (2019). Data gaps and opportunities for comparative and conservation biology. *Proceedings of the National Academy of Sciences, 116*(19), 9658–9664.

DiVincenti Jr., L., McDowell, A. & Herrelko, E. S. (2023). Integrating individual animal and population welfare in zoos and aquariums. *Animals, 13*(10), 1577.

European Association of Zoos and Aquaria (EAZA). (2023). *Position statement on management Euthanasia/Culling*. https://www.eaza.net/assets/Uploads/Position-statements/2023/2023-04-EAZA-Management-Euthanasia-Culling-Statement.pdf [Retrieved August 31, 2024]

EAZA RMG. (2024). *Welcome to the EAZA RMG website. EAZA Reproductive Management Group*. https://eazarmg.org/ [Retrieved August 31, 2024]

Faust, L. J., Long, S. T., Perišin, K. & Simonis, J. L. (2019). Uncovering challenges to sustainability of AZA Animal Programs by evaluating the outcomes of breeding and transfer recommendations with PMCTrack. *Zoo Biology, 38*(1), 24–35.

Gray, S. M., Parsons, A. W., Andrews, J. E., Schad Eebes, K., Faust, L. J. & Che-Castaldo, J. (2023). Institutional networks in cooperative population management: Exploring patterns in transfer fulfilment. *Zoo Biology, 42*(4), 547–556.

Greggor, A. L., Vicino, G. A., Swaisgood, R. R., Fidgett, A., Brenner, D., Kinney, M. E., & Lamberski, N. (2018). Animal welfare in conservation breeding: Applications and challenges. *Frontiers in Veterinary Science, 5*, 323.

Hammerly, S. C., De la Cerda, D. A., Bailey, H. & Johnson, J. (2016). A pedigree gone bad: Increased offspring survival after using DNA-based relatedness to minimize inbreeding in a captive population. *Animal Conservation, 19*(3), 296–303.

Harley, J. (2022). Behavioural biology in animal collection planning and conservation. In: Rose, P. E. (ed). *The behavioural biology of zoo animals*. CRC Press, Boca Raton, FL, pp. 11–19.

Heldstab, S. A., Müller, D. W., Graber, S. M., Bingaman Lackey, L., Rensch, E., Hatt, J. M., & Clauss, M. (2018). Geographical origin, delayed implantation, and induced ovulation explain reproductive seasonality in the Carnivora. *Journal of Biological Rhythms, 33*(4), 402–419.

IUCN/SSC. (2014). *Guidelines on the use of ex situ management for species conservation*. Version 2.0. IUCN Species Survival Commission, Gland, Switzerland.

Martin-Wintle, M. S., Shepherdson, D., Zhang, G., Zhang, H., Li, D., Zhou, X., & Swaisgood, R. R. (2015). Free mate choice enhances conservation breeding in the endangered giant panda. *Nature Communications*, *6*(1), 10125.

Minnesota Zoo. (2021, May 13). *Saying Goodbye to Musk Oxen at the Zoo - Minnesota Zoo.* https://mnzoo.org/blog/saying-goodbye-to-musk-oxen-at-the-zoo/

Mooney, A., Conde, D. A., Healy, K. & Buckley, Y. M. (2020). A system wide approach to managing zoo collections for visitor attendance and in situ conservation. *Nature Communications*, *11*(1), 584.

Mooney, A., Teare, J. A., Staerk, J., Smeele, S. Q., Rose, P., Edell, R. H., & Buckley, Y. M. (2023). Flock size and structure influence reproductive success in four species of flamingo in 540 captive populations worldwide. *Zoo Biology*, *42*(3), 343–356.

Norman, A., Putnam, A., & Ivy, J. (2018). Use of molecular data in zoo and aquarium collection management: Benefits, challenges, and best practices. *Zoo Biology*, 38(1), 106–118.

Pearson, C., Grant, T. D., McKinney, M, & Andrews, J. (2023). *AZA species survival plan® program population analysis and breeding and transfer plan for Jamaican Iguana (Cyclura collie).* Final Report – 5 May 2023. AZA Population Management Center: AZA Population Management Center at Lincoln Park Zoo, Chicago, IL.

Penfold, L. M., Powell, D., Traylor-Holzer, K. & Asa, C. S. (2014). "Use it or lose it": Characterization, implications, and mitigation of female infertility in captive wildlife. *Zoo Biology*, *33*(1), 20–28.

Powell, D. M. & Ardaiolo, M. (2016). Survey of US zoo and aquarium animal care staff attitudes regarding humane euthanasia for population management. *Zoo Biology*, *35*(3), 187–200.

Ralls, K. & Ballou, J. (1986). Captive breeding programs for populations with a small number of founders. *Trends in Ecology & Evolusion*, *1*(1), 19–22.

Rendle, J., Jackson, B., Hoorn, S. V., Yeap, L., Warren, K., Donaldson, R., & Vaughan-Higgins, R. (2020). A retrospective study of macropod progressive periodontal disease ("lumpy jaw") in captive macropods across Australia and Europe: Using data from the past to inform future macropod management. *Animals*, *10*(11), 1954.

Rodriguez-Clark, K. M., Echeverry-Alcendra, A., Penfold, L., Leus, K., Coyle, B., & Sullivan, E. (2022). The Group Management Initiative: Strategic Framework and 2023-2024 Plan. IUCN SSC Conservation Planning Specialist Group (CPSG), Apple Valley, MN, USA. https://cpsg.org/content/groupmanagement-initiative-strategic-framework.

Saint Louis Zoo. (2024). *AZA Reproductive Management Center (RMC).* https://stlzoo.org/conservation/reproductive-sciences/aza-reproductive-management-center [Retrieved August 31, 2024]

Smeele, S. Q., Conde, D. A., Baudisch, A., Bruslund, S., Iwaniuk, A., Staerk, J., & Aplin, L. (2022). Coevolution of relative brain size and life expectancy in parrots. *Proceedings of the Royal Society B*, *289*(1971), 20212397.

Tidière, M., Gaillard, J. M., Berger, V., Müller, D. W., Bingaman Lackey, L., Gimenez, O., & Lemaître, J. F. (2016). Comparative analyses of longevity and senescence reveal variable survival benefits of living in zoos across mammals. *Scientific Reports*, *6*(1), 36361.

Tidière, M., Müller, P., Sliwa, A., Siberchicot, A. & Douay, G. (2021). Sex-specific actuarial and reproductive senescence in zoo-housed tiger (*Panthera tigris*): The importance of subspecies for conservation. *Zoo Biology*, *40*(4), 320–329.

Wark, J. D., Cronin, K. A., Niemann, T., Shender, M. A., Horrigan, A., Kao, A. & Ross, M. R. (2019). Monitoring the behavior and habitat use of animals to enhance welfare using the ZooMonitor app. *Animal Behavior and Cognition*, *6*(3), 158–167.

Young, A. M., Hobson, E. A., Lackey, L. B., & Wright, T. F. (2012). Survival on the ark: Life-history trends in captive parrots. *Animal Conservation*, *15*(1), 28–43.

Zoo and Aquarium Association (ZAA). (2015). *1b.2 -guidelines – Animal Euthanasia.* https://zooaquarium.org.au/common/Uploaded%20files/Portal/Policies/1b_2_G_Euthanasia.pdf [Retrieved August 31, 2024]

24

Animal-centred habitat design as the foundation of zoo animal welfare and the role of welfare prioritisation

JAKE S. VEASEY

24.1 INTRODUCTION

The axiom "prevention is better than cure" is currently a more apt descriptor for the provision of healthcare in zoos than it is for welfare. A key component of any modern zoo is a comprehensive preventative healthcare programme; by contrast, management of an animal's psychological needs often emphasises tactical, reactive interventions to perceived welfare challenges (often centred around environmental enrichment or behaviour training) rather than proactively addressing welfare priorities through effective enclosure design and integrated management. The differences in approach to providing for the physical and psychological needs of animals are potentially so ingrained that they might not at first be obvious; however, a simple thought experiment can illustrate some key differences.

The captive state imposes challenges to the physical health of animals that necessitate routine healthcare interventions which the wild does not. Examples include footcare in captive elephants (Proboscidea), penguins (Spheniscidae), and equids (Equidae), where inappropriate substrates, limited activity levels, and unnatural diets contribute varyingly to poor foot health (Radcliffe et al., 2002; Meehan et al., 2016; Rebstock et al., 2024). Similarly, dental care for captive elephants and walruses (*Odobenidae rosmarus*) (Emily & Eisner et al., 2021) often becomes necessary due to the absence of opportunities to interact with more natural, less abrasive materials. Such proactive, and largely successful approaches to the physical care of zoo animals contrast with the more reactive approach to safeguarding their psychological well-being. Once basic needs like food, water, shelter, and sanitation are met, a key preoccupation of animal care staff is reactively addressing the consequences of inadequate environments through training and enrichment. Fernandez & Martin (2021) recognised the welfare significance of these reactive interventions; "The modern zoo has been associated with two major behavioural welfare

DOI: 10.1201/9781003429425-24

advances: (a) the use of training to increase voluntary husbandry care, and (b) the implementation of environmental enrichment to promote naturalistic behaviours." That they did not recognise enclosure design as a significant driver for advancing "behavioural welfare" is noteworthy. They also point to the fact that the advances they have identified have their roots in "the operant conditioning-centred, reward-based approach to behavioural psychology" rather than our advanced understanding of the behavioural ecology and fundamental needs of zoo animals. With reference to mammal-specific examples, this chapter argues that, despite claims that the most notable advances in the behavioural welfare of zoo animals have been in enrichment and training, rooted in behavioural psychology and operant or classical conditioning, the greatest opportunities to enhance welfare lie in evidence-based, animal-centred enclosure design, guided by a species' behavioural ecology.

The tendency for zoos to prioritise reactive welfare interventions stems, in part, from the dominance of hybrid welfare models that define welfare through a combination of physical and psychological well-being (Veasey, 2022a). This is in contrast to hedonistic definitions favoured by most welfare scientists, where physical health is recognised as an influential, but non-defining, component (Robbins et al., 2018; Webb et al., 2019). These hybrid definitions are ultimately recursive in nature, defining welfare based on how it is measured and managed rather than by what it is. The logic being that, since health can impact how an animal feels (and vice versa), health is both a worthy welfare objective and useful welfare indicator. While this is valid to a point, it is fallacious to assume that health and welfare are synonymous; health's relevance to welfare lies in how it impacts, or indicates, the affective states of animals. This misconception persists because health and welfare management actions are easier to quantify than impacts on an animal's affective state, resulting in the prioritisation of physical health and management practices over actual welfare outcomes. This disconnect between how welfare is evaluated and how it is experienced by animals is problematic because prioritising health can conflict with welfare outcomes. In general, smaller, simpler environments offer greater oversight and control by carers and so are better suited to risk mitigation and healthcare provisioning, whereas larger, more species-appropriate and complex environments are better suited to providing for species-appropriate psychological opportunities (Figure 24.1). This tension was validated by Veasey (2022a) through a facility-wide welfare audit, revealing an inverse relationship between the capacity of habitats to facilitate veterinary care and their capacity to maintain welfare gauged across a range of physical, environmental and psychological parameters.

There are systemic challenges stemming from hybrid welfare conceptions that inevitably prioritise health-based outcomes and reactive interventions, such as enrichment and training, for example, over proactive measures. This results in habitats better suited to risk mitigation and medical management, rather than catering to a species' fundamental needs. Consequently, zoos are heavily reliant on enrichment and training to compensate for environments that ultimately fail to provide species-appropriate opportunities. A more effective approach focuses not on enhancing reactive interventions but on designing environments that better meet species-specific needs. To ensure that zoo habitats genuinely safeguard welfare rather than acting as mere backdrops for their management, we must recalibrate welfare concepts, adopt proactive solutions over reactive measures, and redesign environments accordingly.

24.2 ZOO HABITATS AS WELFARE DRIVERS

If captive environments offered animals the same cognitive and behavioural opportunities as the wild, welfare would be more secure, abnormal behaviours would be minimal, and enrichment and non-medical training obsolete. Therefore, rather than considering welfare-focused training and enrichment as evidence of an institution's commitment to welfare, its requirement can equally be viewed as an indication of inadequate baseline welfare provisioning, with habitat (enclosure) design accountable for much of those shortcomings. To express this in another way, the aim of any facility's habitat should be to render welfare-focused enrichment or training unnecessary by virtue of a habitat's ability to provide for a species' needs and safeguard its welfare.

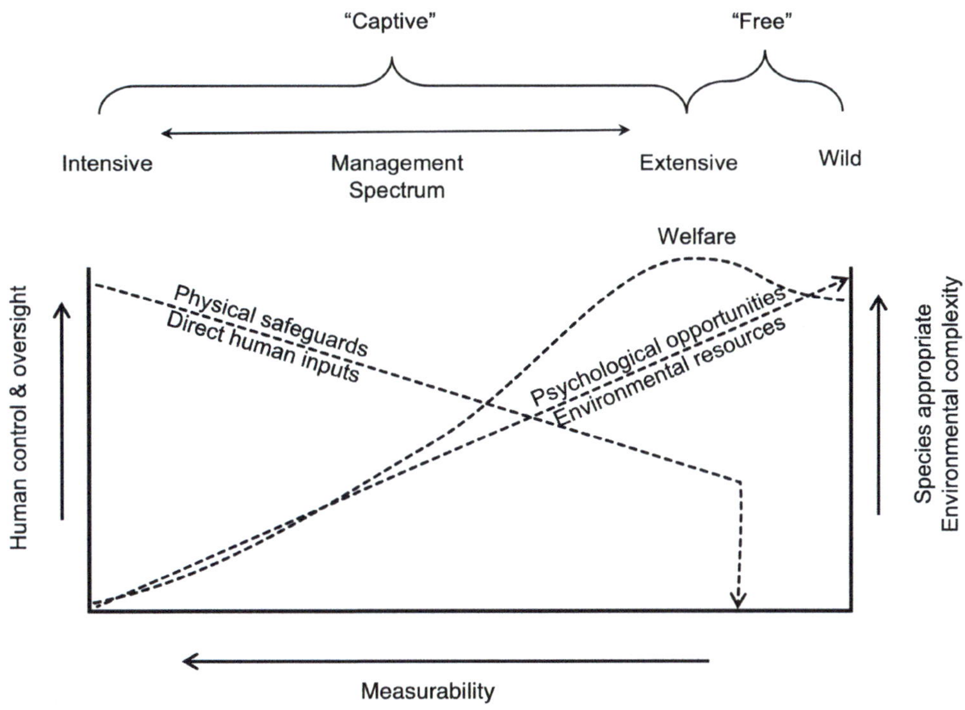

Figure 24.1 Schematic representation of the relationship between population-level animal welfare, efforts to safeguard physical health and the availability of species-appropriate, beneficial psychological opportunities across the spectrum of managed environments – adapted from Veasey (2022a).

However, over the past half century, habitat design in well-resourced zoos has been more successful in transforming visitor experiences than it has been in improving the "behavioural welfare" of animals. By removing bars and immersing visitors in intensely themed environments, the visitor experience is unrecognisable compared to that experienced a generation ago. Despite the outward transformation of zoo habitats and undoubted improvements in enrichment, training, life-support systems, veterinary care, and health-based metrics, improvements in the psychological well-being of animals have not kept pace, with many historic welfare challenges still widespread. This is perhaps most apparent amongst large carnivores, the majority of which still pace for extended periods of time (Bandelli et al., 2023). Currently, carnivore habitat design does little to address the environmental deficiencies manifesting as stereotypic pacing and has focused, instead, on disguising pacing using surfaces and substrates which conceal routes taken, or overhead walkways that turn stereotypic travel into exciting viewing opportunities for zoo visitors.

In understanding how enclosures might better support welfare, it is first necessary to challenge the idea that captivity will inevitably compromise wild animal welfare, regardless of quality. This view is shared by many anti-zoo organisations who argue that even the best zoos cannot prevent mental suffering due to a lack of freedom and natural opportunities. However, it is argued here that it is the quality of habitats provided to zoo animals, together with their management, that constrains welfare, rather than the state of being captive per se.

The logic behind this statement is as follows:

1. Given that all wild animals are constrained by geographical, ecological, and social barriers, it follows that it is the consequences of and the extent to which animals are confined, rather than the nature of that confinement, that is more pertinent to their welfare. Wild lions (*Panthera leo*), for example, confined to

fenced reserves that marginally exceed their natural range, show no apparent behavioural consequences (Hayward et al., 2009). There is no reason to assume that human-imposed boundaries, provided they encompass comparable opportunities to those contained within natural ranges, would have differing welfare impacts.

2. Species' home ranges naturally vary significantly, both between individuals and seasonally within the same individual. For example, tigers (*Panthera tigris*) can range from 16 km² to 10,500 km² (Mazak, 1981; Kroshko et al., 2016), polar bears (*Ursus maritimus*) from 185 km² to 373,539 km² (Mauritzen et al., 2001), and Asian elephants (*Elephas maximus*) from 10 km² to 10,000 km² (Atkinson & Lindsay, 2022). These thousand-fold differences in home range sizes within a species illustrate the inherent flexibility species have in relation to spatial needs, with the smallest wild ranges often being closer in scale to larger captive habitats than with the most expansive wild ranges.
3. There is no reason to assume that a dependence on humans, and resulting insulation from many natural evolutionary pressures, should compromise net welfare, provided that human provisioning meets a species' needs (Rochlitz, 2007). Indeed, given the variability of evolutionary pressures across a species' natural range, the removal of many natural threats such as predation, starvation, and disease – typically absent in the captive state – are arguably more likely to enhance welfare than compromise it.

Despite the variability in spatial utilisation by wild animals, both large and small wild ranges provide the resources necessary for survival, albeit at varying densities and in different forms. And so, while wild animals display significant flexibility in their spatial requirements, their fundamental needs remain largely consistent. There is no compelling reason to suggest that this flexibility should not also extend to the welfare of animals in captivity and provided the resources and opportunities necessary to meet their psychological and physical needs are available, there is therefore little basis to argue that captivity, irrespective of its quality, will inevitably compromise welfare.

Just as the relationship between space and survival resources varies across wild habitats, the relationship between space and welfare opportunities in captivity must also be viewed as distinct but connected. Space undeniably plays a role in captive habitat quality, but larger spaces do not necessarily guarantee better welfare outcomes – smaller, well-designed habitats can outperform larger, less functional spaces in supporting welfare, and increasing habitat size without addressing species' needs directly will not solve welfare issues. It follows, therefore that welfare challenges experienced by captive animals stem from a failure in habitat quality as judged by their ability to meet a species' needs directly, rather than as an inevitable consequence of human-imposed captivity itself. Achieving a point of "peak welfare" requires the recognition that, just as ecosystems provide wild animals with the opportunities necessary for survival, captive habitats must offer animals the opportunities necessary to maintain their motivational equilibrium, and that understanding the former is likely key to identifying requirements for the latter.

Whole-facility welfare audits, carried out by the author, are illustrating the critical role that habitats play in shaping welfare. Principal component analysis of welfare drivers and welfare indicators gathered from multiple audits has so far shown that enclosure quality and the behavioural opportunities they shape account for as much as three-quarters of the variance in welfare states across zoos. These data are graphically summarised in Figure 24.2, where the thickness of the connectors between welfare drivers reflects their impact on the variance in welfare states and behavioural opportunities. It's important to understand that the comparatively modest influence of health, enrichment, and training-related interventions on variances in welfare states may reflect the consistency by which these interventions are provided in the zoos assessed, and the ephemeral nature of many healthcare issues, as much as it does their modest impact on welfare states. Regardless, these assessments confirm the profound impact of habitat quality in shaping behavioural opportunities, and subsequently welfare states.

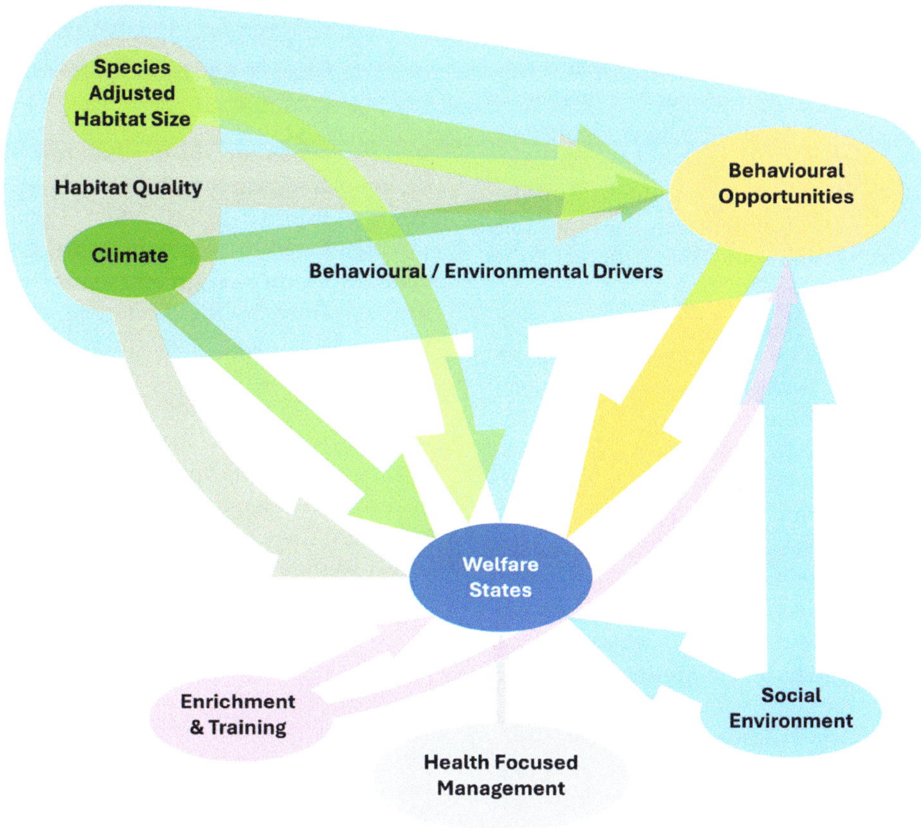

Figure 24.2 The relationships between welfare drivers and welfare states are determined by principal component analysis of data collected from whole facility Animal Welfare Fingerprint™ assessments. The width of arrows reflects the relative influence of factors on the variance observed in subsequent variables.

24.3 BARRIERS TO INNOVATION IN ZOO HABITAT DESIGN

Despite the critical role of habitats in determining welfare states, design-led innovation has lagged behind advances seen in husbandry practices and healthcare (Tidière et al., 2016, 2023; Fernandez & Martin, 2021). Designing zoo habitats that more precisely meet species-specific needs will not only require institutions to reassess how they conceive and judge welfare, but also accept a shift in the role of carers. Instead of focusing on providing for an animal's needs through direct interactions, carers will need to increasingly focus on maintaining environments that empower animals to meet their own motivational and physiological needs, as has been routine in the care of many ectotherms for some time. This conceptual shift may act as an impediment to change, aggravated by the longer lead times and greater evidential threshold required to support costly innovations in infrastructure compared to adjustments in day-to-day husbandry.

Comparative welfare assessments are currently the principal source of evidence used to support changes in facility design. Here, the welfare of animals is evaluated either before or after specific interventions, or by comparing conspecifics in different circumstances and habitats. Aggregated insights from such studies have been used to inform new habitat design concepts, but this approach may inadvertently constrain impactful innovation for the following reasons:

1. Measuring welfare and attributing cause and effect in non-experimental situations such as zoos is challenging; for example, in a scenario

where habitat size correlates with a range of welfare indicators, it is rarely possible to determine whether these changes reflect differences in the volume of space, or changes in variables that may covary with space such as privacy, complexity, or choice. This matters because targeted approaches focusing on the most impactful welfare drivers are likely to be more effective and achievable than increasing enclosure size alone.
2. Comparative welfare assessments can flatter enrichment and training to the detriment of facility design. Although meta-analysis confirms most enrichment studies for mammals report positive outcomes (Shyne, 2006), this could be skewed by a combination of confirmation bias favouring the publication of positive results, and a backdrop of a low welfare baseline. In more optimal environments in which the needs of wild animals are more adequately met, the positive effects of enrichment or training are likely to be less pronounced (Taylor et al., 2023).
3. Benchmarking welfare against the wild presents significant practical and conceptual challenges (Veasey et al., 1996; Dawkins, 2004; Mellor & Beausoleil, 2015) meaning comparative welfare assessments typically reference only captive environments or interventions which may all be deeply suboptimal or lacking in welfare-critical opportunities. Such assessments cannot account for potentially important behavioural opportunities that are typically absent in captivity, such as deep diving, hunting, or migration, and thus offer a flawed basis for facility design.

This is often compounded by the exclusion of field experts, applied ethologists, and welfare scientists from the design process. Although zoo professionals typically provide important input, it is largely based on personal experience and existing best practice guidelines, which are grounded in prevailing husbandry techniques centred around non-hedonistic welfare conceptions, rather than an understanding of the innate needs of the species. This results in an inherently retrospective design process, lacking a truly animal-centred, evidential foundation.

24.4 THE ANIMAL WELFARE PRIORITY IDENTIFICATION SYSTEM

In recognition of the pivotal role that habitat design plays in captive animal welfare, the Animal Welfare Priority Identification System (AWPIS) was developed to empower evidence-based innovation in facility design and welfare management, unconstrained by the limitations of comparative welfare assessments, conceptions of best practice, and current facility design conventions (Veasey, 2020). AWPIS attempts to quantify the significance of opportunities that species encounter in the wild to their welfare in captivity. While the wild serves as a logical reference point for identifying these priorities, there are inherent challenges in making direct welfare comparisons with wild animals, as has already been discussed. It is also neither possible nor desirable to provide captive animals with all the experiences they might encounter in the wild. For instance, while hunting opportunities might improve the welfare of captive predators, offering prey species the opportunity to experience predatory attacks would not improve their welfare, even though both are equally normal and natural behaviours in the wild.

Consequently, in optimising captive welfare by referencing a species' behavioural ecology, the aim is not to replicate the wild but to identify and cater for those naturally occurring opportunities that matter most, recognising that not all are equally beneficial to captive welfare. AWPIS does this by quantifying the importance of naturally occurring behavioural and cognitive opportunities to captive animal welfare by evaluating their evolutionary significance, welfare impacts (where known), and motivational characteristics, using expert assessors and a variant of the Delphi technique. This process relies on the relationship between evolution, motivation, and animal welfare, where the evolutionary significance of a behaviour or cognitive process dictates the intensity and frequency of the corresponding motivation, with animals facing greater welfare challenges when they are prevented from expressing behaviours or cognitive processes of higher evolutionary significance, than those of lesser significance. This relationship is modulated by the origin of the stimulus; internally motivated behaviours or cognitive processes, like sleep, or behaviours associated with reproductive or seasonal cycles for example, tend to be motivated regardless of the environment, while those triggered by external stimuli, such as some display behaviours or the avoidance of acute stressors like predatory attacks, are less likely to arise in the absence of these stimuli (see Figure 24.3). In many instances, such stimuli can be eliminated in captivity, resulting in net welfare improvements.

24.5 EMERGING INSIGHTS FROM AWPIS: THE IMPORTANCE OF INTENTIONAL TRAVEL

The AWPIS approach is yielding striking results that reveal systemic shortcomings in existing conceptions of best practice and facility design, as well as previously unrecognised welfare priorities that require more universal attention. Amongst the most significant findings are the consistently high welfare rankings of physiological necessities (resting, eating, drinking/hydration) and the emergence of intentional/purpose-driven travel as a previously unrecognised welfare priority across a wide range of species (Figure 24.4).

Since AWPIS quantifies the relative welfare significance of behaviours and cognitive processes (based on hedonistic conceptions of welfare), as opposed to evaluating their physiological importance, the high rankings for eating, hydration, and rest emphasise the need for greater consideration of the behavioural and cognitive components of these activities, rather than just their physiological consequences. This mismatch has been highlighted in zoo facility audits where 94% of expert assessments found that zoo animals were provided with nutritionally appropriate diets, but only 30% considered those same diets to be behaviourally appropriate (Figure 24.5).

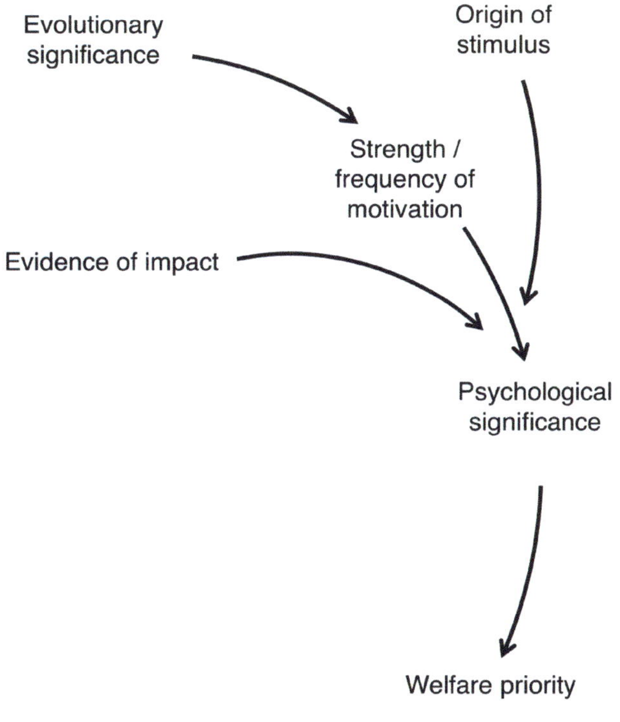

Figure 24.3 Flowchart outlining the principal factors, behaviours, and cognitive processes that are assessed in order to establish their welfare significance in the AWPIS process (Veasey, 2019).

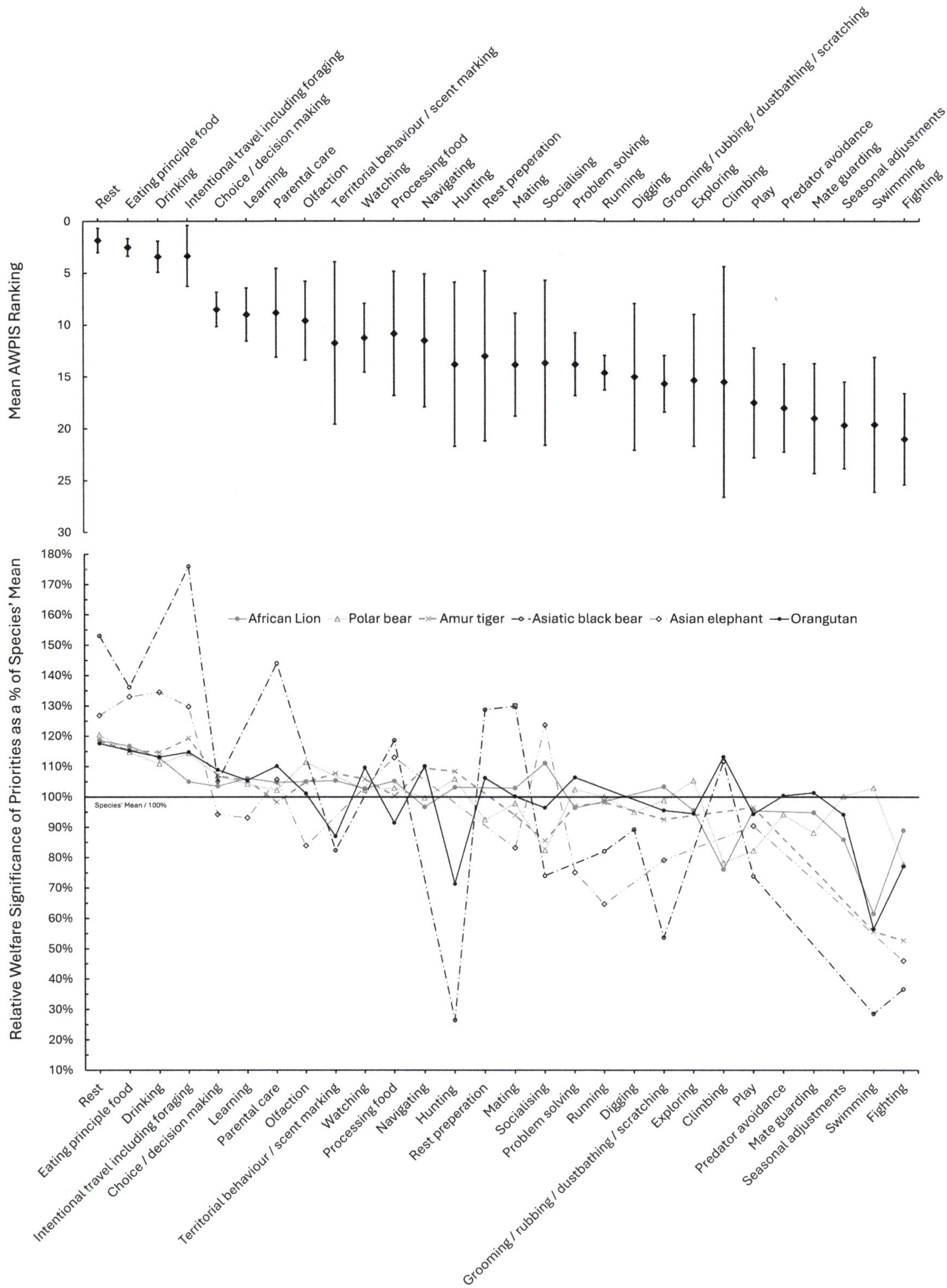

Figure 24.4 Mean AWPIS ranks and scores as a percentage of each species' mean for comparable behaviours and cognitive processes from the following six animal : Asian elephant, polar bear, Asiatic black bear, African lion, orangutan, and Amur tiger.

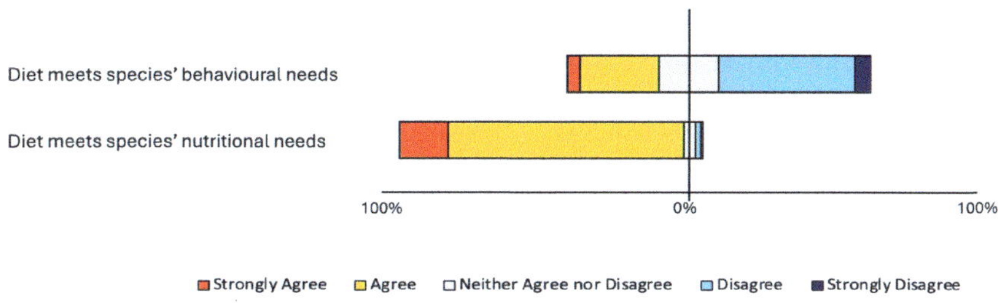

Figure 24.5 Perceived performance in catering to the nutritional and behavioural needs associated with feeding across a diverse zoo, as assessed by zoo staff.

While it is widely accepted that behaviours and cognitive processes associated with food acquisition are important for good animal welfare (Bergeron et al., 2006; Kroshko et al., 2016; Mellor et al., 2021), AWPIS assessments reveal lack of nuanced understanding of the appetitive and consummatory needs of species, resulting in habitat designs and feeding strategies that actively undermine welfare. For instance, although scatter feeding may target important foraging behaviours for omnivorous bear species such as Asiatic black bears (*Ursus thibetanus*), it is inappropriate for obligate carnivores like polar bears. Similarly, while dispersed feeders and scatter feeding may reduce competition among captive elephants for calorie-dense food, it also promotes dispersal, which is at odds with their social structure and natural feeding ecology. The consistency of the AWPIS finding in respect of intentional travel and feeding behaviours, combined with the insights from whole-facility welfare assessments, should compel zoos to pay as close attention to the feeding and foraging ecology of species as they do to their nutrition.

Unlike feeding behaviours, the importance of behaviours and cognitive processes associated with rest and hydration is perhaps less well appreciated. For instance, while captive wild animals are, of course, provided opportunities to remain hydrated, AWPIS assessments show that some species (including polar bears and lions) are not empowered to do so in a species-appropriate manner. Wild polar bears rely on fat metabolism for hydration rather than drinking water, but in captivity, they are provided access to drinking water and are fed diets that lack sufficient fat to support species-typical hydration (Veasey, 2022b). Similarly, wild orangutans (*Pongo* sp.) rarely drink from terrestrial sources akin to drinkers provided in zoos (Veasey et al., 2023). And wild lions travel periodically to specific destinations and drink socially, rather than having free access to drinkers that accommodate a single lion (Veasey, 2023). These differences may seem inconsequential, but the evolutionary importance of hydration, along with associated behaviours and cognitive processes, suggests an underlying powerful motivational need and is likely to play a critical role in a species' welfare. It is also essential to consider such behaviours and cognitive processes within a broader, holistic context, rather than as isolated elements to be considered discretely. For example, social facilitation is a foundation of group organisation for gregarious species such as elephants and lions and has been shown to influence the drinking behaviour of other social carnivores (Glickman et al., 1997). Therefore, it is not inconceivable that the seemingly trivial matter of designing a habitat that precludes lions or elephants from drinking as a pride/herd might undermine cohesion and, consequently, welfare.

Key behavioural and cognitive needs associated with rest are also routinely ignored for many species, particularly the extent to which animals can determine when, where, how, and with whom they can rest. Nomadic species such as elephants and polar bears typically choose to rest in different locations and at different times, and some species, like gorillas (*Gorilla gorilla gorilla*) and orangutans, actively construct resting sites, while others, including elephants and various bear species, may modify areas before resting. Activities such as these are frequently overlooked in facility design and animal management despite their intimate association with powerfully motivated, evolutionarily critical behavioural and physiological needs.

One of the most significant insights from the AWPIS assessments completed so far is the consistently high ranking of intentional travel, which includes foraging, across a diverse range of mammalian species (Figure 24.5). The term 'intentional' is used here to differentiate between the predominant form of travel in the wild and that seen in captivity. In the wild, travel is driven by meaningful, self-selected objectives and it serves as a precursor to achieving motivational equilibrium – whether to access resources or avoid stressors. In captivity, this relationship is broken as care focuses on providing resources and removing stressors, negating the need for a meaningful movement.

Wild animals invest significant time and energy in travel, requiring them to optimise their movements. The decisions they make about when and where to travel, and how this connects to motivated outcomes and activities, highlight the interconnected nature of welfare priorities and reinforce the critical significance of travel. Intentional travel engages complex cognitive processing and decision-making, linking choices to actions that, in captivity, can elevate the welfare value of meaningful choices. For example, in captivity, the decision of an animal to move 5 m towards food type A, or 10 m to food type B may represent a choice, but it is incomparable to those routinely encountered in the wild where an animal might be faced with the choice of a 2 km journey to secure a 40% chance of securing food type A, or a 3 km journey to secure a 70% chance of securing food type B. The difference is not just in the nature of the food, but the cognitive processing required, the time and energy costs associated with that choice, the uncertainty of the outcome, and the positive impacts of good choices. While the captive scenario described here provides a choice, it has little or no consequence and no net caloric impact as the choice is easily reversed. Therefore, it places few, if any, cognitive demands on the animals, eliminates any scope for a true "reward" for a good decision, and so likely has limited value.

Hierarchical cluster analysis of AWPIS rankings, using the Ward method, confirms that travel is one of the most significant determinants of welfare priorities across species (Veasey et al., 2023). Social characteristics and proximity to feeding opportunities, which influence ranging behaviours, leads to surprising similarities in species' needs. For example, in terms of welfare priorities, Asiatic black bears are more like orangutans than polar bears and African lions are more similar to Asian elephants than to Amur tigers (*Panthera tigris altaica*). While the shared social structures of lions and elephants are clear, their nutritional similarities are less obvious, yet both inhabit landscapes where they do not need to travel far between meals compared to Amur tigers and polar bears (Figure 24.6). These examples underscore the significance of behavioural ecology rather than taxonomy in determining welfare needs (Clubb & Mason, 2007; Kroshko et al., 2016; Bandelli et al., 2023) and highlight the cascading significance of intentional travel in shaping a species' needs.

Across a wide range of species, the consistent identification of intentional travel as the most important behaviour that is not a direct physiological necessity, along with its role in shaping welfare characteristics and the prevalence of locomotor stereotypies linked to travel, suggests that intentional travel is one of the most underappreciated welfare priorities in modern animal care. Furthermore, since intentional travel interacts with and influences virtually all other motivated behaviours and objectives for animals, it must be considered a key factor in addressing welfare priorities. Because of its spatial dimension, this can only be effectively managed through animal-centred habitat design rather than interventions such as enrichment and training. Consequently, intentional travel should be established as a discrete welfare objective and is subsequently defined here as the ability of animals to undertake journeys to and from biologically appropriate places, circumstances, and opportunities of their choosing, comparable to those made in the wild.

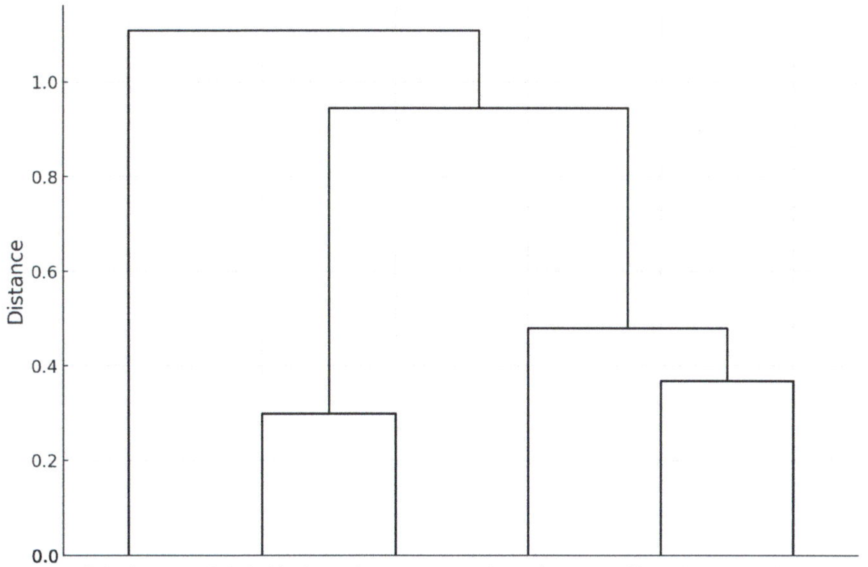

Species	Polar bear	Asiatic black bear	Orangutan	Amur tiger	African lion	Asian elephant
Order	Carnivora	Carnivora	Primate	Carnivora	Carnivora	Proboscidea
Social structure	Solitary	Solitary	Solitary	Solitary	Social	Social
Ranging	Wide	Narrow	Narrow	Wide	Wide	Wide
Diet	Carnivore	Omnivore	Frugivore	Carnivore	Carnivore	Herbivore

Figure 24.6 Dendrogram of Hierarchical Clustering Analysis illustrating the grouping of species based on the similarities of their welfare priorities as determined by AWPIS, and the apparent significance of sociality and ranging characteristics compared to taxonomy.

24.6 THE SIGNIFICANCE OF SUBTLE DIFFERENCES: A CASE STUDY

Amongst the species evaluated through AWPIS, African lions and Amur tigers were assessed specifically to guide new facility developments (Veasey 2020). The assessments revealed nuanced differences in the needs of the two species while highlighting shared priorities, such as intentional travel, and emphasise the need for species-specific, animal-centred habitat design and management strategies, even for closely related species.

Historically, big cat enclosures were clustered together and had shared design criteria regardless of the species – a tiger cage was identical to that of a lion. While zoos have since shifted towards zoogeographic clusters, the tendency to transfer design concepts between related species persists; in the absence of the animals or relevant signage, theming often provides the clearest indicators of which species a habitat is intended for, rather than any species-specific design features. Indeed, prior to these AWPIS assessments, it was originally presumed that African lions and Amur tigers might have similar needs due to their taxonomic, dietary, and ecological similarities, with key differences emerging principally due to differences in sociality. However, the assessments revealed nuanced, yet fundamental, differences requiring entirely different design solutions. The differences between assessments for the species are illustrated in Figure 24.7 which shows the relative ranking of behaviours and cognitive processes shared between the species. That behaviours linked to sociality, including grooming, mating, learning, and fighting, are more important to lions than tigers is self-explanatory, but that travel, foraging, navigation, and to a lesser degree hunting-related behaviours are less important to African lions than Amur tigers is less obvious at first glance (Veasey, 2020, 2023).

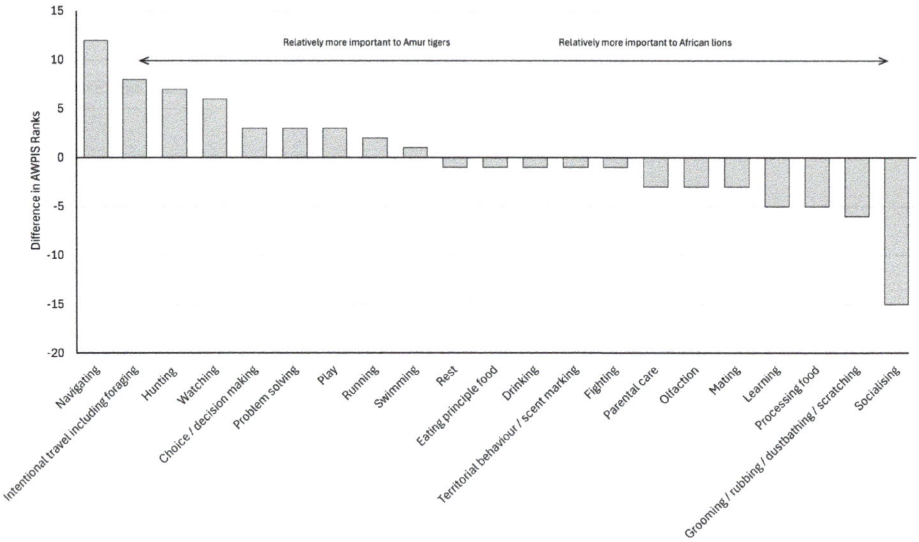

Figure 24.7 A comparison of AWPIS ranks of comparable behaviours and cognitive processes between African lions and Amur tigers.

The moderately lower ranking of intentional travel and foraging for African lions reflects their proximity to feeding opportunities and shared obligation to secure food within the pride. While African lions may need to chase prey, most rarely need to travel long distances to find it, living as they often do amongst some of the highest concentrations of herbivore biomass on Earth. Nonetheless, behaviours linked to intentional travel, which overlap with foraging, ranked highest of the non-physiological necessities for Amur tigers, and for African lions, these behaviours ranked second below sociality. Notably, intentional travel/foraging ranked higher than hunting for both species, even though hunting/feeding-related behaviours have historically been the primary focus of enrichment for captive carnivores (Shyne, 2006). Not only does this emphasis overlook more critical priorities, but it can also be counterproductive for some species. For example, increasing feeding frequency and reducing meal size to maximise enrichment opportunities can result in enriched carnivores being uncharacteristically motivated to seek food after meals that fail to satisfy them in comparison to larger meals secured in the wild (Veasey, 2020).

Amur tigers, living in less productive wild habitats with lower prey densities, range between four and one hundred times more widely than African lions (Mazak, 1981; Hernandez-Blanco et al., 2015). Consequently, behaviours like foraging, intentional travel, navigation, and exploration are likely more critical to individual Amur tigers than to individual African lions, who share navigational and travel demands within a pride in smaller home ranges with higher prey densities. This is also reflected in the greater activity levels of wild tigers compared to lions – 42.75% versus 13% – and may explain why tigers tend to exhibit 36% more stereotypic behaviour in captivity than lions (Kroshko et al., 2016). However, while wild Amur tigers may travel significantly more than wild African lions, the modest drop in ranking of intentional travel for African lions confirms that it remains critical to captive lion welfare.

Species differences in hunting behaviours are also more nuanced than those associated with sociality and travel. While a pride of lions may be able to tackle larger prey than a solitary tiger, typical meal size can be significantly smaller for lions than for tigers. For instance, in the Kruger National Park, impalas (*Aepyceros melampus*) at ~40kg account for over 50% of lion kills (Owen-Smith & Mills, 2008), while wild boar (*Sus scrofa*) at ~100 kg constitute most of the Amur tiger's diet

(Dou et al., 2019). Furthermore, an average lion pride of 15 (Schaller, 1972) means individual lions generally consume smaller, more frequent meals compared to Amur tigers, which not only prey on larger animals but also rarely share kills, except with their dependent young. Consequently, a single kill sustains an adult Amur tiger for significantly longer than it does an African lion (Kroshko et al., 2016). Individual lions also have a more flexible relationship with hunting compared to tigers, as not all members of a pride are required to hunt to eat. Males often do not hunt (Funston et al., 1998), not all females participate in every hunt, and kleptoparasitism of other carnivore kills plays a significant role in lion sociality (Palmer et al., 2023).

Veasey (2020) identified travel as critical to the welfare of captive Amur tigers, because it not only connects tigers to their goals, but also becomes essential for maintaining and monitoring territory. While both African lions and Amur tigers defend their territories, the solitary nature of tigers and their forested habitats leads to a greater reliance on site-specific scent marking, travel, and monitoring. Conversely, the more open savannah environment of lions supports vocalisation and group-based defence, meaning tigers generally would need to cover more ground to defend a typically larger territory, as well as secure prey.

Using the results from AWPIS assessments, a habitat intended to address the needs of tigers is, at the time of writing, nearing completion. This habitat will incorporate a network of paths within a bamboo forest, featuring programmable smart gates, motion sensors, and controllable auditory, visual, and olfactory cues tied to both fixed and variable opportunities. Aiming to replicate the functional complexity of game trails used by this wide-ranging species, intelligent hardware and software are integrated into a dynamic, living landscape, thereby offering tigers biologically relevant opportunities mediated through cognitively challenging, species-appropriate journeys. The enclosure is to be managed by an advanced digital system that interprets the tigers' choices, adapting their journey through the core bamboo forest by adjusting routes based on the destination selected by the tiger. Their subsequent journey is to be augmented by context-specific cues that allow the tigers to learn, navigate, and make complex decisions, leading to species-appropriate journeys with meaningful, biologically relevant outcomes. Cues and journeys linked to mutually exclusive, fixed physical destinations such as those areas providing water, shade, elevation, and shelter will be predictable. Conversely, cues linked to social or feeding opportunities that are not fixed, will be indicative and time-bound, reflecting natural contingencies in the wild.

Whilst the welfare priorities for Amur tigers should be viewed through the lens of providing opportunities for intentional travel, African lions need to be viewed through the lens of their sociality, including fission-fusion dynamics and intentional travel. Subsequently, a habitat devised to accommodate the needs of African lions currently under development is intended to ensure that they can drink together, feed together, travel together, make meaningful group decisions linked to intentional travel, cooperate to achieve motivated outcomes and, importantly, experience fission-fusion/spatial separation as they do in the wild. For lions, therefore, an ideal captive habitat that will enable a pride to travel both collectively and individually to discrete locations provides different biologically relevant, mutually exclusive opportunities. Shade and water, for example, might not be in the same space and, using interactive system technologies adapted from the tiger habitat, the lion habitat will require lions to undertake more meaningful journeys between each location that was previously possible. Configured to recognise individual lions, track their location, and monitor their pride allegiances, the proposed system will allow different prides to overlap in space, not time, replicating natural contingencies. Smart gates will automatically grant or deny access based on pride's allegiances, allowing lions to separate from their pride and return safely to zones with compatible individuals, while still enabling staff to intervene remotely when necessary. This would enable multiple prides, or even different species, to safely cohabit the same expanded space on their own terms, with real-time adjustments guided by the system and staff inputs.

When applied to zoo enclosure design, the variation in welfare priorities among these closely related species as described here results in distinct design concepts for each. This illustrates the need to utilise evidence-based methodologies to identify and then satisfy the unique needs of each species, regardless of perceived similarities or established practices. It also highlights the critical role that enclosure design can play in meeting those needs.

24.7 CONCLUSIONS

While modern zoo welfare practices are well-intentioned, they may unintentionally compromise animal welfare by prioritising welfare interventions such as training, enrichment, and healthcare provisioning over more subtle welfare outcomes which are tied to an animal's affective states. Though these practices offer benefits, they typically fail to meet the core needs of captive animals which, realistically, can only be addressed through habitat design. However, while enhanced habitat design holds the greatest promise for long-term welfare improvements, it is also the least adaptable solution once built. This underscores the need for more thoughtful, collaborative, and multidisciplinary design processes that integrate behavioural ecology, welfare science, and technological innovations into landscape architecture, using tools like AWPIS to provide an animal animal-centred foundation. Ultimately, the success of an animal-centred habitat lies in its ability to meet the behavioural and cognitive needs of each species, rather than relying on established practices or reactive measures to compensate for design flaws. Tools like AWPIS are already pointing the way to transformative, habitat-led solutions to previously misunderstood or overlooked welfare challenges. As habitats better suited to the needs of their species become more prevalent, the role of animal care professionals will also shift from one in which they focus on direct interventions to improve welfare to one where they increasingly focus on managing environments that empower animals to maintain their own motivational equilibrium. Ultimately, though, the speed of this transition depends on the willingness of key stakeholders to recalibrate their welfare conceptions and embrace the necessary changes in both habitat design and welfare management.

REFERENCES

Atkinson, R. & Lindsay, K. (2022). *Expansive, diverse habitats are vital to the welfare of elephants in captivity*. Conservative Animal Welfare Foundation. https://www.conservativeanimalwelfarefoundation.org/wp-content/uploads/2022/07/Digital-CAWF-Elephants-in-Captivity-Report-final-v3.pdf

Bandeli, M., Mellor, E. L., Kroshko, J., Maherali, H., & Mason, G. J. (2023). The welfare problems of wide-ranging Carnivora reflect naturally itinerant lifestyles. *Royal Society Open Science, 10*(9), 230437.

Bergeron, R., Bandell-Waters, J. J., Lambton, S. & Mason, G. (2006). Stereotypic oral behaviour in captive ungulates: Foraging, diet and gastrointestinal function. In Mason, G. & Rushen, J. (Eds.) *Stereotypic Animal Behaviour: Fundamentals and Applications to Welfare* (2nd ed., pp. 19–57). CABI Publishing, Wallingford, UK.

Clubb, R. & Mason, G. J. (2007). Natural behavioural biology as a risk factor in carnivore welfare: How analysing species differences could help zoos improve enclosures. *Applied Animal Behaviour Science, 102*(3–4), 303–328.

Dawkins, M. S. (2004). Using behaviour to assess animal welfare. *Animal Welfare, 13*(S1), S3–S7.

Dou, H., Yang, H., Smith, J. L. D., Feng, L., Wang, T. & Ge, J. (2019). Prey selection of Amur tigers in relation to the spatiotemporal overlap with prey across the Sino–Russian border. *Wildlife Biology, 2019*(1), 1–11.

Emily, P. P. & Eisner, E. R. (2021). *Zoo and Wild Animal Dentistry*. John Wiley & Sons, Inc, Hoboken, NJ.

Fernandez, E. J. & Martin, A. L. (2021). Animal training, environmental enrichment, and animal welfare: A history of behavior analysis in zoos. *Journal of Zoological & Botanical Gardens, 2*(4), 531–543.

Funston, P. J., Mills, M. G. L., Biggs, H. C. & Richardson, P. R. K. (1998). Hunting by male lions: Ecological influences and socio- ecological implications. *Animal Behaviour, 56*(6), 1333–1345.

Glickman, S. E., Zabel, C. J., Yoerg, S. I., Weldele, M. L., Drea, C. M. & Frank, L. G. (1997). Social facilitation, affiliation, and dominance in the social life of spotted hyenas. In Carter, C. S., Lederhendeler, I. I. & Kirkpatrick, B. (Eds.) *Integrative Neurobiology of Affiliation* (pp. 131–140). MIT Press, London, UK.

Hayward, M. W., Hayward, G. J., Druce, D. J. & Kerley, G. I. H. (2009). Do fences constrain predator movements on an evolutionary scale? Home range, food intake and movement patterns of large predators reintroduced to Addo Elephant National Park, South Africa. *Biodiversity and Conservation, 18,* 887–904.

Hernandez-Blanco J. A., Naidenko, S. V., Chistopolova, M. E., et al. (2015). Social structure and space use of Amur tigers (Panthera tigris altaica) in Southern Russian Far East based on GPS telemetry data. *Integrative Zoology, 10*(4), 365–375.

Kroshko, J., Clubb, R., Harper, L., Mellor, E., Moehrenschlager, A. & Mason, G. (2016). Stereotypic route tracing in captive Carnivora is predicted by species-typical home range sizes and hunting styles. *Animal Behaviour, 117,* 197–209.

Mauritzen, M., Derocher, A. E. & Wiig, Ø. (2001). Space-use strategies of female polar bears in a dynamic sea ice habitat. *Canadian Journal of Zoology, 79*(9), 1704–1713.

Mazak, V. (1981). Panthera tigris. *Mammalian Species, 152,* 1–8.

Meehan, C. L., Mench J. A., Carlstead, K. & Hogan, J. N. (2016) Determining connections between the daily lives of zoo elephants and their welfare: An epidemiological approach. *PLoS One, 11*(7), e0158124.

Mellor, D. J. & Beausoleil, N. J. (2015). Extending the five domains model for animal welfare assessment to incorporate positive welfare states. *Animal Welfare, 24*(3), 241–253.

Mellor, E. L., McDonald Kinkaid, H. K., Mendl, M. T., Cuthill, I. C., van Zeeland, Y. R. A. & Mason, G. J. (2021). Nature calls: Intelligence and natural foraging style predict poor welfare in captive parrots. *Proceedings of the Royal Society B, 288*(1906), 1–10.

Owen-Smith, N. & Mills, M. G. (2008). Predator–prey size relationships in an African large-mammal food web. *The Journal of Animal Ecology, 77*(1), 173–183.

Palmer, M., Borrego, N. & Packer, C. (2023). Social strategies of the African lion. In Srinivasan, M. & Würsig, B. (Eds.) *Social Strategies of Carnivorous Mammalian Predators: Hunting and Surviving as Families.* Springer International Publishing AG, Cham, Switzerland.

Radcliffe, R. M. & Osofsky, S. A. (2002). Disease concerns for wild equids. In Moehlman, P. D. (Ed.) *Equids: Zebras, Asses, and Horses: Status Survey and Conservation Action Plan* (pp. 124–152). The World Conservation Union, Gland, Switzerland.

Rebstock, G. A., Wellington, K. P. & Boersma, P. D. (2024). Foot darkening with age in Spheniscus penguins: Applications and functions. *PeerJ, 12,* e17937.

Robbins, J., Franks, B. & von Keyserlingk, M. A. G. (2018). 'More than a feeling': An empirical investigation of hedonistic accounts of animal welfare. *PLoS One, 13*(3), e0193864.

Rochlitz, I. (2007). *The Welfare of Cats.* Springer International Publishing AG, Cham, Switzerland.

Schaller, G. B. (1972) *The Serengeti Lion: A Study of Predator-Prey Relations.* University of Chicago Press, Chicago, IL.

Shyne, A. (2006). Meta-analytic review of the effects of enrichment on stereotypic behavior in zoo mammals. *Zoo Biology, 25*(4), 317–337.

Taylor, A., Edwards, L. & Van Doorn, M. (2023). An effective environmental enrichment framework for the continual improvement of production animal welfare. *Animal Welfare, 32*(e14), 1–11.

Tidière, M., Gaillard, J. M., Berger, V., Muller, D. W. H., Bingaman Lackey, L., Gimenez, O., Clauss, M. & Lemaitre, J. F. (2016). Comparative analyses of longevity and senescence reveal variable survival benefits of living in zoos across mammals. *Scientific Reports, 6,* 36361.

Tidière, M., Colchero, F., Staerk, J., Adkesson, M. J., Andersen, D. H., Bland, L., ... & Conde, D. A. (2023). Survival improvements of marine mammals in zoological institutions mirror historical advances in human longevity. *Proceedings of the Royal Society B, 290*(2009), 20231895.

Veasey, J. S. (2019). *Identifying design priorities for optimal welfare.* Proceedings of the International Zoo Design Conference, Wroclaw, Poland, April 4–7 (pp. 57–71).

Veasey, J. S. (2020). Can zoos ever be big enough for large wild animals? A review using an expert panel assessment of the psychological priorities of the Amur Tiger (*Panthera tigris altaica*) as a model species. *Animals, 10,* 1536.

Veasey, J. S. (2022a). Differing animal welfare conceptions and what they mean for the future of zoos and aquariums, insights from an animal welfare audit. *Zoo Biology*, *41*(4), 1–16.

Veasey, J. S. (2022b). *Bear Necessities: Priorities in Polar Bear Welfare, Management and Facility Design*. Wildlife Foundation, Doncaster, UK.

Veasey, J. S., Armstrong, K., Butcher, F., Embury, A., Grossfeldt, L., Robbins, A. & Thompson, H. (2023). *Orangutan Welfare Priority Assessment*. https://www.carefortherare.com/news/welfare-priority-assessment-for-orang-utan-launched (Access Date 04/03/2025)

Veasey, J. S., Waran, N. K. & Young, R. J. (1996). On comparing the behavior of zoo housed animals with wild conspecefics as a welfare indicator. *Animal Welfare*, *5*, 13–24.

Webb, L. E., Veenhoven, R., Harfeld, J. L. & Jensen, M. B. (2019). What is animal happiness? *Annals of the New York Academy of Sciences: Reports*, *1438*, 62–76.

25

Zoo welfare strategy – from page to practice

JUSTINE K. PARTOON

25.1 INTRODUCTION

This chapter is dedicated to the development and implementation of an animal welfare strategy for a zoological park, marine park, or public aquarium (hereafter "zoos"). The challenges of defining and assessing animal welfare are discussed elsewhere in this book. The approach taken in this chapter is aligned with the definitions of welfare detailed in previous chapters and acknowledges three key components of animal welfare: (1) biological functioning, (2) naturalness, and (3) affective state. Best practice animal welfare science is rapidly evolving with the advancement of both scientific study and changes in public attitudes. Animal welfare is a societal issue, where citizens decide whether to endorse or disengage from animal-related organisations (e.g. zoos), based on their perception of animal welfare. A failure to examine and underpin the ethical foundation of zoos may lead to a loss of the 'social licence to operate' (SLO) (Gray, 2017). This loss of support can have significant impacts on zoos in terms of patronage and therefore financial viability.

To avert a decrease in the SLO, proactive stakeholder engagement and transparency of practice are of paramount importance, especially as they relate to animal welfare (Hampton et al., 2020), requiring continuous and diligent animal welfare evaluation and rectification of any welfare concerns. For zoos to uphold their SLO, they must prioritise the delivery of conservation and education outcomes. These activities should always be conducted in a manner that promotes and enhances positive animal welfare, ensuring a clear and strong ethical foundation of operations.

The gold standard approach to zoo animal welfare is to proactively strive for optimal welfare, as opposed to reactively responding to problems when they emerge (Hill & Broom, 2009). Increasingly, zoos prioritise animal welfare as part of their wider operational management; however, only a few, if any, describe themselves as animal

DOI: 10.1201/9781003429425-25

welfare organisations (Kagan et al., 2018). If the core aim of a zoo is that animals experience positive welfare and continue to thrive, it is essential to have a visionary animal welfare strategy that encompasses the zoo's core ethos and articulates its unwavering commitment to every animal within its care.

This chapter advocates for all zoos to adopt a strategy that embeds animal welfare in every aspect of management, beyond animal husbandry, into the wider realms of zoo policy, from financial planning to educational delivery, from conservation to commercial operations to marketing and beyond.

25.2 WHAT IS AN ANIMAL WELFARE STRATEGY?

An animal welfare strategy is not a husbandry plan. Rather, it is a wider integration of animal welfare into all zoo policies and procedures, with the aim to embed good welfare practice across the entire organisation and beyond to the broader community. This chapter highlights the importance of ensuring positive animal welfare in zoos. There is a significant gap in the process of advocating, monitoring, and ensuring positive welfare across all aspects of zoo operations. This realisation led to the development of an animal welfare strategy for Zoos South Australia, which serves as a potential model for other organisations.

25.3 WHY DO ZOOS NEED AN ANIMAL WELFARE STRATEGY?

An effective animal welfare strategy consists of a robust framework that integrates all key relevant areas of operation, provides clear measurable goals, and embraces a proactive approach to the promotion of positive animal welfare throughout the organisation. The strategy should commit the zoo to conducting ongoing applied animal welfare research, individual monitoring, and regular assessments of groups, identifying both welfare risks and enhancement opportunities wherever they exist (Sherwen et al., 2018).

A comprehensive animal welfare strategy is not a step-by-step guide to improved animal husbandry, but rather a holistic approach to improving all avenues that impact animal welfare. It encompasses a 'big picture' approach to animal welfare throughout the zoo, ensuring consultation with all relevant stakeholders, including animal care and non-animal care staff. Ultimately, zoos have a moral obligation to provide a life worth living for all the animals under their care. A robust animal welfare strategy is a great mechanism to deliver on this objective. It is the intention of this chapter to help zoo personnel create their own animal welfare strategy, committing their institution, at the highest level, to effective management of animal welfare throughout their organisation.

Most zoos have a series of core documents that underpin their operations. These can include plans (e.g. master plan, business plan) and charters (e.g. sustainability charter, accessibility charter), which are typically reviewed every 3–5 years and are bound together by a strategic plan (Figure 25.1). To be effective, it is essential that the animal welfare strategy is seamlessly integrated into this 'ecosystem', especially the strategic plan, and that the documents 'communicate' effectively with one another.

There are key inputs to consider when formulating an animal welfare strategy, which are critical to the development and success of the strategy (Figure 25.2). The strategy must ensure that the zoo adheres to legislative and regional zoo association accreditation requirements. However, it is worth noting that legislative and accreditation requirements are enforced using minimum standards, which are intended to prevent negative animal welfare states but are not always sufficient to provide optimal and/or positive animal welfare (Melfi, 2009). It is, therefore, good practice to strive beyond these standards.

In addition, there should be a person in charge of the development of an animal welfare strategy to ensure its progress and implementation. This person should be determined, knowledgeable about animal welfare science, and an inspiring communicator, able to articulate a clear vision and ensure that the development and implementation of the strategy is successful.

Figure 25.1 A schematic overview of a zoo 'ecosystem' from Zoos South Australia showing core organisational plans and charters. The animal welfare strategy should be integrated within this organisational framework at the highest level.

25.4 REVIEWING CURRENT ANIMAL WELFARE DOCUMENTATION: CREATING A BASELINE

The first step in formulating an animal welfare strategy at a zoo is to compile and analyse all existing documentation and procedures pertaining to animal welfare. Key documents and procedures to consider can include:

1) Strategic plan (and key performance indicators).
2) Animal welfare charter/code.
3) Animal welfare evaluation tools:
    a) animal welfare assessment.
    b) internal/external animal welfare transfer protocol.
    c) quality of life assessment.
    d) health assessment.
    e) nutritional evaluation.
    f) environmental enrichment protocol.
    g) behavioural monitoring program.
    h) animal-visitor interaction protocol.
    i) animal training program.
    j) habitat design protocol.
4) Animal record-keeping data and analysis tools (ZIMS, OERCA, ZooMonitor, etc.).
5) Operational policies and procedures.
6) Animal welfare concern reporting systems (internal and external).

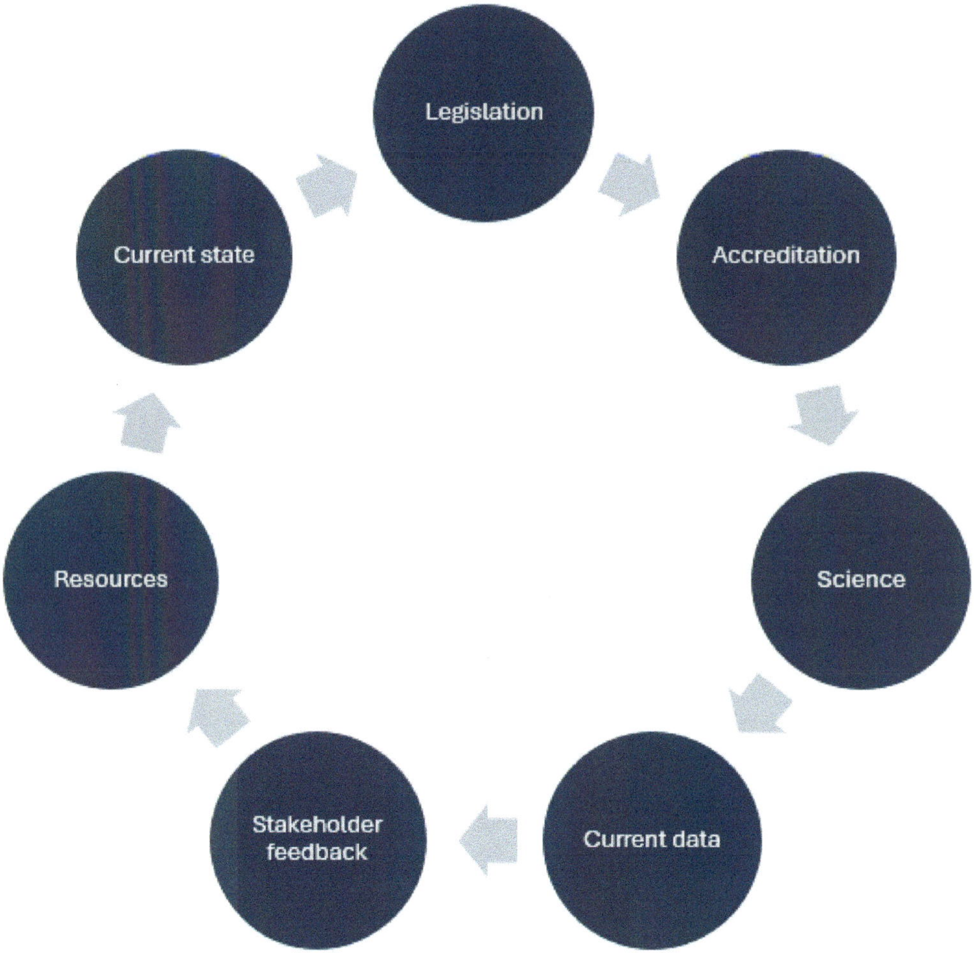

Figure 25.2 A schematic overview of the key inputs to consider when creating a new animal welfare strategy.

7) Husbandry manuals.
8) Research projects (internal and external).
9) Animal Welfare and Ethics Committee Terms of Reference.
10) Collection management and planning processes.
11) Animal welfare budgets.
12) Local legislation and standards.
13) Regional association welfare assurance tools (ZAA, AZA, EAZA, BIAZA, etc.)

Once compiled, a gap analysis should be conducted on the existing documentation and procedures. This analysis will help illuminate areas of disparity between existing practices and protocols, and the future aspiration for positive animal welfare, facilitating the creation of solutions to drive change. The initial gap analysis establishes a baseline from which a new animal welfare strategy can be launched.

When reviewing the key inputs detailed in Figure 25.2, it is important to analyse how the different parts of the animal welfare 'puzzle' overlap, which helps identify areas that may be overlooked or missing. For example: is there a clear policy/protocol that ensures that animal welfare assessments inform annual budget-setting? Is there a clear habitat design policy that ensures all relevant stakeholders are consulted and that the animal's natural habitat and behavioural repertoire are considered? Is this process nimble?

A gap analysis also reveals other less direct avenues for improvement – e.g. the implementation of new technologies or the professional development of staff in support of emerging practices. An animal welfare strategy must be underpinned by science. Using peer-reviewed science as the foundation of an animal welfare strategy assists with communicating and effectively implementing the strategy across the organisation, helping address challenges associated with status quo bias or other impediments arising from the diverse array of stakeholders across the zoo (Ward et al., 2018).

25.5 INVOLVING STAKEHOLDERS: ESTABLISHING THE VISION

After reviewing all relevant documentation, it is beneficial to gather stakeholder feedback to assess their understanding of animal welfare science and to understand the degree to which it is effectively implemented in practice. This feedback can be gathered using the Delphi method (i.e. consultation of experts to gather consensus across a broad base of subjective opinions), online surveys, workshop focus groups, and one-on-one interviews. One benefit of collecting this data through online surveys is that the process can be repeated in the future, providing a quantitative baseline that can be tracked and measured against over time, providing insight into the effectiveness of the animal welfare strategy. While it is preferable that all staff feel at ease when raising animal welfare concerns, allowing staff to remain anonymous may enable them to feel comfortable about sharing their viewpoints.

The composition of the stakeholder group is important, as there may be a tendency to only consider the obvious stakeholders who have direct involvement with the animals, such as keepers, animal management (curators, etc.), veterinary surgeons, and veterinary nurses. However, there are many other key stakeholders in a zoo, such as management, horticulture, technical services, marketing, visitor services, education, conservation, human resources, volunteers, etc. Even engaging members of the public is key for a diverse perspective, transparency, and positive community engagement.

The level of stakeholder involvement in the development of an animal welfare strategy can vary, depending on their role and their level of expertise. Animal care staff may be expected to contribute more than other stakeholders; however, input from stakeholders without an animal care background provides a different and valuable perspective, promoting a culture of participation and inclusion, which is more likely to secure future organisation- and community-wide 'buy-in' of the resulting animal welfare strategy.

Reputable zoos are now employing animal welfare specialists, including animal welfare managers, enrichment coordinators, animal welfare researchers, and behavioural specialists. Employing animal welfare scientists is highly recommended (Ward et al., 2018), as these staff are key to the implementation of overarching and effective animal welfare initiatives. Some zoos opt for external animal welfare consultants, as they can bring an objective, scientific approach to the assessment of animal welfare and any associated prescriptive measures.

If a zoo does not employ specialised animal welfare staff (i.e. animal welfare scientists), it is recommended that, during the stakeholder engagement phase, animal welfare scientists from a local university and/or other zoos are consulted. Additionally, it is valuable to consult with regional accreditation organisations, as they will be able to provide constructive feedback and verify whether your strategy aligns with their framework and with future strategic directions.

If a zoo has an internal Animal Welfare and Ethics Committee (AWEC), they should be included in the stakeholder engagement process. The role of an AWEC typically focuses on providing independent animal welfare advice, investigating animal welfare concerns, and enhancing day-to-day husbandry practices to improve animal welfare. It should be noted that an internal AWEC is different from an external Animal Ethics Committee, which are established to ensure compliance with ethical practices and standards required by federal or state government legislation. If your zoo does not have an AWEC, it is recommended to form one as part of the animal welfare strategy and it should include committee members educated in animal welfare science (Miller & Chinnadurai, 2023).

It is essential that the final animal welfare strategy has 'buy in' from all facets of the organisation, including zoo leadership (i.e. Chief Executive, Chief Financial Officer, Human Resources Director, etc.), managers, and all other staff. The leadership team will provide the resources to implement the strategy and ensure it is aligned with other core aspects of the zoo's operation (the strategic plan, annual budgets, etc.). It is also crucial to have a clear pathway to implementing the animal welfare strategy, including the provision of necessary resources (e.g. staffing, financial, facilities/equipment), which itself will require a gap analysis to ensure that the aspiration of the strategy does not unreasonably outstrip the capacity of the organisation.

25.6 DEVELOPING THE STRATEGY–MAKING IDEAS A REALITY

A key part of an animal welfare strategy is to determine the approach to assessing and improving welfare. Evaluating animal welfare in a zoo is a challenge as there is no universal standardised method of assessment and, as a result, some zoos do not have formalised procedures in place. However, many well-operated zoos, and some regional zoo associations, have taken the initiative and developed methods and strategies to assess, evaluate, and enhance animal welfare. In many cases, these assessments are guided by standards, legislation, and accreditation organisations (e.g. ZAA, AZA, EAZA, BIAZA).

There have been many recent advancements in the domain of assessing animal welfare, including universal animal welfare framework surveys (Kagan et al., 2015), species-general welfare assessments (O'Brien & Cronin, 2023), animal welfare risk assessments (Sherwen et al., 2018), and the creation of species-specific assessments (Clegg et al., 2015). An important part of an animal welfare strategy is not only to educate others about animal welfare science, but also to determine clear and objective methods of animal welfare assessment. This process can include the development of an internal animal welfare assessment tool, or perhaps multiple species-specific or taxa-specific assessment tools for different taxonomic groups (e.g. birds, mammals, ungulates, reptiles). Many zoos use the Five Domains Model (Mellor et al., 2020) as one means of assessment. Zoos South Australia have used this 2020 version of the Five Domains Model to refine their overarching approach to animal welfare assessments (Figure 25.3).

When developing an animal welfare strategy, one of the key aims is to ensure that every staff member of the zoo and every external stakeholder advocates for and upholds the goal of positive animal welfare. Some zoos have external-facing animal welfare documentation on their websites, such as animal welfare charters or codes of conduct for animal welfare and ethics. These documents may be used as a starting point for the development of an animal welfare strategy. Other tools to aid with the development of a strategy include a review of peer-reviewed animal science articles, internal workshops and brainstorming sessions, animal welfare strategies from other zoos, consultation with animal welfare scientists, consultation with animal care staff from other non-zoo facilities, and policies from accreditation bodies, amongst others. Internal collaboration and consultation are crucial, ranging from director to keeper, especially with animal care staff, examining all aspects of the strategy to ensure it is a staff-driven process.

It is also important to verify whether the animal welfare strategy aligns within the organisation. Zoos South Australia's animal welfare strategy informs their commitment to animal welfare and is articulated in the external-facing Animal Welfare and Ethics Charter (Figure 25.4). This charter is an aspirational commitment for all staff, volunteers, and contractors, making them universally accountable for positive animal welfare. The criteria in Figure 25.4 are interconnected'; application of one specific criterion does not guarantee success in the others. For example, a zoo can provide great care for an animal, but an animal can still suffer from poor welfare.

An animal welfare strategy requires a well-defined vision and mission statement, serving as both a long-term aspiration and a short-term objective for the zoo. Incorporating stakeholder feedback obtained through surveys, workshops, and other means is crucial during the development of the vision and mission statement. Involving staff in the decision-making process for the vision and mission, such as through voting on several

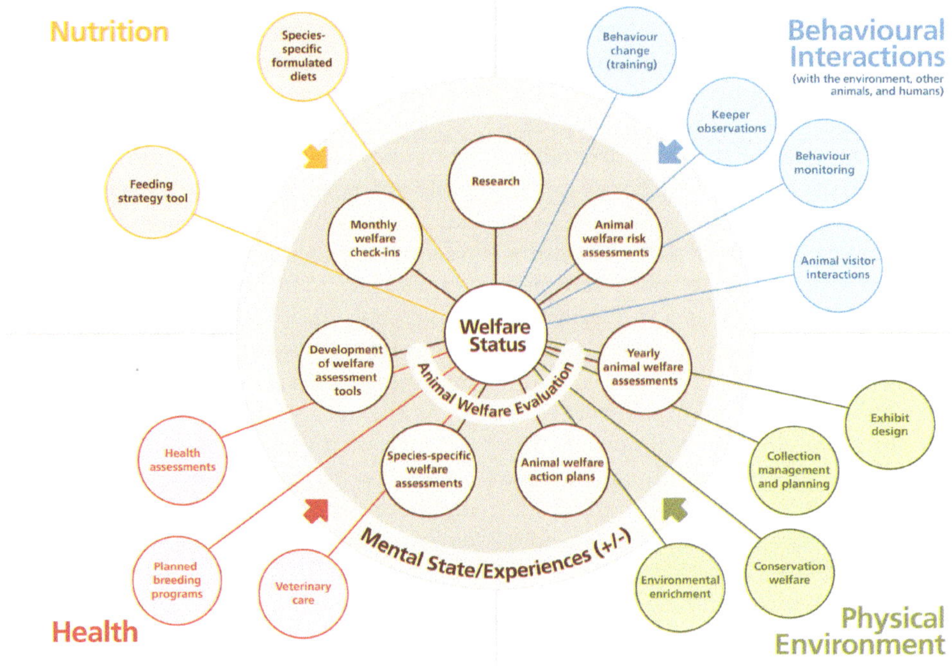

Figure 25.3 An overview of Zoos South Australia's approach to animal welfare assessments, with incorporation of the Five Domains Model.

versions, ensures a sense of inclusivity for everyone involved.

An example of an animal welfare vision and mission from Zoos South Australia is provided below.

Vision: Animal thriving 24/7.
Mission: All animals should experience a thriving, rewarding, and good life.

The comprehensive framework of an animal welfare strategy is different from that of a charter, as it bridges the gap between a high-level guiding document and operationalising animal care. An overview of Zoos South Australia's Animal Welfare Strategy framework is summarised in Table 25.1. This summary provides a planning overview only and does not include all operational details. Zoos South Australia's animal welfare strategy relies on animal welfare assessments, rigorous animal care, ongoing evaluation of the strategy and welfare programme, and continuous communication. The strategy framework is segmented into measurable components that are tested daily, weekly, monthly, quarterly, and annually.

Once the framework of the animal welfare strategy has been established, the next step is to formulate objectives or goals for the strategy. Each core

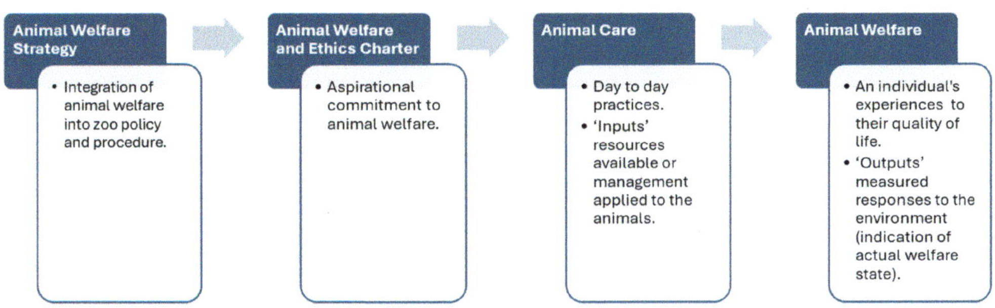

Figure 25.4 Zoos South Australia's overarching approach to animal welfare.

area of the framework should clearly be allocated measurable goals in support of the strategy. A five-year plan for each area can be effective, incorporating a cyclic evaluation process that allows the strategy to be refined and enhanced as needed. The number of goals for each area varies, depending on the contribution of that core area. For example, the core area of 'Animal Welfare Evaluation' will likely have the largest number of goals. To ensure that the goals have achieved the intended outcome, it is recommended to establish key performance indicators (KPIs) against which progress can be measured.

Table 25.2 provides an example of goals and KPIs derived from the overarching animal welfare strategic framework and implemented through a data collection tool such as ZooMonitor (Wark et al., 2019). This example set of goals and KPIs includes aspects of animal welfare evaluation, animal care and communication (internal and external), and is integrated into the routine of zookeepers.

It is expected that there will be hiccups along the way and adjustments will need to be made, demonstrating why having a strategy that is nimble and adaptable is important. Addressing the welfare needs of animals is a continuous endeavour, requiring planning, implementation, evaluation, and adjustment (Wolfensohn et al., 2018). As the exploration of the emerging field of animal welfare science progresses, the strategy will incorporate new learnings.

25.7 IMPLEMENTING THE STRATEGY – STEPS AND RESPONSIBILITIES–

Implementing an animal welfare strategy requires all staff to have a cohesive and clear understanding of its framework, goals, and objectives. The strategy can be communicated face-to-face in meetings, but additional training of key staff will likely be required. A clear implementation plan needs to be created that details the steps and responsibilities of staff, and where applicable, the associated resourcing. This final plan must be endorsed by leadership to actively support its implementation. The strategy and implementation plan must then be presented to all staff and stakeholders to ensure everyone has been appropriately briefed and that they know their respective responsibilities, which will cover the entire spectrum of animal welfare – consideration of good welfare is as important as the consideration of poor welfare.

Communication when implementing the strategy is key. Breaking KPIs into small, achievable goals assists with this process, as does ensuring a fluid model where KPIs are monitored for effectiveness and adjusted when and where required. There is a tendency to focus on instances when things go wrong, but it is crucial to celebrate the wins and recognise positive steps and progress towards optimal animal welfare.

As the poet Maya Angelou eloquently states, 'I did then what I knew how to do. Now that I know better, I do better.' This sentiment reminds us that despite setbacks, each experience provides an opportunity for growth. At many zoos, plans may not always unfold as expected, but such instances present opportunities for continued improvement.

These opportunities for continued improvement could be small initiatives like migrating keeper diary notes to an electronically supported platform. Another example could be to transition paper-based animal welfare assessments onto a platform that allows for digital data collection. Although these initiatives require an initial commitment of time for training, the benefits derived frequently outweigh the initial investment. Such initiatives allow staff to assess welfare more readily and quickly, and they enable the organisation to better analyse and act on collected animal welfare data.

25.8 MONITORING AND EVALUATING THE STRATEGYA JOURNEY OF CONTINUOUS DEVELOPMENT

To be effective, an animal welfare strategy must far exceed a token 'tick box' exercise. There must be support at the highest level to implement the strategy to its fullest capacity, with the intention to aid the delivery of the mission and vision for the organisation. To know whether the animal welfare strategy is effective, it needs to be monitored and evaluated.

Table 25.1 Zoos South Australia's Animal Welfare Strategy framework

	Daily	Weekly	Monthly	Quarterly	Annually
Animal Welfare Evaluation	Animal welfare concerns reported. Observations (i.e. Zims reports, animal behaviour data) recorded.	Individual or species welfare assessments conducted.	Animal-based measures (i.e. weights, body condition scores, etc.) analysed. Welfare risk assessments on new habitats, animal moves or acquisitions conducted.	Enrichment, training and behavioural monitoring programs (i.e. activity budgets from ZooMonitor), evaluated. Animal welfare action plans evaluated. Quality of life assessments (e.g. review of chronic cases) conducted.	Animal welfare assessments on remaining animals completed. Written welfare reports on all animals completed and circulated. Animal Visitor Interaction programs assessed and reviewed.
Animal Care	Animal health checks conducted. Habitat status (e.g. cleaning, maintenance, UV light checks) assessed. Training, enrichment provision assessed.	Animal-based measures (i.e. weights, body condition scores, etc.) collected and recorded.	Nutritional reviews for targeted species established.	Habitat integrity checks conducted. Species Planning conducted.	Research Department (long term goals and priority projects) reviewed.

(*Continued*)

Table 25.1 (Continued)

	Daily	Weekly	Monthly	Quarterly	Annually
Communication (internal and external)	Welfare concerns reported to Curator and Scientist Team. Pro-active promotion and education content related to animal welfare circulated through social media platforms.	Public feedback queries or concerns relating to animal welfare are addressed. Meetings to review welfare priority cases conducted. Animal welfare contributions to newsletter(s) for internal and external communication written and submitted.	Professional development for stakeholders is provided. Communication and marketing plan for animal welfare reviewed.	Quarterly board reports produced. Articles submitted to Zoo Times (bi-annual member newsletter). Staff contributions to Specialist Advisory Groups and Taxon Advisory Group meetings reviewed and encouraged.	Research Department (long term goals and priority projects) reviewed and circulated to all staff. Organisation commitment to welfare evaluation (mission and goals) reviewed and reinforced. Internal Animal Welfare and Ethics Meeting (bi-annual) conducted and minutes with key outcomes and actions circulated. Annual general meeting (members and volunteers) conducted with animal welfare content included. Representation of advancements in animal welfare practises at regional and international Conferences and forums.

(Continued)

Table 25.1 (Continued)

	Daily	Weekly	Monthly	Quarterly	Annually
Evaluation of strategy and programme	Animal care practices evaluated and amended.	Animal welfare science papers reviewed and shared with peers to encourage improvement and broader advocacy.	Animal welfare updates sent to management/ leadership teams.	Annual budget assessed for level/ rate of expenditure related to animal welfare. Standards and legislation related to animal welfare reviewed. Animal welfare strategy goals reviewed. Animal welfare outcomes reviewed.	Animal welfare strategy reviewed against expected KPIs. Ensure compliance and obtain regional member association accreditation (~every 3 years). Animal welfare policies and procedures reviewed. Improvements to animal welfare assessed and celebrated organisation wide.

Table 25.2 One example of a strategic goal is to implement ZooMonitor into the keeper's daily routine

Project	Time Frame	Goal	KPI	Who is responsible for implementation
Implement ZooMonitor into keeper routine for behavioural monitoring and animal welfare enhancement	2024	Ethogram and observation protocol development.	Determine target species and develop ethograms. Determine observation parameters (type of data collection, duration, accompanying question, heat maps). Scientific references and support documentation.	Animal Welfare Officer.
		Project details set up in ZooMonitor.	Set projects up on ZooMonitor. Pilot testing of observation protocol.	
		Staff training tools.	Establish training protocol. Video library of ethogram behaviours. Guidance video of ZooMonitor. Practice observation videos. Inter-observer reliability testing.	
	2025	1 (5 minute) observation per day.	Meeting number of scheduled observations.	Assistant Curator and Senior Keeper.
		Initial analysis of data and recommendations.	Bi-annual analysis of data. Bi-annual reports written including activity budgets and recommendations to improve animal welfare. Review whether recommendations are improving animal welfare.	Animal Welfare Officer.
	2027	3 (5 minute) observation per day.	Meeting number of scheduled observations.	Assistant Curator and Senior Keeper.

(*Continued*)

Table 25.2 (Continued)

Project	Time Frame	Goal	KPI	Who is responsible for implementation
		Analysis of data and recommendations.	Newly trained staff to analyse data and write behavioural reports (Animal Welfare Officer to support) including activity budgets and recommendations to improve animal welfare. Review whether recommendations are improving animal welfare. Review how frequently staff can analyse and write reports. Present findings at conferences.	Animal Welfare Officer and Assistant Curator/Senior Keeper.
	2028	4 (5 minute) observation per day.	Meeting number of scheduled observations.	Assistant Curator and Senior Keeper.
		Analysis of data and recommendations.	Quarterly reports written including activity budgets and recommendations to improve animal welfare. Review whether recommendations are improving animal welfare.	Animal Welfare Officer and Assistant Curator/ Senior Keeper.

As mentioned above, one way to monitor the strategy is through KPIs with set targets. Below are some examples:

1. Animal welfare assessments (numbers completed per year).
2. Animal welfare action plans (percent of action plans completed).
3. Animal welfare incidents (% decrease over time).
4. Animal welfare concern protocol (% decrease over time).
5. Animal training programme (% of trained behaviours for preventative health care increasing overtime).
6. Enrichment programme (structured enrichment program, evaluated, monitored and enhanced).
7. Financial metrics (e.g. increase in financial support directly related to animal welfare).
8. Professional development (goals for staff professional development and resource provision).
9. Real-time animal welfare tracking using suitable computer apps (% of animals where welfare data is collected and tracked over time).
10. Monitoring progress against KPIs (e.g. spreadsheet checklist with both quantitative and qualitative methods of data that can be shared with leadership and the greater team) indicating number of KPIs met.

11. Tracking stakeholder feedback (e.g. surveys, reports – including biannual/annual updates on strategy progress internally and externally).

It is important to have KPIs that indicate whether the zoo is succeeding in their mission. For example, some zoo's mission is for the animals to thrive. Acknowledging the multidimensional nature of animal welfare assessments, zoos may opt to gauge progress by establishing a specific benchmark, with the expectation that scores will increase over a defined period. Some other key points to use for evaluation, include: (1) financial impact; (2) deviations from expected outcomes; (3) stakeholder feedback; and (4) lessons learned.

In the case of lessons learned, these can help to improve development of the strategy, refine objectives, help reallocate resources, aid staying abreast of zoo animal welfare science, and aid in adopting appropriate practices.

25.9 PROFESSIONAL DEVELOPMENT FOR STAFF

A key aspect of an animal welfare strategy is staff professional development, encompassing three basic premises: understanding (foundation of welfare management), evaluation (evidence-based management), and improvement (optimal animal welfare management). Staff can be supported in progressing through these three strata (Figure 25.5). Investing in professional development for staff is essential as it supports continuous learning, job satisfaction, and employee retention. It is important that knowledge transfer on topics such as zoo animal welfare science is supported and encouraged, to ensure that all zoos irrespective of their culture or location, can improve standards of animal care and welfare through evidence-based management ensuring that animal welfare is optimised (Ward et al., 2020).

Zookeepers play a pivotal role in animal welfare due to their daily care and monitoring of animals, allowing them to identify nuances in behaviour. However, it is imperative that staff are provided with opportunities for professional development. Bacon et al. (2023) recommends that zoo leadership

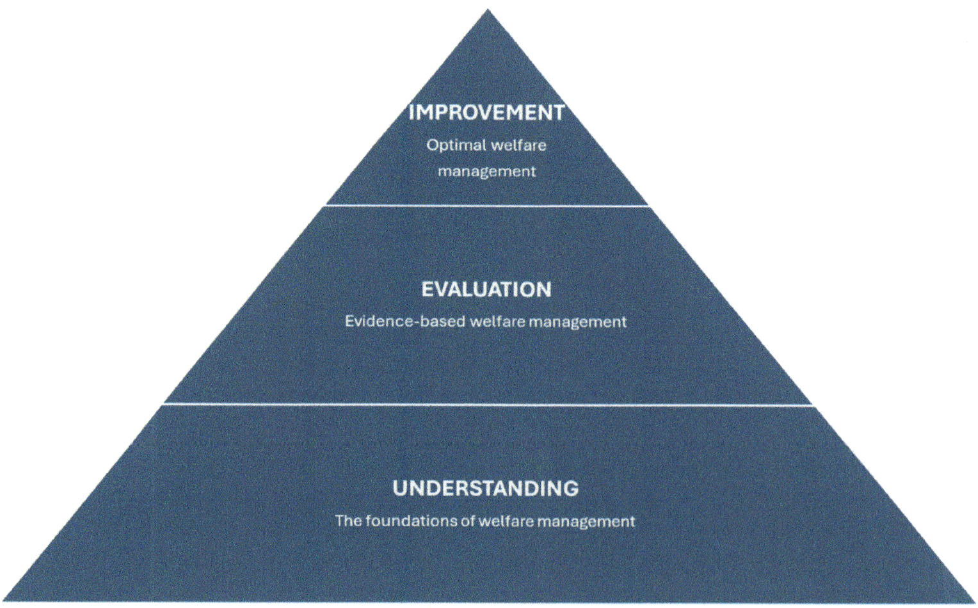

Figure 25.5 An approach to staff professional development.

be actively involved and allocate suitable resources to staff training in animal welfare and behaviour. Furthermore, the authors suggest incorporating comprehensive information on the ecology and behaviour of zoo species into the training of zoo staff, complementing traditional animal welfare education. This goal can be achieved via online courses, courses conducted by zoo staff, university courses, workshops, seminars, on-the-job training, internships, online resources and webinars, cross-training, and research partnership opportunities. Professional development not only fosters career satisfaction but also enhances animal care and welfare.

25.10 CONNECTING WITH OTHER ZOOS AND THE WIDER COMMUNITY

Another part of an animal welfare strategy, and an area where zoos could enhance their practice and improve, is to 'co-evolve' together. Expanding on the recommendations of Ward et al. (2018), a strong case can be made for an enhanced collaborative approach when it comes to animal welfare in zoos. Of course, there is already useful work that highlights zoo collaboration (e.g. inter-organisation research, development of apps such as ZooMonitor, sharing of resources), but this area could be further improved. There are opportunities for zoo organisations to take examples of their work and circulate these with their members. This approach influences and inspires members through improved networking. Sharing work and collaborating with other zoos can also act as a means of continued professional development. This approach loops back into the SLO, further enhancing the credibility of zoos in the eyes of the public.

Many current standards originate from minimum baselines, lacking an explicit direction and mandate for optimal standards (Kagan et al., 2018). Whilst many zoos strive to provide experiences for their animals that contribute to an optimal welfare state, there can be a focus on providing care to minimum standards, often resulting from resource constraints. EAZA's best practice guidelines are creating some useful husbandry care documents, incorporating the latest science for certain species. Various keepers and scientists have educational podcasts on social media platforms, which also facilitate continued learning. Zoos have immense capacity for outreach so they need to promote positive animal welfare far beyond their gates. An infographic designed to enhance understanding of animal welfare in a zoo setting is detailed in Figure 25.6. Its objective is to enhance comprehension and engagement by presenting complex information in a more accessible format for learners.

Considering the role of a zoo within its community is crucial, and as such, it is important that this role also has a part in the strategy. Active engagement in both research endeavours and community initiatives, whenever possible, helps promote positive welfare for all animals. For example, by contributing to a review of state or federal animal welfare acts.

Having clear channels for two-way feedback, focused on both disseminating information to the public and actively listening and responding to community concerns and sentiments, is essential. There are many ways for the public to provide feedback to the zoo – e.g. having surveys and feedback forms, establishing an advisory committee, setting a clear process for reporting animal welfare concerns, including a member of the public on high-level committees such as the Animal Welfare and Ethics Committee and/or the zoo board. Other avenues include community events, public forums, social media engagement, educational outreach programmes, and partnerships. By having an approach like this, it acknowledges the importance of transparency and trust, which assists with upholding the SLO between the zoo and its community stakeholders. Collaborative research between academic institutions is also beneficial, helping engage the next generation of zoo supporters and conservation advocates. Examples of research collaborations from Zoos South Australia are described in Box 25.1.

The continuous evolution of animal welfare science will inform the next steps for zoo animal welfare strategies. It is hoped that many zoos will create and integrate animal welfare strategies into their institutions and continue to collaborate on these endeavours. An increase in technology use is already becoming apparent in the animal welfare sector outside of zoos, and this trend is already informing practices within zoos. Animal

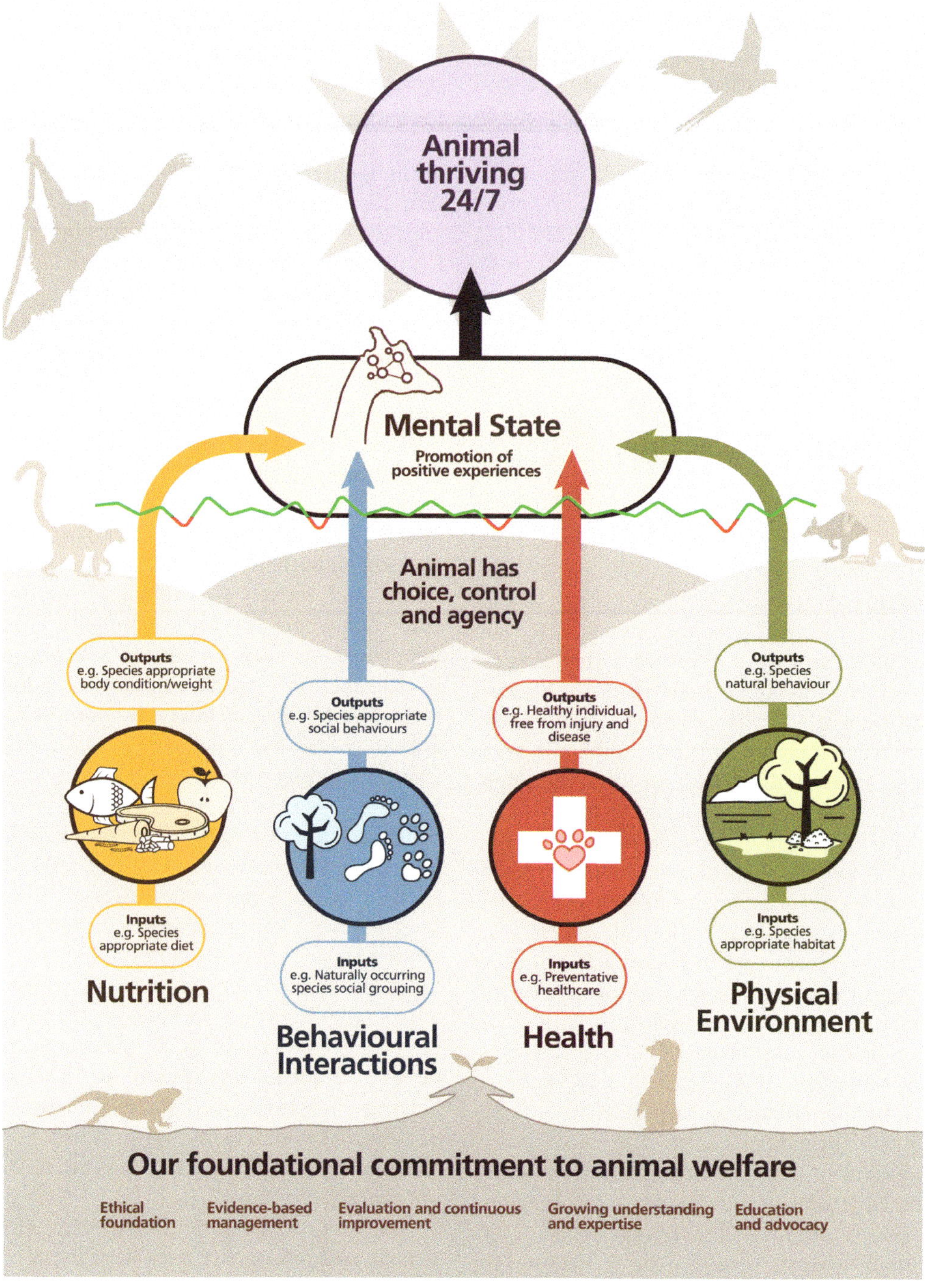

Figure 25.6 An infographic showcasing a zoo's dedication to animal welfare. It emphasises a core commitment to positive animal welfare through the integration of the Five Domains Model and animal welfare assessments, with the overarching objective of promoting net positive animal welfare and striving for animal thriving 24/7.

BOX 25.1: Collaborative research at Zoos South Australia

As mentioned in this chapter, the involvement of an array of stakeholders is important in a strategy. Collaborative research initiatives between academic institutions and zoos can contribute to advancing the conservation and educational objectives of zoos. This relationship extends to the animal welfare domain as well; there are a lot of great initiatives and ideas that can be further supported through engaging with external institutions such as universities.

Zoos South Australia has developed a strong working relationship with Adelaide University. There are several projects currently underway with multi-disciplinary research from the Veterinary Science and Engineering Schools focusing on enhancing animal welfare through technological innovation. One of these projects includes a pilot study using visual object tracking designed to monitor orangutans' (*Pongo abelii*) behaviour and movement (Figure 25.7). The long-term goal of this project is to have a 24/7 activity budget of the orangutans and apply these findings and technology to other species, further informing our understanding of welfare through evidence-based approaches and providing opportunities for welfare enhancement.

Figure 25.7 Study utilising visual object tracking for monitoring orangutans' *(Pongo abelii)* behaviour and movement.

Other projects that Zoos South Australia are collaborating on to enhance welfare include independently engineered feeding devices with cognitive challenges where animals work together to press buttons simultaneously for food to be released. Additionally, efforts are underway to provide animals with more control over their environments, such as an environmental enrichment device being developed for Asian elephants (*Elephas maximus*). This device allows elephants to control water pressure, change shower heads, play with hoses and turn on and off white noise. These projects exemplify the zoo's commitment to embracing technological innovations in animal care.

welfare assessment tools will continue to improve, potentially using automatic remote technology to capture continuous data (24/7), helping to better understand the multidimensional nature of animal welfare over the individual's lifetime. In addition to technological advancements, the future will evolve in relation to societal views. It is important that we actively engage with the community at large to ensure continued public support and involvement of different stakeholders. It will be interesting to observe how zoos evolve their animal welfare strategies in the coming decades, potentially setting new standards in care, conservation and animal welfare science. For further information and resources on animal welfare strategies, see Box 25.2.

25.11 CONCLUSIONS

What is the value of an animal welfare strategy? An animal welfare strategy is the mechanism to do right by the animals. Zoos have a moral obligation and duty of care to ensure all animals in their care have the right to a life worth living for their whole life. A well-created and implemented strategy allows an organisation to go beyond reactive approaches to animal welfare to proactive welfare monitoring, recognising that animal welfare is on an individual continuum and needs to be tracked throughout all life stages. A well-planned animal welfare strategy encompasses 'big picture' planning, all the way down to the operational details. This approach includes having a framework that focuses on promoting positive experiences for animals and providing resources and mechanisms to deliver them effectively. A strategy should be a nimble, flexible document, periodically updating and integrating new information, and evolving to respond to advances in animal welfare science. Equally important is supporting and fostering a culture of continuous improvement, a willingness to learn, to innovate, and ongoing refinement.

BOX 25.2: Further information

World Association of Zoos and Aquariums, Animal Welfare Strategies

Animal Welfare Strategies - WAZA - https://www.waza.org/priorities/waza-code-of-ethics/animal-welfare-strategies/

This World Association of Zoos and Aquariums Animal Welfare Strategy highlights the importance of animal welfare strategies in zoos and aquariums. Their comprehensive approach emphasises the care and conservation of animals, ensuring both their physical and psychological well-being. It covers a variety of aspects, including animal welfare and its assessment, monitoring animal welfare, environmental enrichment, exhibit design, breeding programmes and collection planning, conservation welfare, animal welfare research, partnerships in animal welfare and engagement and interaction with visitors.

Zoos South Australia Animal Welfare and Ethics Charter

Animal welfare - Assess the welfare of your animals with Zoos SA online tool https://www.zoossa.com.au/animal-welfare/

Zoos SA Animal Welfare Charter

https://www.zoossa.com.au/wp-content/uploads/2024/03/Zoos-SA-Animal-Welfare-Charter.pdf

Zoos Animal Welfare and Ethics charter is an example of an aspirational commitment to animal welfare, ensuring all animals have a thriving, rewarding life. Their comprehensive Animal Welfare Charter prioritises species-appropriate care, continuous improvement, and promoting positive experiences for animals. With a strong focus on ethical foundations, staff training, and community engagement, Zoos SA aims to set the standard in animal welfare, backed by rigorous scientific assessments and global accreditations.

REFERENCE LIST

Bacon, H., Vigors, B., Shaw, D.J., Waran, N., Dwyer, C.M. & Bell, C. (2023). Zookeepers – The most important animal in the zoo? *Journal of Applied Animal Welfare Science*, 26(4), 634–646.

Clegg, I.L.K., Borger-Turner, J.L. & Eskelinen, H.C. (2015). C-Well: The development of a welfare assessment index for captive bottlenose dolphins (*Tursiops truncatus*). *Animal Welfare*, 24(3), 267–282.

Gray, J. (2017). *Zoo ethics: The challenges of compassionate conservation*. Csiro Publishing, Australia

Hampton, J.O., Jones, B. & McGreevy, P.D. (2020). Social license and animal welfare: Developments from the past decade in Australia. *Animals*, 10(12), 1–11.

Hill, S.P. & Broom, D.M. (2009). Measuring zoo animal welfare: Theory and practice. *Zoo Biology*, 28(6), 531–544.

Kagan, R., Allard, S. & Carter, S. (2018). What is the future for zoos and aquariums? *Journal of Applied Animal Welfare Science*, 21(sup1), 59–70.

Kagan, R., Carter, S. & Allard, S. (2015). A universal animal welfare framework for zoos. *Journal of Applied Animal Welfare Science*, 18(sup1), S1–S10.

Melfi, V.A. (2009). There are big gaps in our knowledge, and thus approach, to zoo animal welfare: A case for evidence-based zoo animal management. *Zoo Biology*, 28(6), 574–588.

Mellor, D.J., Beausoleil, N.J., Littlewood, K.E., McLean, A.N., McGreevy, P.D., Jones, B. & Wilkins, C. (2020). The 2020 five domains model: Including human–animal interactions in assessments of animal welfare. *Animals*, 10(10), 1–24.

Miller, L.J. & Chinnadurai, S.K. (2023). Beyond the five freedoms: Animal welfare at modern zoological facilities. *Animals*, 13(11), 1818.

O'Brien, S.L. & Cronin, K.A. (2023). Doing better for understudied species: Evaluation and improvement of a species-general animal welfare assessment tool for zoos. *Applied Animal Behaviour Science*, 264, 105965.

Sherwen, S.L., Hemsworth, L.M., Beausoleil, N.J., Embury, A. & Mellor, D.J. (2018). An animal welfare risk assessment process for zoo. *Animals*, 8(8), 130.

Ward, S.J., Sherwen, S. & Clark, F.E. (2018). Advances in applied zoo animal welfare science. *Journal of Applied Animal Welfare Science*, 21(sup1), 23–33.

Ward, S.J., Williams, E., Groves, G., Marsh, S. & Morgan, D. (2020). Using zoo welfare assessments to identify common issues in developing country zoos. *Animals*, 10(11), 1–15.

Wark, J.D., Cronin, K.A., Niemann, T., Shender, M.A., Horrigan, A., Kao, A., & Ross, M.R. (2019). Monitoring the behavior and habitat use of animals to enhance welfare using the ZooMonitor app. *Animal Behavioural Cognition*, 6(3), 158–167.

WAZA. (2023). Animal welfare goal. https://www.waza.org/priorities/animal-welfare/2023-animal-welfare-goal (Accessed: 05/03/2025).

Wolfensohn, S., Shotton, J., Bowley, H., Davies, S., Thompson, S. & Justice, W.S.M. (2018). Assessment of welfare in zoo animals: Towards optimum quality of life. *Animals*, 8(7), 110.

26

Enabling animals to participate in their own care

CAMILLE MUNDAY, UNNAR AEVARSSON AND ZOË NEWNHAM

26.1 INTRODUCTION

Over the past century, zoo animal care and management has made significant strides. This progress is driven by a deeper understanding of animals' ability to experience both well-being and suffering, coupled with a growing desire to meet the needs of all creatures and ensure high levels of welfare. These advancements encompass the handling, training, and all other interactions with animals, from the smallest invertebrates to the largest mammals.

The training of animals in zoos and aquariums (hereafter "zoos") offers numerous benefits. Physically, it promotes exercise and can help facilitate medical procedures with reduced stress for the individual, as animals can be trained to voluntarily participate in their own care, such as through health checks and treatments promoting choice and control (Melfi et al., 2020). Mentally, training provides enrichment and stimulation, helping to prevent boredom and reduce abnormal repetitive behaviours (ARBs) while increasing behavioural diversity (Miller et al., 2020). Training can also help create positive relationships between animals and caregivers, building trust and cooperation across a range of species (Rault et al., 2020).

Additionally, and arguably a less well-known application, is the ability of training to support conservation and reintroduction projects. Training ex situ in zoos can facilitate research used to improve the care of species in situ as well as aid with reintroduction projects, enabling them to return to the wild. Overall, animal training is crucial for ensuring high levels of welfare and quality of life for animals in human care and despite time and resource constraints faced by many animal caregivers such

DOI: 10.1201/9781003429425-26

as zookeepers (hereafter referred to as 'keepers'), it should form an integral part of daily routines.

An in-depth review of the fundamentals and processes of animal training in zoos is beyond the scope of this chapter; for further reading on this topic, please refer to Melfi et al. (2020); however, this chapter aims to present a small selection of case studies illustrating how training can be used to help improve and maintain high standards of welfare in the zoo, how training can be adapted into enclosure design, and how training can positively influence conservation projects both in situ and ex situ.

26.2 LONDON ZOO BIRD TRAINING

Despite the evolving understanding of bird cognition, the practice of training birds via positive reinforcement to take part in their own care is not as widespread in zoos as within other taxa and has historically been based on birds trained for entertainment using outdated practices such as weight management. However, recent advancements in training methods have led to the widespread adoption of operant training techniques, especially across BIAZA (British and Irish Association of Zoos and Aquariums) institutions, playing a crucial role in bird care and management.

London Zoo recognises the immense value of training, which makes animal care less stressful and safer for both the animals and the keepers. On a daily basis, various bird species are trained to actively participate in their own care, including station, scale, and recall training. This approach enhances welfare, ensures effective health monitoring, reduces stress, and contributes to overall well-being, promoting positive welfare states.

26.2.1 CASE STUDY: ENHANCING BIRD WELFARE AT LONDON ZOO USING THE BIRD SELF-MANAGEMENT SYSTEM (BSMS)

London Zoo houses Abdim's storks (*Ciconia abdimii*) in a spacious, naturalistic, outdoor, mixed-species walkthrough aviary. This enriching environment supports their well-being, allowing them to thrive; however, during winter, the storks require extra attention due to their adaptations for warmer climates. In recent years, the Bird Team has closely monitored the storks' behaviour, especially during overnight frosts. Early signs of wing tip oedema were identified, a condition that can occur due to reduced blood flow from vessel constriction in cold temperatures. The team proactively addressed this issue by implementing preventative measures to minimise the risk of the condition worsening, including providing heaters outside, allowing the birds to maintain access both indoors and outdoors on colder days, and a new management system built into the enclosure to help regulate the access birds have if needed. These measures ensured that the storks could continue to perform natural behaviour such as roosting, nesting in high places, and flying together as a group, which is essential for this gregarious species. In some severe cases, wing tip oedema can be challenging to manage, but the team remains vigilant in providing the best care possible to ensure good bird welfare.

26.2.1.1 New management method and training protocol

To further manage and reduce wing tip oedema, it is essential to keep the birds indoors during extremely cold weather. However, this can be challenging in a large outdoor walkthrough aviary because the species prefer to spend time outside due to their ecological niche and the complex topography provided outdoors. Using nets to catch and move the birds to a warmer indoor space can also cause significant stress, reducing welfare. To address this problem, the Tropical Bird Team redesigned and expanded the indoor habitat to make it more enticing. The habitat now includes UV lighting, ample heating, more roosting spots and food provision, providing a comfortable and inviting environment for the storks. While these modifications successfully encouraged the birds to increase their use of the indoor space, when given the choice, the storks still preferred to spend most of their time outside, even during cold weather, necessitating additional measures.

To address this, the "Bird Self-Management System" (BSMS) was designed and implemented in

2023 to facilitate bird moves without causing stress and ensure the storks remain indoors when temperatures drop below 5°C. The BSMS features a row of lightweight, vertical PVC pins that enable storks to freely enter and exit their lofts when the weather is warm enough. However, when temperatures fall, the system is adjusted to function as a one-way entrance, ensuring the storks remain inside. This is accomplished by using a solid horizontal bar that prevents the pins from being pushed outward (Figure 26.1).

Since the storks were not familiar with this system from a young age, a specific training process was necessary for them to enter the loft. This involved gradually introducing the new system in small, manageable steps to help them acclimate (Figure 26.2). Early trials revealed that the storks preferred clear pins over solid ones, prompting a change in the pins.

- Day 0. All pins are raised, allowing free movement between the indoor and outdoor areas of the enclosure.
- Day 1 – Day 6. Each day, an additional pin is lowered, allowing the birds to gradually get used to the sensation of the pins brushing against them as they move in and out of their indoor area.
- Day 7. All pins are lowered. The training is considered successful if all birds are using the system, which is monitored via camera traps. If any birds are hesitant to use the system, the criteria can be relaxed, and pins reset to an earlier stage, giving the birds the opportunity to gain more reinforcement history for the behaviour of pushing through pins.

26.2.1.2 Welfare benefits of training and the BSMS system and further applications

The protocol to train Abdim's storks to use the BSMS has yielded promising results. Previously, during winter's colder temperatures, the birds had to be kept indoors all winter for their well-being. This necessary measure, though essential for their safety, posed challenges for both the keepers and the birds. With the new system, the birds are only secured indoors when necessary, allowing them to spend time both outdoors and indoors during winter, thus reducing stress and improving welfare. The system suits Abdim's stork well, as their gregarious nature naturally leads all the birds indoors.

Figure 26.1 Photos of Abdim storks (*Ciconia abdimii*) using the Bird Self-Management System (BSMS) at ZSL London Zoo taken by camera traps (A & B). When weather conditions permit, access is allowed in both directions, which is an important part of the training process, especially after a period of being shut inside (C).

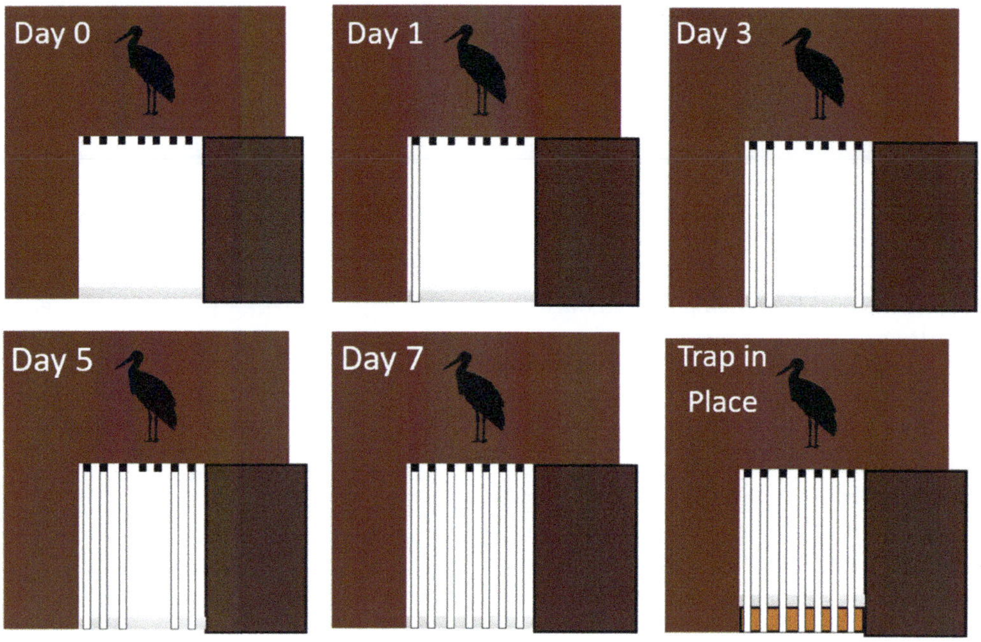

Figure 26.2 Pin arrangements for the gradual, day-by-day introduction of the BSMS. Please note that the "days" are intended as a guide, not a strict target.

However, the benefits of this innovative system extend beyond just Abdim's storks, with the system also used for the scarlet ibis (*Eudocimus ruber*), which share the habitat with the storks. Smaller systems have been successfully trialled with various bird species, including multiple dove (*Columbidae* spp.) and pigeon (*Columbidae* spp.) species, lilac-breasted rollers (*Coracias caudatus*), Vietnamese pheasants (*Lophura edwardsi*), Argus pheasants (*Argusianus argus*), and white-faced whistling ducks (*Dendrocygna viduata*). For these smaller birds, the system is primarily used for vet checks, locating birds in the large aviary, and aiding in counting.

26.2.2.1 Case study: Recall training Eastern white pelicans

At ZSL (Zoological Society of London), recall training is an indispensable practice for bird management. This method teaches birds to voluntarily return to a designated location or respond to specific cues, thereby minimising the need for physical capture and handling, significantly reducing stress. Recall training is particularly advantageous for species that are challenging to access or require frequent handling. Furthermore, this behaviour promotes increased physical activity, encourages natural foraging patterns, and enhances enrichment within the birds' main habitat.

At London Zoo, eastern white pelicans (*Pelecanus onocrotalus*) and greater flamingos (*Phoenicopterus roseus*) cohabit in "Three Island Pond", an artificial enclosure comprising a pool, islands and a smaller land section known as the corral. The corral includes a small water body and indoor housing specifically for bird management. Given that both species show preferences for the deeper waters of the main pond, training them to recall away from this area into the corral is essential and provides a secure and controlled environment for conducting training sessions and other husbandry practices.

Currently, all flamingos and pelicans have been successfully trained to respond to recall cues during feeding times, reliably returning to the corral. Among them, "Strider", an adult male pelican, stands out as the first to master feeding at a designated station within the corral (Figure 26.3). His advanced training can be attributed to a combination of factors, including his extended training period, innate curiosity, and the strong bond he developed with caregivers during this time. Additionally, his competitive behaviour during feeding sessions made him an ideal candidate for targeted training, as station feeding was necessary to reduce instances of him taking fish from conspecifics and to promote equitable food distribution. While Strider was prioritised due to these specific considerations, efforts are underway to extend station feeding training to the remaining pelicans to achieve similar outcomes across the group.

The training plan detailed below consists of structured steps used to recall the pelicans and station-train Strider in the corral. The ultimate goal is to extend station-training and scale-training behaviours to all individuals while also facilitating desensitisation to tactile interactions and handling. It is important to conduct training sessions at various times throughout the day to accustom the pelicans to a flexible schedule and prevent them from associating training with a specific time of day.

26.2.2.2 Training plan: Recall and station training for pelicans in Three-Island Pond

Goal: To train the pelicans to recall into the corral and to station train for a long enough duration to allow caregivers to obtain accurate weights and conduct visual health assessments.

Step A: Establishing a Positive Association

1. The trainer enters the Three-Island Pond carrying a red bucket of fish and positions themselves at the pond's edge. This serves as a visual cue. Once the trainer observes pelicans moving towards them three times upon entry, they advance to Step 2.
2. Employing an audio cue (maracas, referred to as the "shaker"), the trainer continuously shakes it. Pelicans are reinforced with a fish each time they respond positively by swimming towards the trainer. This step is repeated

Figure 26.3 'Strider' the Eastern white pelicans (*Pelecanus onocrotalus*) successfully station training.

until pelicans consistently react to the shaker noise three times in succession.

3. After pelicans reliably associate the shaker noise with food and respond consistently while in the water, this step is considered complete.
4. The trainer then transitions to Step B.

Step B: Recalling into the Corral

1. The trainer positions themselves in the corral, maintaining a suitable distance from the entrance to provide pelicans with a comfortable trust space. They continuously observe the pelicans' behaviour and body language for signs of stress or discomfort, such as hesitating to enter the corral or exhibiting a vigilant posture. The trainer adjusts their positioning and distance as needed to ensure the pelicans remain at ease.
2. The trainer shakes the shaker continuously and reinforces the pelicans with a fish when they enter the corral. At this stage, the distance they enter is irrelevant; crossing the entrance threshold is sufficient for a reinforcer. The pelicans can be reinforced whether they are in the small body of water or on land.
3. The trainer repeats step 2 until the pelicans are comfortable with being recalled into the corral.
4. Once the pelicans consistently recall into the corral, this step is complete.
5. The trainer proceeds to Step C.

Step C: Station training "Strider"

The trainer positions themselves in the corral, keeping a suitable distance from the station, which, in this case, is a blue tree stump.

1. The trainer shakes the shaker and directs Strider to the station by pointing and saying "station". Once Strider stands on the station, he is reinforced with a fish.
2. Once Strider consistently stations, this training plan is complete.

26.3 OPERANT CONDITIONING FOR AVIAN CONSERVATION: TRAINING SIHEKS FOR SELF-FEEDING

Conserving endangered species often requires innovative strategies to ensure their survival and well-being. The ZSL, alongside partners including those within the AZA (Association for Zoos and Aquariums), has embarked on a pivotal project focusing on the Extinct in the Wild Guam Kingfisher (*Todiramphus cinnamominus*), known as sihek by the indigenous CHamoru* people. This project incorporates simple training techniques to improve the success of hand-rearing chicks in captivity, ensuring proper nutrition, minimising stress, and promoting natural behaviours. By addressing these challenges, the project aims to improve the chances of successful reintroduction and long-term survival of the sihek in the wild.

In the wild, mother birds use vocalisations to alert chicks to the availability of food and guide them in locating it once they leave the nest. To replicate this behaviour, sihek chicks were conditioned to associate an audio cue with feeding, as this common avicultural practice encourages chicks to feed during hand-rearing. Immediately after hatching, the chicks were conditioned to associate feeds with a whistle blown immediately prior to handfeeding. When the chicks were still within their nest chamber, the whistle was blown, and food items were promptly presented nearby, typically resulting in the chicks retrieving the food themselves (Figure 26.4). As the birds matured and fledged, the whistle training continued, with food being strategically placed within their aviary, prompting the birds to fly directly to their food.

The training was designed to encourage feeding in chicks, promote self-feeding in fledglings, and was hoped to function as a cue to encourage birds to return to feeding stations post-release. Encouraging the birds to return post-release aimed to provide additional food during the slow-release process and facilitate post-release health checks,

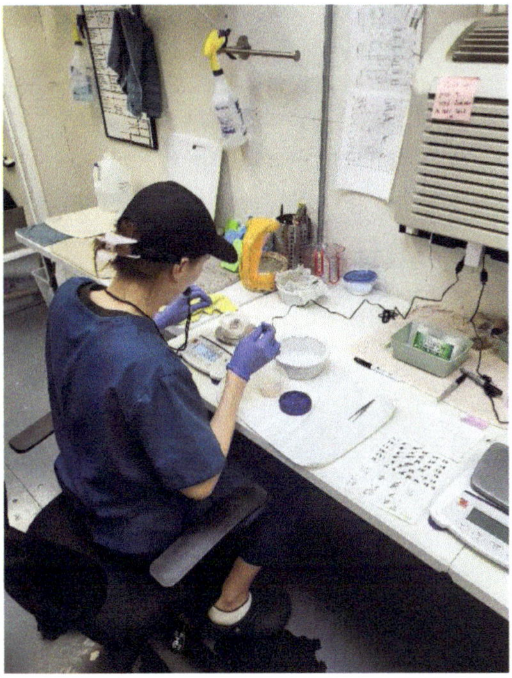

Figure 26.4 To replicate natural feeding behaviours, sihek (*Todiramphus cinnamominus*) chicks were conditioned to associate an audio cue with feeding. Immediately after hatching, the chicks were conditioned to associate feeds with a whistle blown immediately prior to handfeeding.

including weighing on scales integrated within feeding platforms. While the birds maintained the conditioned response in captivity, post-release outcomes did not meet expectations for this species.

It is important to note that not all species may respond positively to recall in the wild, as this behaviour can be highly species-specific and closely linked to their natural habits and instincts. For instance, sihek are solitary birds and, once released, showed no interest in human interaction or additional food provision, quickly adapting to hunt their own prey, consistent with their relatively short post-fledging parental care (Tarwater & Brawn, 2010). In contrast, sooty terns (*Onychoprion fuscatus*) and common noddies (*Anous stolidus*), with their social nature and extensive post-fledging parental care, have been successfully audio-recalled using comparable audio cues post release (McMurdo Hamilton, 2021) which in these examples enabled researchers to monitor fledglings and provide supplementary food and care post release.

Even though it did not yield the expected results post-release, the training proved beneficial during their hand-rearing, allowing for nine chicks to be successfully hand-reared (Figure 26.5). The training also proved beneficial during pre-release captive care, with hand-reared chicks successfully thriving in the wild six months later. Furthermore, valuable information about the species and how to apply behavioural science in future conservation programmes was gained. Future research and adaptations of these techniques could further enhance the rehabilitation and management of various bird species, contributing to their recovery and preservation.

*"CHamoru" (capitalising "CH") is recognised by the CHamoru Language Commission as the orthographically correct spelling.

26.4 REPTILE TRAINING AT LONDON ZOO

Although reptiles have captivated human interest throughout history, modern reptile husbandry only truly began to develop in the 19th century with the creation of specialist reptile houses and the publication of foundational texts. The Reptile House at London Zoo, notably one of the world's

Figure 26.5 Image depicting the nine sihek (*Todiramphus cinnamominus*) fledglings that were successfully hand reared.

first, played a pioneering role by providing suitable environments for reptiles and amphibians (ZSL, 2025).

Despite advancements in understanding reptile cognition, training reptiles remains less common in zoos compared to other taxa. Initially, herpetologists employed traditional methods such as manual restraint, chilling, and sedation rather than cooperative training. Today, London Zoo recognises the value of reptile training in making animal care less stressful and safer for both the animals and the keepers. Various reptile species are now trained in station, crate, target, and recall training to enhance welfare, ensure effective health monitoring, reduce stress, and contribute to overall well-being.

26.4.1 CASE STUDY: CRATE TRAINING A PHILIPPINE WATER MONITOR

Crate training encourages voluntary participation in husbandry practices and medical examinations by temporarily moving an animal to a separate holding space. This method is particularly beneficial for species that pose a danger to keepers working in close proximity to the species. It is also advantageous for species that have learnt the behaviour of door rushing; rapidly approaching the service door when a keeper enters. At London Zoo, a captive-bred female Philippine water monitor (*Varanus cumingi*) (hereafter referred to as "monitor") is trained to enter a smaller separate holding space (hereafter referred to as "crate") connected to the outside of her main habitat.

The training consists of structured steps to ensure the monitor reliably responds to cues and enters her crate. This process gradually trains her to the crate, allowing her to wait inside it while keepers service the main habitat. An audio cue, (in this case and hereafter "whistle"), is used to call the monitor, guiding her to enter the crate and wait until the servicing is completed (Figure 26.6).

Training sessions were conducted at various times throughout the day to ensure the monitor became accustomed to a flexible training schedule. It is crucial to include sessions in both the morning and afternoon routines to prevent monitors from associating training with a specific time of day. All training sessions took place during working hours, as this is a diurnal species.

26.4.2 TRAINING PLAN: CRATE TRAINING FOR PHILIPPINE WATER MONITOR

Goal: To train the monitor to shift from the main habitat to an external crate in response to a specific audio cue and remain measurably calm while the keeper services the habitat.

Step A: Initial crate introduction

- The trainer sets up the crate and places a high-value food item inside.
- The trainer allows the monitor to enter the crate and receive the reinforcer without using the whistle.
- This step is repeated 4–5 times to familiarise the monitor with the crate.

Step B: Associating whistle with reinforcer

- The trainer sets up the crate with a high-value food item.
- The trainer opens the crate door and waits for the monitor to approach before blowing the whistle.
- As the monitor enters the crate, the trainer blows the whistle three consecutive times, reinforcing the monitor by the third whistle. This helps the monitor associate the whistle with the high-value food item.
- This step is repeated 4–5 times before using the whistle to call the monitor from inside the enclosure.

Step C: Whistle cue and reinforcement

- The trainer sets up the crate with a high-value food item.
- The trainer opens the crate door, blows the whistle three times, and waits for a response.
- The trainer waits for the monitor to enter the crate, then reinforces her with the high-value food item. The trainer continues reinforcing this behaviour with additional high-value food items for approximately two minutes.
- It is important at this stage that the crate door remains open, to prevent any negative association with the crate.

Step D: Comfort with the crate

- The trainer repeats Step C until the monitor becomes desensitised with the crate.

Figure 26.6 Ectotherm keeper at ZSL London Zoo successfully crate training the Philippine water monitor.

- Once desensitised, the trainer stops baiting the crate and reinforces recall behaviour using small reinforcer items with tweezers or tongs only.

Step E: Enclosing the monitor

- The trainer repeats Step D but starts closing the crate door to enclose the monitor.
- The trainer frequently repeats this step to reinforce recall behaviour.
- The trainer reinforces any neutral responses while the door is closing or closed. If the monitor shows any sign of discomfort or negative behaviour (e.g. glancing at or moving towards the closing door, freezing, hissing, tail whipping), the trainer stops closing the door to give her control over the situation.

Step F: Reducing high-value food items

- Once the training is well-established, the trainer begins to decrease the quantity and frequency of high-value food items. Gradually shift to using less high-value food items to prevent unwanted weight gain.

26.5 THE IMPORTANCE OF SPECIFIC ANIMAL TRAINING LOCATIONS WITHIN ANIMAL ENCLOSURES: A CASE STUDY FROM MARWELL WILDLIFE

Marwell Wildlife advocates for and promotes the training of all species via operant and classical conditioning using positive reinforcement, promoting welfare and ensuring choice and control. Where possible, specially designed species-specific areas for animal training are built into enclosures when the opportunity to design and build a new enclosure arises, or via modifications to existing enclosures. Having specially designed areas for animal training can promote and facilitate training, allowing keepers to work more efficiently, increasing the likelihood of sessions being carried out during a keeper's busy schedule.

During the design and building process of the white rhino (*Ceratotherium simum*) house, a training raceway was built into the design. The area was built into the middle of the house and consists of a raised raceway separate from the rest of the stalls (Figure 26.7). The raceway is accessed via one of two doors (one either end) and has space for two poles to be added at each end to allow the space to be reduced to safely secure the individual further if needed and allow access both in front and behind the rhino.

The design of the raceway allows keepers and vets to maintain protected contact whilst working in close proximity to the rhino. This allows for multiple behaviours to be trained and for more accurate and thorough health checks to be carried out safely. Training only takes place if the rhinos choose to take part in the session and stops as soon as the individual becomes disengaged or shows signs of wanting to exit. Allowing the rhinos choice and control of whether to engage with training sessions or not prioritises welfare and helps maintain positive keeper and animal relationships.

The raceway is visible from the guest area, allowing training sessions to be viewed. This presents the opportunity to educate the public on the importance and applications of behaviours such as target training, injection and blood draw training, footcare and health checks, and the positive benefits these behaviours have on animal welfare (as well as the importance of having designated areas for training).

When new training areas are built, the first consideration is to desensitise, and then train your animal to enter the training area so sessions can take place. Below is a brief plan of how the rhinos were introduced to the raceway.

26.5.1 CASE STUDY: INTRODUCING RHINOS TO A NEW TRAINING AREA AT MARWELL WILDLIFE

Note: Target training behaviour was already established. For the full target training plan, please contact the author.

Goal: The goal of this training was for each individual to be comfortable standing in the raceway with both front and back bars in place and doors secured for duration of each training session, allowing for in-depth health checks and other behaviours to be trained for.

Figure 26.7 Specifically designed raceway and training area for White Rhino (*Ceratotherium simum*) at Marwell Wildlife. Consisting of a raised raceway separate from the rest of the stalls and accessed via one of two doors (one either end), the raceway has space for poles (as can be seen in front of the raceway on the floor) to be added at each end (four circular metal holes attached to bars, two at each end) to allow the space to be reduced to safely secure the individual further if needed, allowing access both in front and behind the rhino (top photo). This design allows keepers and vets to maintain protected contact whilst working in close proximity to the rhino, allowing for multiple behaviours to be trained and for more accurate and thorough health checks to be carried out safely. Depicted is a keeper carrying out a target training session (bottom photo).

Steps:

1. Use positive reinforcement to move the individual rhino (who will be taking part in the session) to the stall with entry to raceway.
2. Open both raceway doors (entrance and exit) and see how the individual reacts – if positive, the individual can be targeted near the entrance to raceway.
3. Leave individual with free access to walk through the raceway to build up a positive association (individuals can be targeted into and through the raceway if calm - but the aim at start is to not stop individuals in the raceway).
4. Once individual is comfortable entering and exiting the raceway, target behaviour can be used to stop the rhinos in the raceway (still with both doors open). Build up the time individuals are stopped in the raceway in small approximations.
5. Once individual is comfortable being stopped, the entrance door can start to be closed. Individuals will need to be habituated to the sound and motion of doors - a verbal signal of "door" should be used before movement. This can begin by the door only being closed a small way and built up in approximations until it can be left fully closed for the duration of the session. The exit door should be left open to ensure a quick forward exit is available.
6. Once individual is comfortable in the raceway with the entrance door closed behind, the exit door can start to be closed (using the same verbal signal). Like the entrance door, this can be built up in approximations until left fully shut for the duration of the session.
7. When individual is comfortable in the raceway with both doors closed, bars can be added. A verbal signal of "bars" should be used before keepers move/lift a bar. This should be done in small approximations to prevent negative reactions.
8. Once individual is desensitised to bars moving, a single bar can be placed at the lower level in front – inputted at arm's length so if an adverse reaction is seen, the bar can safely be removed. During this step, the raceway exit door will be open to allow the individual to back out of the raceway if needed. Repeat until the individual is comfortable with both bars in place. The exit door can then be closed behind them.
9. Step 8 should be repeated with adding both back bars. To start, the front bars and doors should be left open to allow the individual to exit the raceway if needed. Once the individual is comfortable in the raceway with both back bars in place, the front door should be closed, and front bars to be added. This step may take several sessions.
10. End goal is rhinos being stood in the raceway, with both front and back bars in place and doors secured for duration of session.

26.6 CONCLUSIONS

Zoo training is essential for animal welfare but comes with limitations influenced by various factors. The outcomes depend heavily on species, individual animals, trainers, the environment, and time. Differences in temperament, trust, size, age, and health among animals affect their responsiveness, while trainers' consistency and expertise play a pivotal role. Time constraints at work often limit the duration and frequency of training sessions, and environmental factors, such as distractions or enclosure design, further shape progress.

Overcoming these challenges calls for tailored strategies and approaches (alongside patience and flexibility), which keepers and other animal care staff are instrumental in successfully implementing. Strategies such as the Self-Management System for birds and specifically designed crates for monitors at London Zoo, and the training raceway at Marwell Wildlife show that allowing animals to take part in their own health care thus improving positive welfare states is possible when training is factored into keeper daily routines and enclosure designs.

Regular training sessions not only reinforce desired behaviours but also provide opportunities for physical and cognitive activity and enrichment. The training techniques of station, scale, and recall discussed in this chapter are valuable tools for improving the care and well-being of all animals in zoos and (alongside other behaviours not covered in this chapter), trained based on the principles of behaviour science also foster trust and cooperation between all species and their trainers helping to form positive human and animal relationships further increasing welfare.

The implementation of animal training throughout this chapter demonstrates the effectiveness of training in managing animal health, reducing stress, and ensuring the overall welfare of zoo housed animals alongside its application to conservation as should form an integral role of modern zoo science and keeper routines.

ACKNOWLEDGEMENTS

The Bird Self-Management System (BSMS) at London Zoo was designed and implemented by Adrian Walls (Section Manager of Tropical Birds) and Thomas Lawrance (keeper at London Zoo). Thanks to Gary Ward (ZSL Curator of Birds) and the London Zoo Tropical Bird Team, Ben Tapley (ZSL Curator of Reptiles and Amphibians) and the London Zoo Ectotherms team. Training advice from Jim Mackie (ZSL Animal Behaviour Management Officer and ZSL Evidence-based Animal Care Team). Thanks to Professor John Ewen (ZSL Senior Research Fellow) and partners who contributed to the Sihek Recovery Project, including Guam Department of Agriculture, AZA, The Nature Conservancy, and US Fish & Wildlife Service.

REFERENCES

McMurdo Hamilton, T. (2021). *Making better decisions for the conservation of threatened species: science, values, and New Zealand's rarest bird* (Doctoral dissertation, UCL (University College London), 33–47. https://discovery.ucl.ac.uk/id/eprint/10140698/ (Access date: 24/03/2025).

Melfi, V. A., Dorey, N. R., & Ward, S. J. (Eds.). (2020). *Zoo animal learning and training*. John Wiley & Sons, Hoboken, NJ.

Miller, L. J., Vicino, G. A., Sheftel, J., & Lauderdale, L. K. (2020). Behavioral diversity as a potential indicator of positive animal welfare. *Animals*, *10*(7), 1211.

Rault, J. L., Waiblinger, S., Boivin, X., & Hemsworth, P. (2020). The power of a positive human–animal relationship for animal welfare. *Frontiers in Veterinary Science*, *7*, 590867.

Tarwater, C. E., & Brawn, J. D. (2010). The post-fledging period in a tropical bird: Patterns of parental care and survival. *Journal of Avian Biology*, *41*(4), 479–487.

ZSL. (2025.). *The history of ZSL*. https://www.zsl.org/about-zsl/organisation/history-of-zsl (Access date: 24/03/2025).

27

Zoo animal welfare into the 21st century: Conclusions and future considerations for welfare-focused husbandry

PAUL ROSE, MARK WHITESIDE, ZOË NEWNHAM AND IAN HICKEY

27.1 WHAT HAS THIS BOOK DISCOVERED ABOUT ZOO ANIMAL WELFARE?

Forty years after the definition was first published, animal welfare is still the state of the individual as they attempt to cope with their environment. We now have more information on what these states are (behavioural, psychological, physical) as well as how an individual's internal (e.g. hormones, homeostasis, thoughts and emotions) and external (e.g. resources, climate, presence of humans) environments influence the chances of coping with, and successfully responding to, particular stimuli of importance. We also know to consider affective state, biological functioning, and naturalness in our welfare concept – and this is particularly important to the zoo when we manage animal populations to meet specific aims ("keeping wildlife wild" goes hand-in-hand with understanding – and providing for – the wants, needs, and physical and psychological health of the species that underpin a zoo's conservation, research, education, and engagement aims). This book quite rightly discusses the social licence that a zoo has to operate – a prickly issue that will continue to grow more thorns in the future as acceptable standards of welfare rise and tolerance for welfare compromises lowers. Being transparent with animal management practices that are essential for population sustainability and conservation viability is essential, as such practices may be misunderstood or misconstrued by those outside of zoo operations, and this can lead to mistrust, incorrect assumptions being made, and emotional arguments being presented as fact.

Bias in how species are favoured in research outputs (Rose et al., 2019), or focused on for husbandry developments (Figure 27.1), suggests there is still an

DOI: 10.1201/9781003429425-27

ingrained culture of prioritising the needs of some taxa over others. This may be justified in cases where species are seen to be controversial in the eyes of the public or activists, e.g. elephants (Proboscidea) (Doyle, 2014), and prevailing zoo practices are hard to justify from a welfare standpoint. Thus, in this case, great strides have been made to develop, evolve, and thus improve animal management for elephants based on this intense and integrated applied research effort (Glaeser et al., 2021; Miller et al., 2016; Yon et al., 2019). However, there are likely cases where a species that is not large, enigmatic, or is perceived to have higher cognitive ability still experiences welfare challenges across zoos and aquariums, and, as such, such species is not the focus of welfare reviews and husbandry developments. Identification of gaps in housing and husbandry knowledge, as per Melfi (2009) who, for example, highlighted the large number of facilities housing chelonians yet noted very limited research output to support prevailing management practices, should be undertaken more broadly and not just published on – but actioned. Bridging the gap between the academic and the practitioner to encourage more applied husbandry, housing, and welfare research, which is then utilised in the zoo to make a positive difference to animal care and zoo operations, is a key focus for zoos moving forwards. Further collaboration with individuals external to the zoo that have expertise in less-researched taxa may also be needed to plug holes in ecological or biological knowledge of such animals and help advance husbandry-based research. Such collaborations would feed into the One Plan Approach of integrated conservation between in situ and ex situ populations and the personnel working with them (Byers et al., 2013).

Hearteningly, this book has identified that the implementation of practices that are considerate of zoo animal welfare, both in terms of promoting positive outputs and ensuring that animals can meet challenges and thrive accordingly, is fundamental to the fulfilment of a zoo's aims, the execution of its operations, and essential to the zoo's continued social right to practice. However, what animal welfare looks like for all taxa housed in the zoo, and how it can be accurately inferred, remains challenging. Therefore, a pragmatic and operational approach is needed. Building on evidence for captive care and constantly evolving better practice guidelines to ensure the most current, reliable, and relevant information is embedded into animal management is the surest foundation for good welfare in the zoo. Zoos can eventually make the move towards "husbandry-based evidence" (Rose & Lewton, 2025) when practices align with species needs and the outputs of such practices are, themselves, the evidence that the zoo knows what it is doing to successfully look after all life stages of that species.

Population and animal managers must consider the role of that species to the zoo (more broadly) when providing resources and stimuli for animals to engage with. For example, a species that is housed in the zoo as an ambassador for the wild, with little to no chance of in situ reintroduction, could be provided with a different set of stimuli (suitable for management and safe working around

Figure 27.1 A simple count of taxa-specific husbandry manuals (not including educational or awareness-raising documents or non-husbandry annexes) reveals that mammals dominate the publication of best practice guidelines across the EAZA region, with fish accounting for <3% of available husbandry resources (https://www.eaza.net/BPG/ accessed 04/03/2025).

humans, for example) compared to a species from an insurance population for eventual field-based conservation action (where key anti-predatory behaviour or abilities to outcompete rivals are required for survival). As was introduced in the preface, there are many challenges that zoos have to grapple with, especially when trying to balance meaningful welfare assessments against providing for "wild type" experiences. Figure 27.2 summarises some of the conundrums for the zoos around the common theme of ensuring animals experience positive welfare outputs whilst not creating domesticated versions of their species that have lost key adaptive traits. We would encourage others to take these conundrums and use them as the basis of further welfare-focused husbandry science and investigation.

Measurement of animal welfare moving forward needs to be refined and validated for species, avoiding generic styles of assessment that will not capture the specific outputs shown by individual species that can indicate changes in their welfare over time. Even when zoos are bound by policy or law to undertake animal welfare assessments, this should be completed using welfare assessment tools that are relevant to individual species to provide the most reliable results from such assessments as possible. A welfare assessment tool for birds, for example, is not going to be able to capture the nuances in behaviour and behavioural expression displayed by the multitude of different species managed in zoos and aquariums. Figure 27.2 shows that enabling fitness benefits to accrue from behavioural performance can mean one animal successfully competing with another to better access rewards or utilising sexually selected traits to outcompete rivals for mating rights. These varied behavioural outputs ultimately have positive connotations for some animals to the detriment of others but, holistically, would improve the natural functioning of the social group overall. As "naturalness" is a key component of welfare, we therefore argue for a change in how zoo animals are managed to remove an excess of comfort in some situations to ensure individuals can retain wild-type characteristics into the future. For example, still provide unrestricted drinking water, but make the animal seek this out or work for access to the resource as a wild conspecific would do.

As an example of the conundrum surrounding how to ensure that animals have control, choice and comfort, yet are still challenged to experience new things and thus develop ways of coping with different experiences over the course of their life, the case study in Box 27.1 discusses the balance between providing appropriate care and promoting positive welfare outcomes, while considering that the wild is not always welfare-positive. Box 27.1 gives food for thought on ideas of promoting resilience and the conservation of adaptive behaviours that could be applicable to zoo populations in some way, shape or form.

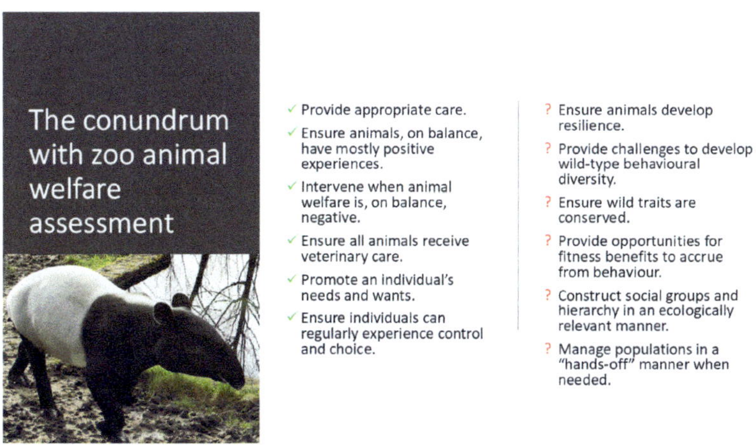

Figure 27.2 The conundrum with the zoo surrounds a balance of comfort (positive welfare) against conserving wild traits that rely on survivability and developing strategies to cope with and overcome adversity.

BOX 27.1: Stress for success? Giving a beneficial taste of the 'real world' to improve long-term welfare for animals in reintroduction programmes

For most animals housed in zoos, positive welfare is achieved via such things as well-designed enclosures, careful monitoring of the animal throughout their life where treatment is easily implemented, and death typically prescribed by defined endpoints and euthanasia by trained practitioners. However, there are situations where animals are only captive for part of their life, such as prior to release into the wild as part of a conservation programme. In these instances, animals need to navigate complex environments, feed efficiently, compete successfully, and survive long enough to breed viably (Armstrong & Seddon, 2008). During this time in the wild, there is often limited ability to monitor welfare and administer treatment if required.

Despite advances in reintroduction biology, most programmes can have a high failure rate (Fischer & Lindenmayer, 2000; Griffith et al., 1989; Gross et al., 2024). A key cause of failure is high post-release mortality that has been attributed to cognitive, behavioural, and physiological deficiencies in captive-reared animals compared to their wild-reared conspecifics (Fischer & Lindenmayer, 2000; Gross et al., 2024). As death is not always quick in the wild, such deficiencies suggest that the captive environments used for such "future-in-the-wild" populations pose a long-term threat to the welfare of the animals involved. However, when these environments are enriched and there have been efforts to make experiences more naturalistic (e.g. via anti-predator training), the fitness of a released population markedly improves (for review see Gross et al., 2024; Reading et al., 2013). There can be a lack of consistency in results across programmes, with reviews highlighting the need for more scientific rigour (Tetzlaff et al., 2019) and more extreme conditions to develop necessary survival characteristics (e.g. lethal exposure to predators) for a viable future in the wild (Zhu et al., 2023). This poses two questions for the zoo: what are the consequences of more extreme naturalistic captive environments and how does this influence an individual's overall quality of life?

Such questions may be difficult to answer in zoo settings. First, zoos may be restricted in how "naturalistic" the environment can be due to ethical or legal obligations. For instance, would a zoo be able to implement a stressful, fearful or distressing environment filled with predation, competition, unpredictability, starvation, parasites and disease that is synonymous with a true reflection of the wild? Secondly, zoos can be constrained by their infrastructure and resources available, or because of limited funding available to improve methods and experimental approaches and therefore scientific rigour – thus a lack of replicates, controls and the ability to follow animals after release can limit the reliability of outputs from such programmes. Because release programmes for conservation typically involve endangered species, there may be little scope for risk if zoo personnel are dealing with the last few individuals of a species or population. There is, however, a surrogate system that could be used to test the application of environmental complexity and challenge on long-term animal welfare to offer valuable insight into conservation programme success: the rearing of game birds (Galliformes).

Released gamebirds show similar mortality to that from conservation programmes, with up to 50% of individuals dying within a few weeks post-release (Madden et al., 2018) – attributed to behavioural, physiological and cognitive deficiencies of captive-reared conspecifics (for review see Madden et al., 2018). These birds are reared in large numbers, in replicated environments, and (with the necessary licences secured) can be subjected to naturalistic environments to test potential means of improving welfare (e.g. Beardsworth et al., 2021). Tags and sensors can also be fitted to monitor behaviour post-release (e.g. Heathcote et al., 2023).

A review of how to improve post-release survival (Madden et al., 2020) shows the complex relationship between post-release survival and pre-release welfare, highlighting four important scenarios:

1) A negative coincidence of interest where captive environments result in poor pre-release welfare and birds that are not suited to life in the wild. For example, a barren, non-naturalistic, industrial system does not allow birds to exhibit normal behaviours, coupled with high levels of aggression, apathy and social withdrawal. This style of environment causes high mortality in a released population.
2) A conflict of interest where environments that promote good pre-release welfare leads to the production of "healthy" individuals that are not prepared for release. This occurs when animal management practices provide *ad lib* resources in sanitary conditions and without experiences of predators. With close monitoring of welfare, animals are treated when there are signs of illness. Overall, this results in significantly lower mortality than age-matched wild-reared conspecifics (Đorđević et al., 2010; Madden et al., 2018) but such lack of preparation for wild challenges means birds suffer high mortality rates upon release.
3) A conflict of interest where environments that improve post-release survival compromise pre-release welfare. For example, the development of anti-predation learning by presenting predator cues improves survival (Gaudioso et al., 2011) but causes stress during the development phase (Rabin, 2003). Dietary provision (e.g. food type or predictability) can cause appropriate dietary behaviour (foraging, food preference) and physiology (gut morphology) that improves survival (Homberger et al., 2014; Whiteside et al., 2015) but can increase competition and monopolisation that results in some individuals experiencing hunger (Rushen, 2003; Stahl & Kaumanns, 2003).
4) A positive coincidence of interest where management improves pre-release welfare whilst allowing birds to develop essential survival skills. For example, the addition of perches to a captive environment increases the likelihood of birds avoiding nocturnal predation by roosting above ground (Whiteside et al., 2016). Not only does access to perching improve post-release survival, but it also reduces pre-release aggression and improves feather coverage (Santilli & Bagliacca, 2017; Whiteside et al., 2016).

These scenarios show the delicate and complex balance between pre-release welfare and post-release survival, highlighting the dilemma that animal managers have when planning captive housing regimes. Do we cosset animals, adhere to all our perceived positive welfare obligations, but release animals that are ill-prepared for life in the wild? Or should we compromise pre-release welfare and implement environments causing stress, fear and disease to then release animals with a better probability of survival? Such are the big and challenging questions that zoos need to grapple with if they are to truly manage "the wild" viably into the future.

27.2 MEANINGFUL BUT SENSIBLE ASSESSMENT OF WELFARE OUTPUTS

Throughout this book, we have stressed the importance of meaningful, relevant and validated welfare assessment approaches and the use of credible, trustworthy evidence. So, we now pose the question, "Do we always need to think about emotions?" This might feel like an uncomfortable question for some, working in environments where interpretation of the mental domain is prized above all. But is it really achievable and realistic to work out how best to record if each leafcutter ant (*Atta cephalotes*) in a colony of thousands is feeling good? Perhaps we can infer what the ant is likely to be feeling (are we ever going to actually know?)

if the overall package of enclosure, husbandry and enrichment (all of these ant-focused inputs) is biologically relevant. An individual ant in a barren plastic tub with nowhere to hide or no conspecifics to interact with is going to outwardly demonstrate actions that suggest a lack of comfort with their immediate situation unlike an individual ant carrying out behaviours indicative of their life stage and role within a colony. If we take this leafcutter ant out of the bare tub and place them back into a functioning, socially structured colony and they resume the behaviours that we would associate with their life stage and social caste, we have strong evidence here that the inputs (environment, social group, resource access) are relevant and appropriate for the species and therefore welfare is likely to be, on balance, positive in these conditions. Such considerations and ideas for developing appropriate inputs to support good outputs are illustrated in Figure 27.3.

Therefore, a key question needs to be, "What is any meaningful measure of emotional state for a species? And does this (attempt at measuring emotion) really matter if the inputs are correct (i.e. ecologically relevant for the species in question)?" This is not a blanket "give up on trying to infer emotional outputs" but instead a checkpoint to encourage scientists and practitioners to first understand the species in question – what we know about it, what it needs, and what we currently can provide under human care. For species where documentation of body language or behavioural expression of moods is more certain, and where species are complex in their processing of external information, we most certainly should include mental states as part of welfare assessment protocols. Continuing with the invertebrate scenario, determining the overall quality of life for captive octopuses (Octopoda) needs to integrate inputs (resources) and emotional, behavioural, and mental outputs in regular welfare assessments due to the intelligence, sentience, and complex neurobiology of these species (Amodio et al., 2019; Hochner, 2012) that would influence changes in their perception of welfare across different times of the day. The review of the biology of these species and the integration of reliable evidence into husbandry continue to enable development of care regimes that should, ideally, support appropriate behavioural, psychological, and physiological welfare outputs more frequently. However, the gaps that are still apparent in our knowledge of fundamental aspects

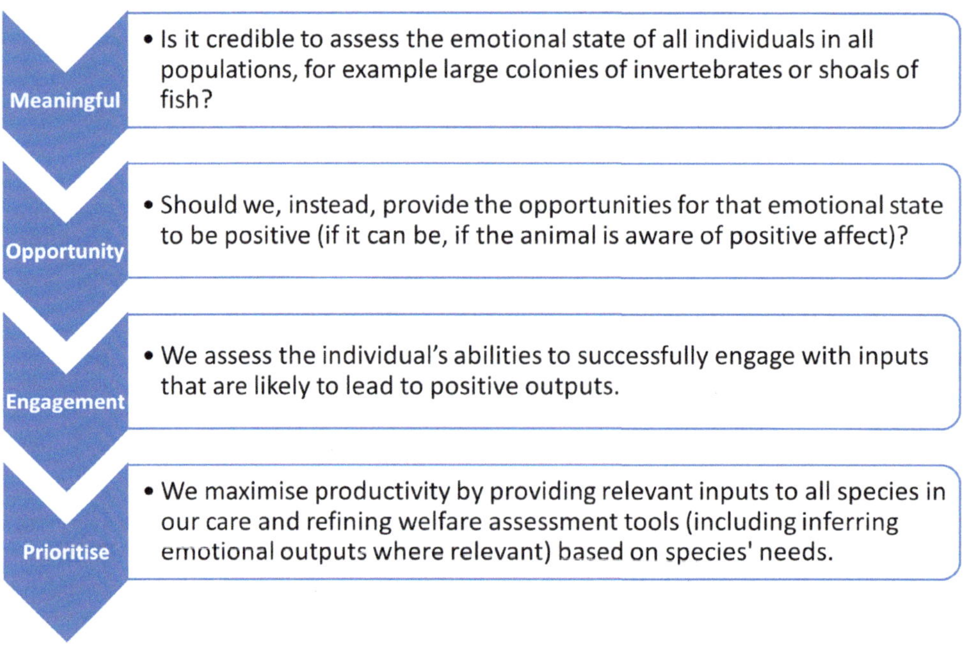

Figure 27.3 If we consider the appropriateness of husbandry as the best proxy for good welfare outputs, we can put renewed efforts into the science behind captive care to evidence as many aspects of husbandry as possible.

of species biology (as mentioned previously in this chapter) must be filled to give us confidence that our animal management practices will ultimately enable positive welfare outputs. As mentioned in this book's chapters on enclosure design, putting the animal at the centre of the design and build phase will ultimately lead to more appropriate welfare outputs because the environment created will be a suitable fit for the animal being housed. When attempting to promote the performance of complex behaviour patterns, such as the challenges surrounding meaningful time spent flying for birds in aviaries (Rose et al., 2024) due consideration must be given to the traits that a species possesses. The concept of "functional ecology" and how this links to zoo animal behaviour and husbandry is covered in detail in "The Behavioural Biology of Zoo Animals" (Rose & Huffman, 2022). The idea that we can look across species, see common traits linked to a specific habitat or niche, and provide for these accordingly is essential. As functional ecology has been useful for understanding conservation needs, population dynamics, and extinction threats, perhaps we can also use this to help us understand which species are likely to suffer. This is an idea that can be found in works that predict the likelihood of suffering from ecological characteristics (Bandeli et al., 2023; Lewis et al., 2022; Mellor et al., 2021). Use of functional ecology concepts alongside phylogenetic comparative approaches – statistical methods that consider the independence of individual species as data points when attempting to model responses to husbandry and housing (Mellor et al., 2018) – could provide a finer scale view into species differences in their responses to a life in the zoo.

Environmental enrichment is a key feature across many chapters of this book. It is heartening to see how enrichment has become a mainstream aspect of daily care across reputable, accredited zoos globally. There are still areas for improvement with this practice, however. When providing environmental enrichment, for example, we should start by questioning what we are aiming to achieve (e.g. what is the behavioural outcome and how will this improve the animal's welfare or quality of life experience in the zoo?) and do this from a strong baseline of species' natural history, evolutionary biology, and behavioural ecology viewpoints. We can then integrate into enrichment practices opportunities for eustress (positive challenges that further develop an animal's cognitive, spatial, and problem-solving skills). The idea of challenge being important to welfare is eloquently examined and evaluated by Meehan and Mench (2007). This approach can work for a range of species, even for species that can be seen to be challenging to enrich, such as aquarium fish (Rose, 2024), Figure 27.4.

27.3 THE WILD IS NOT A PLAYGROUND

"...retaining and encouraging natural abilities" is embedded into The World Zoo and Aquarium Animal Welfare Strategy (Mellor et al., 2015). Consequently, as this book has demonstrated, the backbone of what welfare is, and what equates to evidence-based husbandry, is natural history, ecology and evolution. Evidence of the importance of such traits comes from how a species engages with stimuli and resources in its natural habitat. Figure 27.5 shows that the wild is a harsh and unforgiving place. If we remember that selection pressures have honed certain individuals to be more successful – and possess traits that give them an advantage over others – we can recreate habitat features and opportunities in the zoo that enable this natural process to continue. For example, giving more credence to mate choice in conservation (Asa et al., 2011) is an excellent example of the need to consider selection pressures (and their involvement in the phenotypic traits that species possess) within zoo animal management more widely.

We need to provide elements of natural living where possible, by including, measuring and evaluating enrichment, building enclosures full of challenging and engaging features and resources, and considering the impact of people (caregivers and visitors) on what animals are likely to do. Ensuring individuals can engage fully with ecologically relevant aspects of their environment is crucial. Thus, providing husbandry and housing in line with the role of that species in the zoo (e.g. is this an ambassador species that will not breed in the future, or is this individual due for reintroduction in the near future) has a clear welfare angle. The enrichment and experiences presented for learning, developing resilience and solving challenges

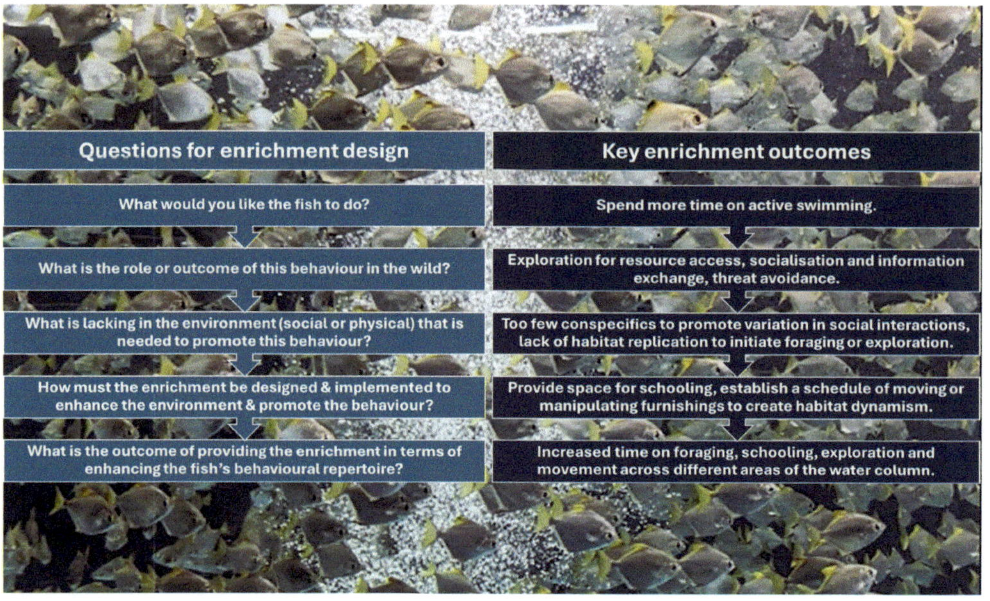

Figure 27.4 Adapted from Rose (2024). Designing and implementing enrichment with the animal in mind. We must start with what we would like to see the animal achieve from interaction with the enrichment and build towards a meaningful enhancement of behavioural and welfare outputs. Learning from solvable challenges (negative experiences that become positive) can, and should, be part of this process.

for the ambassador animal will be very different from those in the reintroduction programme.

By focusing more on the application of evidence that is known to be relevant and accurate, we could further help to determine what welfare outputs present for the many species we manage in zoos. "Naturalness" in the wild compared to the zoo, for example, could be based on specific phenotypic traits that change with season or over time. Welfare scientists could work together to provide behavioural inventories for specific species that give clear identifiers of affective states, or

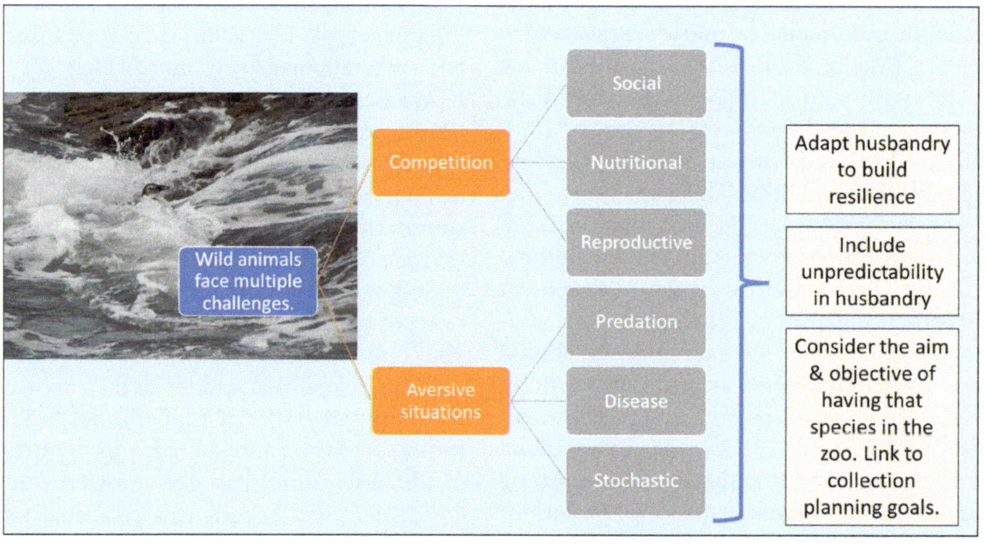

Figure 27.5 What are the challenges of the wild? How can these challenges be filtered, discarded or applied where relevant to each species population under human care?

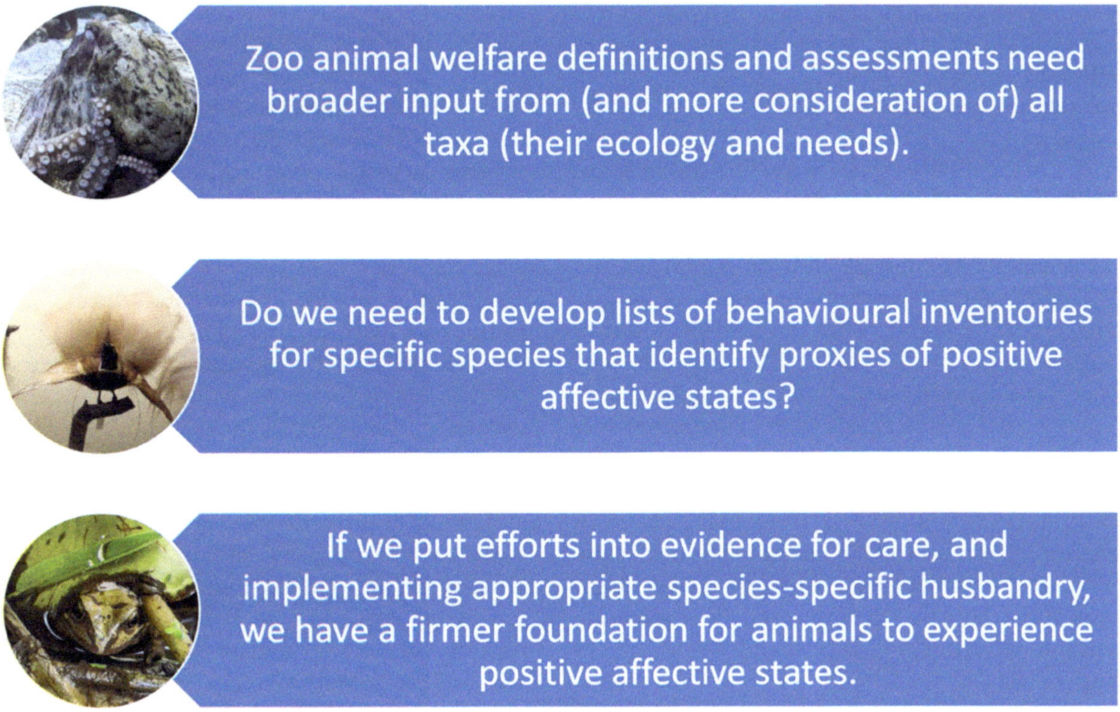

Figure 27.6 Moving forwards from the debate on "what is welfare" to the application of welfare knowledge that has a positive impact.

the proxies for these, that could form the basis of understanding behavioural needs, or how to provide choice, control and challenge, in the zoo. This approach of understanding the wild in all its forms and therefore being able to comprehend how and why non-domesticated species behave in the manner that they do gives us a more secure foundation for the provision of appropriate inputs (Figure 27.6).

27.4 WELFARE AND THE WORDS WE USE

At the time of writing this book, research shows that keeping animals captive in zoos and aquariums is a controversial topic, even with the well-intentioned conservation and education aims that underpin the workings of the modern zoo (Villarroya et al., 2024). The uncomfortable nature of how zoos operate, and the practices that are undertaken for population-level welfare (at the expense of individual welfare), e.g. culling of surplus individuals and then using these culled animals for educational gain (Clauss et al., 2025) stick in the public imagination seemingly more readily than the success stories. Therefore, the way in which zoos engage with their visitors and how they explain what they do needs to be honest and transparent to show why such decisions are taken and how benefits can accrue from them.

At the same time, zoos need to consider how they engage and interact with younger audiences to ensure their education and conservation messages remain valid. With research finding that younger age groups in society are more keen to keep non-domesticated species as pets than traditional ones (Cronin et al., 2022), the role of zoos in disseminating clear and honest animal welfare messaging is even more important. The use of technology such as online livestream cameras gives further insight into the wild world of the animals, their behaviours, and resource usage (Des Pallieres & Rose, 2023; Wood et al., 2020) – such tools could be better utilised for evidence collection to benchmark captive behaviour and environments. But this technology also provides excellent opportunities for wider visitor and public engagement. Zoos could become creative in how they use technology to bring the zoo animal and the free-living animal closer together in the stories and messages they tell their visitors. Accessibility of wildlife on such livestream cameras could, however, be a double-edged sword – will this make the public even more aware of the differences between how animals can live in a natural habitat compared to within a zoo enclosure?

Therefore, we must be clear on our words and their meaning in the zoo, both in terms of how we talk internally and how we explain operations to visitors and the wider public. "Best practice care" is perhaps more realistically "better practice care". We can see this from the number of updates to species husbandry guidelines that are published in a short period of time by the large zoo accreditation organisations. Will we ever reach best practice in the zoo? Or do we instead have a continuous cycle of review and improvement that conveys more truthfully the care, thought, and dedication we put into ensuring animal husbandry is as correct as possible?

Another good example is the growing use of the term "habitat" to describe enclosures and exhibits – something aspirational but maybe not wholly accurate. A habitat is a dynamic, natural system that encompasses landscapes, biotic and abiotic factors, and resources needed to sustain a species and their life processes (Miller & Hobbs, 2007). Can this be true for all the forms of environment created for species under human care in zoos and aquariums (Figure 27.7)? Research on the example provided in Figure 27.7, the common hippopotamus (*Hippopotamus amphibious*), tells us that behavioural needs are not catered for in some zoo enclosures (Tennant et al., 2018).

When a species' zoo enclosure can promote the widest possible array of ecologically relevant behaviours and provide opportunities for challenge, comfort and eustress in a way that hones survival skills and the use of adaptive traits, calling this environment a habitat is a true and honest reflection of how this species is managed and cared for as part of their wider welfare-focused husbandry. For example, we can already recreate managed habitats in captivity for some species, particularly smaller ectothermic species like fish, amphibians and reptiles. Advances in technology now allow us to provide environments for these animals that are almost indistinguishable from their wild habitats, with heating, lighting and precipitation delivery systems able to mimic both daily and seasonal variations within a species' natural home range (Figure 27.8). This is of course largely dependent on a species' size, as the greater the size of an environmentally controlled exhibit, the more financially costly it will be to maintain in terms of energy consumption, for example.

Figure 27.7 Are these habitats? Or are they environments that we have created to provide housing for common hippos in zoological establishments? Do these environments enable the animals to perform their full 24-hour cycle of different behaviours? If we compare the diversity, scale and scope of resources provided and look at the range of resources available in the species' wild habitat, there are marked differences. Therefore, the concept of a "hippo habitat" in a zoo may not be the most relevant way of describing how the species is being displayed.

Figure 27.8 Technological advances can enable us to create a habitat. They allow for the provision of correct lighting, temperature, and humidity requirements of a species' natural home range when that species is small enough to manage in such spaces. This exhibit for the blue poison dart frog (*Dendrobates tinctorius azureus*) and emerald tree boa (*Corallus batesii*) provides UV and LED lighting, various heat gradients, and a precipitation/misting system, all on automated timers to replicate the natural environment of the Amazon and minimise the need for keeper interference.

27.4.1 WHAT WELFARE IS ACCEPTABLE WELFARE?

Zoos need to wrangle with what is acceptable in terms of animal welfare experiences for different species and for specific individuals. Public attitudes towards welfare infringement for large charismatic mammals are becoming less tolerant, and they are aware of signs of poor welfare when these are expressed by individuals of such species at the zoo. There are, though, many examples of where individual animals do not look their best or display outward signs of poor welfare or physical injury that can detract from overall positive welfare outcomes. We, the editors of this book, are not suggesting that animals with healed injuries or behavioural pathologies be instantly euthanased, but rather that care and/or consideration is given to how this animal is displayed to zoo visitors, what message the animal's physical condition may conjure if no context is given, and whether the on-show environment is suitable for the animal or whether better accommodation should be created where more bespoke care can be given. If a zoo has put animals on display that clearly appear welfare-compromised (Figure 27.9) without explanation or interpretation of the individual's current situation or what is being done about this, is the zoo accepting that such welfare compromise is fine for some individuals of some species in captivity?

27.5 WELFARE-FOCUSED HUSBANDRY – THE CONCLUSION OF THIS BOOK

The ultimate concept of this book is to encourage efforts into welfare-focused husbandry. This means aligning animal management regimes in the zoo with species' needs. Not to wrap animals in cotton wool and never challenge them, nor to use the need for resilience as an excuse for excess stress and poor care, but to ensure that wild animals in zoos remain just so and can experience

Figure 27.9 What is an acceptable welfare compromise? Examples of where potentially negative welfare may be visible, or not, to a zoo's visitors. Top left: Poor feather condition and missing upper mandible in a Eurasian spoonbill (*Platalea leucorodia*). Top middle: Damaged and slipped wing in a black stork (*Ciconia nigra*). Top right: Overgrown and misaligned beak in a white-fronted bee-eater (*Merops bullockoides*). Bottom left: Repetitive licking of enclosure furnishings by an adult giraffe (*Giraffa camelopardalis*). Bottom middle: Damaged mouthparts and cloudy and damaged eyes in a *Pangasius* sp. catfish. Bottom right: Small, barren, homogeneous exhibit for an alligator snapping turtle (*Macrochelys temminckii*).

opportunities for important ecologically relevant traits to be utilised or displayed.

Figure 27.10 presents a structured approach to welfare-focused husbandry in zoos and aquariums, emphasising the interconnected roles of scientific outputs, animal-centric assessment, and engagement with the wider zoo world. It provides a framework to ensure that husbandry practices are informed by reliable evidence, assessed for their impact on individual animals, and refined through broader dissemination and relevance when applied in the zoo. Developing operational welfare indicators for specific species (across all life stages of a species so it is clear when completing a zoo welfare

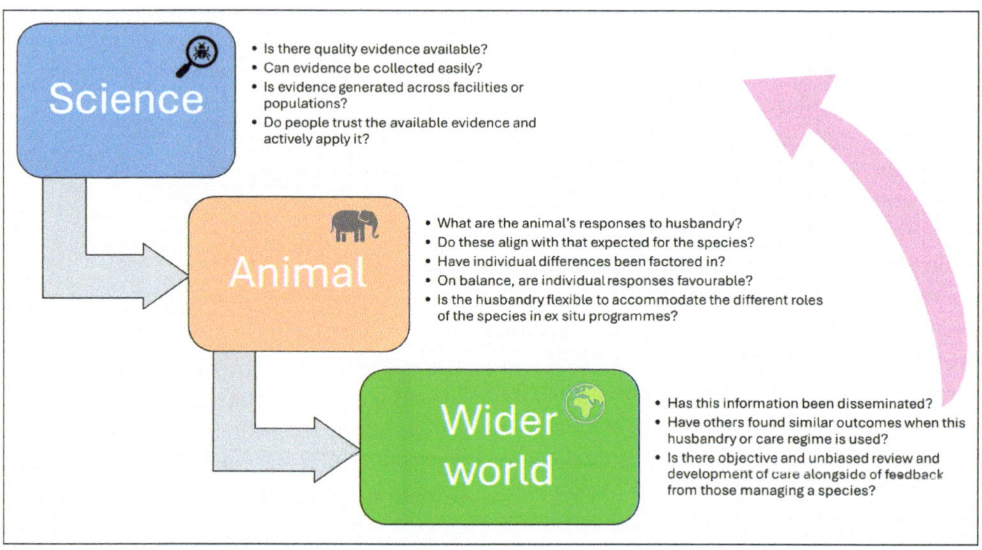

Figure 27.10 The three pillars of welfare-focused husbandry: the science for generating evidence, the animal that this science benefits, and the wider world where science-based husbandry is ultimately applied and evaluated.

assessment what is meaningful to assess at a particular time) would be relevant to include in taxa-specific husbandry guidelines.

Science serves as the foundation of evidence-based husbandry, ensuring that management decisions rely on high-quality, reproducible research rather than anecdotal information. The availability and quality of evidence are crucial, alongside the feasibility of collecting data across multiple facilities to provide a broad coverage of species holdings. Any generalisability of findings is reliant on multiple data points from individuals, populations, and zoos being included in data gathering exercises. This, in turn, promotes more trust in such scientific knowledge and plays a key role in determining whether stakeholders implement such evidence-based approaches accordingly.

Once husbandry practices are informed by scientific research, their direct effects on the animals must be evaluated. Monitoring an animal's responses to husbandry provides insight into welfare, ensuring that care aligns with species-specific evolutionary history and behavioural ecology. It is essential to consider individual differences in responses, as each animal may react uniquely to specific care routines. The overall welfare balance must be carefully assessed to determine whether a practice is beneficial. Furthermore, flexibility in husbandry is necessary to accommodate the different roles that animals may play in ex situ conservation programs.

Beyond individual facilities, the dissemination of information within the wider zoo and aquarium community is vital for continuous improvement in animal welfare. Findings from research and welfare assessments should be shared broadly, allowing for validation across different institutions. Cross-institutional comparisons help determine whether similar outcomes are observed when the same husbandry methods are applied elsewhere. Unbiased review and ongoing development are key components of this process, encouraging independent evaluations and refinements based on feedback from practitioners.

Figure 27.10 highlights a systematic approach to zoo animal welfare, ensuring that husbandry decisions are firmly rooted in scientific evidence, individually assessed for their effects, and continuously refined through collaboration within the wider zoological community. By integrating these three pillars, zoos and aquariums can maintain a high standard of care, ensuring that their husbandry practices remain responsive, effective, and scientifically sound.

To put this framework into practice, zoos and aquariums can adopt several actionable steps:

Develop standardised data collection protocols that ensure consistent methods for gathering welfare-related data across facilities to improve comparability and reproducibility.

Implement routine welfare assessments whereby regular behavioural and physiological evaluations are conducted to monitor individual responses to husbandry practices.

Encourage cross-institutional research collaborations to share findings and methodologies between facilities to enhance uptake and usage of evidence-based husbandry for each species they have been designed for.

Train staff in evidence-based practices by providing ongoing professional development to ensure that caretakers and management teams have dialogue and discussion on husbandry evidence developments that lead to more successful application of the latest research in animal care to the zoo.

Provide staff with opportunities for accessing scientific information and supporting resources by facilitating access to a zoo library or online database of publications that everyone can learn from. From a zookeeper focus, welfare assessment tools need to be practical and applicable.

Adapt husbandry to individual animal needs across life stages by using behavioural observations, health records, and other pertinent methodologies (such as Qualitative Behavioural Assessment, QBA) to tailor care plans for specific animals, recognising individual variability. Zookeepers need to understand the species in their care across the lifetime of the animal. The skills to do this need to be learned, and training is important in differentiating what methods to use, when and why.

Create feedback loops for continuous improvement that support the "better husbandry practice" process. Establish systems where zookeepers, higher managerial staff, and researchers can regularly review and adjust husbandry regimes when needed.

Going beyond firefighting to create positive experiences. Whilst there will always be urgent situations that require quick resolving, encouraging animal care staff to implement correct husbandry practices (and giving them the time and resources to do so) will pay off in the long term and create a more satisfying working environment.

Engage in public and professional outreach to disseminate welfare insights through conferences, publications, and education programmes that foster widespread adoption of best practices and engage as many stakeholders as possible in welfare-focussed husbandry.

Zoos and aquariums are a force for good. They have done, and continue to do, excellent work in researching and conserving the natural world and educating a global population about it. It is an unequivocal fact that, without zoos, some species would be extinct. But, like any institution or organisation, zoological ones are not perfect. And these imperfections are more frequently under the spotlight because they are held to a higher account due to the lives that they manage – those of the animals in their care that did not ask to be housed in human-created enclosures. We should also remember that the wild is not perfect – it is not a Garden of Eden that provides positive experiences for wildlife on a daily basis. We must consider how animals meet these wild challenges when designing captive housing and husbandry, as welfare may be more compromised in the long term if we end up creating domesticated versions of the original species. Ultimately, this means that in all the zoo's operations and wider missions, the animals must come first. Continuing to work for these animals means a constant cycle of research, review, application and evaluation. We need to know what they need by gathering information on their evolutionary history, natural history and ecology. We must scrutinise such evidence to check that it is reliable and then apply such information to species-specific animal care approaches. Then we can truly develop better practice management that enables a whole suite of beneficial welfare outputs to be experienced by the individual under human care. This approach, we feel, will best keep zoo animals wild.

REFERENCES

Amodio, P., Boeckle, M., Schnell, A. K., Ostojíc, L., Fiorito, G. & Clayton, N. S. (2019). Grow smart and die young: why did cephalopods evolve intelligence? *Trends in Ecology & Evolution*, *34*(1), 45–56.

Armstrong, D. P. & Seddon, P. J. (2008). Directions in reintroduction biology. *Trends in Ecology and Evolution*, *23*(1), 20–25.

Asa, C. S., Traylor-Holzer, K. & Lacy, R. C. (2011). Can conservation-breeding programmes be improved by incorporating mate choice? *International Zoo Yearbook*, *45*(1), 203–212.

Bandeli, M., Mellor, E. L., Kroshko, J., Maherali, H. & Mason, G. J. (2023). The welfare problems of wide-ranging Carnivora reflect naturally itinerant lifestyles. *Royal Society Open Science*, *10*(9), 230437.

Beardsworth, C. E., Whiteside, M. A., Laker, P. R., Nathan, R., Orchan, Y., Toledo, S., van Horik, J. O. & Madden, J. R. (2021). Is habitat selection in the wild shaped by individual-level cognitive biases in orientation strategy? *Ecology Letters*, *24*(4), 751–760.

Byers, O., Lees, C., Wilcken, J. & Schwitzer, C. (2013). The one plan approach: The philosophy and implementation of CBSG's approach to integrated species conservation planning. *WAZA Magazine*, *14*, 2–5.

Clauss, M., Roller, M., Bertelsen, M. F., Rudolf von Rohr, C., Müller, D. W. H., Schiffmann, C., Kummrow, M., Encke, D., Ferreira, S. & Duvall, E. S. (2025). Zoos must embrace animal death for education and conservation. *Proceedings of the National Academy of Sciences*, *122*(1), e2414565121.

Cronin, K. A., Leahy, M., Ross, S. R., Wilder Schook, M., Ferrie, G. M. & Alba, A. C. (2022). Younger generations are more interested than older generations in having non-domesticated animals as pets. *PLoS One*, *17*(1), e0262208.

Des Pallieres, C. G. & Rose, P. E. (2023). Two's company, three species is a crowd? A web-cam-based study of the behavioural effects of mixed-species groupings in the wild and in the zoo. *PLoS One*, *18*(4), e0284221.

Đorđević, M., Pekeč, S., Popović, Z. & Đorđević, N. (2010). Influence of dietary protein levels on production results and mortality in pheasants reared under controlled conditions. *Acta Veterinaria*, *60*(1), 79–88.

Doyle, C. (2014). Captive elephants. In L. Gruen (Ed.), *The ethics of captivity* (pp. 38–56). Oxford, UK: Oxford University Press.

Fischer, J. & Lindenmayer, D. B. (2000). An assessment of the published results of animal relocations. *Biological Conservation*, *96*(1), 1–11.

Gaudioso, V. R., Sánchez-García, C., Pérez, J. A., Rodríguez, P. L., Armenteros, J. A. & Alonso, M. E. (2011). Does early antipredator training increase the suitability of captive red-legged partridges (*Alectoris rufa*) for releasing? *Poultry Science*, *90*(9), 1900–1908.

Glaeser, S. S., Shepherdson, D., Lewis, K. D., Prado, N., Brown, J. L., Lee, B. & Wielebnowski, N. C. (2021). Supporting zoo Asian elephant (Elephas maximus) welfare and herd dynamics with a more complex and expanded habitat. *Animals*, *11*(9), 2566.

Griffith, B., Scott, M. J., Carpenter, J. W. & Reed, C. (1989). Translocation as a species conservation tool: Status and strategy. *Science*, *245*(4917), 477–480.

Gross, I. P., Wilson, A. E. & Wolak, M. E. (2024). The fitness consequences of wildlife conservation translocations: A meta-analysis. *Biological Reviews*, *99*(2), 348–371.

Heathcote, R. J., Whiteside, M. A., Beardsworth, C. E., Van Horik, J. O., Laker, P. R., Toledo, S., Orchan, Y., Nathan, R. & Madden, J. R. (2023). Spatial memory predicts home range size and predation risk in pheasants. *Nature Ecology & Evolution*, *7*(3), 461–471.

Hochner, B. (2012). An embodied view of octopus neurobiology. *Current Biology*, *22*(20), R887–R892.

Homberger, B., Jenni, L., Duplain, J., Lanz, M. & Schaub, M. (2014). Food unpredictability in early life increases survival of captive grey partridges (*Perdix perdix*) after release into the wild. *Biological Conservation*, *177*(0), 134–141.

Lewis, K. D., Parker, M. O., Proops, L. & McBride, S. D. (2022). Risk factors for stereotypic behaviour in captive ungulates. *Proceedings of the Royal Society B*, *289*(1983), 20221311.

Madden, J. R., Santilli, F. & Whiteside, M. (2020). The welfare of game birds destined for release into the wild: a balance between early life care and preparation for future natural hazards. *Animal Welfare*, *29*(1), 1–18.

Madden, J. R., Hall, A. & Whiteside, M. A. (2018). Why do many pheasants released in the UK die, and how can we best reduce their natural mortality? *European Journal of Wildlife Research*, *64*(4), 40.

Meehan, C. L. & Mench, J. A. (2007). The challenge of challenge: Can problem solving opportunities enhance animal welfare? *Applied Animal Behaviour Science*, *102*(3–4), 246–261.

Melfi, V. A. (2009). There are big gaps in our knowledge, and thus approach, to zoo animal welfare: A case for evidence-based zoo animal management. *Zoo Biology*, *28*(6), 574–588.

Mellor, D. J., Hunt, S. & Gusset, M. (Eds.). (2015). *Caring for wildlife: The world zoo and aquarium animal welfare strategy*. Gland, Switzerland: WAZA Executive Office.

Mellor, E. L., McDonald Kinkaid, H. K. & Mason, G. (2018). Phylogenetic comparative methods: Harnessing the power of species diversity to investigate welfare issues in captive wild animals. *Zoo Biology*, *37*(5), 369–388.

Mellor, E. L., McDonald Kinkaid, H. K., Mendl, M. T., Cuthill, I. C., van Zeeland, Y. R. A. & Mason, G. J. (2021). Nature calls: Intelligence and natural foraging style predict poor welfare in captive parrots. *Proceedings of the Royal Society B: Biological Sciences*, *288*(1960), 20211952.

Miller, J. R. & Hobbs, R. J. (2007). Habitat restoration - Do we know what we're doing? *Restoration Ecology*, *15*(3), 382–390.

Miller, L. J., Chase, M. J. & Hacker, C. E. (2016). A comparison of walking rates between wild and zoo African elephants. *Journal of Applied Animal Welfare Science*, *19*(3), 271–279.

Rabin, L. A. (2003). Maintaining behavioural diversity in captivity for conservation: Natural behaviour management. *Animal Welfare*, *12*(1), 85–94.

Reading, R. P., Miller, B. & Shepherdson, D. (2013). The value of enrichment to reintroduction success. *Zoo Biology, 32*(3), 332–341.

Rose, P. E. (2024). Understanding the welfare of aquarium fish. *Animal Sentience, 36*(3), https://doi.org/10.51291/52377-57478.51858.

Rose, P. E., Brereton, J. E., Rowden, L. J., Lemos de Figueiredo, R. & Riley, L. M. (2019). What's new from the zoo? An analysis of ten years of zoo-themed research output. *Palgrave Communications, 5*(1), 1–10.

Rose, P. E., Freeman, M., Hickey, I., Kelly, R. & Greenwell, P. J. (2024). Considering what animals "need to do" in enclosure design: Questions on bird flight and aviaries. *Birds, 5*(3), 586–603.

Rose, P. E. & Huffman, B. (2022). Behavioural biology & the future zoo: Overall conclusions In P. E. Rose (Ed.), *The behavioural biology of zoo animals*. Boca Raton, FL: CRC Press.

Rose, P. E. & Lewton, J. (2025). Key concepts for enhancing zoo animal welfare: Coping, comfort, choice, control, challenge and compassion *Journal of Applied Animal Welfare Science*, in press.

Rushen, J. (2003). Changing concepts of farm animal welfare: Bridging the gap between applied and basic research. *Applied Animal Behaviour Science, 81*(3), 199–214.

Santilli, F. & Bagliacca, M. (2017). Effect of perches on morphology, welfare and behaviour of captive reared pheasants. *Italian Journal of Animal Science, 16*, 1–7.

Stahl, D. & Kaumanns, W. (2003). Food competition in captive female sooty mangabeys (*Cercocebus torquatus atys*). *Primates, 44*(3), 203–216.

Tennant, K. S., Segura, V. D., Morris, M. C., Snyder, K. D., Bocian, D., Maloney, D. & Maple, T. L. (2018). Achieving optimal welfare for the Nile hippopotamus (Hippopotamus amphibius) in North American zoos and aquariums. *Behavioural Processes, 156*, 51–57.

Tetzlaff, S. J., Sperry, J. H. & DeGregorio, B. A. (2019). Effects of antipredator training, environmental enrichment, and soft release on wildlife translocations: A review and meta-analysis. *Biological Conservation, 236*, 324–331.

Villarroya, A., Miranda, R., Pino-del-Carpio, A. & Casas, M. (2024). Social perception of zoos and aquariums: What we know and how we know it. *Animals, 14*(24), 3671.

Whiteside, M. A., Sage, R. & Madden, J. R. (2015). Diet complexity in early life affects survival in released pheasants by altering foraging efficiency, food choice, handling skills and gut morphology. *Journal of Animal Ecology, 84*(6), 1480–1489.

Whiteside, M. A., Sage, R. & Madden, J. R. (2016). Multiple behavioural, morphological and cognitive developmental changes arise from a single alteration to early life spatial environment, resulting in fitness consequences for released pheasants. *Royal Society Open Science, 3*, 3.

Wood, K. A., Ham, P., Scales, J., Wyeth, E. & Rose, P. E. (2020). Aggressive behavioural interactions between swans (Cygnus spp.) and other waterbirds during winter: a webcam-based study. *Avian Research, 11*, 1–16.

Yon, L., Williams, E., Harvey, N. D. & Asher, L. (2019). Development of a behavioural welfare assessment tool for routine use with captive elephants. *PLoS One, 14*(2), e0210783.

Zhu, J., Gosnell, J. S., Akallal, L. & Goltsman, M. (2023). Fear changes traits and increases survival: A meta-analysis evaluating the efficacy of antipredator training in captive-rearing programs. *Restoration Ecology, 31*(3), e13674.

Index

P

For Product Safety Concerns and Information please contact our EU representative GPSR@taylorandfrancis.com Taylor & Francis Verlag GmbH, Kaufingerstraße 24, 80331 München, Germany

* 9 7 8 1 0 3 2 5 4 7 3 6 7 *

T - #0243 - 151225 - C410 - 254/178/19 - PB - 9781032547367 - Gloss Lamination